Butterflies of
the East Coast

Butterflies of the East Coast

An Observer's Guide

Rich Cech

and

Guy Tudor

PRINCETON UNIVERSITY PRESS

PRINCETON AND OXFORD

Co-author Guy Tudor and naturalist John Yrizarry afield in Orange County, New York

Library of Congress Cataloging-in-Publication Data

Cech, Rick, 1949–
Butterflies of the East Coast : an observer's guide /
Rick Cech and Guy Tudor.
p.   cm.
Includes bibliographical references (p.   ).
ISBN 0-691-09055-6 (cl : alk. paper)
1. Butterflies—Atlantic States.   I. Tudor, Guy, 1934–
II. Title.
QL551.A7C43   2005
595.7′89′0974—dc22   2004054130

British Library Cataloging-in-Publication Data is available

This book has been composed in Garamond and
GillSans

Printed on acid-free paper

www.pupress.princeton.edu

Printed in Italy by EuroGraphica

10 9 8 7 6 5 4 3 2 1

Butterflies and soap bubbles and whatever is like them amongst us seem most to enjoy happiness. To see these light, foolish, pretty, lively little sprites flit about—that moveth Zarathustra to tears and song. I should only believe in a God that would know how to dance. —Friedrich Nietzsche, *Thus Spake Zarathustra* (1892)

People often come to me with some strange animal they have found. "What is it?" they ask. "Oh, I say brightly, "that is a swallowtail butterfly, *Papilio cresphontes.*" It is curious how happy people are to have a name for something . . . . But other questions follow, "Where does it live?" and "What does it do?" I explain that it is a tropical butterfly, common in Florida, which sometimes gets quite far north in the United States . . . . Almost inevitably there will come another question, "What good is it?" I have never learned how to deal with this question. I am appalled by the point of view that makes it possible . . . . Often my reaction is to ask in turn, "What good are you?" —Marston Bates, *The Forest and the Sea* (1960)

The work of human hands appears more crude and less orderly the more closely one looks at it. In nature the situation is quite different; the orderliness remains and becomes progressively finer and more fascinating. Not until we look through the microscope do we discover how delicately and ingeniously the two tubes of the lepidopteran proboscis are coupled and interlocked. —Friedrich G. Barth, *Insects and Flowers* (1991)

Contents

Preface

From the scattered artifacts that remain of early human cultures, we can already see evidence of an odd but enduring fascination with butterflies. Stylized butterfly imagery appears in Neolithic cave art as far back as 6500 B.C.—interspersed among more utilitarian drawings of hunting parties and game animals (Manos-Jones, 2000). Butterfly decorations can also be found on Minoan signet rings, hairpins from the Sung Dynasty in China, and gold discs buried with Mycenaean kings. Early tribes in the American West sewed pupal shells filled with sand onto ceremonial garments as rattles.

But connections with early culture run deeper than mere display. In preagricultural societies, powerful butterfly goddesses represented the closely linked forces of death and regeneration in the world (Neumann, 1955). Itzpapálotl was an Aztec butterfly-goddess with wing edges of obsidian (the substance used for spears and sacrificial knives). A Cretan vase from around 1400 B.C. depicts a half-woman/half-butterfly deity, with vulture claws. In these hard times, the short-lived beauty of a butterfly, despite its attractions, conveyed an ominous portent.[1]

In classical Mediterranean civilizations, with daily necessity giving way to more leisurely contemplation, the mysterious transformations of the butterfly—from larval "worms" to deathlike pupae, to splendid, reemergent adults—inspired both philosophers and storytellers: Greek myth includes the story of Psyche, a mortal princess who was granted immortality by the gods in recognition of her deeply felt love. Not coincidentally, the ancient Greek word for soul (*psyche*) was also their word for butterfly. Early Christians compared the butterfly's developmental stages to the life of Christ, with the chrysalis representing the tomb (*Verrmis quia resurrexit* [literally, "the worm has risen again"], Pope Gelasius I, 5th century). And to generations of modern poets and novelists, as well as in folk traditions worldwide, butterflies have symbolized a range of intangible and aesthetic concepts, from beauty, freedom, and whimsical frolic to good fortune and spiritual ascent.

Such a deep-rooted, cross-cultural affinity for a group of animals is not unprecedented, of course—although it is unequaled, as far as we can see, among arthropods. A worthwhile question to explore is, why should this be? What is it in our encounters with butterflies that provokes such a lively response from our collective imagination, and so persistently?

We can obtain at least a rudimentary answer to this question by comparing the images of butterflies seen in art and literature with certain key traits that distinguish butterflies as biological organisms. This is tricky ground, since a wide disparity exists between people's naive impressions of butterflies, based on casual observation, and the scientific reality of what butterflies *are* in biological terms. Yet these dissimilar perspectives share some significant and enlightening points of contact:

Beauty. The attribute most frequently associated with butterflies is beauty. Many other insects are visually striking, it is true, and some butterflies are in fact quite plain. Yet few animal groups in any order can boast of such intricate, highly colored, and endlessly variable patterns as we see in the wings of butterflies.

From a biological perspective, the beauty of butterflies is no mere coincidence. Rather, it is a by-product of detailed evolutionary processes, working on several planes at once. The basic visual appearance of a butterfly is produced by *wing scales*—the anatomical structure for which the order Lepidoptera was named (Greek *lepis* [scale] and *pteron* [wing]). With minor exceptions, wing scales occur only in Lepidoptera (moths and butterflies) and in some closely related Trichoptera (caddisflies). Individual scales derive their color from a variety of pigments and intricate refractive structures. Grouped together, scales can create additive colors (as when yellow and black scales together appear green). Or, when arranged in groups, they form a rich variety of eyespots, bars, ripples, or other patterns. The specific layout of scale types on an individual butterfly's wing is determined by an elegant series of chemical processes that commence late in the larval stage. It has been noted that the patterns of butterfly wings tend to be more differentiated than those of moths (Nijhout, 1991, p. 34). *Why* butterflies have invested such a significant degree

---

[1] Although the butterfly was the regenerative half of this yin-yang combination—not the representation of death and destruction at the end of the life cycle—the butterfly still falls victim to guilt by association in some folklore. Thus, a Serbian proverb says that to kill a butterfly is to kill a witch, meaning the early, death-dealing Goddess (Gimbutas, 1989).

of evolutionary energy into pattern development is a different question, which we will try to address later.

Flight Style. Poets have long rhapsodized over the graceful, "merry-winged" flight style of butterflies. Something in their active yet meandering fluttering, and in their smooth, graceful approach to flowers, causes a distinct reaction in human observers. Every gradation of wistful human emotion has been projected onto the flight of butterflies at one time or another.

But here again, our intuitive perceptions have biological correlates. Experts in the evolution of insect flight have concluded that the flight style of butterflies—while not as swift or tactical as that of some other insects—is in fact highly evolved for gliding and low-speed flight. Efficient gliding facilitates long-distance movement, such as the dramatic fall migration of Monarchs. Controlled, low-speed flight is an adaptive skill that is difficult to achieve aerodynamically, and in butterflies (as well as birds) it has been acquired only through specific anatomical adaptations. Any experienced field observer can tell you that it is oftentimes possible to distinguish moths from butterflies at a glance by the comparatively clumsy way in which moths land.

Harmlessness. As long ago as the 5th century B.C., Buddha exhorted his followers to "satisfy the necessities of life like the butterfly that sips the flower, without destroying its fragrance or its texture." A few butterfly species are agricultural pests in the caterpillar stage, but as a whole these creatures are about as nonthreatening as animals can be. The sight of a butterfly perched on an exposed forearm is greeted with delight the world over (even though the visitor's true motive is to acquire minerals, not to form any particular emotional bond with the appendage's owner).

The biological foundations of the butterfly's gentle lifestyle, of course, rest on the fact that it consumes only liquids, drawn up through its proboscis. The proboscis is another unique and remarkable feature of the Lepidoptera. (Other insect groups have separately evolved sucking mouth parts, but they generally are involved with invasive puncturing.)

All of these considerations raise a paradox. Specifically, we generally accept without argument the Darwinian principle that in nature only the fit survive. Yet while butterflies have existed for at least 50 million years—longer than lions or tigers or bears—they possess none of the traits we typically associate with fitness. They are neither strong nor fierce nor armored nor cunning nor stealthy (at least in flight). Indeed, the three defining traits just discussed (beauty, gracefulness, and harmlessness) scarcely seem a prescription for competitive survival. What are we to conclude? In short, either "survival of the fittest" does not apply in this corner of the animal kingdom, or else we need to rethink seriously what it means to be "fit." The authors recommend the latter alternative as highly edifying. Characteristics we initially see as weak or unfit may, on closer view, end up as badges of strength in the butterfly's ongoing struggle for survival.

∘ ∘ ∘ ∘

A few words about our objectives in writing this book. The principal topic is butterflies, naturally, but there is a strong secondary emphasis on the U.S. East Coast itself—in particular, its diverse ecology and significant natural areas. To convey fully the story of butterfly life in any specific region, the environment itself must be included as a principal character. Our target readership includes field observers in the moderate to proficient range who are interested in butterfly identification and ecology, although we have included information and images meant to be useful regardless of experience.

The key denominator of our approach is that the book is written for "field observers." This perspective is apparent in the topic headings of the Introduction: "Where to Find Butterflies"; "When to Find Butterflies"; and "What to Look For" (once butterflies are found). We use the term "field observer" in a broad sense, to include anyone who studies butterflies in a free and unrestrained setting. It is important to recognize that rewarding and useful observations of butterflies can take place in backyard gardens or city parks. Indeed, in some cases populations of rare species can be materially enhanced by conscientious home gardeners.

On the other hand, backyard enthusiasts should realize that only a portion of a butterfly's life cycle is apt to be visible at a typical garden nectary. More important, many butterfly species can be found only in certain specialized or remote habitats. To appreciate fully the butterfly life of a region, it is necessary to step out into the local environment, and we encourage the exploration of these exciting and important areas.

Readers will quickly note a distinct emphasis on hostplants in this book. The authors are both active field naturalists with a longstanding interest in botanizing. We believe that, apart from learning species identification marks, the most important incidental skill that a beginning field observer can acquire is the ability to identify hostplants. With experience, an observer will also want to locate and identify caterpillars. But in general, butterfly caterpillars tend to be secretive and difficult to locate. Many caterpillars are nocturnal, moreover, and a number even burrow into flower heads. (Those seeking to rear such larvae are often forced to collect the flowers and wait for the caterpillars to emerge when the blossoms wilt.) For these reasons, we have placed primary emphasis on hostplants and habitats in this text, more than on describing early stages. In cases where larvae are likely to be seen in the field on a regular basis, nonetheless, we have made efforts to include them.

The scientific literature on butterflies, finally, is replete with technical terminology, much of it frankly abstruse. While a serious student of butterfly biology needs to learn these terms, we have found that they form an impermeable barrier to entry for many otherwise interested observers. For this reason, we have used "plain-language" terms wherever possible in this text—though also noting technical terms in parentheses from time to time so that they will begin to feel familiar. Thus, we will generally use "emergence" to describe the departure of an adult butterfly from its chrysalis, rather than "eclosion," and "leading edge" to describe the front boundary of a wing, rather than "costal margin." In referring to subfamilies, we may use a common group name such as "hesperines" rather than the scientific Hesperiinae. We hope this level of informality will not unduly disturb more studious readers.

○ ○ ○ ○

A last preliminary question is, why the U.S. East Coast states? We use this term to refer to the Appalachian Mountains and Atlantic coastal plain environments extending eastward. For convenience, we have used 17 states plus the District of Columbia as a rough proxy for the region (Maine, New Hampshire, Vermont, Massachusetts, Rhode Island, Connecticut, New York, New Jersey, Pennsylvania, Maryland, Delaware, District of Columbia, Virginia, West Virginia, North Carolina, South Carolina, Georgia, and Florida). About 250 butterfly species have been recorded in this area. Of these, we include full accounts for 234 full species that are either established breeders or recurring colonists. Rare strays are noted from time to time (and occasionally illustrated), but not given full accounts.

Ecologically, the U.S. East Coast region is largely self-contained, with many links in terms of geological history (also a few disparities, admittedly, such as the presence or absence of recent glaciation). The region is biologically unified by its principal geographic feature, the Atlantic coastal plain. This flat expanse of low-lying territory allows unrestricted north-south movement of individual butterflies and migratory populations. At the same time, the Appalachians form a substantial barrier to movement in and out of the region from the west. For many species, it is quite accurate to think of the "East Coast population" as a single demographic assemblage that maintains more or less continuous genetic contact.

The East Coast region is not hermetically sealed, of course, and many of its interesting butterfly "events" occur along contact zones with adjoining ecological regions, especially the Caribbean, Gulf Coast, Ohio River Valley, and Canadian boreal zone. In recent years, the colonization of Acacia Blues in Florida and the appearance of exotic strays in the Northeast such as Marine Blue and Dainty Sulphur on Long Island (of presumed midwestern origin) are examples of noteworthy extraregional occurrences.

Another reason to study the U.S. East Coast is that, unlike the West Coast, it has seldom been treated as a contiguous region in natural history guides. Most equivalent nature books cover the entire eastern United States—often using an arbitrary geographic boundary line, such as the 100th meridian—or else split the territory into Northeast and Southeast subregions, thus overlooking many important linkages. A broader perspective would enhance our appreciation of biological interconnections that draw the region together.

A final reason to consider the U.S. East Coast as a contiguous region is its human demography. The East Coast states occupy just 12.5% of the total U.S. land area (about 461,000 square miles). Yet they are home to more than 37% of the U.S. population, according to 2000 census figures. This results in a population density of nearly 230 persons per square mile—as compared with only 77 persons per square mile nationwide. These numbers help us to visualize the tremendous development pressure that has built up throughout our region. And this pressure shows little sign of abating: of nearly 11 million people added to the United States population during the past decade, more than a third settled in the East Coast region. With specialized regional habitats and interlinked butterfly populations under threat up and down the coastal plain, there is a pressing need to develop regional conservation priorities. Much work remains to be done to avoid losing a wealth of nearby resources that are being held together, in many cases, by a strained or fraying ecological thread.

ACKNOWLEDGMENTS

A great deal of generous assistance is needed to complete a book of this sort—in helping to locate rare species and hostplants, sharing knowledge and experiences, and simply extending encouragement and friendship. We have been fortunate in receiving valuable help from many people in the course of our work, and together we have shared many unforgettable experiences. A number of ineffable moments have involved mud, as the day that Alana Edwards's State Park vehicle, running on bald tires, became mired in a Florida "muck swamp," requiring a fire engine rescue (the fire engine also got stuck at first); or the time Rick wrecked a newly purchased pair of trousers in an acid bog at Ft. Bragg while stalking a mated pair of Mitchell's Satyrs with Eric Hoffman; or the day a shoe was lost exploring the unknown depths of a wet savanna in New Jersey in search of Arogos Skippers with Dale Schweitzer. Then there was emerging from a Suffolk County, New York field after a trek with Steve Walter, covered with a three days' supply of ticks; having Walt Gould tell Rick to be more careful after he noticed him standing over a coiled Pygmy Rattlesnake (resting between his feet) in the Everglades, as he distractedly photographed a Ruddy Daggerwing; Roger Hammer, strolling casually through a mosquito-infested Florida hardwood hammock in shorts, explaining which of the tropical species we were brushing against might cause a serious rash; and John Yrizarry suggesting we not look for Compton Tortoiseshells in the back yard when Black Bears were exploring the bird feeders. Or Tom Allen, driving his pickup over a narrow, slippery wood-plank bridge in Fork Creek WMA in Boone County, West Virginia, remarking calmly that the bridge was not dangerous as long as you stayed on the planks. There were also sublime moments, as when our first Cofaqui Giant-Skipper alit on a rock outcrop just before dusk in 1999 on a trip with Derb Carter and Harry LeGrand, the 45-second "window" when a Polixenes Arctic presented itself on top of Mount Katahdin, halfway through an 11-hour round trip hike to the Tablelands, rediscovering Bog Elfin in an acid bog in New Hampshire with Emily Peyton, a North Carolina native who does not relish wading thigh-deep in the chilled waters of a Black Spruce bog in May (but did). And many more.

We are particularly indebted to the following people for special assistance or information: Tom Allen, Joanna Burger, John Calhoun, Derb Carter, Brian Cassie, Buck and Linda Cooper, Harry Darrow, Bob Dirig, Alana Edwards, Tom Fiore, Walt Gould, Jeffrey Glassberg, Mike Gochfeld, Roger Hammer, Eric Hoffman, Jeff Ingraham, Harry LeGrand, Harry Pavulaan, Peter Post, Robert Pyle, Don Riepe, Dale Schweitzer, Vallerie Vallely, David Wagner, Kristine Wallstrom, Steve Walter, Marc Weinberger, John and Mary Yrizarry, and Harry Zirlin.

Also to Lyn and Brook Atherton, Barb Barton, Skip Blanchard, Mike Bochnik, Donald Burgy, John Burns, Paula Cannon, Fabrice Delacour, Tom Dodd, Jim Dowdell, Edna Dunbar, Jim Duquesnel, Tom Emmel, Steve Hall, Scott Hartley, Karen Hsu, Peter Joost, Rich Kelley, David Kunstler, John Lawrenson, Andy Miller, Jacqueline Miller, Marc and Maria Minno, Tom Neal, Gil Nelson, Dave Norris, Bob

Pemberton, Jeff Pippin, Eric Quinter, Bob Robbins, Jane Ruffin, Mark Salvado, Jim Springer, Don Stillwaugh, Bo Sullivan, Pat and Clay Sutton, Ann Swengel, Mike Thomas, Dick Walton, Wade and Sharon Wander, Reggie Webster and David Wright. Also numerous additional members of the New York City Butterfly Club, Connecticut Butterfly Association, and NABA Chapters of New York City and New Jersey not named individually. As well as the staff and managers of many reserves and public lands throughout the region, including Castellow Hammock (Miami-Dade County, FL), Crocodile Lake National Wildlife Refuge (Miami-Dade County, FL), Deering Estate (Miami-Dade County, FL), Florida State Parks Department, Ft. Bragg Military Reservation (Cumberland County, NC), Ft. Indiantown Gap (Lebanon County, PA), Jamaica Bay National Wildlife Refuge (NY City), Jonathan Dickenson State Park (Martin County, FL), New Jersey Audubon Society, New Jersey Dept. of Fish and Wildlife, Sandhills Game Land (Scotland County, NC), The Nature Conservancy, Ward Pound Ridge Reservation (Westchester County, NY), West Virginia Department of Natural Resources, and Weymouth Woods Sandhill Nature Preserve (Moore County, NC).

The origins of the recent butterfly observation movement can be traced in significant part to the New York City Butterfly Club, which has held monthly meetings at the apartment of the second author in Forest Hills since the mid-1980s. Observers owe a significant debt to Harry Darrow, an outstanding naturalist and pioneer in photographing butterflies in the field. His work still stands out in an era of automated macrophotography.

A majority of the photos in this book were taken by Rick Cech. We have been pleased to also include photos taken by a number of excellent photographers. Photos were provided to us by Derb Carter, Jaret Daniels, Harry Darrow, Jeffrey Glassberg, Roger Hammer, Peter Post, and Don Riepe. (Credits for specific photos are listed in the text.). Specimens used for photos were loaned by Jeff Ingraham, David Wagner, and Harry Zirlin.

Finally, Rick extends deepest thanks to Emily Peyton for every imaginable form of support and assistance throughout the project. Field observations in the text attributed to "we" often include Rick, Emily, and a variety of field companions. Also to Clinton Bartow and Russ and Jane Kinne, unselfish mentors to a young naturalist-photographer in Connecticut.

Introduction

I. WHERE TO FIND BUTTERFLIES

BUTTERFLY RANGES

To observe butterflies in the field, we first need to locate them. This task is often not easy, as any experienced observer knows. A few abundant species, such as Cabbage Whites and American Ladies, can literally be found almost anywhere—roadsides, gardens, parks, even vacant city lots. But most others require at least some degree of specific knowledge to locate. And a surprisingly large number of East Coast species—more than a quarter, by our count—reside in limited areas within our region, some with just a few known sites or colonies. To further complicate matters, butterfly ranges are not static. A dozen or more East Coast species have experienced significant range reductions in recent decades, often without clear reason (e.g., Grizzled Skipper and Regal Fritillary). And a smaller number of species, sometimes just as unaccountably, have undergone pronounced extensions (e.g., Common Ringlet and Red-banded Hairstreak). Even species with stable ranges may repeatedly extend their boundaries northward in mild years, only to be pushed back when a harsh winter intervenes (e.g., Shapiro, 1991).

A range, stated most simply, is the geographical region in which a species *usually* lives or occurs. Some variant of this simple definition is the starting point for even the most sophisticated computer models that simulate distribution patterns (e.g., Rapoport, 1982, p. 3). Most ordinary butterfly books (including this one) illustrate ranges as color blocks on a standardized map, each with well-defined boundaries. Such drawings can easily create the impression that butterflies are dispersed uniformly through the designated area, as if in a giant corral. This approach is useful as a means of visually condensing a great deal of complex information; but it can easily mask underlying biological dynamics, limiting our ability to appreciate the ecological forces that shape a species' distribution.

Researchers suggest that butterfly populations exist in constant flux. This may seem to overstate the case somewhat, from an observer's perspective, since we know of many stable colonies that persist for decades in one location. Yet population densities do regularly vary within butterfly ranges, often from year to year, and the precise location of the edge of a distribution, which is important to know, can be vague and shifting.

In recent years, *metapopulation theory* has been invoked to explain the range dynamics of many butterfly species. First articulated in the 1950s, metapopulation theory holds that many organisms live in a series of largely isolated local populations, linked only by periodic dispersals, or cross-migrations (*see* Hanski, 1999). The fate of a species in any area depends on a set of key factors, including the viability of individual colonies, the rate of successful cross-dispersal, the "patchiness" of the environment, etc. Individual colonies may become extinct, but if conditions are favorable an area may later be recolonized (Ehrlich, 1985). As one author has phrased it, individual colonies blink off and on over time, "like lights on a Christmas tree" (Douglas, 1989, p. 100).

This is all a bit theoretical, however, if our main purpose is to locate butterfly populations in the field. For this, we need to focus on more tangible factors. With respect to butterflies, three key influences predominate: hostplants, habitats, and climate. These factors are not fully independent, to be sure, as climate affects the distribution of habitats, habitats define the range of hostplants, and hostplants (in large numbers) can modify climates. Still, each element has independent effects that can be explored separately.

CLIMATE

For cold-blooded (*exothermic*) creatures such as butterflies, temperature is the most critical aspect of climate, followed closely by precipitation. Species living in the North naturally require physiological adaptations to survive regular hard freezes. But other, less obvious effects are also apparent. In the South, for instance, longer flight seasons and faster caterpillar growth rates combine to increase the average number of yearly broods. (Arctic species, by comparison, may require two or more full summers as a caterpillar to reach adult status.) Climate can even affect subtle traits such as wing size: it has been proposed that medium-sized butterflies have a selective advantage in northern climates, regardless of family. The reasons for this interesting fact are discussed below.

The influence of climate is sufficiently pervasive that entire biological communities can be identified in climate-specific terms—such as *holarctic* plants and animals, found in northern territories around the world. In our mostly temperate region, though, no single family (and only a few subfamily groups or tribes) can be described as "climate specialists." Instead, most groups have representatives scattered across the various climate zones. Climate does remain a significant component of *habitat,* on the other hand, and is discussed in that connection. For the moment, however, we proceed to a second factor—the most important single determinant of butterfly distributions, as well as many other aspects of their lives—specifically, hostplants.

HOSTPLANTS

Caterpillars are highly fussy about the foods they will accept, and adult female butterflies are even choosier in selecting target plants on which to lay eggs (*oviposit*). Evolution has been refining these preferences for tens of millions of years. Disagreement remains as to whether or not butterflies and hostplants *co-evolved*—with new defensive plant capabilities met by improved methods of caterpillar exploitation, prompting additional plant defenses, and so on (*see* Ehrlich & Raven, 1964; *but* Courtney & Chew, 1987; Scoble, 1995, pp. 174–76). Whatever the case, we know that present-day interrelationships between hostplants and butterflies are very close indeed. Scientists and observers alike try to make sense of these complicated interconnections.

Let us start at the tangible end of the process—the act of egg-laying by a female butterfly. This is among the most subtle and intricate of butterfly behaviors. Accurate oviposition is critical because the resulting caterpillar has little capacity to recover from a parental mistake. Even if caterpillars were more mobile, they would still lack the sensory apparatus and intelligence to seek out a suitable host over any significant distance. It is true that in some species females lay eggs on surfaces *near* the hostplant, rather than directly on it, seemingly to reduce predation; but in these situations hostplants are within the "crawl range" of instinctually guided hatchlings (e.g., Gannon, 1986, Brown, 1981; MacNeill, 1964).

Adult females select hostplants methodically. The initial contact is usually visual, based on leaf shape or various other specialized cues. Some species prefer isolated hostplants, or hostplants of a particular size or shape (*see,* e.g., Douglas, 1989, p. 180, Chew & Robbins, 1985, pp. 71–72). Approaching the target host, the female lands and tastes or smells the leaf with chemical sensors on her feet. Often, she drums at the leaf surface, evidently to dislodge chemicals for sampling (or to detect roughness). The chemical

cues that induce egg-laying are often highly specific (e.g., *see* Haribal et al., 1998). And some chemicals deter ovipositing rather than induce it.

Butterflies are exquisite taxonomists. They can often locate and identify plant species more reliably than trained human botanists. Naturalist Roger Hammer tells of the day he followed a female Polydamas Swallowtail around Elliot Key (in Biscayne Bay off Florida), until she led him to an endangered Marsh's Dutchman's-Pipe (*Aristolochia pentandra*) that he had been unable to find himself, despite diligent searching. Female butterflies do make oviposition errors, especially early in an egg-laying session (Stanton, 1984), though it is difficult to rule out the possibility that some of these errors may be part of a genetically programmed strategy for sampling alternate hostplants, or, as already noted, a strategy to avoid placing eggs in predictable locations that predators can target. Perhaps for this reason, caterpillars can often survive on a wider range of hosts than females are prone to lay on (Janz, 2003).

Having located a proper hostplant, the females of some species next check the plant for existing eggs, to avoid overcrowding or cannibalism on the part of another female's earlier-hatched larvae (Raucher, 1979).[1] In some groups, such as checkerspots, eggs are laid communally. Here, the female may actually *seek out* existing eggs rather than avoid them. Finally, in *myrmecophilous* (ant-loving) lycaenids—certain blues and hairstreaks whose young are protected by ants in exchange for producing sweet secretions—the female may refuse to oviposit unless ants of the appropriate species are present. Females actually touched by these ants are prone to lay the most eggs (Douglas, 1989, p. 134).

The ovipositing female often lays only on particular portions of the hostplant. The concealed underside of a leaf may be selected rather than the exposed upper surface. Or, in species whose caterpillars feed on flowers, the female may lay only on unopened buds. Greater fritillaries lay eggs near the base of the hostplant in the fall, so that young caterpillars will not have far to look for a sprouting hostplant come spring. Females often lay eggs near the site of newly emerging foliage, which may offer desired concentrations of certain chemicals (either very low or very high, depending on the objective), or which may simply be more tender, and thus easier for newly hatched caterpillars to chew. Whatever determinations a female needs to make prior to ovipositing must be concluded quickly, in any case, as the moment of egg-laying is a vulnerable one for her (Hanski, 1999, p. 212).

In sulphurs, egg-laying begins as soon as a day or two after emergence and persists for up to two weeks. Female

---

[1] Few caterpillars are carnivorous per se, but many larvae will eat others of their own species that they happen to encounter. The purpose of larval cannibalism may be territorial more than nutritional, but this subtlety is largely immaterial to the victim.

I-1. In Florida, trees that shed their leaves during the winter drought refoliate in the spring. Here, a female Ruddy Daggerwing lays her eggs near the newly emerging leaves of a Strangler Fig, which will provide newly hatched caterpillars with a tender meal.

sulphurs lay an average of 700 eggs in laboratory conditions, though probably a fraction of this number in the wild. Other species are estimated to lay roughly between 200 and 1,000 eggs, occasionally more. Only a few of these need to reach adulthood in any locality for the species to persist.

HOSTPLANT USE: STRATEGIES AND PATTERNS

All these details beg a fundamental question. How did female butterflies acquire the genetic programming to seek out and identify specific hostplants in the first place? Unfortunately, it may be impossible to reconstruct the exact origins of early plant associations, since they were formed millions of years ago, when plants and butterflies were both still quite different. Some intelligent guesswork has been done to reconstruct these early events (e.g., Scott, 1986, p. 64), but it is still most practical for the moment to understand present-day circumstances from a functional perspective.

Researchers pay attention to the number of hostplants a butterfly species utilizes. Is there a single hostplant per species (*monophagy*), a limited number (*oligophagy*), or many (*polyphagy*), and are the hosts closely related? Polyphagous species may use multiple hostplants across their range, or they may be polyphagous at the population level only, keeping to a single host at any one location. Or, they may go through polyphagous episodes while shifting hostplant use. Much remains to be learned here.

Another approach is to analyze the hostplants used by all the butterfly species in a *particular region*. Table I lists 62 plant families with known primary butterfly associations

in our area (excluding some incidental uses). This list includes a wide variety of plant types, but what may be most telling is the large number of common plant families that are *not* included—maples (Aceraceae), mints (Lamiaceae), lilies (Liliaceae), orchids (Orchidaceae), and, for the most part, conifers (just two species feed on pines, another two on cedars). Even composites (Asteraceae) are rather modestly represented if we consider the size of the family and the large number of butterflies that use this family for nectar. Moth larvae, by comparison, have relatively diverse tastes, consuming fungi, lichens, nonvascular plants (moss/liverworts), algae, and ferns, as well as flowering plants in many families. Butterflies outside our region also use some of these additional food types, but Table I shows plainly that the process of hostplant selection among East Coast butterflies is not random. On the contrary, our species have made some very specific choices in affiliating themselves with particular hostplant families over time.

The hostplant choices of East Coast butterflies can be assigned to a concise set of descriptive categories, which we will call *hostplant-related lifestyles*. See p. 4. For the most part, these lifestyles amount to nothing more—nor less—than a suite of integrated adaptations that enhance the survival prospects of species using a particular class of food plant.

One common lifestyle strategy involves concentrating on a set of nutritious and digestible hostplants, exploiting them aggressively to achieve a high reproductive rate. This strategy may enable a species to outpace the many mor-

Table I. Hostplant Families Used by East Coast Butterfly Species

# of Butterflies Using Family for Principal Hostplants	Hostplant Family
15 or more	Grass (56), Pea (31), Sedge (16).
10–14	Aster (Composite) (11), Beech (10), Violet (10), Willow (10).
6–9	Birch (9), Rose (9), Cassia (8), Elm (7), Heath (7), Mustard (Crucifer) (7).
4–5	Mallow (5), Nettle (5), Acanthus (4), Buckwheat (4), Mimosa (4), Passionflower (4), Rue (4), Spurge (4).
2–3	Goosefoot (3), Figwort (3), Milkweed (3), Quassia (3), Cashew (3), Laurel (3), Vervain (3), Agave (2), Amaranth (2), Bayberry (2), Birthwort (2), Buckthorn (2), Buttercup (Crowfoot) (2), Cedar (2), Currant (Gooseberry) (2), Holly (2), Olive (2), Palm (2), Plantain (2), Pine (2), Saltwort (2), Soapberry (2), Walnut (2).
1	Arrowroot, Black Mangrove, Bursera, Cacao, Caltrop, Canna, Caper, Custard-Apple, Cycad, Dogwood, Flax, Hemp, Honeysuckle, Leadwort, Locustberry, Magnolia, Mistletoe, Mulberry, Parsley (Carrot), Red Mangrove, Sweetleaf.
0 [No principal hosts]	*Examples:* Arum, Cactus, Evening-Primrose, Gentian, Geranium, Iris, Lily, Maple, Milkwort, Mint, Morning-Glory/Bindweed, Nightshade, Orchid, Saxifrage, St. John's-Wort, Sycamore, all lower plants & ferns.

Note: "Plant Family" listings include number of butterflies using the family.

tality factors that beset butterfly populations. Nonwoody vascular plants (*herbs* in botanical parlance) are often well-suited to this strategy. Scott believes that primitive butterflies pursued an "herbal" lifestyle built around early members of the pea family (Scott, 1986, p. 64; *compare* Janz et al., 2001). A drawback to specializing on easily exploited herbs is that many such plants are ephemeral, growing in disturbed or transitional habitats. This injects an element of uncertainty, especially in prehistoric times, when these preferences evolved, since disturbed habitats were probably rarer in those times than human activity has made them today.

It was most likely the stable, widespread availability of two other plant groups, trees and grasses, that placed them at the center of their own strategic lifestyles—this despite their sophisticated chemical defenses, frequent indigestibility, and inconsistent nutritional content. A different set of factors would be required to account for the origins of root-feeding, as engaged in by giant-skippers, or the development of a newly acquired taste for flower parts, which is thought to have sparked the evolution of lycaenids (blues and hairstreaks) as a separate taxonomic group (Scott, 1986, p. 99).

Nutrition alone is not the sole basis for defining hostplant strategies, however, often not even the primary basis. Butterflies use hostplants as a source of diverse chemical substances, employed in defense, mating, synthesis of wing and body pigments, and other special life functions (Brower, 1985). Examples include alkaloids, cardenolides, flavonoids, glucosides, tannins, coumarins, organic cyanides, and many others. The main function of these so-called secondary plant chemicals seems to be the deterrence of herbivores, in one way or another. Some are actively toxic, others may simply be unpleasant, causing rejection because of bitter taste or unfamiliarity. A number of hostplants that are described in plant books as "aromatic" or "medicinal" may fall into this category. Some defensive plant chemicals, lastly, may protect butterflies indirectly, by making them hard to digest. There is experimental evidence that nestling birds fed on caterpillars with high tannin concentrations develop more slowly than chicks fed on caterpillars low in tannin (Brower, 1985, p. 121).

Ironically, secondary plant chemicals—once tamed—become primary targets for ovipositing female butterflies, especially those that can be extracted while feeding, thereby conferring toxicity during the butterfly's life cycle. In some cases, particular defensive chemicals occur in a number of unrelated plant families, and this can result in unlikely seeming hostplant combinations. Cabbage and Great Southern Whites, e.g., feed on plants in four genera (three related, one not), but all contain mustard oils, or *glucosides* (Scott, 1986, p. 65). Similarly, Dorcas Coppers feed across families and genera (docks, smartweeds, and

Major East Coast Butterfly Lifestyles Based on Hostplant Use: An Overview	
Hostplant Type	**General Strategy Factors**
Vascular Plants ("Herbs")	• Relatively easy to digest, offer ready nutrition • Fast caterpillar growth, numerous broods • Transient lifestyle, since many herbs are ephemeral • May be difficult for predators to target shifting populations
Grass	• Hostplants widespread and abundant, available throughout warm season • But often difficult to digest, nutritionally incomplete • Slow caterpillar growth, fewer broods per year (or smaller individuals, e.g., grass skippers) • Difficult for predators to target in large feeding areas • Hostplants not toxic, so adults are often cryptic (satyrs) or evasive (skippers) • Caterpillars often live in protective leaf rolls, feeding at night
Woody Plants (Trees, Shrubs & Woody Vines)	• Many trees have well-developed chemical defenses, with leaves hard to digest • Seasonally targeted broods may coincide with "leaf out" periods • Some caterpillars can detoxify a range of defensive chemicals (tannins, etc.) • But this slows development, reducing average brood frequency (one per year in *Satyrium* hairstreaks) • Tree feeders are often polyphagous, and may use a number of hostplant families • Exposure exists to arboreal predators, "false heads" are common
Flowers	• Flower parts and fruits relatively easy to digest; eaten by certain blue and hairstreak larvae • Seasonally targeted broods coincide with peak flower periods • Flower parts may lack strong defensive chemicals found in leaves; allows use of unrelated hostplants • Caterpillars cryptic and nocturnal, may hide or burrow in flower heads • Caterpillars often ant-tended
Insectivorous	• Harvester caterpillars eat Woolly Aphids; some blues eat ant larvae • Available nutrition from insect protein is relatively high • Cycle time short; from egg to adult just 3 weeks in Harvester • Adult Harvesters feed on "honeydew" secreted by aphids
Roots	• Hostplant for two giant-skipper species only • Yucca root is a rich food supply for burrowing caterpillars • But highly difficult to digest • Slow larval development, food energy stored as fat • Concealed lifestyle limits exposure to predation • Adults do not feed
Toxic Plants	• Some species sequester noxious chemicals ingested by caterpillars, rendering them unpalatable. • Bright, aposematic colors deter predators • This strategy can overlap with others above; much remains to be learned about the use of plant toxins by butterflies

cinquefoils), but all contain a specific flavonoid that stimulates egg-laying in the female (Douglas, 1989, pp. 178–79). In such cases, defensive chemical considerations in host-plant selection may outweigh nutritional ones.

USING HOSTPLANTS TO LOCATE BUTTERFLIES

In locating butterflies, we have noted the importance of hostplant associations. Equally important, however, is the recognition that hostplants *themselves* have specialized ranges, often local and patchy. In some cases, hostplants are a strong and immediate indicator of a butterfly's likely presence, making it worthwhile to begin searching whenever the plant is found, even if no local colony is known. For example, Coinvine (*Dalbergia ecastophyllum*) is tightly associated with Statiras in southern Florida, Dutchman's Pipes (*Aristolochia*) with *Battus* swallowtails (Pipevine and Polydamas), and hackberries (*Celtis*) with American Snout and emperors.

This approach is heartening when it succeeds. On a birding "big day" in 1995, the first author paused to inspect a patch of Wild Indigo (*Baptisia tinctoria*) in a pinewoods clearing in southern New Jersey, and was excited to find a previously unknown colony of Frosted Elfins. (His birding companions, more impatient than enthused, were puzzled at this sudden loss of focus on warbler chirps.) Elsewhere, in an awe-inspiring display of serendipity, West Virginia state biologist Tom Allen happened to notice some tiny, near-invisible eggs on the underside of Dwarf Cinquefoil (*Potentilla candadensis*) leaves while doing fieldwork in the late 1990s. Returning the next April, he was rewarded with a colony of Appalachian Grizzled Skippers, among the rarest of East Coast butterflies.

Associations between butterflies and hostplants are not always obvious in the field, although with practice many subtle links will become apparent. Zebra Swallowtails, for instance, range widely in their habitats, seldom lingering in the immediate vicinity of pawpaws, except when ovipositing. Yet it is rare to see the swallowtail more than a few hundred yards from some pawpaw species. When everyday butterfly walks are redefined to include routine observation of local flora (including potential hostplants), new vistas materialize.

Changes in butterfly distribution can sometimes be linked to shifts in hostplant use. This is not especially common, since the ranges of most hostplants are well-established and a majority of hostplant-butterfly associations date back many millennia. But abrupt shifts do occur, often in connection with introduced plant species. In recent decades, Wild Indigo Duskywings have "jumped" from Wild Indigo (*Baptisia tinc-toria*) to Crown Vetch (*Coronilla varia*), an invasive member of the pea family that is widely planted on interstate right-of-ways to control erosion. Likewise, it has been suggested that the range of Hayhurst's Scallopwing extended north from its original boundaries because of Lamb's Quarters (*Chenopodium album*), an alien weed whose range exceeds those of the Hayhurst's native hosts (Opler & Krizek, 1984).

Not all exploitations of non-native hostplants lead to range expansion, however. In one of nature's more perverse gambits, some introduced hostplants are (1) attractive to native species, yet (2) fatally toxic to their caterpillars (e.g., Chew, 1975). The rapid decline of West Virginia Whites in the northeastern United States in recent decades has been linked to a fatal attraction on the part of female whites to chemical markers in the highly invasive European weed, Garlic Mustard (*Alliaria petiolata*).[2]

Occasionally, hostplant analysis provides unexpected insights into butterfly ecology. In conducting their classic studies of the Edith's Checkerspots in California, Paul Ehrlich and his students made a curious observation: checkerspot colonies at the Jasper Ridge site, near Stanford, were found only on serpentine soil, even though the primary hostplant (*Plantago erecta*) also grew on sandstone-based soils nearby. Subsequent research found that many caterpillars starve in dry years, when *Plantago* dries out before caterpillars finish feeding (Singer, 1972; Douglas, 1989, p. 102). In these years, populations survived by using a backup hostplant, Owl's Clover (*Orthocarpus densiflorus*). It so happens that Owl's Clover lives only on serpentine soils, explaining the checkerspots' limited distribution. Similar discoveries undoubtedly can be made regarding East Coast species.

But before striding into the field, flush with newfound confidence, we must acknowledge that using hostplants to locate butterflies is not always easy. Sometimes the hostplant is difficult to locate, such as Virginia Snakeroot (*Aristolochia serentaria*), a widely distributed but elusive native hostplant of Pipevine Swallowtails, or various northern currants (*Ribes* spp.), hostplants for the northern Gray and Hoary commas.

At other times, a hostplant is too common to be of much use in narrowing our search. This is especially likely with widespread trees and grasses. Or, a particular species may use a number of host species in a particular family (e.g., the Pea Family, Fabaceae), all of which may be found locally. Finally, there are cases in which the range of the hostplant extends well beyond that of the butterfly. Examples include Dusky Azure, 'Appalachian' Grizzled Skipper, and Milbert's Tortoiseshell. Why this discrepancy should exist is a difficult question to be explored at another time.

---

[2] Garlic Mustard is toxic to caterpillars of both West Virginia and Mustard Whites, usually in the 1st or 2nd instar (Bowden, 1971). But one population of Mustard White (a species that eats Garlic Mustard in Europe) is apparently already becoming acclimated to the new host (Courant et al., 1994). The ultimate fate of West Virginia White remains uncertain, but recent population declines are foreboding. *See also* Casagrande and Dacey, 2001 (Monarchs and Black Swallowwort).

We are far from having a complete understanding of all hostplant associations, even for East Coast butterflies, which are comparatively well-studied. Much information in the literature comes from informal trial-and-error testing, in which captive-reared caterpillars were offered a variety of different food plants. Such studies are most conclusive when a particular plant is firmly rejected, or when it is eaten but ends up being toxic. Otherwise, caution must be advised, since the number of plants accepted in captivity frequently exceeds the number chosen voluntarily in the wild (Scott, 1986, pp. 60–61).

Opportunities abound for individual observers to make new discoveries. In the spring of 1999, for instance, natural history educator Alana Edwards noted that Gulf Fritillaries in a Lake Worth, Florida, nursery greatly preferred a native herb, Piriqueta (*Piriqueta caroliniana*), to nearby passionvines, the universally cited hostplant. An herbaceous member of the Turnera Family (*Turneraceae*), Piriqueta is closely related to passionvines (*Passifloraceae*). And, like passionvines, Turneras contain cyanogenic glucosides (Olafsdottir et al., 1990). Edwards suspects that use of Piriqueta is related to their dense arrangement in nursery beds (vs. a widely dispersed growth pattern in the field), and possibly to the fact that the female did not detect hostile ants on the Piriqueta, which can deter ovipositing on the passionvines.

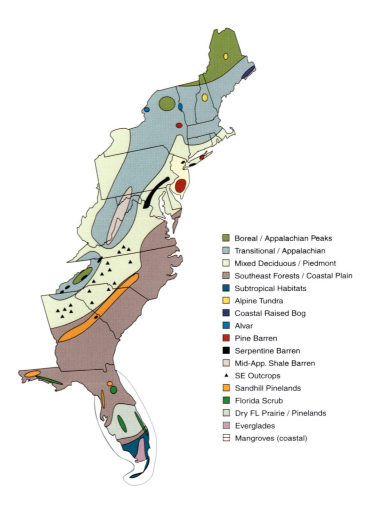

Boreal / Appalachian Peaks
Transitional / Appalachian
Mixed Deciduous / Piedmont
Southeast Forests / Coastal Plain
Subtropical Habitats
Alpine Tundra
Coastal Raised Bog
Alvar
Pine Barren
Serpentine Barren
Mid-App. Shale Barren
SE Outcrops
Sandhill Pinelands
Florida Scrub
Dry FL Prairie / Pinelands
Everglades
Mangroves (coastal)

HABITATS

No one doubts the importance of habitat in determining where butterflies live. But apart from being a nursery for specific hostplants, what *direct* role does habitat play in defining butterfly ranges? Quite a significant one, actually. To begin with, habitats are shaped by major *physical* and *climate factors,* such as susceptibility to recurrent storms and flooding, local patterns of seasonal rainfall and temperature fluctuations (including the length of the flight season), susceptibility to fire, proximity to coastal salt spray or tidal inundation, specialized soil composition, and so forth.

A long and bewildering list of *ecological factors* also play a role, including habitat density and patchiness; availability of patrol corridors or hilltopping sites; availability of hostplants; availability of adult food sources; proximity of adult food to hostplants—and of both to shelter; availability of damp soil for puddling; habitat color (relative to camouflage strategy); density and variety of local predators and parasitoids; incidence of lepidopteran diseases (viral, bacterial, and fungal); availability of protected overwintering spots; availability of protected surfaces for pupation; presence of attending ant species for those species requiring it, etc.

Just how specialized habitat requirements can be is apparent from the findings of Dave Norris's prodigious work on Northern Metalmarks. A Connecticut field biologist,

Norris began to locate metalmark colonies in 1989. Initially, he would overlay geological survey maps (which show surface topography) with USDA soil maps, in order to locate calcareous outcrops. After screening more than 500 potential sites with off-season visits, and later with in-season fieldwork, he determined that the presence of open cedar woods was a key indicator of potential metalmark habitat. But even sites with suitable hostplants and topography were not used if the nearest suitable nectar source (normally Butterflyweed or Black-eyed Susan) was more than 50 meters away. Indeed, adults rarely stray more than 40 to 50 meters from either hostplant or nectar locations. Over more than a decade of work, Norris has found and monitored 30 previously unknown metalmark colonies in the Connecticut-New York-New Jersey area.

SPECIFIC EAST COAST BUTTERFLY HABITATS

Low-Growth Habitats: Transition and Disturbed

Although definitive statistics are not available, it is reasonable to suggest that people most frequently see butterflies in *disturbed* or *transitional habitats*. This is where most people occur, for one thing, but in addition more than 90% of East Coast butterfly species use a "disrupted" habitat during some portion of their life cycle.

I-2. Transition habitat: field and woodland edge

Disrupted habitats include two general subtypes, disturbed and transitional. *Disturbed habitats* are significantly altered (often permanently) by human activity. Examples include settlements, waste areas, and agricultural fields. *See* p. 75. *Transition habitats,* by comparison, are less severely impacted, with intermittent or reversible disturbance, and retain a higher degree of ecological integrity. Examples include fallow fields and pastures, clearings, edges,[3] second-growth scrub, power line cuts, rural roadsides, etc. The line between these two habitats is not always clear, given the rich and unpredictable variety of disturbances that human activity engenders in the environment (nor is a precise delineation usually important).

As a group, butterflies specializing in disrupted habitats are not the rarest or most sought-after. But if some are "over-familiar" it merely reflects the fact that conditions in these habitats may be ideal for butterfly propagation. Disturbed and transitional environments are widespread, comparatively warm, and sunny, and each supports a diverse array of fast-growing, opportunistic hostplants and nectar flowers.

It was recently shown that caterpillars of the Cabbage White—a premier disturbed habitat specialist—emit sticky secretions on their body "hairs." These secreted chemicals (*mayolenes*) repel predatory ants and probably other attackers (Smedley et al., 2002). This helps explain how Cabbage Whites are able to survive in open agricultural areas, reproducing at rates 2 to 3 times higher than those of the closely related Mustard White (Ehrlich, 1985, p. 39). In our area, Mustard Whites are found in shadier field edge habitats than Cabbage Whites. It has not been determined whether the Mustard White's "edge player" status is purely a matter of wired-in behavioral predisposition, or whether it arose from differences in effective chemical defenses. Disturbed habitats by their nature highlight many such probing questions.

Among the numerous species closely linked to disturbed/transitional habitats (in terms of hostplants as well as nectaring) are Black Swallowtail, Checkered White, Cabbage White, Clouded and Orange Sulphur, Little Yellow, Dainty Sulphur, American Copper, Gray Hairstreak, Eastern Tailed-Blue, Variegated Fritillary, Phaon and Pearl Crescent, American and Painted Lady, Red Admiral, Common Buckeye, White Peacock, Monarch, Long-tailed Skipper, Wild Indigo Duskywing, Common and Tropical Checkered-Skippers, and Common Sootywing, plus a significant percentage of the "grass skippers," most notably European, Fiery, Peck's, Baracoa, Sachem, Monk, Eufala, Brazilian, and Ocola.

An interesting subgroup in this category is the *subtropical scrub* specialists, which frequent tropical-style second-growth plant communities in Florida. Burgeoning development over the past century has greatly increased the supply of second-growth scrub habitat there, and this may help account for successful colonizations (or range extensions) of several species on the lower peninsula, such as Orange-barred Sulphur, Statira, Soldier, Dorantes Skipper, and Monk (Opler & Krizek, 1984, p. 20).

Climax Low-Growth: Barrens, Outcrops, and Alpine Tundra

Climax, low-growth habitats are specialized and distinctive. Though biologically diverse, they share an important defining characteristic: they retain a low-growth profile even when left undisturbed for long periods of time. The exact reasons for this vary, but climax low-growth nearly always results either from a locally harsh climate, a peculiarity of underlying geology, or both. To thrive in such habitats, resident plants must specialize. And specialized plants attract specialized herbivores—including the caterpillars of several of our most range-restricted butterflies. For this reason, we examine these environments is some detail. (*See* Anderson et al., 1999; FFWCC, 1999; Myers & Ewel, 1990.)

Barrens are a rare but widely occurring form of climax, low-growth habitat. There is no precise, agreed-upon definition of the term "barrens," but it generally refers to areas that experience so-called edaphic climax, i.e., a limited-growth ecological steady state that is restricted not by local climate as such, but rather by soil deficiencies or other limitation. Barrens are generally best for specialty butterflies in the spring, before summer heat and dehydrating exposures deplete available nectar sources.

Mid-Appalachian shale barrens are found primarily in the "Ridge and Valley Province," between the Allegheny

---

[3] About 20 to 25 mainly woodland species use edge habitats on a regular basis. If these were removed from the list, the percent of species using disrupted habitats would decline to the low 80s—but this is still the highest use rate for any habitat class.

I-3. Shale barrens often support thin woodlands, with surface minerals appearing at outcrops and in erosion areas, such as along streams or roads.

Plateau in West Virginia and the Blue Ridge Mountains of western Virginia, Maryland, and Pennsylvania (Braunschweig et al., 1999). The geology here is dominated by sedimentary rock, folded into a series of rolling hills and steep-sided, picturesque valleys. Shale barrens usually occur along south-facing ridges with a relatively steep slope (> 20%), many times in places where a stream has undercut the hillside, hastening erosion. The habitat consists of rough, mineral soils, acidic and of low organic content (pH 4.0–5.0). There is usually a thin overstory of trees, interrupted by bare (or nearly bare) shale outcroppings. One butterfly hostplant, the Shale Barren Rockcress (*Arabis serotina*), spends its first year of life after germination growing a deep tap root. It blooms only in its second year, after a reliable source of moisture is ensured.

Shale barren soils are botanically uninviting because of poor water retention and a marginal nutrient supply. But their most detrimental feature is prolonged exposure to direct sunlight: midday soil temperatures in summer are comparable to those of southwestern deserts. Plants adapted to survive here must be heat- and sun-tolerant (*obligate heliophytes*).

The rarest butterfly species associated with shale barrens in our area are Olympia Marble and 'Appalachian' Grizzled Skipper. A number of other otherwise uncommon species also appear here regularly.

East Coast *serpentine barrens* occur in discontinuous patches along the Appalachian ridgeline (bypassing the Blue Ridge area of western Virginia, where shale predominates). Notable examples occur in eastern Maryland and Pennsylvania (Soldier's Delight, Nottingham Barrens, Goat Hill, etc.). Others are scattered loosely from the southern Appalachians to New England. Typical flora consists of open grassland intermixed with a variety of scrub oaks and low-growing scrub pines. (*See* Tyndall & Hull, 1999; Brooks, 1987; Gatrelle, 2001e.)

The origins and geology of serpentine barrens contrast sharply with those of their shale counterparts, but the result is often the same: infertility. While shale is a shallow, sedimentary rock, serpentine has deep, igneous roots. It develops well below the surface and is pushed upward by tectonic activity. Both shale and serpentine soils dehydrate rapidly—in the case of shale because it is nonporous and sheds water, in the case of serpentine because it is highly porous and lets water drain through. Shale soil is low in inorganic nutrients, while serpentine is extremely high in such components. (The drawback here is that serpentine typically contains high levels of *toxic* elements, such as chromium, nickel, or magnesium, while being deficient in important *nutritional* elements, such as calcium.)

Early settlers recognized serpentine barrens as odd lands, nearly devoid of mature timber. While poor for agriculture, they were actively grazed, and later mined for chromite. Through the mid-1950s, scientists sought to determine why these barrens (like many other serpentine habitats in the world) were relatively infertile. But then, as grazing declined, the "barrens" suddenly began succeeding into full-sized Virginia Pine (*Pinus virginiana*) woodlands, a process now more than 90% complete. Evidently the mid-Atlantic serpentine barrens are less infertile than many serpentine barrens elsewhere. It was ultimately concluded that the original low-growth character of mid-Atlantic serpentine barrens likely resulted from intentional fires set by precolonial hunters to improve access to game.

Surviving serpentine barrens are appealing butterfly sites, hosting a number of "barrens specialists" such as Mottled Duskywing, Edward's Hairstreak, Common Roadside-Skipper, and several *Andropogon* grass-feeding skippers (Cobweb, Dusted, and Leonard's). In western North Carolina, isolated populations of Tawny Crescent and Gorgone Checkerspot are associated with high-altitude serpentine barrens, and a former population of Arogos Skippers (now extirpated) once lived on a serpentine outcrop in Staten Island (Shapiro & Shapiro, 1973, p. 104).

Limestone barrens (*alvars*) are odd, intriguing habitats. Globally imperiled, they barely reach into our area from the Great Lakes, with scattered examples near Watertown, New York, and around Lake Champlain. Unlike shale or serpentine barrens—which are often thinly wooded—alvars truly *look* barren, with flat, exposed, pavement-like rock surfaces thinly covered by moss, lichens, and small shrubs, but little tall vegetation. (*See* Catling & Brownell, 1999.)

Alvars consist of horizontal limestone or dolomite bedrock, scraped clean by retreating glaciers. Climate and geology have not allowed a deep base of organic topsoil to become reestablished. What soil there is is calcareous (alkaline). The word "alvar" is Swedish for "limestone barren" (the habitat also occurs in Scandinavia).[4] Where the limestone bedrock has weathered, the pavement surface becomes pocked with distinctive, crevasse-like *grykes,* some quite deep. Ecologically, alvars represent a peculiar mix of boreal and prairie influence. Newer examples, such as those in New York, show more of a prairie imprint, including colonies of locally endemic grassland plants such as Prairie Smoke (*Geum triflorum*).

The Chaumont and Limerick barrens in northwestern New York are best visited in spring, when Hoary Elfin, Columbine Duskywing, and Olympia Marble fly simultaneously. Tawny Crescent, an alvar specialist in Canada, has not been reported from New York sites but should be sought there in early summer.

Pine (or sand) barrens occur in large, scattered patches in the Northeast, mostly along the outer coastal plain—in southern New Jersey, eastern Long Island, and Cape Cod—but also at a few sites inland (*see* Boyd, 1991; Gibson et al., 1999; McPhee, 1967; habitat photo, p. 243.[5] Though found in large, individual tracts, sand barrens are still a niche habitat, covering less than 1% of the East Coast's land area (about 4,000 square miles). A number of other pinewoods habitats in the region could also justly be called "pine barrens," as they too grow on sandy, acidic, well-drained soil and maintain their ecological character with periodic fires. But at present the term "pine barrens" is usually reserved for a set of northern communities in which the dominant conifer, Pitch Pine (*Pinus rigida*), grows in stunted, or "dwarf," form, in places no more than 9 feet tall. (Away from pine barrens, they reach heights of 50 to 60 feet.) Pitch Pines are adapted to frequent burns, sprouting new foliage from their roots after fires.

The pine barrens themselves are a harsh and at times monotonous environment, somewhat poor in faunal diversity (except for a few groups, such as reptiles). Many noteworthy butterflies live in one of the distinct *subhabitats* found in the pine barrens, including low, swampy ponds or bogs, tundra-like Bearberry flats, scattered Atlantic White-Cedar swamps, etc. Species to be sought in pine barrens (all habitats) include elfins (mainly Brown, Eastern Pine, and Hoary, but locally Frosted and Henry's), 'Olive' and Hessel's Hairstreaks, Atlantic Azure, Georgia Satyr, Bog Copper, Sleepy Duskywing, Cobweb Skipper, Leonard's Skipper, Dotted Skipper, Arogos Skipper, and Two-spotted Skipper.

An important *inland* pine barren site is the Albany Pine Bush, located in the upper Hudson River Valley. Originally about 40 square miles, this small habitat is now less than one-tenth that size. The Pine Bush was formed when the

I-4. Limestone barren, or alvar

sand-laden meltwater of late Pleistocene glaciers emptied into a large glacial lake that covered the site of modern-day Albany. The Pine Bush is famous as a stronghold for the Karner Blue, a rare and threatened race of Melissa Blue. It also supports colonies of Mottled Duskywing, Frosted Elfin, Cobweb Skipper, Sleepy Duskywing, Common Roadside-Skipper, and others. Unfortunately, the area has proven difficult to manage for the Karner Blue, even with carefully prescribed burns, since colonies of Wild Lupine (*Lupinus perennis*), on which the blue depends, is difficult to reestablish once shaded-out by successional foliage.

Barrens of the southeastern states include *granite outcrops* and *balds,* both of which occur at higher altitudes in the central and southern Appalachians. *Balds* are treeless habitats, usually found above 4,000 feet (Wiser & White, 1999). There is no climatic timberline in the southern Appalachians, since no summits there exceed the altitude beyond which trees can grow. But the *heath balds* that form in deforested areas such as avalanche zones can be slow to grow back into forest—thwarted by locally harsh climatic conditions and, in places, by inhospitable, serpentine soils. Balds are surrounded by cool, northern-style mountain forests. Given their northern affinities, balds and their associated woodlands have become southern refugias for relict populations of boreal butterfly species such as Pink-edged Sulphur, Atlantis Fritillary, Tawny Crescent, Gorgone Checkerspot, and Gray and Green Comma.

Granite outcrops of the upper Appalachian Piedmont are stark, awesome habitats. (*See* Shure, 1999; habitat photo, p. 319.) Though situated well below the timberline, they are almost totally treeless, giving the impression of an immense parking lot, with nothing in view but solid granite as far as the eye can see. The reason for this appearance is

---

[4] There is no apparent connection with early Hindu mystics, also called "alvars."

[5] Coastal pine barrens formed on or near the southern terminus of Wisconsin

Stage glaciers (which retreated in the late Pleistocene, about 10,000 years ago). Some are situated atop *terminal moraines* (Strahler, 1966).

simple: the Precambrian granite surface layer is hard enough to resist erosion, and most soil that does manage to form is washed away before it can accumulate. A number of sparse but extremely specialized ecological communities have developed in flat outcrop pools and other niches, where meager growth is possible. Outcrops are interesting to butterfly observers largely because of yucca plants that grow in dry, thin-soil woodlands around the edges. These yucca plants can be the host of a Yucca or Cofaqui Giant-Skipper.

At the northern edge of our region, far from these Appalachian balds and outcrops, a few mountain ridges pass the timberline and support true *alpine tundra*. This tundra habitat has close biological affinities with arctic tundra farther north—indeed, historically these are remnant patches of "true" northern tundra, stranded at high altitude when the glaciers last retreated. (*See* Steele, 1982; Harris et al., 1964; habitat photos, pp. 169, 222, 223.) Virtually the only trees in this environment are a few low-growing conifers and birches, gnarled and twisted into bizarre shapes by the cold and wind (*krummholz*).[6] Three butterfly species in our area are alpine tundra specialists: Purplish Fritillary, Melissa Arctic, and Polixenes Arctic. The first two are found in this region solely in the Presidential Range in New Hampshire (on Mt. Washington and adjacent ridgelines), the third on the Tablelands of Mt. Katahdin, in Maine.

Open Woodlands: Savanna, Sandhills Pines, Pine Rocklands

Southeastern *pine savannas* are open, grassy habitats interspersed with clusters of trees. These clusters may vary greatly in density, but do not form closed canopies.[7] Southeastern savannas are ancient environments, never glaciated, whose origins are believed to extend back more than a million years. They are an extremely important regional habitat, with an exceptionally high rate of endemism. Interestingly, many savanna plants and animals have close ancestral ties with the arid Southwest—evidence of drier climatic times, when widespread xeric habitats across the southern rim of the United States allowed dispersal of plant and animal species directly to the Southeast and Florida. (*See* Platt, 1999.)

Pine savannas maintain a delicate ecological balance, in that trees and grassland coexist in long-term proximity. The durability of this unlikely equilibrium can be explained by two primary factors: sand and rainfall.[8] Over the

eons, countless tiny rock fragments (mostly quartz) have washed or blown from the aging Appalachian Mountains onto the coastal plain and nearby continental shelf. As ancient sea levels rose and fell during the Ice Ages, these deposits formed beaches and river deltas. Today, their sandy remains underlie modern, inland savannas. Savanna soils are largely dry and sterile, their nutrients leached away by draining surface water. An exception is where clay or some other hard substrate materials create a "lens" that holds water near the surface. In such cases, specialized habitats such as clayhills, "seepage savannas," or pocosins result, as discussed below.

Mature hardwood forests still might overrun savanna habitats, except for seasonal variations in precipitation. Rainfall levels in the Southeast are the highest in our region (40 to 65 inches), yet they are highly seasonal, with a distinct winter drought and usually a late-spring dry period. Tinder-like drought conditions spawn frequent fires, as often as once a decade in some places. This cycle favors fire-resistant pines over hardwoods, and allows grasslands to persist in areas that would otherwise be overgrown with forest.

In many respects, open pine savannas are ideal for butterflies. They are sunny, high in plant diversity, and offer a good deal of shelter. Seasonal aridity and regular burns are drawbacks, on the other hand, that require special adaptations. Uncommon butterflies associated with open, grassy savannas include Georgia Satyr, Arogos Skipper, Dotted Skipper, Meske's Skipper, and Reversed Roadside-Skipper.

Sandhill pines (or high pines) are a handsome form of upland savanna that once covered some two-thirds of the southeastern coastal plain. Much of this original habitat has been lost—mostly logged, developed, overgrown because of excessive fire suppression, or converted to commercial "pine plantations" of little ecological value. Large remaining stands can be found along the dividing line between the Piedmont and coastal plain in the Southeast, e.g., in the Sandhill Gamelands of North Carolina, and separately in northern Florida, e.g., at Ocala National Forest. (*See* LeGrand, 2000; Platt, 1999; Myers, 1990.)

Sandhills have an overstory of tall, straight Longleaf Pines (*Pinus palustris*) and an "herb layer" of thin Wiregrass (*Aristida stricta*). There is only a sparse understory of patchy shrubs, giving the habitat a spacious, parklike appearance. (*See* habitat photo, p. 306.) Ecologically, sandhills form on elevated, especially well-drained tracts of sandy soil. Frequent, low-intensity fires are needed to maintain the habitat, and Longleaf Pines are superbly

---

[6] Technically, the large assemblage of miniature heaths that grow in "alpine gardens" above timberline qualify as "woody plants." But they are almost universally treated as wildflowers, rather than as trees or shrubs, in natural history literature.

[7] Savannas also occur in northeastern pine barrens. Some are seasonally wet, kept open by regular inundation. See photo and discussion on p. 284.

[8] A third, less regular factor is strong wind, which occurs frequently in the Southeast, in the form of hurricanes and tornadoes. Savanna pines fare better in high-wind situations than either hardwoods or tall, forest pines.

I-5. Pine savanna I-6. Regular fires are needed to maintain the ecological balance of savanna habitats.

adapted to this exposure. Their long, moist needles resist low-intensity fires. Even the saplings are fire-resistant, a distinctive trait among our conifers. Individual trees can reach an age of 500 years or more.

Sandhills are a key habitat for a number of the rarest or least-often seen butterflies of the East Coast, including Confused Cloudywing, Mitchell's Satyr (extremely local in sandhill bogs), Meske's Skipper, Dusky Roadside-Skipper, and Yucca Giant-Skipper. Also present are Palamedes Swallowtail, Little Metalmark, Gemmed Satyr, Mottled Duskywing, Zarucco Duskywing, Dotted Skipper, Arogos Skipper, and Lace-winged, Carolina, and Reversed Roadside-Skippers.

Florida sand scrub (or Florida scrub) is among the most specialized of East Coast habitats, also perhaps the driest of the sand-pine environments. It grows on thoroughly drained, highly infertile sandy soil (*quartzipsammet*). Early settlers referred to sand scrub as "concealed desert." The dominant overstory plant is Sand Pine (*Pinus clausa*), a quick-growing, short-lived tree (average life just 50 to 70 years). The understory is an "impenetrable mass" of scrub oak, Florida Rosemary (*Ceratiola ericoides*), and miscellaneous other shrubs.[9] (*See* Menges, 1999; Profant, 1989–90; Myers, 1990.)

Florida sand scrub grows on the remains of ancient dunes and beaches. *Inland sand scrub* grows on high ridges that became islands when glaciers melted in earlier eras,

raising water levels worldwide. These islands served as refugia for local plants and animals, many of which developed into separate species during their isolation. Endemism is especially high in the Lake Wales Ridge scrub, near Sebring. This is now a highly endangered ecosystem. The largest remaining inland tract is the Big Scrub at Ocala National Forest. *Coastal scrub* occurs on the panhandle and the Atlantic Ridge. This type can be seen at Jonathan Dickenson State Park.

No butterflies are endemic to Florida scrub, but it is a regional stronghold for several species such as Zebra Swallowtail, which uses certain scrub-dwelling pawpaws—including the endangered southeastern coast Four-petaled Pawpaw (*Asimina tetramera*)—in place of the common Pawpaw (*A. triloba*), which does not grow in southern Florida. Southern Dogface also specializes in this habitat in our area. It feeds on Prairie Clover (*Dalea fayei*), a scrub pea.

Pine rocklands (also called rimrock pines) are in a class by themselves among southeastern savannas. (*See* Platt, 1999; Snyder et al., 1990; habitat photo, p. 204.) They grow precariously atop exposed limestone outcrops in southernmost Florida and on certain of the lower Keys. The surface is perforated with sharp-edged, ankle-twisting "solution holes" created where rainwater has gradually dissolved soft areas of limestone. There is extremely little organic soil. The rimrock pine overstory consists of Slash Pines (*Pinus elliottii*), a local form of Longleaf Pine. Most understory plants are of West Indies lineage, and the incidence of endemism is high. Pine rocklands are fire-maintained, although only the scrubby understory typically burns, not the widely spaced crowns. Most herbs are perennials that resprout after fires. Indeed, many plants dedicated to this habitat rarely bloom at any other time.

Pine rocklands are an extremely important habitat for butterflies—even allowing for a number of species that have been lost in recent decades. They include Florida's only fully endemic butterfly species, the Florida Leafwing, also Bartram's Hairstreak (an endemic race), Miami Blue

[9] The habitat is resilient after fires, as many plants maintain large underground root systems from which they resprout after blazes. The inland race of Sand Pine is *serotinous,* dropping its seeds only after fire (which may kill the adult tree). Many sand scrub plants have relatively long regeneration cycles; burns that occur too frequently can kill vulnerable saplings, destroying the habitat altogether.

I-7. Boreal forests extend across the northern tier of the continent and southward in pockets at high altitude in the Appalachians. Low-lying boreal forests are crisscrossed with streams and bogs.

(also an endemic race, now restricted to a single known population on Bahia Honda Key), Acacia Blue (recently established), Florida Duskywing, and southern forms of Meske's and Palatka Skipper (declining, widely extirpated).

TRUE WOODLANDS

True woodlands in our region (those with closed canopies) do not seem ideally suited for butterflies, especially in the North. They tend to be dim, cool, and low in available nectar. Offsetting advantages include a relative degree of shelter, from both predators and adverse weather, plus a ready supply of leaves to eat (albeit often difficult to digest). Such benefits evidently outweigh disadvantages on balance, since more than 40% of East Coast butterflies use woodlands during some part of their life cycle. After emergence, the adults of many tree-feeding species linger in the canopy, or else nectar in nearby clearings or fields. A few groups, notably pearly-eyes and wood-satyrs, inhabit interior woodlands as adults, feeding mainly on dung, droppings, sap, or decaying matter, as opposed to nectar.

Ecologists traditionally divided North American habitats into a series of "life zones," based in large part on dominant forest types (Sutton & Sutton, 1985). These zones are still useful as habitat guides for woodland butterflies. They include Arctic/Alpine; Hudsonian (or taiga); Canadian (or boreal); Transition; Upper Austral (mixed deciduous); Lower Austral (southeastern); Subtropical (south Florida).

Northern Forests: Boreal (or Canadian) and Transitional

The boreal and transitional climate zones—in which northern forests occur—occupy roughly 40% of the East Coast

region. Outside of Maine, *boreal forest* is found only in scattered patches along the Canadian border and at high altitudes in New England, New York, and the Appalachians. Boreal forests are composed of northern conifers (fir, spruce, tamarack, etc.) and northern hardwoods (aspen, alder, birch, maple). Interspersed among these forests stands are patches of second-growth northern scrub thickets, overgrown with various woody shrubs and saplings. Characteristic boreal butterflies include the "northern commas" (Green, Gray, Hoary, and Satyr), Pink-edged Sulphur, Harvester, Silvery Blue, Atlantis Fritillary, Milbert's Tortoiseshell, White Admiral, Arctic Skipper, and Common Branded Skipper. A number of boreal bog specialists also occur in the vicinity, as discussed below.

Northern pinewoods probably support fewer butterfly species than any major habitat in our region. They have dense canopies and dark, cold interiors. In addition, there is little by way of caterpillar food, since few East Coast butterflies have learned to feed on pines.

Transition forest contains a variable mixture of conifers (White Pine, Eastern Hemlock, etc.) and northern hardwoods (Red and Sugar Maple, American Beech, Red Oak, etc.). One transition zone specialist in our area, the Early Hairstreak, is most common in higher-altitude American Beech forests. Other regular transitional forest species include Canadian Tiger Swallowtail, Compton Tortoiseshell, and Northern Pearly-eye.

Mixed Deciduous Woodlands (Upper Austral)

The Upper Austral climate zone, in which *mixed deciduous forests* occur, covers about 25% of our region. It extends from southern New England through the mid-Atlantic states, then southward across the upper Piedmont and lower mountain provinces of the Carolinas and northern Georgia. As the name implies, mixed deciduous forests consist almost entirely of various hardwoods (elm, hickory, oak, maple, cottonwood, dogwood, etc.). *Cove forests*, which grow at altitudes between 3,200 and 5,000 feet in the southern Appalachians, are an especially handsome and biologically diverse form of mixed deciduous forest. They form part of the rich, southern Appalachian forest system—often classified as a separate habitat zone, since nearly every eastern forest type can be found here in relatively close proximity. Significant tracts of this forest were preserved in the 1930s, with the formation of the Great Smoky Mountains National Park.

Cove forests illustrate an interesting ecological distinction between the Northeast and Southeast. Specifically, in the northeastern United States, deciduous trees tend to grow at lower altitudes and conifer forests on mountaintops. The reverse is typically true in the Southeast, where mountain deciduous forests occupy all but the highest

I-8. Cove forests of the southern Appalachians, high in species diversity, served as refugias for deciduous forest species during past glacial epochs.

I-9. Pine flatwoods

ridges, towering over a vast expanse of pine flatwoods that blanket the outer coastal plain.

About 20% of East Coast butterfly species have some connection with mixed deciduous forests. Characteristic species includes Eastern Tiger Swallowtail, Spicebush Swallowtail, *Satyrium* hairstreaks (Banded, Hickory, and 'Northern' Oak), Appalachian and Spring Azures, Northern Metalmark, Diana (Appalachians only), Silvery Checkerspot, Question Mark, Eastern Comma, Mourning Cloak, Red-spotted Purple, Little Wood-Satyr, Juvenal's Duskywing, Hobomok Skipper, and Bell's Roadside-Skipper (the last being extremely local, mainly found in the western South Carolina Piedmont). Some resident species, such as Eastern Tiger Swallowtail, feed on hostplants from different tree families, allowing them to take advantage of the botanical diversity contained in these forests.

Southeastern Woodlands (Lower Austral)

Southeastern woodlands form a broad band along the lower Piedmont and southeastern coastal plain. They grow in the Lower Austral climate zone, which covers nearly a third of the total East Coast area. Lower Austral woodlands include a diverse and extensive variety of pinewoods, plus several specialized types of deciduous forest. Taken as a whole (including swamps, savannas, and other specialized habits), the Lower Austral zone contains an exceptionally large number of endemic butterflies. But the forests themselves support rather few distinctive species.

Pine flatwoods, the most common forest type in the Southeast, are so-called because they occupy low-lying sites along the coastal plain where topographical variation is minimal. (*See* Abrahamson & Hartnett, 1990; Nelson, 1995). Flatwoods have an overstory of pines (including Slash, Pond, and Longleaf) plus a dense shrub layer (e.g., Gallberry, Fetterbush, Staggerbush, Wax Myrtle, Saw Palmetto in the south, etc.).[10] Like all southeastern pinelands, pine flatwoods are fire-maintained. But because of recent fire suppression, the shrub layer in many pine flatwoods has lost much of its original character. Typical butterflies include Palamedes Swallowtails, Sleepy Orange, Southern Broken-Dash, and—much less commonly—Yucca Giant-Skipper.

Many ecologists consider *Florida dry prairie,* found in the central peninsula, to be a treeless form of pine flatwood. (*See* habitat photo, p. 161.) Butterflies can be common in dry prairies—provided they are neither flooded nor drought-stricken, both of which conditions occur frequently. Some butterfly species reach notable concentrations on dry prairies, including many of the specialty skippers of the Florida peninsula. An amazing 29 skipper species have been found at Kissimmee Prairie SP in a single day (B. & L. Cooper, pers. comm.).

[10] Sandhill pines, discussed earlier, generally occur on higher ground than flatwoods and have better drainage. They also have an understory of wiregrass, with few shrubs. This said, the two habitats are closely allied, and even "trade places" over time at particular sites. The fact that some flatwoods lack a closed canopy further complicates the classification of these habitats.

Apart from the mixed deciduous forests that grow along the Piedmont, upland hardwood forests are comparatively uncommon in the Southeast. Many southeastern hardwoods grow in wooded wetlands (swamps or pocosins), as discussed below, or in maritime forests along the coasts. Among deciduous habitats that do occur are *oak mottes*, productive for Southern Hairstreaks, and limited tracts of rich, temperate hardwood forests. Those in north Florida are primary habitat for the 'Viola's' Wood-Satyr.

Coastal Woodlands: Maritime Forest and Subtropical Hardwood Hammocks

The stresses on maritime forests are daunting: fierce storms, tidal surges, abrasive sand particles, caustic salt spray, and thin soil with few nutrients. In addition, the habitat is under constant threat from development almost everywhere. In the North, coastal forests of American Holly, Northern Red-Cedar, and American Hackberry provide habitat, respectively, for Henry's Elfin, 'Olive' Juniper Hairstreak, and the "hackberry butterflies" (American Snout, Tawny Emperor, and Hackberry Emperor). In the South, coastal groves of Southern Red-Cedar and maritime live oak forests are prime habitat for Juniper and Oak Hairstreaks. In all these cases, resident species are joined by coastal migrants at various times. (*See* Johnson & Barbour, 1999.)

Subtropical hardwood hammocks are a uniquely important coastal (or near-coastal) habitat for East Coast butterflies. (*See* habitat photo, p. 70.) These exotic woodlands are limited to the southernmost counties of mainland Florida and the Keys. More than 150 trees and shrubs of Caribbean lineage grow here, most found nowhere else in the United States. The hammock overstory is typically dominated by Wild Tamarind, Pigeon Plum, Gumbo Limbo, Jamaican Dogwood, or Poisonwood (a tree-sized Poison Ivy relative). Understory shrubs are enormously varied, and include such butterfly hostplants as Limber Caper and Cat's Claw.

The most celebrated butterfly resident of tropical hardwood hammocks is the federally endangered Schaus' Swallowtail. Other important species linked to the habitat or its edges include Florida White, Mimosa and Dina Yellows, Amethyst Hairstreak, Silver-banded Hairstreak, Miami Blue, Zebra, Dingy, and Florida Purplewings, Ruddy Dagger Wing, Hammock Skipper, and Three-spotted Skipper.

WETLANDS

For butterflies, wetlands are an acquired habitat. None of our species is aquatic at any life stage, save perhaps the Bog Copper, whose overwintering eggs can survive prolonged submersion. Wetlands offer definite advantages to adult butterflies, since plants with "wet feet" can produce a consistent supply of nectar and lush larval food, notably grass and sedges. But wetlands are a difficult habitat in which to find shelter during floods and storms, or to secure a protected niche for overwintering. Indeed, it is not clear where the caterpillars of some marsh-dwelling skippers (e.g., in genus *Euphyes*) spend the winter.[11]

Freshwater Wetlands

Among wetland habitats, *damp meadows* and *water edges* are used most commonly by East Coast butterflies. This habitat type is really an intergrade between upland habitats and wetlands. It combines the advantages of ready moisture with the security of a generally nonaquatic landscape. At least 20% of East Coast species regularly exploit such "friendly" wetland settings—and the count would be even higher if we included incidental use by nectaring adults of nonwetland species. Among the many butterflies regularly found in wet meadows or along the edges of streams or drainage ditches are Bronze Copper, Acadian Hairstreak, Regal Fritillary (prairie swales), Meadow Fritillary, Harris' Checkerspot, Texan Crescent, Baltimore, White Peacock, Viceroy, Appalachian and Eyed Brown, Least Skipper, Long Dash, Delaware Skipper, Byssus Skipper, Black Dash, Berry's Skipper, and Dun Skipper.

The next most commonly used fresh water butterfly habitat is *wooded wetlands,* or *swamps.* Swamps may be defined as wetland habitats dominated by trees—or, alternatively, as forests growing in water. In the North, *Red Maple swamps* are widespread but rather little used by butterflies. This is not surprising, since none of our species feed on maples, the major available hostplant.

Matters are very different in *southern hardwood swamps,* which are a haven for butterflies. Ecologists recognize many varieties of southern swamp habitat, including *river swamps* and *still-water swamps.* The latter are especially productive for butterflies. In late summer, blooming sprays of Hercules' Club (*Aralia spinosa*) in the Great Dismal Swamp of southeastern Virginia, their lush flowerheads laden with nectar, can create an amazing feeding frenzy. Characteristic species in these storied habitats include Zebra and Palamedes Swallowtail, Great Purple Hairstreak, Striped Hairstreak, Carolina Satyr, and Clouded Skipper. (*See* Roble et al., 2000; Glassberg, 1993b; Simpson, 1990; Ewel & Odum, 1984.)

Canebrakes occur commonly in or near southern hardwood swamps. These grassy enclaves are exceptionally

---

[11] The wintering habits of certain marsh-dwelling skippers are poorly known. Some may endure submersion for short periods, but not indefinitely based on recent evidence (Konvicka et al., 2002).

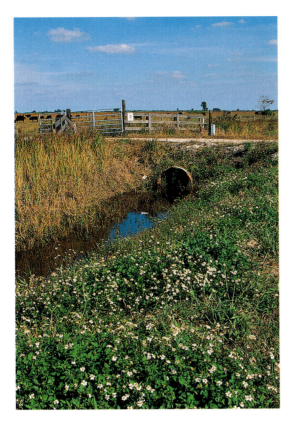

I-10. Wet meadows and roadside ditches provide a lush habitat, combining many of the best features of grasslands and wetlands.

I-11. Southern hardwood swamps host a great variety of specialized butterflies.

important for butterflies. A number of highly sought-after "southern swamp specialists" are actually obligate cane-feeders. They inhabit swamps only because their hostplant grows there. These species include Southern Pearly-eye, Creole Pearly-eye, and three species of roadside-skipper (Lace-winged, Carolina, and Reversed). Yehl's Skipper also eats cane, though not necessarily exclusively. (*See* Hall, 1996; Platt, 2002.)

A number of East Coast butterflies utilize *bogs* and *fens,* often called *peatlands* since both are characterized by an accumulation of undecomposed organic matter in a continually wet environment. Water is held in the habitat by an underlying, impermeable substrate, or by a high water table or local climatic conditions that restrict evaporation. (*See* Cassie, 1994; Johnson, 1985.)

Bogs are highly acidic peatlands. Northern "kettlehole" bogs are the footprints of retreating glaciers, created where large, slow-melting ice blocks were left behind. The remaining indentations now experience little inflow except for rainwater. Decomposition is slowed by the buildup of acids, and from a lack of inflowing nutrients.

Not all northern bogs were formed in this characteristic manner, however. Some were created by rising water tables (many times associated with beaver dams), which drowned formerly mesic upland habitats. And *raised bogs* in New England formed along foggy coasts with regularly high humidity, where evaporation was insufficient to eliminate standing moisture. (*See* habitat photo, p. 146.) The distinctive flora of northern bogs includes *Sphagnum* moss, a myriad of low-growing heaths, exotic bog orchids, and carnivorous plants, such as Northern Pitcher-Plant and sundews.

However formed, bogs have a cadre of unusual butterflies, many found nowhere else. In the boreal north, Black Spruce-Tamarack bogs are primary habitat for Dorcas Copper, Brown Elfin, Bog Elfin, Bog Fritillary, and Jutta Arctic. Farther south, bogs also host Georgia Satyr, Mulberry Wing, Dion Skipper, and Two-spotted Skipper. Northern Blues are found in raised coastal bogs of eastern Maine.

Fens differ from bogs in that they are fed by an inflow of groundwater, and are generally alkaline rather than acidic.[12] This groundwater inflow increases available nutri-

---

[12] It is sometimes categorically stated that bogs are acidic while fens are alkaline. In fact, the precise acidity of a fen depends on the soil through which in-seeping waters have percolated. Northeastern fens vary from somewhat acidic to highly basic (Johnson, 1985, p. 27). Interestingly, "kettlehole" bogs lying below the water table also receive groundwater inflows, and thus technically meet the definition of a fen. Such anomalies cause some ecologists to define both habitats as "peatlands," with other features discussed separately.

I-12. Sawgrass (*Cladium jamaicensis*) marshes are an important butterfly habitat in our region. They occur as far north as southeastern Virginia.

ents, boosting species diversity. Sedge rather than *Sphagnum* moss is the dominant ground cover, and trees and shrubs can be more common and varied.

The best-noted fen specialist in our area was the northern race of Mitchell's Satyr (*Neonympha mitchellii mitchellii*), which lived in calcareous fens in western New Jersey until extirpated in the 1980s. A surviving East Coast population of Mitchell's Satyrs uses bogs and artificial, boglike habitats, notably at Ft. Bragg, North Carolina.

The pocosin ("swamp-on-a-hill") is an endemic, boglike habitat of the Southeast. (*See* Kirby, 1995.) The name is at times applied to a number of rather different kinds of habitats. By most definitions, pocosins are thick, deciduous "shrub bogs" that form on nonporous substrates (e.g., clay) along the coastal plain or in pinewoods habitats. Some have trees, usually a few pines or a stand of Atlantic White-Cedars. But as most strictly defined, a true pocosin is purely shrubby, without trees. Near the coast, pocosins form in intergrade zones between hardwood swamps and open marshes. But in the sandhills, they occur along seep lines or in low-lying sites where the water table is near the surface. Pocosins are highly acidic, which retards the decay of organic matter. Over time, a thick layer of peat forms, up to 60 inches deep. Such deposits are thickest near the center of the pocosin, forming a hill that gives the habitat its name. Pocosins are thought to require periodic, low-intensity fires during dry seasons to maintain their character.

Many butterflies visit pocosins when Sweet Pepperbush (*Clethra alnifolia*) is blooming. Others use pocosin plants as hosts. The Palamedes Swallowtail uses Red Bay (*Persea borbonia*) growing in pocosins or in hardwood swamps. But few species are true pocosin specialists. Among them are southern populations of Hessel's Hairstreak (in Atlantic White-Cedar stands—if shrub bogs with trees are considered pocosins!), and probably King's Hairstreak, the most elusive of *Satyrium* "woodland" hairstreaks. King's Hairstreak cater-pillars feed on Sweetleaf (*Symplocos tinctoria*), a shrub that occurs in (but is not limited to) pocosin thickets.

Apart from brief visits to sip nectar, a limited number of butterflies are equipped to inhabit *open marsh* habitats. Formally, a marsh is a treeless wetland with emergent vegetation (i.e., vascular plants rooted under water but growing above the surface). Not surprisingly, most butterflies that specialize in marshes are grass- or sedge-feeders. Examples include Appalachian and Eyed Brown, Georgia Satyr, and a variety of skippers such as Least Skipper, Arogos Skipper (some populations), Delaware Skipper, Byssus Skipper, Mulberry Wing, Broad-winged Skipper, Dion Skipper, Black Dash, Two-spotted Skipper, Palatka Skipper (sawgrass marshes), and Twin-spotted Skipper. Some of these species also live in bogs.

Saltwater and Brackish Wetlands: Coastal Habitats and Migration Corridors

When tidal flooding, salt spray, and blowing sand are added to the challenges of wetland existence, the ranks of resident butterflies thin dramatically. *Open salt marshes,* though among the most productive of all ecosystems, have only one fully dedicated butterfly species in our area, the Salt Marsh Skipper. *Brackish marshes* and *backwaters* fare somewhat better, since tidal activity and weather exposure are less severe away from the immediate coast. (*See* photo, p. 295.) Several distinctive backwater species include Rare Skipper, Aaron's Skipper, and Dukes' Skipper. Some colonies of Bronze Coppers also dwell on brackish marsh edges.

Slightly inland from the coast, *salt flats* flood periodically, forming shallow pools. As this overflow evaporates, a highly saline residue remains. Only the most highly salt-tolerant of plants, such as glassworts and saltworts, can survive on such "salt pans." Few butterflies have adapted to salt flats as their primary habitat, but those that do can develop enormous populations. Tiny Eastern Pygmy-Blues, are often seen by the tens of thousands (even millions) on large flats in the Everglades, fluttering weakly among the glassworts. Great Southern Whites can also occur in enormous numbers in similar settings.

A number of butterflies inhabit *coastal scrub* and *dunes,* especially in the South. Examples include Great Southern White, Miami Blue (presently limited to Bahia Honda Key in Florida), Cuban Crescent, Obscure Skipper, and Yucca and Cofaqui Giant-Skippers. In eastern North Carolina, a recently discovered form of Dusted Skipper lives on scrubby, coastal habitats (including dredge spoil flats).

Mangroves are saltwater swamps that occur along warmer coastlines around the world. They have only a few associated butterfly species, the two most obvious examples being Mangrove Skipper (on Red Mangroves) and

Mangrove Buckeye (on Black Mangroves). Great Southern Whites also fly in this habitat, and Cuban Crescents and Obscure Skippers can be found along the separation zone between mangroves and inland habitats.

Finally, eastern coastlines are important as *migration routes*. From mid-summer through fall, a strong northward movement builds along the coast, from the Carolinas through New England. Some northward migrants, such as Cloudless Sulphur and Common Buckeye, can be readily seen in flight by field observers. Others, such as Fiery, Long-tailed, and Ocola Skippers, are less conspicuous, evident only when they pause for nectar along the way. (The latter two species also travel *southward* during the fall from the Carolinas, reaching as far south as the Keys.)

The most conspicuous coastal migration, however, is definitely the southbound fall flight of the Monarch and of certain adult-overwintering nymphalids. It is a dramatic sight to see thousands of Monarchs settling down to rest on coastal foliage at day's end during September and October, bound for wintering grounds in Mexico, Florida, etc. Among the nymphalids, Question Marks, Eastern Commas, and American Ladies move south each fall. Other species, such as the Compton Tortoiseshell and northern anglewings, are usually seen at the coast only in peak years, which occur irregularly. Some hawk watches have begun to record seasonal butterfly activity.

Butterfly migration is not limited to the coasts. Mountain ridges and river courses are also important seasonal movement corridors. In addition, many species engage in dispersal (vs. migration), and these may turn up almost anywhere. Details of butterfly seasonal movement are discussed in more detail below.

SPECIALIZATION: A KEY TO UNDERSTANDING DISTRIBUTIONS

Hostplant lifestyles (discussed above) can explain many aspects of butterfly ecology, especially in relation to specific habitats, where customized survival strategies are most obvious. But even a detailed knowledge of habitat use does not tell the full story of butterfly distributions. For butterflies themselves bring additional dimensions to the process, for example, by being either generalists or specialists.[13]

For discussion purposes, we have assigned East Coast species to categories according to *specialization strategy*: specialist, generalist, and for some purposes intermediate (or medium). There is no universally accepted definition for these terms, and even if there were they represent points on a continuum rather than well-defined "buckets."

Overall, we count 121 generalists and 113 specialists on the East Coast. The same ratio is found in skippers (45:41), and for "true" butterflies (76:72).

The *generalist strategy* stresses diversification, mobility, and opportunism, avoiding exclusive reliance on single hostplants or specialized habitats. Generalists can make do in a number of different settings—and, if not, they are prepared to move elsewhere, reinforced by their genetic adaptability. Common components of a generalist strategy include use of multiple hostplant species or families; regular seasonal movement (dispersal or migration); and opportunistic breeding cycles (often fitting in as many broods per season as weather permits). Generalists are commonly found in disturbed habitats, where they may achieve high reproductive rates compared with their more specialized relatives. Still, some generalists focus their energies on broadening the range of hostplants or habitats accessible to them, rather than on increased reproductive rates per se. Eastern Tiger Swallowtails sacrifice quick growth in exchange for the ability to digest and detoxify leaves from a variety of tree families (Douglas, 1989, pp. 184–86).

Specialists, by comparison, emphasize focus and control, amassing a suite of biological adaptations that enhance their ability to exploit particular niche habitats. Key elements of a specialist strategy may include use of a limited number of specialized hostplants (especially toxic or indigestible ones that would be difficult for a generalist to exploit); low dispersal rates and high site fidelity; and "targeted" brood cycles that coincide with the availability of specific hostplants or seasonal weather patterns. Additional physiological adaptations, such as the ultra-cold tolerance of alpine satyrs, can further enhance the strategy.

It bears restating that specialization strategy is *not* the same thing as hostplant lifestyle. Within each lifestyle, there are both generalists and specialists. One common toxic species, e.g., the Monarch, is a consummate generalist. But Baltimores and Atalas, also both toxic, are in the medium and specialist categories, respectively. Indeed, Figure 1.1 indicates that there is little consistent difference between generalists and specialists based on hostplant use.

Given the limits of our current insight, a significant degree of judgment is involved in assigning East Coast butterflies to specialization categories. But that does not mean that the task is beyond objective analysis. It was shown nearly three decades ago that certain digestive enzymes (*esterases*) produced by North American sulphurs vary genetically in a manner consistent with ecological lifestyle (Burns, 1975). Specifically, species with *high esterase variability,* such as Clouded and Orange Sulphurs (with 25 to

---

[13] See Futuyma and Moreno (1988). British geneticist Julian Huxley referred to generalists as "euryplastic" and specialists as "stenoplastic" (Huxley, 1942, pp. 444, 519). Generalism was then considered the ultimate product of evolutionary advancement. *See* Ford (1977, pp. 252–53); but *see* Levins (1968) (optimal strategy depends on environmental factors).

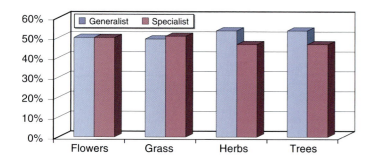

Figure 1.1. East coast butterfly specialization by hostplant type

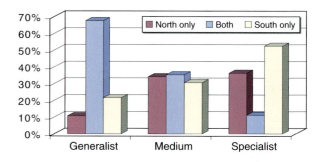

Figure 1.2. Life strategy by latitude zone

32 ESD alleles), were found to be generalists, whereas *low esterase variability* was found in specialists such as the blueberry-feeding Pink-edged Sulphur (3 to 4 alleles). A greater diversity of digestive enzymes would be useful to a species feeding on a wide and unpredictable variety of plants. Undoubtedly, the lifestyle strategy "choices" made by East Coast butterflies have many physical correlates, and it may ultimately be possible to trace their origins genetically.

But back to the original point: What about the relationship between specialization strategy and ranges? If habitat and hostplant use are critical determinants of butterfly distributions, then specialization strategies (which reflect both factors) should be linked to distribution. To explore this question informally, East Coast butterfly species were assigned to one of three *life zones*—"North Only," "South Only," or "Both Zones," as shown in Table II.

When life zones are mapped against specialization, a clear pattern emerges. Figure 1.2 shows that 68% of generalist species were also geographically widespread, living in both the North and South. At the other extreme, only 11% of *specialists* were widespread. The remaining 89% were either "North only" (36%) or "South only" (53%).

This analysis suggests that a butterfly's level of specialization, which reflects its use of hostplant and habitat, is a useful way to describe butterflies ecologically and can be

helpful in understanding range dynamics. In exploring brood cycles and additional life strategies in the following chapters, this will prove to be a recurring source of insight.

II. WHEN TO FIND BUTTERFLIES

BROODS AND BROOD TIMING IN GENERAL

Having learned where to look for butterflies, it remains to determine *when* to look for them. Arriving at a stake-out site a day or two before the beginning of a flight period (or a day or two after one has ended) is a truly vexing experience—especially in those cases where it is difficult to distinguish between a missed flight period and a species that is simply hard to locate. In our experience, an hour or two in prime habitat is usually long enough to determine whether or not a target species is on the wing. But whatever the correct interval, the ability to cope gracefully with such incidents is a valuable field skill.

As noted earlier, many butterflies occur in local population clusters. If the adults are to reproduce, their flight periods must coincide, at least to some degree. Butterflies address this basic imperative with coordinated adult flight cycles, or *broods*. Broods are a purely statistical phenomenon, the result of many quasi-independent subevents. Yet they have distinct regularities (Shapiro et al., 2003). Their "shape" can be represented graphically by plotting the number of adults seen during a fixed period of time on frequency charts called *phenograms*. (*See*, e.g., Cech 1993.)

A basic dimension of butterfly broods is *concentration*. In *synchronous* broods, adults emerge over a relatively short period. *See* Figure 2.1(a).[14] This strategy facilitates mate-finding, but it also tends to put all of the population's eggs—or rather all of its adults—in one basket, vulnerable to predator concentrations, adult food shortages, and weather mishaps. Many elfins and hairstreaks have synchronous broods. Butterflies with *diffuse broods,* by com-

Table II. East Coast Butterflies by Life Zone

	Life Zone Criteria	Number of Species	Percent
North Only	Above VA-NC state line; also Appalachian highlands	66	28%
South Only	Below VA-NC state line, plus extreme southeastern VA	89	37%
Both	Any combination of northern and southern range	84	35%
Total		239	100%

Note: Strays are excluded.

[14] Phenograms are not actually neat, symmetrical bell curves, as illustrated. They might look roughly this way if each adult emerged just long enough to be counted, then vanished. But adults live on after emergence, and are thus recounted in later surveys during the brood period. After an abrupt buildup of numbers early in the brood, the declining "right tail" of an actual phenogram tapers off much more slowly. The longer-lived the adult butterfly, the greater the skewing effect.

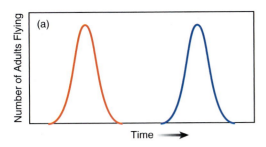 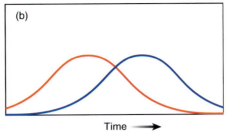

Figure 2.1. (a) Discrete, synchronous broods. (b) Diffuse, overlapping broods

parison, emerge more gradually. This approach is better diversified, but strategically less focused.

A second basic dimension of brood "shape" is *overlap.* In species with *discrete broods,* each adult flight period ends entirely before the next begins. This isolates each brood from those coming before and after it (except by ancestry, of course). In species with *overlapping broods,* there is no interval at all between successive flights. Here, complex genetic mixing can take place across generational lines. Indeed, the level of cross-over can be so great in late summer broods of species like Orange Sulphur that clear lineage patterns can break down altogether. *See* Figure 2.1(b). Field observers are unlikely to miss a target species in such cases, since some individuals will always be on the wing.

Most species have a predictable number of broods per year. A butterfly is *univoltine* if it has one annual brood, *multivoltine* if it breeds continuously, and so forth. Brood frequency is *obligate* in many species—i.e., the number of broods per year is fixed, regardless of environmental conditions. In species with opportunistic (*facultative*) brood frequencies, the number of generations is variable. American Coppers produce a full or partial 4th brood during the fall when seasonal conditions are favorable. (*Partial broods* usually occur when some of the individuals in an earlier brood continue developing into next-generation adults while others cease activity and spend their winter as earlier-generation larvae. At the population level, this strategy hedges the species' bets, ensuring that some individuals will survive the winter even if an early cold spell kills all of the adults in the final generation.)

Some butterflies have evolved highly specialized brood cycles. The caterpillars of many arctic sulphurs and satyrs require two or possibly even three years to reach pupation weight, as they feed on difficult-to-digest hostplants during the cold, short arctic summer. The same may be true of alpine satyrs in our region.

In other cases, the *pupal* stage can extend for more than one year. This occurs routinely in some desert species, apparently as a hedge against severe drought stress during the normal flight period. In addition, some East Coast species, such as the Schaus' Swallowtail and Olympia Marble, are known to remain in the pupal stage for a number of years, in case of drought or other environmental stress (Opler & Krizek, 1984, p. 35). This ability reportedly extends to swallowtails and pierids generally, and perhaps to other species that overwinter as pupae.

BROOD TIMING FACTORS

How long does it take for a butterfly to complete the cycle from egg to adult? Experienced field observers often use a 45-day "rule of thumb" to project the interval between brood peaks—based on an average of 4 to 10 days for the egg stage, 21 to 28 days as a caterpillar, and 7 to 14 days pupating. This rule produces a workable first approximation in a surprising number of cases.

Among rules meant to be broken, however, the 45-day rule "brood cycle rule" certainly ranks high. As already noted, some species take two or more years to become adults. And the Harvester, which feeds on relatively nutritious insect protein, can complete the cycle in just three weeks. Differences in "cycle speed" generally may be explained by a small number of core factors: (1) nutritional content of food, including the ease with which it may be digested; (2) ambient temperatures during the larval stage; (3) extreme weather conditions, such as prolonged raininess, flood, or drought; and (4) the size of the adult butterfly (small species are generally able to complete their cycle faster than large ones). "Runt" adults can occur when full caterpillar growth is inhibited, e.g., by cold weather or prolonged droughts.

In most butterfly species, emergence dates are *sexually asynchronous* (i.e., they exhibit *protandry*). The first males in any given brood emerge before the first females, generally by just a few days, but sometimes much longer (1 to 2 weeks in greater fritillaries). Male and female populations can be thought of as having their own distinct brood curves. These curves must overlap to some extent, naturally, but we still see populations with many different combinations of fresh and worn individuals. The adaptive value of asynchronous emergence remains speculative, but its principal values may lie in (1) maximizing a male's chances of being present to intercept unmated females, while (2) minimizing the time required for a freshly emerged female to mate and begin ovipositing. Prompt female breeding is linked to overall fecundity (Wiklund, 2003).

Voltism is linked to several lifestyle characteristics of East Coast butterflies:

- Multibrooded species are more likely to be generalists (56%) than specialists (44%).

- Single-brooded species are more likely to be specialists (58%) than generalists (42%).

- Nearly 52% of many-brooded species (three-plus broods per year) in our area feed on herbs as caterpillars (this number increases to nearly 57% among multibrooded generalists).

Figure 2.2. Number of broods by life zone

VARIATIONS IN BROODS BY YEAR AND LATITUDE

Brood dates often vary substantially from year to year, although most evidence on this point is anecdotal. In one set of time-series observations from the New York City area, brood peaks in the warm summer of 1991 occurred as much as two weeks ahead of normal. One year later, during the much colder summer of 1992, peaks were recorded two to four weeks *later* than in 1991 (Cech, 1993). In our experience, the volatility of brood dates seems to increase with latitude. Many northern species are missed entirely on weekend visits because of wide swings in brood dates. The timing of early spring broods may also be impacted by adverse weather. An "average" year for emergence dates should be one that is moderate in all key factors: temperature, precipitation, plant growth, etc.

Brood frequency is strongly influenced by latitude. Simply put, brood frequency decreases as latitude increases. (Even more simply, fewer annual broods occur in the North than in the South.) This can be seen clearly in Figure 2.2 (using an estimated average number of broods per species). Opler states that the tendency for brood frequency to increase southward is offset to some degree by longer gaps *between* broods in southern species (Opler & Krizek, 1984, p. 29). Many Deep South species breed more or less continuously, nonetheless, except in mid-winter, when a longer-lived, sedentary brood occurs (*see* discussion in connection with overwintering, below).

Even within a single species, northern populations often have fewer broods than southern ones. Glassberg (1999) shows boundary lines for brood frequency in many eastern species. These geographic "break points" are often quite sharp. Average size varies in an interesting fashion around these dividing lines. Starting in the far north, individuals of a particular species will generally be smallest, reflecting cooler, less supportive growth conditions. Size increases steadily as we move south. But then, just beyond the dividing line between brood frequencies, average size suddenly *decreases,* since the species is now trying to squeeze in an extra brood each season, leaving each generation less time to grow.[15] Moving south from this point, average size begins to increase once again, and the largest members of the species found anywhere generally occur near the southern range limit. This pattern has been carefully documented in Pearl Crescents and Northern Broken-Dashes (Porter & Mueller, 1998; Roff, 1983; Burns, 1985, pp. 21–22).

Brood diagrams showing emergence dates by latitude often look roughly like a pyramid. At the base of the pyramid are southern broods, which emerge earliest and finish latest. As we move north, first emergence dates become later, until, around the break point, the southern population begins to lap its northern counterpart, commencing a second brood while the first northern brood is still finishing up. At some point, the temporal separation of broods at different latitudes can create genetic isolation. But as long as any overlap remains between broods across latitudes, there is a possibility of genetic exchange, and only a modest amount of such exchange is needed to maintain species continuity. Other isolating factors aside, it may take a very long time before genetic isolation based on brood sequence alone will cause a full species break to occur.

BROOD PHENOTYPES

On seeing the first male Black Swallowtail of the season, northern observers may be surprised to have forgotten over the winter how small these butterflies appear. But 1st-brood northern male Black Swallowtails *are* smaller than individuals from later generations. The genetic repertoire of many butterfly species includes the ability to produce distinct seasonal *phenotypes*.[16] Some phenotypes are so distinct that they have separate names, such as the "marcia" form of Pearl and Phaon Crescents, the winter "red

---

[15] Hostplant selection also may change at such locations, focusing on nutritious species that will sustain the most rapid growth, thus allowing completion of the requisite number of generations (Janz, 2003). This strategy would be viable, of course, only in species that utilize multiple hostplants.

[16] Technically, a "phenotype" is the complete "expression" of an organism at any moment in time, including all traits and characteristics. Phenotypic traits can reflect a combination of genetic and environmental influences.

phase" of Barred Yellow, the "dusky" Great Southern White, etc. In many cases, the basal hindwing of off-season broods (from fall through spring) is darker beneath than in summer individuals. The forewing tips of winter-brood Goatweed Leafwings are thinner and more sickle-shaped than in the summer. Changing day length (*photoperiod*) during the larval stage is believed to trigger most seasonal phenotypic changes, though temperature or other factors often play a role as well. In a number of cases, such as the light-colored (or so-called albino) fall sulphurs, seasonal phenotypes are largely or entirely restricted to females.

Monarchs illustrate just how far-reaching seasonal phenotype differences can be. Fall migrants possess a number of very specialized traits not found in summer broods, including the ability to "shiver" in response to cold, directional navigation skill, increased fat stores, and suppression of sexual maturity (*reproductive diapause*) during migration (Douglas, 1989, p. 116).

I-13. Dense cloud cover can bring a swift halt to butterfly activity, especially on cool spring days.

DAYTIME BEHAVIOR CYCLES

The question of when to find butterflies is a function of the clock as well as the calendar. Despite the random appearance of their behavior, butterflies actually stage their daily activities quite purposefully (subject to weather variations). The process begins each morning at the roost. For most species, the roost is nothing more than an ad hoc protected site, away from foraging predators and direct exposure to the elements. Most butterflies roost alone, perched head-up with their wings folded and forewings retracted. In this position, only the ventral hindwing is displayed, usually the best-camouflaged wing surface. Some species roost communally, such as migrant/overwintering Monarchs. Zebra Heliconians roost communally throughout the year, returning to the same site each evening.

TEMPERATURE MANAGEMENT (THERMOREGULATION)

Except in the sultriest weather, the first order of a butterfly's day is warming up. For these cold-blooded creatures, *thermoregulation* is a critical skill. The ideal temperature zone for butterfly activity is between 80 and 100 degrees F. If temperatures fall too low, the butterfly's thorax muscles cannot contract, creating effective paralysis as far as flight is concerned. In some cases, butterflies have flown in laboratory conditions with body temperatures as low as 60 to 65 degrees F, but this is unusual. On the other hand, field observers *often* note scattered flight activity when *air*

temperatures are in the mid-60s, sometimes lower. This is possible because butterflies use various techniques to raise their body temperatures above those of the surrounding environment, and some species are quite adept at this (especially those that live regularly in cold surroundings). In general, we do not expect to see regular flight until air temperatures reach the mid-60s F, but in places with calm, sunny microhabitats surprises are apt to occur with some regularity.

What techniques are used for warming? Some butterflies have the capacity to shiver, i.e., to tense their muscles isometrically. The heat generated by this contraction warms the body substantially, to temperatures as high as 15 to 25 degrees F above ambient levels.[17] Mourning Cloaks perform this behavior in their hibernation crevices on warm winter days, before venturing out for a brief flight (Douglas, 1989, pp. 75–76). We have seen Green Commas shiver on cloudy days in northern New Hampshire in late summer.

A more familiar means of obtaining warmth is *basking*. Most species adopt characteristic basking poses after walking from their roost to a sunny perch in the early morning. Many true butterflies and spread-winged skippers *dorsal bask* with their wings fully extended, a rapid method for gaining heat. Most sulphurs, satyrs, and hairstreaks *ventral bask* with their wings held together, often angled perpendicular to the sun's rays, like a keeled-over sailboat. Grass skippers adopt a characteristic jet fighter pose when basking, in which the hindwings are held flat open and the forewings angled at 45 degrees. It is intuitive to suppose that body fluids, warmed in the veins of a butterfly's wings, will carry their warmth into the body, on the same general principle as a radiator. But repeated careful experiments

---

[17] Though butterflies are cold-blooded, their muscle contractions generate heat, just as in warm-blooded vertebrates. In very warm weather, excess midday activity can result in overheating, even for butterflies, leading to reduced activity. Flying can also create cooling, but this effect is easily outweighed by heat-generating factors at midday in summer.

have shown that heat absorbed by the wings while basking is transferred either mechanically or by radiation to the thorax, not by circulation (Douglas, 1989, p. 79). Heat absorption in the basal one-third of the wings matters most with this form of heat transfer.[18]

Basking is supplemented, especially in cooler climates, by additional techniques and adaptations. Northern and early spring species often have hairlike scales that act much like mammal fur in forming an air pocket to maintain body warmth.[19] Dorsal basking species may perch on a sunlit rock or reflective patch of sand to capture escaping radiant heat, and some press their bodies against the warm surface. Black Swallowtails bask with their abdomens elevated when it is cold, then lower them into the shade of the wings when it is warmer. In many ventral-basking species, the inner surface of the lower hindwing is darkened by scales containing the pigment melanin. In temperate-zone species, darkened underwing surfaces are most common in early season phenotypes (in pierids, Spring Azure, etc.). This is a permanent feature in northern species, such as arctic sulphurs, and it can increase total flight time during the day (Watt, 1968).

In far-northern habitats, butterflies with medium-sized wings enjoy a selective advantage across families. Caterpillars develop more slowly in cold surroundings, making large size an expensive luxury. Extra wing surface is also a disadvantage in negotiating cold, heavy winds that often sweep through low-growth northern habitats. At the other extreme, wings that are too small cannot generate enough warmth from basking to warm the body effectively, thus limiting critical flight time on chilly days (Douglas, 1989).

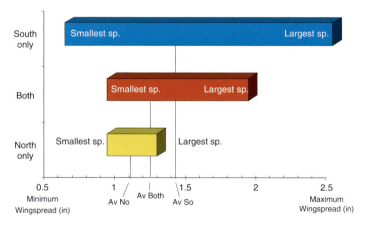

Figure 2.3. Wingspread of East coast skippers: range and variation by region

I-14. Sleepy Duskywing assumes characteristic hiding position during a cloudy interlude.

This extreme case illustrates certain general size relationships that hold throughout our region. Using skippers as an example, Figure 2.3 shows that average wingspread is greatest in the South, and that southern skippers also have a much greater *range* of sizes than northern ones, since they are less climate-constrained and can experiment more freely along this dimension.

In terms of the daily thermoregulatory cycle, basking is normally complete by mid-morning. Active flight commonly begins around 9:30 to 10:00 A.M., weather permitting.[20] Thermoregulatory adjustments continue throughout the day, however. In cool weather, butterflies must maintain their body heat, sometimes by flying low to the ground, where there is less wind and more reflected warmth. Or, they may stop regularly to bask in warm microhabitats. If the weather suddenly cools, as with the passing of a thick cloud, butterflies can disappear into sheltered clumps or thickets with remarkable speed. Duskywings wrap their wings around a twig, resembling moths.

Warmth can easily become too much of a good thing, however. At temperatures above 105 to 110 degrees F, *heat shock* occurs with prolonged exposure. In one study, even

---

[18] In *reflectance basking*, the wings are held 60 to 120 degrees apart, deflecting solar energy onto the body. In pierids with light-colored forewings, the angle of the wings is increased in species that have extensive darkening at the forewing point (*apex*) (Kingsolver, 1983; Schappert, 2000, p. 210).

[19] Even ordinary flat scales are filled with air, and can assist in insulation. Thoracic temperatures of butterflies without scales and hairs are reduced by 15%–20% (Douglas, 1989, p. 80).

[20] On inclement days, butterflies may not fly at all. And in hot, dry conditions they may sometimes fly as early as 8:00. Basking (vs. flying) can begin even earlier. It is surprising, nonetheless, how often active flight commences around 9:30 to 10:00, even in the South. Harry LeGrand (pers. comm.) notes that butterflies may be reluctant to fly in the morning until moisture has evaporated from roosting foliage, even in mid-summer.

modest heat shock reduced the lifespan of male sulphurs by 40% and female egg production by a factor of 4 (Kingsolver & Watt, 1983). Butterflies avoid heat stress by lowering activity rates at midday or by seeking cooler, shadier habitats. Many also hold their wings over their backs, often pointed straight at the sun, reducing heat absorption.

At day's end, some butterfly groups take shelter earlier than others. Most grass skippers roost early, often by 3:30 or 4:00, as do many pierids. Nymphalids often fly relatively late, sometimes until near dusk. An occasional butterfly (often a Red Admiral) is recorded flying after dark in warm weather, e.g., at moth lights.

ACTIVITY BUDGETING

Adult butterflies divide the active portion of their day among three primary core behaviors: foraging/feeding; mate-seeking/reproduction; and host plant location/ovipositing. Each of these activities involves a search phase, followed by some set of completion behaviors. Many butterfly species pursue particular actions only at specified times of the day, and male and female activity schedules often differ significantly. Much remains to be learned. Though clearly important, these behavior patterns are difficult to study systematically in the field. It is known that behavior schedules vary widely among species, and even within species at different times and places.

Not all butterflies feed on nectar as adults. Sap-feeding species are reportedly attracted by chemical esters wafting from tree wounds (Scott, 1986, p. 68), and dung-feeders presumably use corresponding cues. Among nectar-feeders, up to 50% of daily activity is spent locating suitable flowers. The process involves more than merely landing on a flowerhead and sipping nectar (although that process itself is quite remarkable). For one thing, a good deal of searching may be required. Some flowers help the process along by "advertising" themselves with distinctive colors or "nectar guides," visible only in ultraviolet light (Barth, 1991). And in some flower clusters individual blossoms change color as they age, signaling the end of nectar production (Weiss, 1995). Other flowers use varied approaches to *conceal* nectar from all but desired pollinators (which may *not* include foraging butterflies). Lastly, the level of nectar production itself frequently peaks at particular times of day, requiring would-be users to arrange their schedules around supply peaks.

In seeking mates, butterflies fall into two general camps: perchers and patrollers (Scott, 1974). Males of *perching species* establish display territories, often at specific times of day, and wait for a female to wander by. Display perches tend to be located at strategic spots, such as hilltops or hostplant concentrations, or other locations in good light where females are likely to appear.[21] We have noticed that male skippers often establish display territories in the center of formations with a naturally circular shape, such as low, partially bare grass patches in the middle of a taller-grass field.

In *patrolling species,* males fly around the territory more or less continuously, hoping to intercept females as they search for food. Male Falcate Orangetips seldom pause in their frenetic, low-altitude patrols in early spring until the first females emerge, at which point they begin to linger near potential mates. Patrolling species tend to mate opportunistically, whenever they happen on a suitable and receptive female. Many perching species, by comparison, have restricted mating times during the day, reflecting the limits of the male's display schedule.

The dichotomy between perching and patrolling is not absolute. Both behaviors may coexist within a single population, and in some species individual males may alternate styles in different circumstances. For the field observer, moreover, it can be difficult to distinguish display behavior from basking, or patrolling behavior from nectar foraging. Males of a common European satyr, the Speckled Wood (*Pararge aegeria*), come in two general forms, pale-colored perchers and dark-colored patrollers (Van Dyck, 2003). Most patrollers seek mates by flying through the woods close to the ground. Although they lose significant body heat while cruising shaded woodlands, their dark color allows them to warm up quickly when pausing in sunlit openings. The pale color of the perching males, by comparison, allows them to remain in sunlit patches for some time without overheating. Perchers have relatively large thorax muscles needed for explosive flight in territorial encounters, and by remaining at an optimal temperature they are more likely to defeat interlopers in such encounters.

While males divide their time between mate-seeking and foraging, females spend much of the day searching for hostplants on which to deposit eggs. Many pierids are reported to oviposit at midday, and many skippers in the afternoon. It has been said that few species lay eggs in the morning, but we have observed females among diverse butterfly groups ovipositing in the morning on quite a few occasions.

SEASONAL MOVEMENTS

Some butterflies are sedentary, inhabiting local colonies that persist for decades in a single location. At the other ex-

---

[21] When perches are located away from nectar sources or hostplant concentrations (e.g., on hilltops or at leks), the process will work only if receptive females themselves approach the display area. Sometimes only unmated females elect to do so (Wiklund, 2003). Perches that afford a view of approaching females silhouetted against the sky offer males the greatest visibility (Rutowski, 2003).

treme, Painted Ladies disappear almost entirely from North America each fall, repopulating the region the following summer (except in "crash" years, when few return). Between these two extremes, almost every conceivable class of intermediate strategy is pursued by one species or another.

Sedentary butterflies, broadly speaking, are contented butterflies, able to find hostplants, adult food, shelter, and a suitable climate, all nearby. Still, the term "sedentary" must be used advisedly, even in connection with apparently static populations. In recent years, it has been learned that many local colonies of European checkerspots, once considered "closed" by ecologists, actually have significant interchange with other regional colonies (Hanski, 1999, p. 220). Likewise, in contrast with large pierids, such as Cloudless Sulphurs, which have long been singled out for their migratory activity, many medium-sized pierids were traditionally regarded as sedentary. But observers now point out streams of smaller pierids, such as Cabbages, flying *northward* during the fall, e.g., over Delaware Bay.

Nonsedentary butterflies have many different movement patterns. *Dispersal* usually refers to random or irregular movements away from a host population with no clear destination, often by single individuals. *Migration,* by comparison, is more direct and seemingly purposeful. At a minimum, migration requires a regularly recurring, directional movement that ends with a relocation of breeding territory. It usually involves a group of individuals, sometimes entire populations, and it may have a very specific end destination, such as winter Monarch roosts in Mexico and California. One-way migration (i.e., without a return trip, either by the same individual or its immediate descendants) is usually called *emigration* rather than migration.

Dispersal is an inconspicuous process—if indeed it is one process at all, as opposed to several. Butterflies are known to quit their colonies occasionally in apparent response to environmental factors. Although a number of field studies have been conducted, much remains unclear. Dispersal often appears genetically predisposed, especially among disturbed habitat specialists such as Checkered White and Dainty Sulphur.[22] Their tendency to relocate spontaneously is consistent with a generalist lifestyle. But dispersal also occurs in normally sedentary species, often apparently in response to environmental stress or increased population density. In semi-arid western habitats,

Ehrlich and his students noted increased dispersal in exceptionally dry years, especially when large caterpillar populations decimated local hostplant stocks (Ehrlich, 1985, pp. 27–28). Lack of available nectar also may trigger dispersal, and females may depart a crowded colony when harassed by overfrequent mating attempts. In general, small colonies are prone to greater net outmigration and lower reproductive success than larger ones. This detrimental combination tends to hasten their demise (Hanski, 1999, pp. 31–32 ["Allee effect"]). The pattern and prevalence of dispersal may be affected by habitat characteristics, such as the existence of "corridors" between suitable patches (Haddad, 2000).

Migration is a more structured process. It is most advantageous ecologically when (1) a species can enhance its overall reproductive success by sending migrants north during the summer each year to form new breeding colonies, but (2) the colonies cannot withstand winter freezes in their adopted northern territories. At least some southward return movement is necessary in the fall if this strategy is to meaningfully supplement the base population in warmer latitudes (where most migrants originate).

But how common is a southern return migration among butterflies? It is now well-known that virtually the entire northern population of Monarchs moves south each fall, in a mass exodus to southern wintering grounds. But apart from this single dramatic example, which may have relatively recent origins (Shapiro, 1993), the evidence for wholesale return migration among East Coast butterflies is sporadic. There are reliable sightings of consistent southbound movements in Painted Ladies, of winter buildups of Red Admirals in central Florida, and of a variable, partial exodus of certain other northern species that overwinter as adults, primarily anglewings and related nymphalids. Directional, southward movements of some mobile species have also been documented along the Florida coast in fall (Walker, 1991, 2001; Swanson & Monge-Najera, 2000), but this evidence is somewhat equivocal.[23] From all appearances, true migration is by no means a common butterfly lifestyle in our area.[24]

To migrate, a butterfly must be able to maintain a directional course despite weather conditions and physical barriers. Also important for most distance migrants is gliding flight, which allows an individual to cover long distances with reduced effort.[25] Investigators continue to study the

---

[22] In the Dainty Sulphur, only one of the three geographic races disperses seasonally. The eastern race, in our area, remains in the Deep South, but members of the midwestern race wander northward extensively each summer.

[23] Southward movements in Florida do not necessarily involve returning northern migrants. They may just as easily be emigrants from large southeastern population centers that happen to be spreading south rather than north. Indeed, the butterflies involved tend to be warm-weather species that could have originated nearby: Cloudless Sulphur, Gulf Fritillary, Common Buckeye, and Long-tailed Skipper (*see* Scott, 1986, pp. 45–46).

[24] In the tropics, many butterflies are known to migrate over mountain ranges, seeking to avoid dry-season droughts. It is not known whether any East Coast butterflies engage in similar behavior.

[25] In Monarchs, gliding flight is 25 to 30 times more energy efficient than flapping flight. A migrating Monarch with a 140 mg supply of stored body fat can travel for an estimated 1,060 hours gliding, but only 44 hours flapping. Thus, the success of long-distance Monarch migrations seems to depend on gliding (Gibo & McCurdy, 1993).

exact triggers and mechanisms of butterfly migration. Directional navigation seems to involve a number of processes working in parallel, including solar orientation, perception of polarized light, reading the earth's magnetic field, etc. (e.g., Rutowski, 2003). Migrants are skillful at using local wind current to assist their progress (Srygley, 2001). The onset of migratory behavior appears to be triggered mainly by day length, and perhaps by temperature.

One-way, directional flights are achieved by a number of mobile species that do not truly migrate, such as Great Southern Whites and a number of other pierids. Every few years, millions of Great Southern Whites fly northward in massive emigrant groups, maintaining their northward direction despite wind and physical obstacles (Nielsen, 1961). This is technically not a migration—though it is often referred to as such—because the northward movement stalls in the Carolinas and there is no equivalent return flight. American Snouts sometimes darken the skies as they fly north, yet again there is scant evidence of a full migratory cycle.

Among the most perplexing mysteries in all of butterfly behavior is why some species, such as Cloudless Sulphur, fly northward most years with great determination, only to perish in the cold winter months that follow. This may allow a species to constantly test the northern limits of its survival range, or it may be a way to discard excess individuals from southern populations, thus avoiding a ruinous increase in density, or it may be nothing more than an atavistic throwback to some moment in evolutionary history when this behavior made sense and therefore became programmed into the species' genes. Assigning cause to insect behavior is a tricky undertaking at best, and here it presses our ability to understand the adaptive balance of nature's rhythms.

OVERWINTERING (DIAPAUSE)

Except in the southernmost parts of our region, butterflies cannot remain active all year. In most locations, they require some degree of *cold hardiness* to survive the winter, such as "antifreeze" chemicals in the blood that reduce their freezing point. Many butterflies also undergo *diapause,* an insect resting period, analogous to hibernation in mammals. Technically, diapause is a hormonally induced period of inactivity, marked by reduced metabolism and a complete cessation of feeding, reproduction, growth, and development. It can occur in summer or winter, although our primary focus here is on winter diapause.

Butterflies can diapause at any life stage. Most species use a single stage exclusively, or very occasionally two. Even eggs may diapause. Instead of hatching in 4 to 10 days, a diapausing egg will remain dormant for months, producing a caterpillar only in the following spring. Our

I-15. At northern latitudes, a significant portion of a butterfly's adaptive energies must be focused on surviving the winter.

interest in the topic is heightened by the fact that strategies for overwintering are tightly integrated with a butterfly's overall survival strategy—so tightly integrated, in fact, that it can be difficult to understand either in isolation.

Multibrooded species in the Deep South often lack a full-fledged diapause. Instead, they may produce a long-lived winter adult generation that remains mostly inactive, except for occasional feeding activities (Opler & Krizek, 1984, p. 29). Monarchs engage in this behavior, called *reproductive diapause,* even though they largely withdraw from our area during winter. Reproductive diapause is seen widely in tropical butterflies during the dry season, and the practice most likely originated in that setting.

More northerly species cannot ride out the winter so easily. Persistent cold is lethal to butterflies, causing death when it crystallizes body liquids (thereby destroying cell structures), or when loss of liquid water creates toxic chemical concentrations or denaturing. Many insects can avoid such lethal damage by burrowing into the soil, out of cold's reach, but very few butterflies have this capability. For the most part, butterflies must face the winter with meager protection.

Though most temperate zone butterflies are cold hardy, few are truly *freeze tolerant*. In order to survive the winter, they produce one or more *antifreeze chemicals* in their body fluids, such as glycerol or sorbitol, that lower their freezing point and prevent ice crystals from forming on particles within the cell (*nucleators*). Supercooled body fluids can sustain an insect during all but the most severe winter weather without solidifying. A butterfly is considered freeze tolerant, on the other hand, *not* merely because it can survive when the air temperature falls below 32 degrees F, using the techniques just described, but rather because it can tolerate having its *entire body frozen solid*. In very cold weather, the body fluids of freeze tolerant species enter an unusual, glasslike solid state, called *vitrification.* Miraculously enough, frozen larvae are able to awaken from this state in the spring, with no apparent harm (Leather et al., 1993, pp. 94–95). Relatively few of our butterflies are freeze tolerant, the best examples being alpine satyrs.

Diapause is an extremely complex process, with many variations and numerous intricate control mechanisms. Almost every imaginable variation is used by some kind of insect in the world (not all by butterflies). Such a thoroughgoing diversification of life strategies makes it difficult to generalize about the process, but it is valuable in illustrating an essential strength of insect evolution, namely, the dauntless pursuit of unexplored possibilities.

In temperate zone butterflies, diapause is most often triggered by day length (*photoperiod*)—either by absolute day length or by changes in day length. Temperature also frequently plays a role—though usually not a primary one—in determining the length and "depth" of the diapause.[26] In some species, such as Cabbage White or American Copper, elevated temperatures can indefinitely delay the onset of diapause, even after it has been "triggered" by day length (Scott, 1986, p. 28). Some species, such as Viceroys, are *facultative* diapausers; those living in southern Florida fly throughout the year, while their northern cousins overwinter as caterpillars.

Butterflies generally do not enter diapause immediately when initially induced (*triggered*). Rather, they first proceed through a series of preparatory phases. In species that overwinter as caterpillars, the initial response may be to produce an extra larval stage (e.g., a sixth larval molt in Silvery Checkerspot, vs. the usual five). This final instar may be endowed with special adaptations for overwintering, such as a thicker outer skin (*cuticle*) and sometimes a different color. Emperors and Viceroys create a hibernation nest (*hibernaculum*) by joining leaves together with silk threads. Skippers do the same by folding over leaf flaps or grass blades. Overwintering Cabbage Whites produce a higher percentage of brown pupae than usual (vs. green ones), evidently improving off-season camouflage. The diapause trigger also causes the subject to begin synthesizing antifreeze substances and to commence other prediapause behaviors (such as dehydration and evacuation of the gut). Adult diapausers accumulate fat deposits in their bodies that will sustain them during their long, winter fast, then search for a dark, dry cavity in which to take shelter.

Eventually, diapause commences in earnest, induced by hormonal change.[27] Once begun, diapause continues indefinitely unless halted. Diapause end cues are as complex and variable as inducement cues, and the two need not be the same, even in a single individual. Temperature and photoperiod are dominant triggers, often acting in combination—although it is possible for diapause to be programmed at the outset to run for a fixed period of time, regardless of external cues.

Ironically, diapause will not end in many species without a prolonged cold spell. Why a cold spell? Because in species that begin diapause in the summer or early fall, long days and warm temperatures will continue to occur during the present season. Only after a cold spell of some duration can a diapausing butterfly "know" for sure that winter has truly occurred, and that it is safe to respond to day length or temperature as a sign of spring. Of butterflies that do not successfully survive the winter, some are killed directly by exposure, some taken by predators, and some may lose the capacity to respond to "end diapause" triggers, and thus remain dormant until their fat stores are depleted.

What determines the life stage at which a butterfly diapauses? As noted above, most species diapause exclusively in one stage, or occasionally two. Distinct patterns are apparent within butterfly families, indicating deep, ancestral predispositions. Yet despite these genetic biases, there is a clear ecological order to the process. First of all, a species' diapause stage must fit its spring emergence schedule. Figure 2.4 shows that for northern, non-grass-eating species, there is a clear relationship between the first date of adult emergence and diapause stage.

All northern, non-grass-feeding species whose first annual flights occur in late winter (before March 15) overwinter as adults. Not coincidentally, emerging adults feed primarily on sap, dung, or rotting matter after awakening, which is typically the only food reliably available at that season. Species whose first adult flight occurs in the spring (March 15–May 15) tend to overwinter as pupae. And most species with a first adult flight in late spring (May 15–June 15) diapause as caterpillars. Finally, a large majority (86%) of species whose first flight begins after June 15 spent the winter as eggs. Among this late-flying group, more than 90% feed on nectar, which is by then readily available.

At one level, Figure 2.4 merely confirms an obvious fact,

---

[26] Though it is beyond the scope of this discussion, there is evidence that diapause uses two largely separate trigger mechanisms: first, the photoperiod trigger that initiates a comparatively primitive set of "basic" diapause responses; and second, a temperature-related trigger that controls a more complex suite of diapause responses, including the synthesis of protective chemicals. Assuming this model is true, the two mechanisms nonetheless become thoroughly integrated. To further complicate matters, evidence from other insects suggests there may be additional independent diapause trigger mechanisms at work, which are keyed by hostplant maturity, moisture levels, overcrowding, etc.

[27] The exact hormonal mechanism that controls diapause depends on the life stage at which a species overwinters (Scott, 1986, p. 29). Adult diapause is associated with reduced levels of juvenile hormone (the substance that causes reproductive organs to mature in the adult). Larval and pupal diapause result from a reduction in eclosion hormone. Eclosion hormone normally stimulates molting. Its absence during diapause arrests the molting process. Egg diapause is controlled by a hormone produced in the subesophageal ganglion.

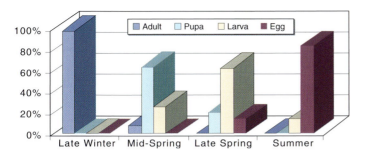

Figure 2.4. Diapause stage by first adult flight date (non-grass eating northern species)

namely, that it takes longer for an overwintering egg that hatches in the spring to reach adult status than is required by an overwintering pupa. But this overlooks the fact that diapause stage and general lifestyle are frequently integrated in subtle and nearly inseparable combinations:

Egg Diapause/Northern Tree-Feeders: Tree leaves and flowers are at their tenderest and most edible when they first emerge in the spring. To capitalize on the spring "leaf-out" requires some coordination, however. First, a butterfly must overwinter near the dormant leaf buds. For effective timing, the diapause stage needs to be either egg or caterpillar. Caterpillars are at risk on outer tree branches during harsh winter weather, however, where they can easily be blown off by winter winds or eaten by birds. Eggs, by comparison, are natural treetop diapausers. Even the eggs of warm-weather species can often withstand extremely cold temperatures, as low as –60 degrees F in some Lepidoptera (Leather et al., 1993, p. 98). The greatest risks faced by diapausing eggs are predators and parasitoids (as described below).

Over 90% of East Coast butterflies that overwinter as eggs are single-brooded. For tree feeders, the peak moment of the year ends after the first rush of spring foliage has matured. Beyond this point, it is adaptive for maturing caterpillars to complete a single adult flight, then enter diapause in the egg stage, beginning in mid-summer. *Satyrium* hairstreaks (Banded, etc.) exemplify this life strategy.

Larva and Pupa Diapause/Herb Generalists: As just noted, most caterpillars and pupae cannot withstand winter cold as readily as eggs. Many herb-feeding generalists spend the winter in one of the more advanced life stages, sheltered as best as possible from the elements. They frequently seek shelter in leaf litter at the base of a hostplant (or sometimes attached to it). Many of these species overwinter at a certain larval stage, or instar. Baltimore caterpillars, for example, always enter diapause in the 4th instar.

Larval Diapause/Grass-Feeders/Tundra Specialists: Grass-feeders were excluded from Figure 2.4 because their dietary habits impose a protracted, idiosyncratic adult emergence pattern. A large majority of East Coast grass-feeders (89%) overwinter as caterpillars. Even so, many of them must feed until late spring or summer to emerge as adults. As noted earlier, few large, temperate grass-eating species that undergo diapause can produce a large number of broods per year.

All of our *alpine tundra* species overwinter as caterpillars. To survive arctic conditions in winter, these caterpillars must be freeze tolerant. This capability allows them to feed during more than one summer until they reach full pupation weight. Interestingly, on Ellesmere Island in the high arctic, *all* insect species reportedly diapause as larvae, not just Lepidoptera (Leather et al., 1993, p. 152).

Pupal Diapause/Flower-Feeders: More than 80% of East Coast species whose caterpillars feed on flowers overwinter (if at all) as pupae. Adult females emerging early in the season lay their eggs on flower buds—which commence blooming actively in the spring. Many species pursuing this strategy are single-brooded spring butterflies, such as elfins and certain azures.

Adult Diapause/Crevice Roosters/Fall Migrants: In the North, adult diapause occurs only in nymphalids, especially anglewings, tortoiseshells, and ladies. In single-brooded adult diapausers, such as Mourning Cloaks, diapause technically begins as soon as fresh adults emerge in early summer. The diapause induction trigger in such cases is set to the longest day-length the species will experience in its environment, and thus is immediately "tripped" (Leather et al., 1993, p. 67). Mourning Cloaks are inactive during much of the summer (*estivation*), then fatten up in the fall before finding a dark crevice in which to pass the winter.[28] Female reproductive organs do not mature until the following spring. Adult diapausers must take care to find a dry crevice, for studies have shown that external moisture greatly reduces an adult butterfly's cold tolerance (Leather et al., 1993, p. 119). As noted above, on awakening in the early spring, these species tend to feed on food types other than nectar. Some adult diapausers withdraw from the North before becoming inactive. Monarchs withdraw by the millions to Mexico each fall, for example, among other winter havens.

III. WHAT TO LOOK FOR

Having determined where and when to locate butterflies—and by now presumably having found some of them—it

---

[28] Winter survival rates of adult diapausers in Europe is related to the amount of body fat accumulated during the fall. The length of time available for feeding is an important determinant of total accumulation. An overwintering European Peacock (*Inachis io*) requires 0.15 mg of lipids per day (Leather et al., 1993, p. 170).

will greatly enhance the encounter if we have spent time beforehand learning a little about the objects of this long search. Observers may sometimes be tempted to short-change this last, important step, once a target species is found and identified. We strongly advocate the study of butterfly biology, yet must admit that a brief introductory chapter such as this provides room for only a cursory overview. We will focus on questions of greatest likely interest to field observers.

BUTTERFLY ORIGINS AND EVOLUTION

The Order Lepidoptera (moths and butterflies) is among the most recently evolved in the insect world. The fossil evidence is fragmentary, but moths seem to have originated roughly 200 million years ago—give or take a few tens of millions of years. It is believed that several modern insect orders, including moths and the closely related caddisflies (*Trichoptera*), split off independently from a primitive common ancestor, probably an early scorpion-fly (*Mecoptera*) (Emmel et al., 1992, p. 2).[29] But whatever the exact path, Lepidoptera appeared at an interesting juncture in earth history, as fully terrestrial plants were first arising. Only nine of 29 insect orders have learned to feed on plants, but these *phytophagous* groups have enjoyed disproportionate success, accounting for nearly half of all modern insect species. The Lepidoptera are prominent among their ranks, having far more herbivorous species (163,000) than any other insect group (Schappert, 2000, p. 48). Lepidopteran caterpillars do not convert their food to protein as efficiently as carnivores, but they are quite efficient relative to other plant-eating insects, which are their main competitors (Scoble, 1995, p. 172).

The first butterflies probably appeared sometime between 145 and 60 million years ago, though here again the fossil record is sparse.[30] It is generally believed that butterflies evolved from day-flying moths.[31] The first definite fossil butterflies, found in Texas and Colorado, date back 48 million years. They include a primitive, swallowtail-like species, a metalmark, and a nymphalid. By about 38 million years ago, all modern families can be traced in the fossil record. During the late Jurassic/early Cretaceous period,

The first question asked by beginning butterfliers is invariably, "What is the difference between a moth and a butterfly?" Taxonomists would be pleased to have a quick and definite answer to this persistent query. The name "butterfly" is usually applied to the members of two Lepidopteran superfamilies—*Papilionoidea* ("true" butterflies) and *Hesperioidea* (skippers). Both groups are in the section *Macrolepidoptera*, suborder *Ditrysia*, which includes higher moths and butterflies. Butterflies were formerly placed in their own suborder, but this approach is no longer accepted, as substantial taxonomic similarities have been identified across Macrolepidopteran groups. In short, taxonomists have not found a "convincing systematic framework" that cleanly separates higher moths from butterflies. Still, butterflies are regarded as at least a "compact group" within the Macrolepidoptera, and have been assigned to their own taxonomic subsection, the *Rhopalocera* (Ackery, 1985, p. 9; Scott, 1986, p. 97; Scoble, 1995).[32]

The detailed anatomical criteria used to distinguish butterflies from moths are obscure to nonspecialists. Ackery (1985, p. 9), e.g., described butterflies as "ditrysian Lepidoptera with mesal fusion of the dorsal laminae of the secondary metafurcal arms," where in nearly all cases "the mesothoracic lateral dorsal muscles [are] twisted." Butterflies also have a characteristic brain form, with an enlarged optic lobe and reduced deutocerebrum, as well as distinctive aortal anatomy. Other features cited include an inflated antennal scape, lack of frenular bristles in the female, and the presence of tibial spines (Scoble, 1995, p. 306).

Such criteria are obviously challenging to apply in the field. Fortunately, butterflies can usually be distinguished by their clubbed antennae (vs. the nonclubbed, often "feathery," antennae of moths), their diurnal habitats and their relatively bright colors and patterning. Butterfly wings do not fold and, with the exception of one Australian skipper thought to be ancestral, the forewing and hindwing are not physically joined together by a coupling structure (*frenulum*).[33] As noted elsewhere, the latter trait allows butterflies to land more gracefully than moths, an ability that can be observed in the field.

Moths outnumber butterflies in terms of species by nearly 10 to 1, and their diverse adaptations can help place butterfly lifestyles in context. Moths use several hostplant types not exploited by butterflies, and they sequester (synthesize) a number of toxins that butterflies do not. Butterflies also tend to locate prospective mates visually, whereas moths use long-distance pheromones for this purpose, effective at distances of up to a half mile.

Day-flying moths are fairly common in butterfly habitats, and field observers should become familiar with the more prevalent ones at least. The brightly colored Virginia Ctenucha contains cardiac poisons toxic to vertebrate predators.

---

[29] Douglas (1989) suggested the likely ancestor was a stonefly (Plecoptera). He also noted an alternate theory, namely, that caddisflies may have first split from the Plecoptera, then moths later branched off from the caddisflies.

[30] As of 1992, only 44 butterfly fossils had been officially described (Emmel et al., 1992). That is less than one fossil for every million years that butterflies existed! The scarcity of fossil evidence is not surprising, given the ephemeral nature of Lepidopteran anatomy.

[31] Proposed ancestral families include *Castniidae, Sphingidae,* and *Hedylidae.* The *Castniidae* are a mostly tropical family of 150 brightly colored day-flying moths with clubbed antennae that closely resemble those of butterflies. The *Sphingidae* (or hawkmoths) are a large family with more than a thousand species, some of which are fully diurnal, hovering at flowers like hummingbirds. *Hedyloidea* are a macrolepidopteran group, formerly classified among the geometrid moths (Scoble, 1995, p. 392).

[32] Some taxonomists, applying cladistic principles, include *Hedyloidea,* a group of tropical "butterfly-moths," in *Rhopalocera.*

[33] Grass skippers are often referred to as "fold-winged," but this simply means that they hold their wings vertically over their heads, as opposed to outstretched from the sides, as in open-winged skippers.

when butterflies were first evolving, the ancient southern continent of Gondwanaland was still largely interconnected, and so modern butterfly families are represented on nearly all modern continents. Certain subfamilies, however, such as metalmarks and fritillaries, apparently evolved later, and are confined to more limited regions (such as the Americas, the Holarctic North, etc.).

The evolutionary branching that produced modern butterfly families appears to have been driven mainly by changes in hostplant use—either by specialization on new plant types, such as grass, or by learning to tolerate new secondary plant toxins, such as aristolochic acids in certain early swallowtails. At the time these splits were occurring—in the early Cretaceous, when *T. Rex* stalked the land—the flowering plants (*angiosperms*) were undergoing a massive, worldwide diversification that eventually resulted in more than 200,000 separate species. On a different level, the original shift from nocturnal activity to diurnal may have been prompted, or at least reinforced, by the evolution of a fearsome new order of nocturnal predators, the bats (Fullard et al., 2001).

BUTTERFLY NAMES AND TAXONOMY

There is disagreement among those interested in butterflies as to the proper role of common vs. scientific names. Many traditionalists prefer to use scientific names exclusively. This does have advantages. First, it keeps taxonomic identities in the foreground. Second, a species may acquire a number of local common names within its range—as well as an additional set of non-English names if its distribution crosses international borders, e.g., into Latin America or French-speaking Canada. There is an entire book devoted to compiling alternate English common names for U.S. butterflies, both species and races (Miller, 1992). And yet, we must also recognize that scientific names have not been especially stable, either. Reclassifications and technical changes have occurred quite frequently, often to correct Latin grammar. As a result, in some species scientific species names have actually changed *more* often than their English equivalents!

In this book, we use common names when discussing butterflies in the text. We follow the NABA 2nd ed. checklist.[34] The availability of standard common names lists, such as NABA's, greatly reduces the potential for confusion, and we believe that most observers find common names more accessible. In addition, common names can

I-16. The Virginia Ctenuchia (*Ctenucha virginica*) is a common day-flying moth throughout most of the East Coast region.

be colorful and interesting, adding an extra dimension to the literature. Obviously, both common and scientific names have a place in butterfly study. And as long as underlying species references are clear we think that choosing to emphasize one or the other is largely a matter of taste.

Issues around species definitions and taxonomy can be considerably less tractable. A few decades ago, some commentators went so far as to suggest that most "big" issues in taxonomy were already resolved, leaving only scattered "clean up" work to finish. But with the advent of genetic biochemistry, taxonomic questions have taken on new life. Like all organisms, butterflies are classified using binomial nomenclature, under the system developed by Carolus Linnaeus in the mid-1700s and now administered pursuant to the International Code of Zoological Nomenclature (ICZN). Some investigators have recently proposed new, entirely numeric classification methods, but to date the traditional approach has not been displaced. Higher-level taxonomic groupings used for East Coast butterflies in this book are listed in Figure 3.1.[35] Family characteristics are discussed along with species accounts in the main text.

There are about 18,000 species of butterflies in the world, depending in part on one's definition of the term "species." Scoble (1995) recognized 14,500 true butterflies and 3,050 skippers, while Scott (1986) estimated 14,750 species in total. Naturally, the discovery of new species can increase this total. We know, e.g., that even within our relatively well-explored East Coast region, completely new species of moth can still be found.[36] But the same does not

[34] The second author is a member of the NABA English Names Committee. In one instance, we have chosen not to follow the NABA nomenclature. This is discussed in the applicable species account.

[35] In the absence of a broadly authoritative taxonomic list for butterflies (such as the American Ornithologist Union's *Check-List of North American Birds*), it can be difficult to decide which subgroupings to include in a reference list of this type. By

including or excluding specific categories in Table III, the authors intend to present a general, relatively conservative listing of taxonomic classifications—not to render a considered scientific opinion on disputed taxonomic questions.

[36] In the summer of 2000, a highly motivated group from the Lepidopterists' Society conducted a one-day blitzkrieg in the Great Smoky Mountains, and discovered a number of entirely new species of moth.

appear true for East Coast butterflies. In fact, it is not clear that *any* completely new butterfly species remain to be discovered in our region today—excluding possible taxonomic splits of existing, sibling species. The only source of significant potential growth may be introduced or immigrant species from outside the East Coast area, such as the Acacia Blue arriving on the Florida Keys in the 1990s, or the Greenish Blue, which colonized northern Maine in the mid-20th century.

Taxonomists use many physical and biochemical criteria in tracing the path of butterfly evolution. Two traditional yardsticks are wing vein patterns and the shape of male genitalia. But a wide range of additional characteristics have been employed as well, including caterpillar anatomy, hostplant relationships, characteristic behaviors, the arrangement of specific leg "hairs," egg shape, eye pigments, and even the shape of the *sphragis,* a thick plug left behind by some males after mating to prevent additional copulations. With the advent of modern biochemistry, cell-based techniques have been developed to explore hereditary affinities at a microscopic level. An early method of this type was enzyme electrophoresis (e.g., Geiger, 1988; Burns, 1975). More recently, investigators have begun analyzing genetic materials directly, including mitochondrial DNA derived from a metabolic organelle (e.g., Sperling, 2003; Caterino et al., 2001). Some taxonomists have recently adopted a "total evidence" approach, factoring in evidence from a variety of sources (Nylin et al., 2001). Numerous other approaches have academic proponents as well.

A complete taxonomic theory needs a clear definition for the term "species," plus methods to determine the biological relationships among species groups. For much of the 20th century, the *biological species concept* of Ernst Mayr and his contemporaries went nearly unchallenged. Mayr's theory proceeded from the now-familiar proposition that adjacent populations are "full species" only if they cannot successfully interbreed on a sustained basis, producing generations of fertile offspring. But in recent decades a rival school has gained influence, the *phylogenetic species concept.* Its adherents consider a species to be the smallest "diagnosable cluster" of organisms sharing a parental pattern of ancestry and descent. Phylogenetic species theory draws heavily on the analytical technique of *cladistics,* first described in 1950 by German entomologist Willi Hennig. Cladistic routines assign taxonomic relationships based on aggregate similarities among organisms (the "best" family tree is one that maximizes the clustering of similar traits within groups, while segregating dissimilar traits as much as possible in different groups). In the end, a synthetic approach drawing from each school of thought may prove most effective (Sperling, 2003).

If species boundaries are difficult to distinguish, the same is doubly true for subspecific categories, e.g., *races*. Some bi-

Table III. Major Taxonomic Classifications of East Coast Butterflies

Phylum:	Arthropoda (Arthropods)
Class:	Insecta (Insects)
Order:	Lepidoptera (Moths and Butterflies)
Suborder:	Ditrysia (Higher Moths and Butterflies, plus some other moths)
Section:	Macrolepidoptera (Higher Moths and Butterflies)
Subsection:	Rhopalocera (Butterflies)

Family	Subfamily Group
"True" Butterflies	**[Superfamily Papilionoidea]**
Swallowtails [Papilionidae]	Swallowtails [Subfamily Papilionidae]
	Aristolochia Swallowtails [Tribe Troidini]
	Kite Swallowtails [Tribe Leptocircini]
	Fluted Swallowtails [Tribe Papilionini]
Whites & Sulphurs [Pieridae]	Whites [Subfamily Pierinae]
	Sulphurs [Subfamily Coliadinae]
Hairstreaks & Blues [Lycaenidae]	Harvester [Subfamily Miletinae]
	Coppers [Subfamily Lycaeninae]
	Hairstreaks [Subfamily Theclinae]
	Blues [Subfamily Polyommatinae]
Metalmarks [Riodinidae]	Metalmarks [Subfamily Riodininae]
Nymphalids [Nymphalidae]	Snouts [Subfamily Libytheinae] *
	Heliconians & Fritillaries [Subf. Heliconiinae]
	Longwings [Tribe Heliconiini]
	Fritillaries [Tribe Argynnini]
	True Brushfoots [Subfamily Nymphalinae]
	Checkerspots/Crescents [Tribe Melitaeini]
	Typical Brushfoots [Tribe Nymphalini]
	Admirals et al. [Subfamily Limenitidinae]
	Leafwings [Subfamily Charaxinae]+
	Emperors [Subfamily Apaturinae]
	Satyrs [Subfamily Satyrinae]
	Milkweed Butteflies [Subfamily Danainae]
Skippers	**[Superfamily Hesperioidea]**
Skippers [Hesperiidae]	Spread-winged Skippers [Subfamily Pyriginae]
	Intermediate Skipper [Subf. Heteropterinae]+
	Grass Skippers [Subfamily Hesperiinae]
	Giant-Skippers [Subfamily Megathyminae]
	Yucca Giant-Skippers [Tribe Megathymini]

* Sometimes considered a separate family

+ No longer generally recognized as separate taxonomic groups (Charaxinae are placed in Apaturinae, Heteropterinae in Hesperiinae).

ologists question the role of formal subspecies, but others devote much of their time to describing distinctive local forms. In some cases, stable geographic "hybrid zones" exist between species that complicate the process of local diagnosis—and may themselves be a source of new forms and ultimately new species (Sperling, 2003). Recent evidence shows that numerous "sibling species" can exist within complexes formerly considered monospecific (Herbert et al. 2004).

Interestingly, many genetic traits in butterflies that define species identity are controlled by the sex-linked X-chromosome. In butterflies, unlike humans, females are the heterogametic sex, having one X- and one Y-chromosome. This is thought to make them more susceptible to nonviability or sterility as hybrids than males.

Such technical considerations may seem remote from the

experience of an individual field observer. But even here it is possible, with experience, to establish a degree of personal contact—e.g., by studying local population differences that occur in many butterfly groups. An example is the confusing Spring Azure complex. A primary issue with such variants is knowing whether local differences are the result of true, stable genetic differentiation, or merely of reversible environmental influences (such as variations in local hostplant chemistry or other phenotypic triggers). Populations whose appearance is determined by environmental factors rather than genetic ones are referred to as *ecotypes.* With many new taxonomic classifications being proposed, it is difficult—yet also important—to balance the acceptance of new taxonomies with a sober respect for the elusive complexities of evolutionary process. The incorrect recognition of a new species by Petvier in 1717—the "Albin's Hampstead-Eye," which was later shown to have been a misdescribed variant—is still cited as a faux pas by commentators more than 250 years later (Ford, 1977, p. 20).

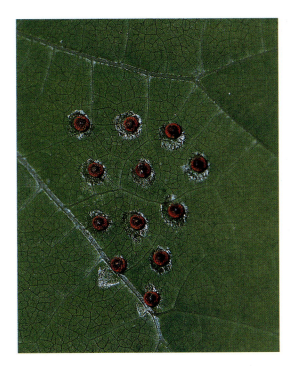

I-17. Eggs of the Pipevine Swallowtail are placed on the underside of *Aristolochia* leaves in groups or clusters.

BUTTERFLY BIOLOGY: EARLY STAGES

Although metamorphosis is a marvelous process, we usually take the four life stages of fully metamorphic (*holometabolous*) insects pretty much for granted. Not all insects have four life stages, however. There are also 150,000 or so *hemimetabolous* insects, such as grasshoppers, which undergo *gradual metamorphosis.* Their external appearance and general lifestyle remain essentially unchanged as their reproductive organs slowly mature while they grow. Full metamorphosis is about five times more common among insects than gradual metamorphosis. This is undoubtedly because full metamorphosis allows insects to key in on specific, targeted niches at various stages in their life cycles, thus optimizing their prospects for survival at each stage. In butterflies, the caterpillar is a food ingestion specialist, with a minor in remaining alive while feeding. Adults, by comparison, generally do not accumulate body mass (other than fat deposits for migration or overwintering). Rather, they eat only enough to maintain themselves while fulfilling their own specialized life functions, primarily dispersal and reproduction. Metamorphosis is a biologically "expensive" process, and yet we can easily appreciate its value for developing butterflies.

Early Stages: Egg, Caterpillar, and Pupa

Butterfly eggs come in varied shapes. They generally show a strong family resemblance.[37] The outer "shell" (*chorion*) consists of several layers of durable protein that resist damage. As many as 14,000 tiny breathing pores dot the outer surface, allowing passage of oxygen and carbon dioxide for respiration while preventing dehydration. A tiny pinhole on the egg's surface (the *micropyle*) lets sperm enter the egg for fertilization (this occurs inside the female, shortly before oviposition). Eggs are cemented to leaves or twigs by an adhesive produced in the female's abdomen. In some lycaenids and skippers, rough-looking anal scales are "glued" to the eggs as they are laid, offering a degree of camouflage (and perhaps chemical protection).

Though naturally tough, butterfly eggs are not invulnerable. Molds and fungi can kill eggs in damp conditions, as can extreme temperatures, physical crushing, or abrasion. And many animals feed on butterfly eggs, from arthropods (such as beetles and true bugs) to foraging birds. The most serious source of egg mortality, however, is undoubtedly the legions of tiny wasps and flies (*parasitoids*) that use their needle-like ovipositors to pierce the chorion, laying their own eggs inside. The emerging parasitoid larva devours the butterfly embryo before it can mature. Parasitoids also attack caterpillars and pupae. Levels of parasitoid infestation in local butterfly colonies vary significantly from year to year—from 0% to 100% in one series of studies in Finland (100% infestations cause extinction of local colonies) (Hanski, 1999, p. 214). One reason for this volatility is that butterfly parasitoids have parasitoids of their own, and this causes their numbers to fluctuate unpredictably.

---

[37] Swallowtail eggs are generally smooth and spherical; pierid eggs are long and tapered, with deep, longitudinal ribs or creases and delicate cross-barring; lycaenid and metalmark eggs are comparatively squat, with intricate, geodesic patterns on the surface; nymphalid and danaid eggs are variable, often oval or barrel-shaped, some with pronounced longitudinal ridges; and skipper eggs are roundish, often with some elongation.

I-18. Feeding on toxic passionvine leaves, a young Gulf Fritillary attains defensive chemicals along with nutrition.

Some plants, when attacked by insect herbivores, emit chemicals that attract parasitoids (Mattiacci et al., 1995).

Most butterflies lay eggs singly, but a number deposit their eggs in clusters, sometimes numbering in the hundreds. Female Baltimores prefer to oviposit near existing egg clusters laid by other females. Regular cluster-layers in our area include aristolochia-feeding swallowtails (genus *Battus*), some whites, Atala, crescents and checkerspots (including Baltimore), Question Mark, Mourning Cloak, tortoiseshells, emperors, and two skippers (European and Golden-banded). Egg clusters are sometimes laid in stacks, as many as three or four deep, or in strings suspended from a leaf. The eggs at the end of such strings, and those on the surface of clusters, reportedly hatch before those laid earlier, which are now rest at the bottom of the stack (Weed, 1917).

Egg masses make an inviting target for parasitoid wasps and flies, but deep clusters do afford some protection—at least for those eggs buried in the center! Brightly colored eggs are frequently *aposematic,* warning predators that they are toxic to consume.

Inside an egg, the embryonic caterpillar feeds on its yolk, growing geometrically. When ready to hatch, the young caterpillar chews its way out of the egg. In many species, the emerging caterpillar consumes the remainder of its egg as a first meal.[38] Noncommunal caterpillars may also look for other, unhatched eggs nearby to consume, contents included. This act of cannibalism may have more to do with reducing intraspecific competition than obtaining nutrition—though from a bioenergetic standpoint it is a meal all the same.

Caterpillars are designed to eat voraciously and bulk up as rapidly as possible, increasing their body mass several thousand times before pupating. Anatomically, they can be pictured as fluid-filled sacs with hardened heads and jaw parts. The outer layer of skin is divided into 13 segments, three of which will become the adult thorax. The other 10 are abdominal. The exoskeleton is not fully hardened (*sclerotized*). Its surface tone is maintained largely by the pressure of internal fluids. A layer of muscles below the skin will tighten to seal off any small punctures. The danger from a puncture wound is not so much that the caterpillar will bleed to death as that it will deflate (Douglas, 1989, pp. 18–19).

Many of a caterpillar's bodily traits are, in effect, "souped-down" versions of the equivalent adult features. This is true of the simplified antennae, eyes, circulatory system, etc. But caterpillars also have some comparatively well-developed capabilities. For example, they produce a variety of enzymes for digestion, whereas adults commonly produce just one (to digest sugar). Caterpillars also have some features that are entirely absent in adults, such as *prolegs,* the large, fleshy appendages located on five rear abdominal segments—behind the true legs that project from the thoracic segments. Hooklike *crochets* on the tips of the prolegs allow a caterpillar to grasp the substrate firmly, even while resting. The remains of a caterpillar's diet after digestion are excreted as *frass.* Some pierid and skipper larvae use "anal combs" to remove frass from their feeding areas. This act of basic hygiene prevents the build-up of fungi or bacteria, especially in the confines of a skipper's leaf roll. It also removes a potential attractant for parasitoids and keeps the domicile unclogged.

Caterpillars move in an intricate manner, rhythmically expanding and compressing their abdominal segments in a wavelike motion. They release the crochets on their prolegs and then refasten them synchronously at the end of each "step." This coordinated series of motions is orchestrated by impulses from a simple nerve cord that runs the length of the body. Caterpillars secrete silk from glands in their mouths. Caterpillar silk is a complex protein, consisting mostly of an elastic substance called *fibroin.* Caterpillars use silk trails to relocate feeding areas after retreating to a sheltered location to rest. Some European swallowtails are known to leave scent-marked silk trails, apparently for signaling purposes (Scott, 1986, p. 63).

Developing as rapidly as it does, a caterpillar outgrows its fixed-size exoskeleton several times before pupating. This necessitates a series of interim *molts,* the exact number of which varies by species. Molting begins when pressure-sensitive nerve endings, located in joints between the body segments, fire in response to stretching of the exoskeleton. This triggers the brain to release molting hormones, called *ecdysones* (Douglas, 1989, pp. 20–22). The caterpillar then begins to grow a soft, new exoskeleton be-

---

[38] There are suggestions that the chorion may be a source of digestive microbes or of key nutrients, or that eating the egg remains removes a potential signal for predators.

neath its current skin and secretes enzymes that chemically weaken the existing outer shell. Then finally, at the critical moment, it draws air into its body, splitting the old skin apart. The new exoskeleton is "pumped up" by blood pressure and hardened chemically (though not fully sclerotized, as in an adult). Each caterpillar generation between molts is referred to as an *instar*. Successive molts will continue to result in new caterpillars as long as the level of *juvenile hormone* remains high in the body fluids; this hormone suppresses advancement to the pupal stage.

Caterpillar lifestyles vary greatly, but most species are either *exposed feeders* or *concealed feeders*. Exposed feeding caterpillars venture into the open to consume leaves or other foliage. They have relatively thick prolegs, allowing them to grip tightly onto twigs and leaves (Scoble, 1995, p. 126). Day-feeding species avoid vertebrate predators with cryptic patterns, or by mimicking inedible substances in the environment (such as bird droppings).[39] Some brightly colored caterpillars feed fearlessly in the open, flaunting toxic body chemistry. Species in our region with aposematic larvae include Atala, Monarch, Zebra, Baltimore, and Gulf Fritillary. Because they are brightly colored and behave conspicuously, these caterpillars are among the most familiar to general field observers.

Concealed feeders may burrow into flowerheads (some lycaenids) or build a leaf nest made of plant materials bound together with silk thread (most skippers, some swallowtails, etc.). Many concealed caterpillars feed nocturnally, and thus have little need for cryptic patterning. Night feeding offers a degree of protection from vertebrate predators, and also helps to avoid overheating and dehydration. On the other hand, nocturnal feeders must cope with cooler nighttime temperatures, which may slow activity. There is evidence that the caterpillars of *Colias* sulphurs can tolerate cooler temperatures than adults, and perhaps this is a more general adaptation (Scott, 1986, p. 42).

It is estimated that caterpillars suffer the highest mortality of any butterfly life stage.[40] They are susceptible to a host of perils, including cold, drought, storms, fungi, disease (both viral and bacterial),[41] parasites, parasitoids, and predators. Caterpillar predators include spiders, ants, robber flies, assassin and ambush bugs, lizards, birds, small mammals, etc. Even hostplants themselves fight back against caterpillar foraging. As already noted, many hostplants produce toxic or irritating secondary chemicals that deter herbivores—at least until the herbivore learns to tolerate

I-19. Giant Swallowtail caterpillar brandishes its osmeterium.

them—and some produce chemicals that attract parasitoids (Mattiacci et al., 1995). In addition, certain plants develop thick or waxy outer cuticles that are difficult to chew. Others, notably the passionvines, grow sharp hooks, called *trichomes,* that can snag or puncture a caterpillar. Some also grow decoy tendrils that attract ovipositing females, then drop off before maturing, carrying any eggs they have accumulated with them. Still other passionvines produce false eggs that may deceive a female into thinking the plant is already occupied by the eggs of another ovipositor, or nectar glands that attract ants likely to prey on eggs or small larvae (Gilbert, 1972).

Against this host of ills, caterpillars have developed a battery of defenses. As already noted, camouflage and aposematic coloring provide some measure of protection, although visual disguises would seem of little use against invertebrate predators, especially at night. Evidence is accumulating that some caterpillars use various chemical excretions to deter attackers. Swallowtail caterpillars thrash vigorously when disturbed, brandishing their *osmeterium*— a forked organ, usually concealed in the thorax. This organ releases a foul-smelling chemical mixture into the air that thwarts some predators, including ants that have already begun to attack (Scoble, 1995, p. 121). Some caterpillars lay trails of silk over dangerous trichomes, neutralizing their spines.

Aristocholic acids sequestered by early stage Pipevine Swallowtails have been shown to deter potential parasitoids, and apparently to kill any that try to develop within them (Sime, 2002). And some Lepidopteran caterpillars,

---

[39] It has been suggested that the "real" purpose of the patterning on early larvae that resemble bird droppings is to protect them from sun damage (Timmermann, 1999). White areas of the "dropping" pattern are derived from urea, a chemical that can function as a caterpillar sun blocker. Of course, there is no reason that the white color patch cannot serve both purposes.

[40] All adult butterflies die, of course, and thus adult mortality is technically 100%. When used in connection with adults, the term "mortality" usually refers to *prema-*

ture mortality, i.e., when death occurs "unnaturally" before an individual has completed its key life functions, specifically dispersal and reproduction. The high mortality of caterpillars may reflect in part the fact that this is the stage of longest duration in most species.

[41] Viral infections such as "wilt disease," which causes caterpillars to deflate and hang limply from foliage, are well-known to those trying to raise caterpillars indoors.

lastly, can detect sound through hairlike sensors on their bodies. Certain moth larvae freeze motionless if exposed to sounds having the same wavelength as the wingbeats of parasitoid flies. Others writhe frenetically in response to the same stimulus, evidently to disrupt the attacker's approach (Scoble, 1995, p. 145). Recent research has shown that certain moth caterpillars make scraping noises with their tails and jaws that deter intruding members of the same species (Yack et al., 2001).

Caterpillar defensive strategies can vary from instar to instar. Many caterpillars switch color as they grow, often in connection with changing lifestyles. For example, the color of Mexican Fritillary caterpillars in Jamaica darkens in late instars, as the caterpillars shift their feeding site from the undersides of hostplant leaves to the darker upper surfaces (Schappert, 2000, p. 27). In Europe, early instars of the Large Blue (*Maculinea arion*) are well-camouflaged on the flowers of Wild Thyme. Later instars lose this protective coloration—but are not disadvantaged since they are ant-tended when older and no longer require camouflage (Thomas & Lewington, 1991, pp. 105–8; Douglas, 1989, p. 135).[42] Along similar lines, spring brood Zebra Swallowtails wriggle to avoid parasitoids and rely on their osmeteria for protection against ants and small spiders. But these defenses are not effective against large spiders and parasitoids that attack the summer brood, so older individuals move away from the hostplant when not feeding (Damman, 1986).

A number of East Coast butterfly caterpillars feed communally, at least in early instars. Communal feeders must suppress the instinct to cannibalize nearby eggs and young larvae. But there are also some more active components to this behavior. For example, communal caterpillars will regroup if artificially separated, and they often travel together and forage on leaves in coordinated "team" attacks. (If a plant produces toxic or irritating secondary chemicals in response to foraging, then finishing off a leaf quickly may be the best way to avoid a rush of toxin.) Groups of communal Mourning Cloak caterpillars react to sudden noises by thrashing violently in synchronous fashion, for up to a minute at a time. The sheer number of individuals participating in this display may help to startle or confuse a predator.

The caterpillars of some communal butterflies weave various forms of dense, silken webs in which they can feed in relative safety (Hanski, 1999). But the net value of this strategy is uncertain, since the very existence of a large caterpillar cluster must create an irresistible lure to many predators and parasitoids, as well as a vector for disease and other density-dependent mortality factors. One the-

I-20. Chrysalis of Tawny Emperor

ory—that communal feeding allows short-lived adults to maximize ovipositing on sparsely distributed hostplants—is interesting, but does not fit the facts with many of the species involved (e.g., Mourning Cloak, with an 11-month adult life span, longest in our region, that lays on common trees, or the Harris' Checkerspot, which uses one of the commonest northern asters as its host).

A specialized subgroup of communal feeders are the ant-tended (*myrmecophilous*) lycaenids (Pierce, 1987). These species have the ability to enlist some of nature's most voracious predators as ecological bodyguards—and even at times to cast them as unwitting victims. They do this by secreting *honeydew*, a sweet liquid, from a nipple-like gland on their abdomen. The ants eagerly consume the honeydew, soliciting meals by stroking the caterpillar with their antennae. Sometimes, caterpillars emit sounds that are thought to attract ants, others may emit chemicals that act as "ant tranquilizers," defusing the hosts' native aggression. In some species, mainly outside our region, later-instar caterpillars are carried into the ant colony, where they feed on ant larvae while masquerading as colony members.

As the larval stage progresses, internal changes occur that presage the transition to adulthood. Beneath the larval exoskeleton, features of adult anatomy begin to form in separate pupal casings, including the wings, proboscis, antennae, adult compound eyes, reproductive organs, and legs. Production of juvenile hormone (which earlier caused the caterpillar to remain in the larval stage) abruptly declines. During its final instar, the caterpillar at first gains weight rapidly, then stops feeding altogether and expels undigested food from its gut. Shortly before pupation, exposed-feeding caterpillars begin to crawl energetically for

[42] Indeed, the older caterpillars change to resemble an ant larva, both in appearance and in chemical "signature." This fools worker ants into carrying them home to the colony, where they prey on real ant larvae. This species became extinct in England in 1979, after changing agricultural practices led to local extirpation of the host ant species (Thomas & Lewington, 1991).

hours, moving away from the larval feeding site. Concealed-feeding skippers tend to remain near the original hostplant, often pupating in the same leaf fold they used as caterpillars.

Once ready to pupate, the caterpillar spins a silk "mat" on the pupal substrate and attaches itself to this base using its *cremaster*—a pointed projection at the posterior end of the body, tipped with crochet hooks. Swallowtails and pierids position their pupae upright, using a silk thread, or "girdle," to secure themselves, like a telephone linesman steadied by a thick leather belt while climbing. Others attach themselves with silk threads in different positions. But most true butterflies simply hang upside down from their silk pad. Skippers, meanwhile, lash themselves in place within their leaf roll, producing a water-repellent powder from an abdominal gland that keeps the outer pupal skin dry. In some species, the color of the pupal shell itself, developing beneath the exoskeleton, may be determined by the wavelength of reflected light falling on the mature caterpillar as it nears pupation. By altering its pupal color to blend with the immediate surroundings in this way, a developing butterfly can significantly enhance its chances of survival.[43]

Finally, with the mature caterpillar firmly attached to its silk mat, the larval exoskeleton strips away, exposing the pupa. At first, the pupa consists of a series of separate casings, each corresponding to principal adult feature (wings, proboscis, antennae, etc.). But within hours these separate compartments fuse into a single structure, the *chrysalis*. Though generally rigid, many pupae are jointed and have some capacity to flex.[44] To our eyes, the pupa now looks inert, even dead, but inside the shell tremendous activity is taking place, as features of the larval anatomy are disassembled chemically and new, adult features take final shape. The thick, chitinous pupal shell prevents water loss while these processes occur. Within the first 48 hours after pupation begins, basic elements of wing pattern become fixed, including shape, color, and venation patterns. The formation of scales is determined in the next 22 hours, and individual scale structure within the next 10 hours. Overall, for nondiapausing pupae, this final preadult stage lasts between four days and two weeks.

While precise mortality statistics are lacking, many pupae are killed by adverse weather conditions and fungal growth. Pupae are also usually defenseless if discovered by predators or parasitoids, and they are also a mainstay in the winter diets of many shrews and mice (Brower, 1985). Although some pupae do have the capacity to thrash actively when disturbed, the main defensive strategies in this vulnerable resting stage are camouflage and inconspicuous placement (the latter being accomplished, if at all, by the final-stage caterpillar). Some pupae also actually make noise, emitting various chirps, hums, or clicks (*stridulation*). The noises of ant-tended lycaenids may constitute a means of communication with their ant hosts (which may thwart attacks by their former tenders). But the exact function(s) of stridulation remain speculative. It is not even known for sure whether it serves a defensive function at all.

As emergence nears, the pupa darkens. Production of juvenile hormone now ceases entirely, being replaced with the production of *eclosion hormone*. The critical moment of emergence often occurs near dawn (Scott, 1986, p. 23). The mature pupa draws in air and expands its muscles, splitting the chrysalis. The moist adult emerges head-first, then clings to some nearby object, often the pupal shell itself, usually hanging upside down. The adult then increases its blood pressure (by as much as 15 times) to pump body fluids into its limp, crumpled wings, which slowly inflate over a period of about 15 minutes. Its wings unfurled, the adult rests motionless for several hours to let them harden.

Mishaps during emergence can be costly. For one thing, the eclosing adult is flightless and vulnerable to predation. And crippling wing damage can result from a simple slip or loss of balance that ruptures the wing veins before they are inflated, or from snagging on the pupal case while escaping. If damage of this type inhibits the capacity for flight, the adult's remaining life expectancy will be short indeed.

While the wings of an emerging adult are hardening, or just afterward, the young adult voids its accumulated liquid waste (*mecomium*), sometimes in a fine spray that may deter intruders. It then cleans its antennae, usually by dragging them quickly through comblike hairs in the crook of its legs, then fastens together the two halves of its proboscis, which arrive unlaced. Finally, with its wings dry and its other preliminaries complete, the butterfly takes flight to begin the adult phase of its life.

BUTTERFLY BIOLOGY: THE ADULT (IMAGO)

Most butterflies spend only a small portion of their lives as adults. Yet we notice this stage most readily, and it is, by any standard, an exceptionally interesting period. The biology of adult butterflies is relatively complex, in keeping

---

[43] Black Swallowtail pupae vary in color, from green to yellow, brown, white, or blackish. In general, prepupal exposure to monochromatic light of various wavelengths produces different pupa color distributions (Hazel & West, 1996; Wiklund, 1972). Separately, Bernath (1982) showed that cryptic Cabbage White pupa—i.e., those matched to their substrate—had improved survival rates. Note, however, that factors other than light exposure can also affect the color of pupae (*see* Douglas, 1989, pp. 162–63).

[44] This capacity varies taxonomically. The pupae of giant-skippers are able to move up and down in their larval tube, using their cremaster hooks for traction rather than attachment.

I-21. A female Dina Yellow hangs motionless from her chrysalis shortly after emergence while her wings dry.

I-22. Failure to exit the pupa cleanly can prevent the wings from fully extending, often a fatal mishap.

with the diverse and shifting behavioral demands to which they must adapt.

Exoskeleton. Insect exoskeletons provide strategic attachment points that magnify the strength of tiny insect muscles, giving them disproportionate strength. Butterflies are seldom noted for muscular prowess, but their chitinous exoskeletons do absorb significant mechanical stresses associated with flight. Critical flexibility is provided by *resilin,* a rubber-like substance found between adjoining plates (*sclerites*).

Head. The head is among the hardest (or most highly sclerotized) parts of a butterfly's body. It houses a well-muscled suction chamber used to imbibe nectar, plus a cavity for the brain, with its distinctively large optical lobe. The head is generally covered with scales. When these form hairlike tufts, the head is referred to as *rough-scaled.*

Legs. The capabilities of insect legs are not widely appreciated. One need only watch a butterfly cling quietly to a blade of grass as it is thrashed about in a raging windstorm, or observe a Hammock Skipper dart under a leaf and fix itself upside down in one uninterrupted motion, to appreciate the great tenacity of these minute appendages. Most butterflies walk on six legs, like other insects. (Each leg has five jointed segments: trochanter, femur, tibia, tarsus, and pretarsus.) But nymphalids walk on their "back fours" only. The front legs still exist, but are reduced in size and usually held against the body, where they are used mainly for sensory purposes. In male nymphalids, the forelegs are covered with long scales, giving rise to the common family name "brushfoot."[45]

Some butterflies have a small, down-hanging flap attached to their forelegs (the *tibial epiphysis*) that is used to clean the antennae. In groups that lack this feature, some use brushes on their middle legs instead (lycaenids, metalmarks, nymphalids), and others may neglect antennal hygiene altogether, e.g., pierids and American Lady (Robbins, 1997). Tibial hairs are often useful in taxonomy.

Respiration/Circulation/Excretion. Adult butterflies have reasonably sophisticated respiratory and circulatory systems. While cold-blooded, with a low basal metabolism, butterflies engage in prolonged muscular activity that requires substantial metabolic support. Still, they can oxygenate their small bodies without need for specialized blood cells such as hemoglobin. Their primary body fluid (*haemolymph*) is greenish rather than red. Air is drawn into the body through a series of air tubes (*tracheae*), connected to the outside world via exoskeletal openings called *spiracles.* When not in use for respiration, they can be closed off to prevent dehydration. Air intake is enhanced by rhythmic body movements, which create a basic form of breathing. A butterfly's circulatory system is "open," meaning that the heart works like an indoor room fan—creating continuous air movement within an open space—rather than like a traditional pump, which forces liquid through a set of closed pipes. There are a number of structured circulatory conduits, nevertheless, including an aorta. With these, the system can achieve certain fairly specific circulatory objectives, such as ensuring adequate oxygenation of the head, thorax muscles, legs, certain areas of the

---

[45] Male snouts and metalmarks also walk on four legs, but the females on six. Male lycaenids have modified front legs (with some fused segments and no claws), but they still use all six legs, including the modified ones, in walking.

wing, etc. The butterfly's heart reverses direction periodically, changing the direction of fluid movement.

Body wastes are filtered from the haemolymph into *Malphigian tubules,* the functional equivalent of kidneys. These wastes empty into the intestine, where they are excreted along with general digestive residue. Adult butterflies produce dry urea, which reduces water loss. In some groups, as discussed below, a portion of the larva's uric acid waste is retained for use in wing pigments.

Butterfly Senses:Vision. It is easier to describe the details of a butterfly's visual system—intricate though they may be—than it is to know for sure what a butterfly actually *sees.* Some degree of visual acuity is necessary for basic adult life tasks, including flight control, location of food and hostplants, and reproduction (Rutowski, 2003). Researchers agree that butterflies can perceive large color fields, including colors we do not see, such as ultraviolet, and that they can detect form outlines and movement. But their ability to see detail beyond short range appears limited (Rutowski & Kimball, 2000). Zebra Heliconians, which have exceptionally large optical lobes, can recognize leaf shapes, identify local landmarks, and avoid spider webs in flight. There is evidence that at least some butterflies can perceive subtle wing pattern details, similar to the field marks used by humans to identify them (Fordyce et al., 2002; Rutowksi, 2003). But our knowledge on this point is limited, making it difficult to interpret many aspects of butterfly display behavior with confidence.

Butterflies have focal, compound eyes, composed of numerous individual eye segments, or *ommatidia.* Each ommatidium points in a different direction, affording an aggregate field of view as wide as 340 degrees, but with limited resolution in any one direction.[46] Each eye facet is covered by an octagonal *cornea,* composed of a layer of clear chitin. Males generally have larger eyes than females, with more ommatidia, presumably to assist in locating potential partners or territorial adversaries at a distance.

Interestingly, the visual styles of skippers and "true" butterflies are very different. In "true" butterflies, each ommatidium acts as a sealed "silo," transmitting visual information independently to the central nervous system (*apposition vision*) (Scoble, 1995, pp. 26–31). In skippers, light from one silo can leak into adjoining ones, creating a composite image at the retinal level (*superposition vision*). It is thought that the latter arrangement results in sharper vision, allowing skippers to engage in the fast, tactical flight from which their name is derived (Scott, 1986, pp. 37–39).

Butterfly eye pigments are species-specific, and can some-

times be used to distinguish similar species in the field. Examples (as discussed in associated species accounts) include Carolina vs. Little Wood-Satyrs (Bernzweig, 1999) and Aphrodite vs. Atlantis Fritillary (Glassberg, 2000). These differences were not discovered using specimens, however, because eye pigments quickly deteriorate after death.

Butterfly Senses:Taste and Smell. A butterfly's senses of taste and smell are closely related, as in humans (when blindfolded, humans often confuse smell and taste stimuli). In butterflies, some individual receptors detect both odors and taste. And even the antennae, which are usually classified as odor detectors, are frequently rubbed against objects the butterfly is trying to sense, as if "tasting" their surface. Apart from this one very general similarity, however, the chemical sensing mechanisms of humans and butterflies have little in common. While our sensors are focused in the nose and mouth, a butterfly's entire body is studded with tiny, specialized *sensilla*—on the antennae, palps, proboscis, legs and feet, wing bases, even the tip of the female ovipositor. Scientists have traditionally classified sensilla by their shape. The recent trend is to classify them by function, but this approach has been slowed by incomplete knowledge as to what each receptor actually senses.

Reflecting on these facts, there may be some foundation after all to the common impression that butterfly behavior is rapturous. An adult almost literally tastes and smells its way through the world, bombarded with gustatory and olfactory information from all over its body. The ecstasy of a French chef might reach new heights if he could taste the dishes he was preparing with his fingers while handling them.

The *antennae* are a butterfly's primary chemical receptor. Most sensory receptors are located on the bare antennal club (*nudum*), typically densest on the front-facing portion. A single female Monarch has been estimated to have some 13,700 antennal sensors (Scott, 1986, p. 38), used primarily to locate nectar and to detect male pheromones. A butterfly's proboscis also contains numerous sensilla. One type (the *sensilla styloconica*) appears to detect optimal levels of sugar concentration. Chemical receptors on the feet (*tarsi*) are also important. When these receptors come in contact with a sugary liquid, the proboscis uncoils reflexively to begin feeding. In the female, tarsal sensors read chemical signatures of hostplants as a cue to oviposit (or, in some cases, as an inhibitory cue).

Butterfly Senses: Hearing. Butterflies generally have a weak sense of hearing, but it is possible that most (maybe even all) adult butterflies can hear sound at least to some

---

[46] Because of overlapping coverage, forward vision is usually superior to vision from behind (where there is a small visual gap). Field observers should remember this fact in approaching wary subjects (Rutowski & Kimball, 2000). The direction in which individual eye cells are pointed can be discerned from the dark pseudo-pupil

that appears in bright light on the eye surface. This darkness is caused by pigments in the individual ommatidia facing the observer; these pigments absorb light that would otherwise be reflected directly outward (Rutowski, 2003).

extent (Scott, 1986, p. 40). The ability to hear has been demonstrated in tropical heliconians, and some nymphalids detect sound using air-filled chambers at the base of the wings. The *subgenual organ* at the top of each tibia can also sense vibrations (as opposed to airborne sound), thus providing early warning of nearby disturbance.

BUTTERFLY BIOLOGY: ADULT FEEDING AND NUTRITION

We have already discussed the most important aspect of butterfly nutrition—the consumption of hostplants by caterpillars. Adult feeding is little more than a complex footnote to the larval feeding story. The two tales are closely interrelated, nonetheless, since many adult feeding behaviors exist to compensate for shortcomings in caterpillar diets. As noted earlier, a caterpillar's main life task is to add biomass, and hostplants provide the raw nutrients required to bulk up. But adults also need additional specific substances—such as particular amino acids, pyrrolizidine alkaloids, even simple sodium—to support various key life functions, including courtship, migration, egg production, manufacture of pigments, and, in some cases, diapause. Larval hostplants often lack one or more of these essential elements—perhaps "on purpose," as an adaptive measure to deter browsing. Herbivores at all levels face equivalent problems. A shortage of dietary sodium, e.g., draws plant-eating mammals such as deer almost magnetically to saltlicks.

To some degree, adult butterflies make up for inherited nutritional deficiencies in the course of their normal feeding. This is especially true for dung-/sap-feeders, whose regular diet provides a significant supply of organic compounds and minerals. And even nectar-feeding adults can partially redress inherited shortages, by choosing "dirty" nectar sources, which contain traces of amino acids, proteins, and other useful substances, rather than "clean" nectar, composed of little but water and dissolved sugars. (Bees, which are not herbivores as larvae, prefer "clean" nectar [Scott, 1986, p. 67].)

But many butterflies still find it necessary to supplement their ordinary diets, often by *puddling*. It is common on quiet country lanes to see butterflies perched on wet sand or muddy puddle edges, sometimes in large groups, all busily sipping. They may do this in some cases simply because they are thirsty. But research has shown that puddling behavior is induced by the presence of specific chemicals, particularly sodium (Scoble, 1995, pp. 20–21). It seems likely that different species may have their own individual objectives for engaging in puddling, even when sipping directly alongside one another. Significant quantities of water may be pumped through a butterfly's body in order to extract dilute target substances. In some Lepidoptera, water is forcibly ejected from the rear of the abdomen during this process.

Oddly, a majority of puddling butterflies (often more than 95%) are males. It has been argued that vigorous mate-seeking activities impose disproportionate nutritional demands. But females surely also expend enormous energy searching for hostplants and producing eggs. A more plausible explanation is that male butterflies specialize in acquiring extraneous nutrients as an indirect part of their larger reproductive role. Specifically, when males mate, they transfer a flexible sac (the *spermatophore*) to the female. In addition to the sperm itself, the spermatophore contains a "gift pack" of nutritional substances that the female absorbs (Watanabe & Sato, 1993). In the case of Monarchs (which migrate and then overwinter), the spermatophore also contains an extra ration of fat (Scoble, 1995, p. 73). Nutrients in the spermatophore extend the lifespan and fertility of female butterflies (Wiklund, 2003).

Butterflies that do not feed on nectar as adults (or that do so rarely) tend to use a variety of alternate foods, often more or less opportunistically, including sap, dung, carrion, etc. As noted earlier, most of these species live in woodlands, where nectar is scarce, or else overwinter as adults, and thus require alternate foods in the early spring. Adult Harvesters may subsist solely on honeydew secreted by aphids—most likely from the same colony that they earlier marauded as larval predators. Non-nectaring butterflies have characteristically short proboscises. When compelled to visit flowers in times of stress, they may find that some blossoms are too deep to negotiate successfully.

Most adult butterflies, of course, do feed primarily on nectar. As with other aspects of butterfly lifestyle, some nectar-feeders are generalists, visiting a wide variety of flowers, while others are quite selective. In either case, nectar flowers are chosen initially by color and shape.[47] Giant sulphurs (genus *Phoebis*) often select red flowers, smaller pierids often choose white or yellow ones, etc., although without knowing how butterflies experience color it is difficult to be sure what the exact selection criteria may be. Some flowers display ultraviolet patterns that assist butterflies in locating their nectary; these patterns are referred to as "nectar guides" (Barth, 1991).[48] There is some evidence that floral scents attract specific butterfly pollinators,

---

[47] In research on nectar preference, flower shape has received less emphasis than color. But studies of butterfly learning (discussed below) have shown that complex flower shapes are often difficult to negotiate, and require practice. Specialized shapes may be yet another strategy utilized by plants to ensure that pollinators will tend to concentrate on desired species.

[48] In general, the nectar-bearing portion of the flowerhead absorbs ultraviolet light, and thus would appear dark to the butterfly, whereas the non-nectar-bearing petals reflect ultraviolet light, and would appear light.

I-23. Most adult butterflies obtain their primary nutrition from nectar.

as they do many night-flying moths (Raguso & Willis, 2003). Butterflies are sometimes important pollinators, but other insect groups tend to play a larger role, specifically bees, flies, and beetles. Among Lepidoptera, Scoble considered moths more significant than butterflies (Scoble, 1995, p. 178).

Nectar is drawn up through the proboscis using suction created by muscles in the head. Contrary to earlier belief, capillary action is not involved (Douglas, 1989, p. 41). Investigators have spent a great deal of time exploring the mechanisms used to coil and uncoil the proboscis. In brief, coiling results partly from the curved shape of elastic tissues embedded in the upper half of the proboscis (*resilin*); the coil is further tightened by muscles located within the shaft. To uncoil the proboscis, a hydraulic mechanism in the head forces blood into hollow channels (*septa*) that run the length of the appendage, causing it to stiffen; meanwhile, muscles in the head raise the now-extended coil, like the arm of a crane. The outer part of the proboscis, now dangling downward, is dipped into the nectar.

Researchers originally believed that butterflies prefer dilute nectar solutions (20%–25% concentration). Thinner solutions would better quench thirst, it was thought, and could be drawn through the proboscis more easily. It was inadvertently discovered, moreover, that pure solutions of sucrose (a disaccharide) could crystallize in the proboscis, killing the butterfly, whereas solutions of glucose (a simple monosaccharide) did not (Scoble, 1995, p. 24). Later studies suggested that butterflies actually tend to prefer higher-concentration nectar sources than originally believed, perhaps 40% sugar on average. European Skippers visit Alfalfa

flowers with a sugar concentration of 40%–65% (Pivnick & McNeil, 1985).

A number of factors may combine to determine nectar preferences. One important attribute, viscosity, is affected by chemical composition and ambient temperature. Habitat structure can also be significant, since there is often a strategic trade-off between searching exhaustively for "ideal" nectar plants vs. accepting lower-quality nectar that is more readily available. It is difficult to state a single, concise formula that will account for the many detailed contingencies involved.

Among the most interesting examples of adult feeding among East Coast butterflies is the Zebra Heliconian (Gilbert, 1972). This mainly tropical species has adopted a lifestyle that minimizes the time it spends in the larval stage. Although its caterpillars are toxic to vertebrate predators by virtue of cyanogenic compounds derived from the passionvines they eat, they are vulnerable to certain invertebrate predators in their tropical forest homes. The main drawback to accelerating quickly through the caterpillar stage is that Zebras enter adulthood with many nutritional requirements unfulfilled. To sustain their long adult lifespan (which may extend for six months or more), longwings have turned to *pollen* as their primary food.[49] First they liquefy the pollen with drops of saliva extruded from the proboscis, then they suck the resulting mixture back in as food. This protein-rich diet extends longevity, and allows the female to produce eggs "from scratch." Adult Zebras also use pollen derivatives to synthesize deadly aliphatic cyanogens that make them toxic. Zebras roost communally, issuing a joint warning to potential predators with their bold colors. Given low adult mortality, Zebras have been able to evolve a sophisticated adult society (for a butterfly), in which older individuals show younger ones where to find local pollen resources. This is one of the few species in which adult food resources outweigh hostplants in determining home range and lifestyle.

BUTTERFLY BIOLOGY: WINGS AND FLIGHT

In all of the animal kingdom, only three existing groups of organisms have the capacity for true flight—birds, bats, and insects. Among them, insects learned to fly first, approximately 300 million years ago, during the Carboniferous period. Wings may have initially evolved as heat-accumulating appendages,[50] but once insects learned to use them for flight they gained an immediate and imposing

[49] Actually, Zebra Heliconians feed on both nectar and pollen, usually pollen in the morning and nectar in the afternoon. Pollen is an actual reproductive germ cell (gamete) produced by the anthers of angiosperms whereas nectar is a sweet liquid whose exclusive function to lure pollinators.; unlike pollen, it has no direct reproductive function.

[50] Brodsky (1994) argues that if wings had evolved to fulfill an important func-

tion, they would still be locked in to that use, and thus unavailable to be wings. It is therefore likely, he reasons, that wings evolved in connection with some relatively transient purpose, such as signaling or securing eggs on the parent's back (Brodsky, 1994, p. 81). But this does not rule out thermoregulation as an initial precipitator, since even today butterfly wings continue to play a critical role in thermoregulation, despite having become specialized for flight. *See* discussion in Section II.

ecological advantage. Explosive diversification ensued. Currently, insects account for more than three-quarters of all living animal species.

The butterfly wing is an unlikely invention. It bears little resemblance to the wings of birds or airplanes, either mechanically or aerodynamically. It has no internal musculature or moving parts, for one thing. Nor does it achieve flight solely through an airfoil effect, in which lift is generated by the flow of air over a curved upper surface (Scoble, 1995, p. 68). By way of a physical analogy, we might compare a butterfly's wing mechanics to the propulsion of a rowboat using oars, or the setting of sheets on a sailboat. Like a sail, a butterfly's wing consists of a thin membrane stretched over a fixed-shape frame, which can be positioned at various angles to the flow of air. And like an oar, it is attached to the rigid outer frame of the exoskeleton, using it as a fulcrum to drive wing movements.

A butterfly's thorax is composed of separate upper and lower plates, joined by a rubbery seal of resilin. As shown in Figure 3.1, the upper plate is the narrower of the two. The upper thorax is the attachment point for the inner base of the wings. This connection is analogous to a rower's hands firmly grasping the tips of a pair of oars. There is also a second connection, slightly farther out, to the wider bottom half of the thorax. This connection acts like the oarlock on a rowboat's gunnel. During flight, relaxation of internal muscles allows the upper thorax to flex upward. As the edges of the upper thorax rise, the wing "handles" are drawn up with them, forcing the wings themselves down by lever action. When the internal muscles contract and the upper thorax is reflattened, the "handles" are pulled down, forcing the wings up.

But a butterfly cannot achieve flight merely by flapping its wings in the manner just described. The angle of attack of the wings changes during the course of each stroke, for one thing. On the upstroke, the wings are angled upward, allowing air to roll off as they rise. On the downstroke, they are likewise angled down, trapping air beneath them and creating a downward and backward push.

Furthermore, mechanical tension builds in the lower half of the thorax during each wing beat, deforming it out of shape, like a spring that is being compressed. This tension reaches its peak when the wings are held straight out to the sides, and can be released entirely only when they come together, either above or below the body. During each wing stroke, thoracic torque builds as the wings reach the midpoint of their trajectory, then quickly discharges as they pass the midpoint. This abrupt release of tension creates a "snap" that contributes to thrust. High-speed photo-

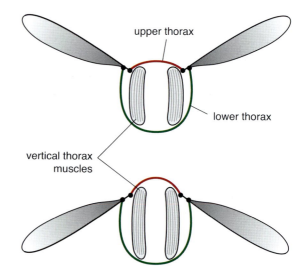

Figure 3.1. Wing Mechanics. The lever-fulcrum action of the wings drives a series of complex movements that ultimately propel a butterfly in flight.

graphs reveal that most butterflies "clap" their wings overhead—and often again below their bodies—on each wing stroke (Scoble, 1995, p. 67). After clapping, the wings peel apart flexibly, beginning at the apex, rather than stiffly separating. As noted below, this detail ends up being significant aerodynamically.

Despite its intricacies, the flapping motion just described is still insufficient, by itself, to produce adequate lift for sustained flight. Rather, as documented in recent decades, the movement of air over the butterfly's wings and body during each wing beat sequence creates a complex series of structured airflows and eddys (*vortices*). The net effect of these flows is to lower air pressure over the body, producing nearly continual lift (Brodsky, 1994, pp. 181–85).[51] The position of these vortices (and even the direction in which they rotate) changes abruptly from upstroke to downstroke, causing air to be thrust from the wings at a variety of different angles. These varying propulsion vectors—some created by the "clap and peel" style of wing flapping described above—help investigators explain the butterfly's ability to remain aloft. It also helps explain the jerky, irregular appearance of a butterfly's movements in the air. A recent study has shown that free flight in butterflies consists of an acrobatic series of individual flapping moves strung together in a random-seeming order (Srygley & Thomas, 2002). Each move is tailored to the aerodynamic needs of the moment and adds lift in its own, particular way.

Butterflies remain less distinctive for their "flapping

---

[51] Traditional aerodynamic theories, based on steady-state or quasi-steady-state models, explain lift in terms of airflows over a mostly fixed wing surface. Such theories can help to account for the lift generated during a butterfly's downstroke. The subtle aerodynamics created by wings *in motion* is the subject of "unsteady-state" models. This newer approach is used, among other things, to explain the lift a butterfly wing achieves on the upstroke (Brodsky, 1994, pp. 184–85).

flight," nonetheless, than for *gliding*. In the long evolution of insect flight, there has been a trend toward ever-increasing wingbeat speeds. This enables faster and more agile flight maneuvers, as epitomized in certain bees and flies (*Hymenoptera* and *Diptera*), whose wings can beat up to 1,000 times a second.[52] In order to achieve such rapid oscillations without battering their wings to pieces, advanced flying insects have evolved a "two-wing" flight style, in which the forewings and hindwings are locked together and move synchronously. Moths have a mechanical wing-coupling mechanism (the *frenulum*) that physically clasps the forewing to the hindwing.

Butterflies have abandoned this evolutionary trend, however, developing instead a flight style that capitalizes on slower wingbeats, averaging just 5 to 12 beats per second, with a maximum of about 20 beats per second in some skippers (Scott, 1986, p. 18). This style facilitates both long-distance migration and efficient local foraging and courtship flights. Not all butterflies are specifically well-adapted for distance flight, but some, such as the Monarch, are world-class masters of the technique. Even butterflies that do not regularly glide have lost their fixed wing-coupling mechanisms. This allows them to separate their wings at low speeds (e.g., when landing), creating an air slot that deters stalling (Scoble, 1995, p. 67).[53]

Gliding requires a number of specialized physical traits (Brodsky, 1994, pp. 66–70). Only reasonably heavy butterflies can glide efficiently, because a significant body mass is needed to maintain a constant, diagonal glide path without disruptive buffeting. Small insects lack the needed mass. The best gliders have exceptionally *low wing loading*, i.e., their bodies must be light relative to their wing surface area. Taken together, these two requirements imply a fairly hefty butterfly with an ample wing surface. In addition, butterflies' wings are relatively thick at the base, where they connect to the thorax; this feature helps prevent rolling in flight.

When gliding, four distinct "vortex braids" form over the upper wing surface. These extend the range of the glide by allowing a butterfly to increase the angle of attack of its forewings—i.e., to achieve more lift without disrupting the smooth airflow over its back (which would cause a stall). Irregularly shaped wing scales on the ventral hindwing, moreover, tend to neutralize disruptive air currents that otherwise would form beneath the wing, creating drag. Research suggests that the absence of wing scales would reduce a butterfly's overall lift by around 15% (Scoble, 1995, pp. 67–68; Nachtingill, 1968). And, finally, by holding the wings in a slightly cocked (or V-shaped) position, a gliding butterfly reduces wingtip turbulence, also lowering drag.

The optimal shape for gliding is reportedly that of the Heliconians, i.e., long, thin wings, like those of a human-built glider. Swallowtails have the least advantageous shape for gliding (among "true" butterflies, at least), but compensate for this with long tails, which catch trailing wing vortices and extend their effect—allowing them to maintain a straighter course and to increase their angle of attack without stalling (Betts & Wootton, 1988).[54]

Butterfly wing shapes are fairly well-correlated with flight style, although unrelated selective factors can also influence wing design, such as thermodynamic requirements and predator considerations. Butterflies with short, broad wings generally tend to engage in slow, "fluttery" flight. Males in this category typically locate mates by patrolling. Narrower wings with pointed tips are better for agility and tactical flight. When combined with large thorax muscles, such wings are well-adapted to meet the performance needs of male perchers, which often engage in dramatic aerial dogfights (*see*, e.g., Van Dyck, 2003; Bitzer & Shaw, 1979). Females often have rounder wings than males, better suited for long, exploratory hostplant forays (Scott, 1986, p. 49). The fastest-flying skippers in our area (*Hesperia* and *Megathymus*) can attain flight speeds for short periods exceeding 50 mph, but "true" butterflies rarely exceed 20 mph.

With a modest degree of field experience, observers can begin to classify butterflies in the air based on flight style (initially by family). Blues typically fly near the ground with a busy, fluttering flight style. Satyr flight is jerky and erratic, with a short hop after landing. Skipper flight is fast and relatively direct. And so on. Because of their complex, "shed vortex" flight mechanics, described above, butterflies seldom fly as smoothly as birds. With frequent changes in direction and flight attitude, their trajectory can be highly unpredictable. And many species mix in occasional glides, causing them to drop rather suddenly from time to time. All of these idiosyncrasies and "imperfections" in flight style actually seem to confer significant protection on flying adult butterflies, however, by making them far less predictable, and thus a less vulnerable target to aerial predators such as fly-catching birds (Brodsky, 1994, p. 186; Syrgley & Dudley, 1993 [flight style of palatable species better for avoiding predators]).

---

[52] Bees and flies have evolved "fibrillar" flight muscles that can oscillate rapidly, producing many contractions for each motor nerve impulse. Among insects *lacking* this oscillating capacity, the fastest known wingbeats occur among certain sphinx moths, at about 100 beats per second (Brodsky, 1994, p. 166; Douglas, 1989, pp. 3–4).

[53] Although they lack a frenulum, butterflies usually hold their wings together during flight, relying on the thorax muscles to maintain an overlap between forewings and hindwings. This is termed *amplexiform coupling* (Scoble, 1995, p. 61).

[54] Monarchs have a somewhat fuller wing shape than that of heliconians, but this allows them to catch warm, rising air currents (*thermals*) in daytime migrations. The benefits of this thermal boost are worth a marginal loss of gliding efficiency.

Learning to Look at Wings

Few of a naturalist's many experiences in the field are any more bewildering than his or her first close look at a butterfly's wings, trying to sort out all of their chaotic intricacies. With practice, it is possible to key in on one or two telltale characters that work as field marks for any given species. But even experienced observers find it difficult to describe a butterfly's overall wing pattern in simple terms. Beginners should not be overly frustrated by this, however, since it took dedicated scientists the better part of a century to work out a general descriptive model for wing patterns, and to apply that model consistently across families.

The vast majority of East Coast butterflies can be identified in the field by sight, provided one has an opportunity to view them satisfactorily. Nonetheless, butterfly identification is a complex, acquired skill, the difficulty of which is not to be underestimated—whether carried out in the field or the museum. A butterfly's scales begin to wear from the moment it emerges, causing some patterns to fade quickly. Others, such as the hyaline spots on skipper wings, may be obscured initially by overscaling, but become larger and more visible with wear. Worn wing surfaces can exhibit a bronzy sheen that is absent in fresh individuals. And wing patterns themselves, finally, show tremendous individual variation in many species.

The biology of wing pattern development is as complicated as the patterns themselves. Indeed, readers may find the following materials to be relatively challenging. But the foundations they establish are important. Starting with the basics, we can see that a butterfly's left and right wing sets are usually identical (*bilateral symmetry*).[55] Yet the forewings often look different from the hindwings, and the upper (or dorsal) wing surface often bears no resemblance whatsoever to the lower (or ventral).[56] In classifying wing patterns, researchers distinguish *background elements* from *pattern elements,* since they have different developmental origins. It is not always easy to tell one from the other. On a Zebra Heliconian, the bright yellow markings are actually "background" while the large dark patches that dominate most of the wing are "pattern" (Nijhout, 1991, pp. 75–79). Background patterns are usually (but not always) dull. They may include color gradients, sometimes even large patches of bright scaling, but they never achieve the detail or complexity seen in pattern elements. Major pattern elements include bands, bars, eyespots, ripples, discal spots, chevrons, etc. (*see* generally Beldade & Brakefield, 2002).

Virtually all of the colors on a butterfly wing are the

I-24. Pattern variations in American Copper

product of scales. *Wing scales* are the remnants of specialized epidermal cells (called scale-building cells, or *trichogens*) that grow on the developing wings of the pupa. After these cells die, their membranes dissipate, leaving only the scales themselves—flat, hollow structures with a variable and often highly detailed latticework of grooves, ridges, internal trusses, and other complex forms. *Structural colors,* including iridescence, are produced by interference and refraction patterns as light passes through scales of particular shape. Iridescent tones on butterfly wings are among the most intense and sophisticated colors to be found in nature. Iridescent colors may extend into the ultraviolet range, invisible to humans. Some specialized scales (*androconial scales*) disperse pheromones used in courtship.

Pigments are synthesized in the developing wing scales of the pupa, generally a day or two before emergence (Nijhout, 1991, p. 11). These substances absorb all wavelengths of light falling on them, *except* for the wavelength of the color that is reflected. There are four principal families of butterfly wing pigments—melanins, ommochromes, pterins, and flavenoids—plus a variety of secondary groups. Pigments produce a wide range of colors, from black and brown (melanins) to white and yellow (pterins, derived partly from uric acid). Some are synthesized by the pupa from available metabolic materials; others, such as flavenoids, are derived from chemicals obtainable only from specific plants.

A general blueprint for the wing pattern of all true butterflies, called the "nymphalid ground plan," was first described in the 1920s (Schwanwitsch, 1924; Suffert, 1927; Figure 3.2).[57] The ground plan consists of several pattern systems running across the wing surface, in roughly concentric zones, at varying distances from the body. These include the *basal symmetry system, central symmetry system, discal spot, border ocelli, parafocal element,* and *marginal*

---

[55] In *bilateral gynandromorphs,* on the other hand, one wing has female coloration, while the other carries the male's patterning. This particular aberration results when an egg cell is produced with two nuclei (Scribner & Evans, 1988; Douglas, 1989, p. 166; Connors, 2002). This oddity may be more common than we know,

since it is apparent only in species whose wing patterns differ by sex or some other trait (*dimorphic*).

[56] Single genes control both of these variances (Carroll, 1997), but that is little comfort to a new observer trying to sort out the many possible variations.

[57] The "ground plan," it should be noted, does not apply to skippers.

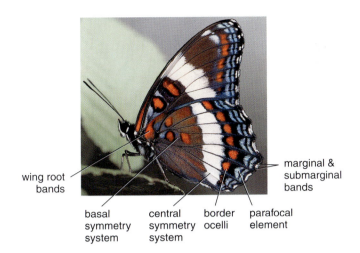

wing root
bands

marginal &
submarginal
bands

basal   central   border   parafocal
symmetry   symmetry   ocelli   element
system   system

Figure 3.2. The Nymphalid Ground Plan. Illustrated here in a White Admiral. Individual species do not visibly incorporate all features of the ground plan.

and submarginal bands. Nijhout (1991) describes the subtle variety of these pattern systems in significant detail.

Since it predates the advent of modern cell biology, the original nymphalid ground plan was not an outgrowth of evolutionary genetics. Rather, it is merely a descriptive guide to regularly occurring pattern elements that appear to have consistent developmental origins (called *homologies*) across species. No single species exhibits all elements of the ground plan, not even nymphalids, nor was it ever considered primitive from an evolutionary standpoint. Indeed, investigators have only recently begun to decipher its full, biological significance.

At first brush, the nymphalid ground plan seems an ideal tool for analyzing wing patterns in the field. But its expression is actually quite obscure in many families. What looks like a single wing line may in fact be a hybrid (or composite) pattern, made up of segments borrowed from two or more different symmetry systems, e.g., bits of the basal and central systems may be joined into what appears to be a single line or marking. This phenomenon is called *pierellization* (after the tropical satyr genus *Pierella,* in which it was first observed). Rearrangements of this type can be detected only by closely examining a wide variety of species from the target family, especially more primitive ones, to discover how the pattern gradually evolved over time.[58] In some species, specific elements of the nymphalid ground plan have become highly distorted or irregular. Two bands can merge together into a single color field. Or a pattern element may assume the same color as its background, effectively disappearing (Nijhout, 1991). So while the nymphalid ground plan is a powerful tool for scientific investigation, it has sharp drawbacks in the field.

The first steps in adult pattern development begin in the caterpillar stage. At this point, the future adult wings are nothing more than small cell clusters called "wing buds" (*imaginal discs*). As these discs grow within the larva, they are invaded by a network of air tubes (*tracheal fibers*) extending from the respiratory system, which create a series of tunnels called *lacunae.* The pattern formed by these tunnels has roughly the same shape as the future adult wing. Eventually, the areas lying *beyond* the borders of the lacunal network undergo programmed cell death, leaving just the wing outline itself intact. The process is similar to using a cookie cutter to define patterns in dough.

With pupation underway, investigators point to a series of chemicals (*activators* and *inhibitors*) secreted from the pupal wing veins. These chemicals cause certain areas within each cell to be marked as *sources* or *sinks* (Nijhout, 1991, p. 217). Nothing visible happens at this point, but later in the pupal stage the sources become active, secreting *morphogens* that diffuse across the wing cells. These morphogens instruct each scale-forming cell what pigment to produce (and possibly what iridescent form to take). In the course of their journey, these morphogens may interact with morphogens emitted by other sources, or they may be neutralized by morphogens emitted from *sinks,* located elsewhere in the cell. On a cell-by-cell basis, these chemical interactions gradually define the ultimate pattern of the adult wing. All that remains is for the actual wing scales to develop, each assuming its designated shape and color pursuant to its instructions. Investigators have been able to replicate the majority of observed butterfly wing patterns with this "source and sink" model (Nijhout, 1991, p. 215), using computer simulations.

But the basic "source and sink" model does not account for all observed wing patterns. It has difficulty representing "wing clouds" of the type found on swallowtail hindwings, or the "ripples" seen in many satyrs, or the random, multicolored flecking on the ventral hindwing of many pierids. Nor does the model readily explain how the formation of certain independent pattern features, such as marginal bands or the discal spot, is controlled. And, lastly, it cannot resolve a simple and longstanding puzzle, namely, how it is that an Eastern Tiger Swallowtail's forewing and hindwing stripes happen to align almost perfectly when the wings are at rest (Nijhout, 1991, p. 236; Oudemans, 1903). Still, advances in recent decades have been enormous, giving promise of additional fundamental insights—not just into wing patterns themselves but also into the genetic mechanisms that control them.

The developmental processes just described rely on chemical reactions, and like any such reaction they can be

---

[58] If the pattern has become highly specialized in all modern species, investigators may lack developmental "stepping stones" to guide them back to the original divergence. This has been a problem with swallowtails and heliconians, among other groups. Investigators must often look to rare, primitive members of the family for clues.

I-25. Aberrant Banded Hairstreak

I-26. Praying Mantis devours an immobilized Silver-spotted Skipper.

disrupted by extreme weather conditions. In some cases (as discussed earlier), regular seasonal temperature changes are a cue for the development of standard seasonal phenotypes. But extreme temperature shock (heat or cold) can disrupt normal pattern formation, producing *aberrations*. Observers often describe aberrations to be genetic mutations, but most probably occur when developmental processes are interrupted at a critical juncture by abnormal temperatures. Indeed, "standard aberrations" occur in particular species, probably as a result of recurring disturbances during certain critical periods of development. It has been noted that record warm temperatures during the summer of 1976 were associated with an unusually large incidence of aberrations in Britain (Russwurm, 1978, p. 12).

BUTTERFLY BIOLOGY: ADULT LIFESPAN AND POPULATION LEVELS

An adult butterfly's *maximum lifespan* is normally no more than a few weeks, even among individuals that do not earlier succumb to predators, disease, etc. And *average life expectancy*—taking all mortality factors into account—barely exceeds a single week in most species. Mark-and-recapture studies indicate that female butterflies outlive males, though only by a few days in most cases, and that large species generally outlive smaller ones (Scott, 1986, p. 25). The longest-lived butterflies by far are those that overwinter as adults (or that have a largely inactive cold-weather adult generation). The Mourning Cloak, a single-brooded species, may live as an adult for up to 11 months after emerging in mid-summer—although it will be inactive for much of this time, estivating in the late summer and hibernating by winter. To our knowledge, Zebra Heliconians have the longest *continuously active* adult lifespan among our species, sometimes exceeding six months.

Adult butterflies are subject to numerous mortality fac-

tors, including disease, predators, parasites, foul weather, and lack of sufficient nutrition. These losses can affect overall population levels, since only 1%–2% of eggs ever reach adulthood in most species to start with (Schappert, 2000, p. 155). Throughout the life cycle, climatic factors and changing weather patterns can have far-reaching effects. In semi-arid habitats with sparse vegetation, caterpillar populations are likely to explode during transient rainy periods. When normal drought conditions return, this can lead to mass starvation, as oversized larval populations decimate the available hostplant stock (Ehrlich, 1985, p. 28). To avoid this result, mass emigrations may occur in "good times." This has been observed in the boom-and-bust life cycle of the American Snout, which streams northward out of Mexico by the billions in years when favorable conditions allow most of the year's caterpillars to reach adulthood (Schappert, 2000, p. 107). In general, temperate zone butterflies are subject to dramatic population shifts across seasons, sometimes even across broods within a single year. These fluctuations can occur with eerie synchrony, appearing in numerous, seemingly unconnected populations over a broad geographic range. Our insights into the drivers of such mass population dynamics are scant indeed.

BUTTERFLY BIOLOGY: SIGNALING, DISPLAY, AND DEFENSE

Butterflies are not always passive victims of circumstance. Researchers have investigated strategies they use to improve their odds of successfully reproducing. Some have taken a fresh look at the nymphalid ground plan, arguing that there are adaptive reasons *why* butterflies have evolved independent pattern systems on different parts of their wings. Markings on the inner (basal) third of pierid hindwings, e.g., seem to change in response to thermoregulatory needs, whereas the outer two-thirds (medial and marginal) change in response to signaling or communication priorities (Kingsolver & Wiernasz, 1991; Nijhout, 1991, pp. 134–36). We do not fully understanding the details of this intricate adaptive process. For one thing, we often do not know which of a butterfly's many natural enemies a particular display is meant to distract (or which potential mate to inveigle), nor do we know in many cases how the message is perceived by the intended recipient (or how clearly). An equation with this many unknowns is difficult and time-consuming to solve.

Still, the approach does hold promise. Researchers from the Netherlands have studied certain tropical butterflies with different wet- and dry-season forms. The wet-season form of one tropical satyr (*Bicyclus anynana*) has large eyespots in the wet season, but very small ones during the dry season. Adults feed actively during the wet season, making good use of their eyespots to divert predators. In the dry season, however, when the satyrs are mostly inactive (in "dry season diapause"), the eyespots could act as a "bull's-eye" for predators. In one experiment, a population of wet-season individuals was released during the dry season, and all were quickly eaten. This result, while not conclusive by itself, is consistent with the possibility that conspicuous markings are disadvantages in the dry season because they increase visibility. *See also* Thomas and Lewington (1991) (Great Heath account).

The researchers in the prior discussion took an additional, illuminating step. Specifically, they reared *Bicyclus* under controlled temperatures in the laboratory, producing successive generations of entirely wet-season or dry-season forms. In these experiments, it took *only 20 generations* for each population to lose its genetic capacity to produce the alternative phenotype! That is, caterpillars from the wet-season population (with large eyespots) would no longer produce a small-eyespot phenotype *even if* exposed to low temperatures during pupation (*see* Carroll, 1997, p. 32). Such rapid alteration of genetic coding suggests a more direct link between evolutionary change and short-term ecological adaptation than was previously suspected.

More generally, significant pattern differences often occur on dorsal and ventral wing surfaces, indicating evolution at work (Nijhout, 1991, p. 239). In many species, the ventral surface (which is normally exposed at rest) tends to have disruptive or cryptic scale patterns, helping to avoid detection. This strategy reaches an epitome in the leafwings, whose brown ventral surfaces closely resemble large, dead leaves, down to the simulated leaf veins on their hindwings and the lifelike fungal damage marks that dot the wing surface. Nijhout (1991) describes the extensive modifications needed in the basic nymphalid ground plan to achieve this "dead leaf" effect.

The bright patterns on a butterfly's dorsal surface, by comparison, are often designed to communicate rather than to conceal. In many species, the nymphalid ground plan becomes so distorted on the dorsal surface that its elements are nearly impossible to trace. Strong selective forces are involved in shaping these colorful displays.

The messages that a butterfly conveys with its wings—as we currently understand them—are fairly basic. Yet our interpretations of these displays have changed over time, and may change again. Not even a century ago, it was thought that alternating shows of "dazzling color" and "eclipsing color" in flight (from on the dorsal and ventral wing surfaces, respectively) allowed species such as the Eastern Comma to elude predators. Weed (1917, p. 27) explained the reasoning behind this theory:

> The eye, having once fixed itself upon an object of a certain form and color, conveys to the mind a corresponding impression, and, if that impression is suddenly found to be unreliable, the instruction which the mind conveys to the eye also becomes unreliable, and the rapidity with which the impression and consequent instruction can be changed cannot always compete successfully with the rapid transformation effected by the insect in its effort to escape.

Most researchers now believe that ventral hindwing patterns evolved primarily to reduce visibility when a butterfly is resting, not to "jam" the perceptual apparatus of pursuing predators. It is certainly plausible, though, that bright dorsal colors may create a vibrant "search image" for predators during flight, which conveniently vanishes when the butterfly lands.

One remnant of the "dazzle-eclipse" theory does survive, however, in the concept of "startle effect." When a drab butterfly, resting with its wings closed, suddenly takes flight, the ensuing flash of color is thought to disorient an unsuspecting predator. Or, a resting adult may distract a predator by momentarily lifting its forewing, suddenly revealing an eyespot that resembles the gaze of an even larger predator, such as an owl. Far-fetched though these ploys may seem, they evidently do succeed at times (Blest, 1957; Waldbauer, 1998, p. 85). Also apparently effective are the large eyespots on late-instar swallowtail caterpillars, which are thought to make them resemble snakes—a ruse that is reinforced when the forked osmeterium is extended, resembling a snake's tongue. More conclusive research is needed, however, for a fully grounded understanding of butterfly distraction displays.

A problem with categorizing wing patterns as cryptic or noncryptic is that camouflage is not always drab. Like a tiger, whose bold stripes provide excellent concealment in tall grass, many bold or busy butterfly scale patterns confer protection by interrupting the visual outline of the wing. It is sometimes difficult for human investigators to determine whether the patterns of a specific butterfly act primarily (1) as disruptive camouflage, (2) as a bold flag to

I-27. Cryptic patterning in Florida Leafwing

warn of toxicity, or (3) as a pattern that mimics the appearance of toxic species. It is sometimes possible to work backward in addressing this issue, starting from the premise that edible butterflies tend to evolve cryptic scaling, whereas toxic ones seek to send a conspicuous signal. Knowing whether a butterfly is edible, i.e., may help decode its "wing message"—but this technique is far from foolproof, as discussed below.

Aposematic coloration, in which a species advertises its toxicity with loud colors, has been the subject of extensive discussion in the literature, especially because of the degree to which nontoxic species are thought to mimic these patterns to gain protection. As with visual displays, toxic defenses come in many varieties and can affect specific predators in different ways.[59] Brower (1985) differentiated between chemicals that are actively toxic to predators (Class I) vs. those that produced a noxious or deterrent effect without being actively harmful (Class II). Many Class I (toxic) chemicals are *emetic* (induce vomiting) below a lethal dose. These include aristolochic acids (found in *Battus* swallowtails), glucosides (the "mustard oils" found in pierids), and cardenolides/cyanogenic glycosides (found in Monarchs and heliconians). Predators can develop an aversion to emetic chemicals before being killed by them. This spares both the initial intended victim and future potential victims. If a Class I chemical were a "silent killer," it could eliminate a

concentrated population of local predators, but would cost the initial victims their lives, and would rarely create a generalized aversion response.[60]

Butterflies deploy defensive chemicals in a number of ways. Caterpillars of the Common Buckeye derive protection from iridoid glycosides ingested with their food, but they evidently digest these chemicals rather than sequestering them, since the adults are edible, a fact evidenced by their cryptic adult patterning, apparently evolved to avoid predation (Schappert, 2000). Monarchs and Atalas are toxic throughout their life cycles, because of cardiac glycosides they acquire as caterpillars. Zebra Heliconians manufacture poisonous aliphatic cyanogens from chemicals derived during adult pollen feeding. These chemicals are similar to those toxins caterpillars have learned to tolerate when feeding on passionvines. (It is not known which came first, the larva's tolerance or the adult's ability to synthesize.) Some species undoubtedly use a combination of these strategies to mount a chemical defense.

Uncertainty persists over the exact effects of defensive chemicals on various butterfly predators. As Brower noted, even substances that are scrupulously avoided by predatory birds may be merely irritating rather than toxic—or may be toxic only when combined with other chemicals, or after metabolic alteration, etc. There is no doubt that some secondary chemicals have direct, toxic properties. Cardiac glycosides (cardenolides), for example, are known to induce arrhythmia, muscle spasms, and eventually death. In nonlethal doses, they can damage mammalian embryos, causing deformities or miscarriage (Douglas, 1989, p. 181). But defensive chemicals can also work in a variety of less obvious ways. Tannins, for instance, appear to make caterpillars harder for predators to digest, and thus less favorable as a food selection item.[61]

A particularly interesting class of defensive chemicals is the pyrrolizidine alkaloids (PAs). These highly toxic substances damage the liver and other organs, but are slow-acting, causing death as long as several months after a lethal dose is ingested (Brower, 1985, p. 121). While some moths obtain PAs from their food, butterflies that use these substances (including Monarchs) must obtain them in the adult stage, evidently from wilting leaves or the nectar of certain plants, mainly heliotropes (*Boraginaceae*), but also asters and peas (Brown, 1984). The best-established use of PAs by adult butterflies is in the synthesis of pheromones for courtship. But they also may act (by themselves or in

---

[59] One difficulty for nonprofessionals is the sheer number of chemicals involved. More than 3,000 kinds of alkaloids alone have been identified, from an equally large number of plant species. Some of the plant families with the greatest diversity of alkaloids (nightshades and amaryllises) are not hostplants for any butterflies in our area, suggesting their level of toxicity may simply be too great to tolerate.

[60] Many butterfly predators are wide-ranging generalists, such as birds, that would be difficult to kill off entirely using undetectable toxins. In addition, it is known that

birds can learn to avoid toxic predators and pass this knowledge along to others of their species "culturally" (Brower et al., 1970). Small mammals may do the same. So an aversion-based strategy has much to recommend it. Still, some defensive chemicals do have unannounced effects, such as substances found in certain hostplants of the Spurge Family (Crabwood, crotons, etc.) that are carcinogenic (Marsh et al., 1985, p. 135).

[61] Nestlings of European Blue Tits were found to grow more slowly when fed larvae with a high tannin content (Brower, 1985, p. 121). The role of tannins in adult defenses, if any, is less clear.

combination with cardenolides) as defensive chemicals. One report indicates that PAs shielded certain tropical butterflies (Ithomiines) from attacks by orb spiders (Brown, 1984; Scoble, 1995, p. 81). The only Ithomiines eaten by the spiders were freshly emerged ones that had not had sufficient time to acquire a supply of these chemicals from nectaring.

To act as a deterrent, a defensive chemical must be immediately obvious to the target predator. Many plant-based toxins are bitter-tasting to humans (Brower, 1985, p. 117), and presumably to other vertebrate predators as well (even birds, whose sense of taste is weak, can sense bitterness). But this requires tasting, which means an attack. Pipevine swallowtails have particularly tough body cuticles, which can withstand an initial bite—long enough, ideally, for the would-be predator to taste their aristolochic acid (Scott, 1986). In a similar vein, studies of beak marks on the wings of various butterflies suggest that aposematic species tend to be bitten and then voluntarily released in a relatively large number of cases (Scoble, 1995, p. 80). Many defensive chemicals have distinctive odors as well, which may give advance warning of their toxicity, at least to predators such as mammals, whose sense of smell is acute.

Sometimes a pungent or unusual-tasting chemical is adequate for deterrence, as already noted, since many predators are relatively *neophobic*. But even Class I toxicity may not deter an exceptionally hungry attacker. Eating poisonous prey may appear suicidal, but this is not necessarily true, since the actual concentration of toxins can vary significantly among individuals. Monarchs are less toxic at the end of their winter diapause than in the fall. And some predators have learned to manage butterfly toxins even at full strength. At overwintering sites in Mexico, Black-backed Orioles prey on diapausing adult Monarchs, eating just those parts of the body that have relatively low levels of cardiac glycosides, such as the thorax muscles, while avoiding the toxin-laced wings, abdomen, and exoskeleton. Black-headed Grosbeaks needn't be so careful, since they have developed a biological tolerance for relatively large doses of the toxin (Waldbauer, 1998, pp. 54–55).

Studies of butterfly palatability, using birds as "tasters," have provided insights into the edibility of several species in our region. Monarchs are almost universally rejected, at least by experienced individuals, as are a number of other aposematic species, such as Pipevine Swallowtails, Baltimores, and Zebra Heliconians.[62] At the other end of the spectrum, a number of groups were eaten without hesitation, including satyrs, *Eurema* sulphurs, most *Papilio* swallowtails (Eastern Tiger Swallowtail, etc.), Malachite, White Peacock, etc. (Brower, 1985). Note that many of these species are boldly patterned, and might be assumed to be aposematic without specific palatability information. In between these extremes, a number of groups exhibited intermediate (or inconsistent) degrees of toxicity, especially among the whites. It must be recognized that these findings, while valuable, do not yet provide a comprehensive guide as to toxicity and deterrence applicable to all developmental stages and predator types.

A final twist in the drama of butterfly display is *mimicry*. This topic has been discussed at length in other sources, enough so that we will limit ourselves to highlights in this introduction (*see* species accounts for some additional details). Mimicry exists in many kinds of organisms, but it was originally studied by Englishman Henry Walter Bates in connection with South American butterflies in the mid-1800s. *Batesian mimicry* occurs when a nontoxic species adopts the appearance of a toxic one, thereby gaining vicarious protection from predators. Oft-cited examples in our area include the female Diana, female Eastern Tiger Swallowtail, Red-spotted Purple, and (possibly) Black and Spicebush Swallowtails—all of which are thought to be edible mimics of the Pipevine Swallowtail, a toxic "model."

A second form of masquerade is *Müllerian mimicry*, named after Fritz Müller, a German naturalist of the 1870s who also worked in the Amazon. In Müllerian mimicry, two species—both toxic—copy each other's appearance, thus providing additional protection for both. In tropical forests, elaborate, multispecies Müllerian "mimicry complexes" have been studied, sometimes involving moths as well as butterflies (e.g., Gilbert, 2003). In our region, it currently appears that Monarchs and Viceroys may be Müllerian mimics, since both seem to have some degree of toxicity. In Florida, Viceroys have a darker background color than in the North, more similar to Queens and Soldiers. An ironic fact, often noted, is that the heliconian butterflies studied by Bates in developing his original theory of Batesian mimicry were actually Müllerian mimics, since all members of the complex were toxic.

BUTTERFLY BIOLOGY: REPRODUCTION

Reproductive behavior in adult butterflies occurs in three distinct stages: location, approach, and completion. Like many other goal-oriented behaviors in lower animals, the beginning phases of the cycle ("appetitive" behavior) vary significantly across species, while the completion phase ("consummatory" behavior) tends to be more routine and

---

[62] Actually, Monarchs reared on cabbage plants in the laboratory were eaten readily by Blue Jays not accustomed to avoiding them based on learned response (Brower, 1985, p. 123). Likewise, Baltimores were palatable if raised solely on plantain, usually an alternate host in late instars. The results of palatability studies can be affected if the part of the butterfly used for testing (e.g., the abdomen) is low in toxins, whereas some other body part not used in testing (e.g., the wing scales) contain secondary chemicals.

stereotyped. As discussed in section II, male butterflies seeking a mate generally either perch or patrol. Perchers in our region include all of the hairstreaks and coppers, most skippers and nymphalids, and some swallowtails. Patrollers include all of our pierids, most blues and swallowtails, and some skippers and nymphalids (such as crescents) (Opler & Krizek, 1984, pp. 24–25). Individuals often converge on hilltop locations to locate mates (or sometimes in gully bottoms, although this has been recorded more in western species). Certain ringlets living in open fields converge near large, isolated trees (Rutowski, 2003).

There is lingering debate as to whether perching males are territorial in the true sense of holding a defined geographical space against rival males. Some argue that they are merely programmed to intercept any moving object that enters their perching zone, with the aim of ultimately encountering a mate. On the other hand, many perching males do regularly defend specific patches of territory, often from a single perch, throughout their adult lives, and some have been shown to attack males of their own species more aggressively than those of other, closely related species (Brown & Alcock, 1990–91; Bitzer & Shaw, 1979). Aerial dogfights, moreover, include bumping and other forms of direct, physical contact (Van Dyck, 2003; Wickman & Wicklund, 1983). Much perching behavior in butterflies appears to involve lek formation (Lederhouse, 1982), moreover. While not specifically territorial, this is hardly a random, hit-or-miss courtship process. Whatever terminology is ultimately agreed upon to describe these various activities, it is clear that butterfly mate-finding behavior includes a wide variety of highly structured and ritualized activities, many not as yet fully explored (see, e.g., Cofaqui Giant-Skipper account).

As noted earlier, most moths locate mates at a distance by means of pheromones.[63] But the initial mating contact among butterflies is normally visual, initiated by the male. This can be a somewhat haphazard process, since the male's vision—while apparently more acute than the female's—is still weak at a distance, prompting many false starts.

In some groups, males may shortcut the visual location process entirely and seek out a female's chrysalis, waiting there for her to emerge. Indeed, one of the few easy ways to locate pupae in the field is to watch for clusters of males, swarming insistently around an unseen object. Male Zebra Heliconians locate pupating females by scent (Scott, 1986, p. 46), and will even mate with females not yet emerged from the pupa. Hybrids between some closely related species (such as Orange and Clouded Sulphurs)

I-28. Female Falcate Orangetip, perched on Poison Ivy, rejects courtship.

may occur when a libidinous male approaches a newly emerged female who cannot yet reject his advances.

Having located a female, the male flies insistently above her. If she is also flying, he vigorously flaps his wings above her to force a landing. At this point, it is the female's prerogative to either accept or reject courtship. There are several techniques for rejecting an advance, including rapid wing-fluttering, raising the abdomen straight upward (common in pierids, a gesture often misconstrued by observers as solicitation), engaging in a ritualized, zigzag "rejection dance," or flying rapidly upward in a dogfight-type spiral (this often in response to advances from a perching male).

If the female allows the male to approach—and we have only a general understanding at present of what drives her decision in this regard—he then begins to direct pheromones toward her to induce receptiveness. Sometimes he approaches closely, so that her antennal tips come in contact with the "scent patches" on his wings (which contain aphrodisiac pheromones). It is not entirely clear whether these pheromones elicit a reproductive response in the female or merely calm her so that she does not leave.

Behaviors at this stage are still quite variable across phylogenetic groups. Danaid males, e.g., eject a "hair pencil" from the rear of the abdomen (composed of hairlike, modified scales) which releases pheromones embedded in a gluelike substance that adheres to the female's antennae (Scoble, 1995, p. 166). Scent cells in butterflies are located in varying folds or crevices on the wings or body, or in the overlap zone between the wings—all places from which the chemical can be released strategically, yet where it is protected from premature evaporation. In groups with many similar-looking species, females may also produce

---

[63] In butterflies, males produce most pheromones. Early naturalists were conscious of the pleasant smell of male butterflies, which varies in strength and character according to species (see, e.g., Ford, 1977, pp. 101–2).

I-29 to I-32. Dorsal female and male Clouded and Orange Sulphurs, under natural and ultraviolet light

pheromones, evidently to allow males to identify them at close range (Scott, 1986).

Part of the female's decision to accept a particular mate depends on her recognizing him as a male of her own species. Mimics may be distinguished from "the real thing" at close range by scent (Douglas, 1989, p. 152). And in a number of species, especially among the pierids, ultraviolet or polarized reflections off the male's dorsal surface can play a critical role. The most commonly cited example is Orange vs. Clouded Sulphurs. In Clouded Sulphurs, the dorsal wings of both the male and female *absorb* ultraviolet light, making them appear dark (Plate 1-31 and 1-32). Clouded Sulphur females recognize males by their chemical signature rather than by appearance (Scoble, 1995, p. 86). In the Orange Sulphur, by contrast, the male's dorsal surface *reflects* UV, and the female can recognize a potential suitor on the basis of his visually flashy topside (Plate 1-29 and 1-30). If a male Orange Sulphur's UV reflectance is eliminated, by painting the wings, females will not accept him (Silberglied & Taylor, 1978).

When seeking mates, male Orange Sulphurs look for dark counterparties (i.e., ones that absorb UV light dorsally). This keeps them from expending energy in approaching other male Orange Sulphurs. It also prevents courtship with female *whites,* which (unlike female sulphurs) *reflect* UV light from the dorsal surface (Scott, 1986,

p. 54; Scoble, 1995, p. 85).[64] Interestingly, it has been determined that just one gene controls the UV reflectance characteristics of the pierid forewing (Nijhout, 1991, p. 15). This illustrates how a single, small genetic shift can have a defining impact on critical behaviors at the species level, and even throughout an entire family, such as the pierids.

Once the formalities of courtship are complete, copulation usually ensues quickly. Sex among butterflies, it should be recognized, is more like a docking of spacecraft than any sort of romantic interlude. The shape of the interlinked genitalia tend to be highly species-specific, and thus have become a principal tool for taxonomic identification. The male's genitalia are among the most durably constructed of his anatomical features—aside from the highly sclerotized head capsule. This reflects the importance of his correctly and reliably executing the reproductive function.

Copulation often begins with the two partners standing alongside each other. The male bends his abdomen around to meet hers. He clasps the female's abdomen with his *valvas* (male abdominal structures that facilitate the coital linkage), and inserts a projecting structure (the *uncus*) into a pocket below her ovipositor, thereby forming a primary junction with the female. The two partners are now standing tail-to-tail, facing away from each other. The *aedeagus* (penis) enters the female's mating tube below the uncus. The male deposits sperm and various nutrients into a

---

[64] Male whites and male Clouded Sulphurs both absorb UV light, and so may be approached by male Orange Sulphurs, but this type of error is less problematical, since it cannot result in undesired hybridization. It is true that female Clouded Sul-

phurs absorb UV light and thus could be confused with a female Orange Sulphur, but the male Orange Sulphur lacks the appropriate chemical signature to gain acceptance from a Clouded Sulphur female.

I-33. Mating Giant Swallowtails

durable white sac, the *spermatophore*. The female stores the spermatophore in a special abdominal chamber, the *bursa copulatrix*. She digests the spermatophore over time and utilizes its nutrients. The sperm, meanwhile, can be kept alive inside the female, sometimes for months, until needed to fertilize eggs (Scott, 1986, pp. 34–45).

Mated butterflies remain connected to each other for a considerable period, from around a half hour up to two or three hours. If disturbed during this time, the couple flies off together. In these "nuptial flights," one member of the pair does the carrying, while the other hangs limply behind. In most species, only one sex does the carrying—in swallowtails, satyrs, checkerspots, hairstreaks, and skippers it is the female; in danaids, heliconians, and blues it is the male (Scott, 1986, p. 50). Either sex may carry in coppers or other nymphalids.

As copulation ends, some species take steps to prevent remating. Male heliconians transfer an *antiaphrodisiac* substance to the female, which will deter unions with other males (Douglas, 1989, p. 153). Female fritillaries also pick up repellant pheromones during mating, but these work only at closer range (Scott, 1986, p. 59). The rejection posture of female pierids (in which the abdomen is raised) may facilitate the dissemination of antiaphrodisiac substances, chemically reinforcing her gesture of refusal. Finally, in some families (especially parnassians, not found in our region), males block the female's mating tube after copulation with a thick black plug (the *sphragis*).

Absent these measures, either sex may remate (although *monandrous* species mate only once, regardless of opportunity). The female usually waits until she has digested her last spermatophore. Remating allows males to extend their genetic legacy, and females to gain additional nutrients for increased productivity and a longer personal life (Wiklund, 2003). The tendency to mate opportunistically may be greater in nomadic species with thinly distributed populations, which need to capitalize on encounters to the great-

est extent possible (Burns, 1968; *but* Cordero, 2000). In crowded situations, females may have difficulty feeding as they fight off unwanted advances.

BUTTERFLY BIOLOGY: LEARNING AND INTELLIGENCE

If someone used the expression "bright as a butterfly," we would assume they were referring to coloration rather than intelligence. Even among entomologists, honeybees have normally been touted as the "brains" of pollinator corps, able to learn colors on one trial, navigate back to their hive, and provide their hive mates with directions to the site via symbolic dance movements. Butterflies were long relegated (albeit without much evidence) to the lightweight end of the IQ scale.

This traditional view of butterfly intelligence—that, to put it simply, there really is no such thing—has two shortcomings. First, as we have seen repeatedly in this introduction, butterflies have wired-in, or instinctual behaviors that are often quite complex and outwardly intelligent. Second, even in terms of the traditional definition of intelligent behavior, including the ability to learn, etc., adult butterflies have at least a modicum of ability that is worth understanding.

It should be recognized at the outset, however, that even the limited rehabilitation we propose here is reserved for adults. Caterpillars, we must acknowledge, appear to be outright dullards in the learning department. Efforts to train them to perform even the simplest of tasks (such as turning a particular direction to obtain food) have foundered, as the subjects were either extremely slow (*Vanessa* caterpillars) or completely unable to learn (danaids) (Scott, 1986, p. 60). It is true that caterpillars at times exhibit some quite sophisticated instinctual behaviors, and perhaps we will someday find that they have certain highly focused learning abilities. But this possibility remains mainly hypothetical at present.

Recent controlled studies have shown that adult butterflies do possess distinct learning abilities. When put to the test, Cabbage Whites were able to associate a particular color with nectar on one try in 82% of cases—about the same as honeybees (Weiss, 1997; Milius, 1998). Changing this association once it is learned is also possible, although this may become difficult after several days using the learned color. Butterflies can even keep multiple associations in mind at once for different behaviors, e.g., one color association for nectaring and a different one for ovipositing (Weiss, 1997).

Female sulphurs can likewise improve their accuracy in ovipositing on legumes (vs. similar-looking nonlegumes) during an egg-laying session, but they backslide after taking a "nectar break" (Stanton, 1984). In various other experiments or field observations, butterflies have learned to select particular plants on which to oviposit, to choose certain trees for feeding, to recall physical features of the hilltop where they had been displaying, to avoid old lantana blossoms whose colors signal reduced nectar output, to avoid experimental "catch-and-release" sites for several days afterward, and to find nectar more quickly in irregularly shaped flowers (Scott, 1986, p. 61; Milius, 1998).

Our most intelligent butterflies appear to be the heliconians. This long-lived group has a large "mushroom body" in the brain thought to be associated with learning. Zebra Heliconians can return to a common roost each evening, visit pollen flowers and hostplants during the day in a regular sequence (*trap lining*), at the same time showing younger adults the route, distinguish hostplant leaf shapes, and even adjust to florescent lights in a room (which other butterflies would beat against incessantly) (Scott, 1986, p. 62).

Life experience can alter the anatomical characteristics of a butterfly's brain. It was recently reported that wild Variegated Fritillaries, exposed to natural stimuli, had larger olfactory lobes than individuals raised under laboratory conditions (Kroutov et al. 2002).

These focused learning abilities allow adult butterflies to make a one-time, rapid adjustment to the surroundings in which they emerge—and in which they will usually spend only a short period before expiring. With so little time to waste, even a small investment in learning capability is worthwhile. And for those few species with a longer active adult life, the payoff from learning increases. Heliconians have even attained a meager level of society.

BUTTERFLY BIOLOGY: ARE BUTTERFLIES FIT?

We ended the preface of this book by raising the paradox of butterfly fitness. How can butterflies, which lack nearly all of the customary indicia of fitness—strength, speed, and ferocity—have survived for more than 50 million years, all the while meandering distractedly and from flower to flower? This puzzle has long fascinated us. Part of the key to butterfly success, it should now be apparent, is defensive chemistry. By tasting badly and in some cases carrying deadly toxins in their bodies, many butterflies have built a zone of deterrence around themselves, which occasionally persists throughout the life cycle.

But what about the butterflies that lack this toxic armor? The bounding, erratic flight style of many palatable butterflies makes them difficult for vertebrate predators to pursue efficiently. Some have adopted a generalist strategy, relocating frequently before predators can key in on their presence. Others have enlisted the help of ants to bolster their natural defenses. But most of the success of nontoxic species comes from a variety of clever subterfuges they

have evolved to deflect harm, including camouflage; disruptive patterns; eyespots and "false heads" that cause predators to strike at the trailing edge of their hindwing; caterpillars that resemble snakes or bird droppings; feeding at night as larvae; building leaf rolls in which to conceal themselves; etc. Like gentle magicians, they project an image to the world that allows them to elude danger in a myriad of small ways.

But evolutionary longevity involves more than simply avoiding predators. True fitness requires the ability to exploit a growing range of habitats in an increasingly skillful and productive manner, with adequate flexibility to accommodate inevitable periods of change and upheaval. Lepidopteran caterpillars are among the most efficient of herbivores, able to exploit a vast array of different hostplants, some barely digestible without significant specialization, and to do so in many different surroundings. With cold hardiness and sophisticated diapause strategies, butterflies can live in nearly any climate, including the high arctic and (in our area) New England mountaintops. They have learned to tolerate or synthesize a broad array of complex, often toxic organic chemicals, which they employ for defense, courtship, communication, and the creation of display pigments. And in their intricately structured wing scales they have evolved one of the animal kingdom's most refined and flexible pattern-generating mechanisms. They have acquired migratory capabilities and seasonal movement patterns that allow them to implement complex lifestyles, spanning thousands of miles in some cases. They have learned to glide with little effort over great distances and to maneuver skillfully at low speeds.

So butterflies have developed a range of adaptations that make them fit. Their unique lifestyle has grown in breadth and sophistication over the millennia, all without clashing with other forms or seeking domination. They chart their own rather peaceable course, deflecting natural pursuers with feint rather than force, and yet well enough to persist. More than an example of fitness, they are in certain respects a model.

IV. HOW TO LOOK

ACTIVITIES INVOLVING BUTTERFLIES

I go and come with a strange liberty in Nature. . . . The bullfrogs trump to usher in the night, and the note of the whippoor-will is borne on the rippling wind from over the water. Sympathy with the fluttering alder and poplar leaves almost takes away my breath; yet, like the lake, my serenity is rippled but not ruffled.

—Henry David Thoreau, *Walden* (1854)

An interest in butterflies can take many forms: artistic, recreational, scientific, even horticultural. We personally enjoy discovering new links between butterflies and their natural communities. Individual contacts on a small scale drive entire ecological systems, and by studying butterflies we gain a close-up perspective on the process.

But conscious interests such as these may really be something of a diversion, keeping our minds occupied while our spirits enjoy the many pleasures of a day in the field. There is a universal element to this response: at times, we can sense an unstated, even unconscious appreciation of nature in people whose declared motives for being in the field differ greatly from our own. In a society accustomed to frenetic, quick-cut images, bombarded with activity, the value of personal time in nature, with its deliberately unfolding rhythms, remains undiminished, though widely neglected.

Field Observation

People who do not observe nature on a regular basis may have trouble seeing the point of it. Observers poke around curiously in a field or woodland, with occasional flashes of excitement, but otherwise do not appear to be *doing* very much. Appearances can be deceiving, though. Field study may not be calculated to impress bystanders, but it is still an active pursuit, often a demanding one.

A number of particulars require attention when planning a field trip. The first is a clear itinerary. At times, it is valuable to wander into the field with no fixed agenda, just to see what may turn up. Many interesting discoveries are made at such unhurried moments. But most successful outings have an organized plan of attack. Butterflies are best observed during the six or seven hours between mid-morning and late afternoon. This leaves a surprisingly short span of time for travel and observation, especially if one stops for lunch (we commonly do not).

A successful agenda is based on well-selected targets. Target butterflies are species that fly at limited times or places, or which are new to the observer; but an observer may also target a large seasonal concentration of commoner species. Around New York City, where we live, early July is the time when the greatest number of species can be seen in a single day (more than 50). Clusters of target species fly at other times as well, such as late April and mid-June. It is important to know where to go at each time in the year. Local lists and personal contacts are important in getting started.

Flight times vary from season to season, mostly in response to differing weather conditions. In Florida, the timing of seasonal wet and dry periods can be pivotal. The most useful resource available to an experienced observer in sorting out these variations may be his or her own personal field notes. The first author has used a pocket dictator to record notes in the field on each outing since 1981. Over

time, this accumulation of seasonal notes, contact names, directions to particular sites, and so forth, has become a highly valuable compendium. (The second author does well with a low-tech, but effective alternative, the index card.)

The best circumstances for butterfly observation can be less than comfortable for human observers, requiring visits to habitats that are hot, exposed, and infested with gnats, chiggers, ticks, mosquitoes, deer flies, prickly plants, tangles, etc. Several basic precautions should be taken on a standard basis, such as a wide-brim hat and sun block, with repellent spray close at hand and even mosquito netting for extreme circumstances. Contrary to what we see in print, we have never noticed that wearing insect repellent affects butterflies we are trying to observe, even when making a close approach for photography. Swatting at mosquitoes and deer flies, on the other hand, *can* easily scare off a skittish subject that one is trying to approach. Typical repellents contain strong chemicals that many individuals prefer to avoid, but applied judiciously (i.e., not directly on the skin, etc.) they can play a useful role.

Many observers wear the sturdiest, and thus most expensive footwear available when watching butterflies. This is a wise choice in many habitats, such as rocky woodlands or mountain slopes. But in swales and other wetlands, an alternate strategy is to buy an inexpensive pair of thin canvas sneakers, and then wade in. Not only does this reduce anxiety over wrecking good boots with too much soaking, but sneakers dry out quickly in warm weather and can be pleasantly cool on the feet. (One drawback to this strategy, as pointed out to us by veteran naturalist and photographer Harry Darrow, is that canvas sneakers "won't be much help against a Water Moccasin," in habitats where this is a risk.)

A good pair of *close-focusing* binoculars is very important for successful field observation. By "close focusing" we mean 5 feet or less, although 6 to 8 feet may be tolerable, depending on individual preference. Whether high-quality optics are essential depends on intended use. More expensive binoculars tend to be relatively well-constructed, likely to stand up to hard use and in unfriendly environments. Also, if a pair of binoculars is to be used for several purposes—especially birding, where looking up, into the sun's glare is required—quality multicoating is a near-necessity, justifying the higher price tag.

In addition to a field journal, as discussed above, many observers engage in the time-honored practice of *listing,* a devotional pursuit that seems to have originated with birders. Most butterfly watchers keep a "life list," detailing the species they have and have not seen. Many also list the species they see each year, the earliest and latest dates for a particular species or seasonal brood, the highest number of a particular species seen in a day, the number of species seen in the back yard, the number of species that have landed on them, etc. At some point the marginal utility of additional lists diminishes. But anything that creates a detailed record of field experiences can prove useful, sometimes in unexpected ways.

Having company in the field is enjoyable. But for beginners it is almost a necessity, since verifying one's early identifications with experienced observers will accelerate the learning process and prevent mistakes from becoming ingrained. Organized trips with a formal leader are a good way to learn, if available. When visiting fragile habitats, either alone or especially in a group, extreme care should be taken not to trample valuable hostplants or nectar sources.

Many butterfliers participate in the annual "Fourth of July" counts co-sponsored by the North American Butterfly Association (NABA) and Xerces Society. These counts actually may be held whenever species abundance reaches its peak in a particular area, from June through August. Each count territory is defined by a circular, 15-mile diameter around a fixed central point. Participants list the total number of butterflies seen, both species *and individuals*. Regular censuses such as this provide valuable longitudinal data. In recent years, an even more ambitious form of survey, called "bio blitzing," has appeared. Pioneered by naturalist Peter Alden in Massachusetts, a "bio blitz" involves a multidisciplinary effort to catalog all of the species *in all biological categories* that can be found in a single day in one defined region. In addition to butterflies, counters survey mushrooms, amphibians, even slime molds.

Inclement weather can bring a quick end to a planned butterfly trip. Rain and unseasonable cold are usually responsible. But an unfavorable weather report should not automatically deter a field trip. Even on rainy days, there may be lull periods when butterflies emerge to feed or bask in the sun. And their movements are less energetic in cool weather, allowing better looks. Indeed, chilly weather is the only time one is likely to see a Spring Azure expose its dorsal surface to view.

Around Home

Butterflies can be observed from the comfort of one's garden lawn chair, or in a "butterfly house" where live individuals from around the world can be observed at leisure. Many people elect these options—although they offer a narrower range of experiences than active field study.

Butterfly gardening is a quickly growing and specialized pursuit. As city apartment dwellers, neither of the authors is well-situated to develop expertise on this topic, and we readily defer to those more knowledgeable than ourselves. A list of published materials is cited. As popular interest in butterflies increases, it seems inevitable that dedicated home gardeners will offer valuable assistance to uncommon species stressed by development or weather events. To play such a role, however, it will in many cases be necessary for

I-34. Butterfly observers in the field

gardeners to cultivate hostplants as well as nectar flowers. Already, in Florida, where massive habitat destruction is occurring on a daily basis (as discussed below), cultivated cycads have allowed Atalas to rebound from near extirpation, and availability of exotic aristolochias has allowed Polydamas Swallowtails to extend their range considerably. This is only possible, of course, if the plants in question are left unsprayed so that they can be safely utilized by butterflies.

Photography

> If you catch a butterfly outside and bring it in you will be likely to find that it is by no means a docile subject. The sunlight shining through the nearest window will be a call which you cannot countermand and your butterfly will constantly respond to it in a most vexing manner.
>
> —C. M. Weed, *Butterflies* (1917)

It is more difficult to obtain high-quality butterfly photos than at first appears, despite a new generation of compact digital cameras that greatly simplify the process. A number of "special techniques" are sometimes used to boost the likelihood of obtaining a clear shot. Yet no matter how sharp the resulting image, we take little pleasure from a photo that does not meet our principal governing criterion: that the subject of the photos be *free-flying and unrestrained* at the time the photo was taken. Many writers on butterfly photography argue that satisfactory photos are prohibitively difficult or time-consuming to obtain unless one captures an individual and chills it for a while to slow down its movement. After this intervention, the immobilized subject is relocated on some attractive background—but rarely one that it would have selected of its own accord. To us, the value of a field photograph is the wealth of accurate information it conveys about the butterfly's life in nature, especially evidence of naturally exhibited behaviors, uncorrected by the photographer's aesthetically skilled (but still heavy) hand.

Except for a limited number of specimen shots, which are obviously not free-flying individuals—and likewise the studio photos of Tawny Emperor caterpillar and pupae, which were taken indoors—all of the individuals photographed by the first author for this book were wild and unrestrained. That is not to say that all had full mobility. The Silver-bordered Fritillary ventral shot is of an individual stunned by an encounter with an automobile, the dorsal female Black Swallowtail was too cold to fly as she basked in the early morning sun, etc.[65] The guiding principle for valuable photos is to exercise one's best efforts to reflect nature in a natural form.

It is important to note that we have no objection to arranged photos, as long as the circumstances under which they were taken are *clearly disclosed*. A friend takes excellent available light photos of sleeping butterflies, e.g., often moving them for better composition and illumination. Sometimes the individuals are coated with dew and look remarkably beautiful. As long as these shots are presented with adequate disclosure (so that a field worker will not use them as a guide to the surroundings in which a roosting butterfly is likely to be found) we think they are wonderful.

Similarly, we have no issue with photos taken in butterfly houses, though disclosure is exceptionally important here, because the backgrounds in these photos usually bear no resemblance whatsoever to a species' natural habitat. The leaf on which a Malaysian butterfly is perched may well belong to a plant endemic to the neotropics. (It is not all that easy, incidentally, to obtain quality photos of fresh specimens in a butterfly house, given restrictions on the photographer's movements, disturbance from other observers, and the rapid rate at which captive butterflies become damaged and worn.)

A few brief comments on photo technique:

Equipment: Even in this rapidly advancing age of digital photography, a serious butterfly photographer should do best with a single-lens reflex camera equipped for macrophotography (of course, the imaging now may be digital). Specialized macro lenses generally produce the best results, but are comparatively costly. A lens with life-size (1:1) capabilities is optimal for photographing small butterflies and grass skippers. Diopter lenses that fit in front of an ordinary lens to allow close-ups have improved in quality in recent years and may be a valid alternative for some users. Telephoto lenses with extension tubes (*telextenders*) also allow the equivalent of macrophotography without the need for an overly close approach to the subject. This rig can sometimes be heavy and unwieldy, however, and may require a tripod, all of which limits flexibility.

Lighting: Almost all of the butterfly photos in this book were taken with strobe lighting. Some photographers strongly prefer natural light photos. We respectfully disagree. In our view, strobe lighting provides even illumination and clear colors that are far more representative of what we see in the field. Natural light photos frequently suffer from uneven lighting, blurriness, and variations in the color temperature based on differing ambient light conditions, all of which make subjects look unnatural. (Our eyes tend to correct for these lighting differences unconsciously, but film does not.) That being said, available light photos can be very handsome, and many photographers have become quite adept at turning out quality images using this technique.

Film: Most of the photos in this book were taken with Kodachrome 25, a film that has unfortunately been discontinued. Indeed, this may be among the last nature books to be published with newly taken Kodachrome 25 images. Some shots were taken on Kodachrome 64, Velvia, Provia 100, and a number of other films. Most professional nature photographers use slide film, and most publishers still prefer (or require) slides, although scanned images on photo-CD are gaining increasing acceptance. Many photographers have recently begun taking digital cameras into the field, and rapid advances are taking place in this medium. Digital photos facilitate quick ID verification and rapid posting to the Internet. Our best advice overall is to select a single film (or medium) to one's liking and use it consistently for a period of time, so as to become better aware of its strengths and limitations. All films and imaging techniques have both.

Stalking: Widely differing techniques are required when approaching different butterflies. Some open-country skippers and satyrs are especially "jumpy" and hard to stalk. (Both groups, not coincidentally, feed on grasses and sedges that do not confer toxicity.) Remember that butterflies can see you. Not as well as you can see them, but their small eyes and sensory systems act like multiphasic motion detectors. Painstakingly slow movements often pay off, in dropping a knee to the ground or in making an initial approach. Be aware, too, that many butterflies are very sensitive to any object that breaks the horizon line. So you may get better results by approaching in a crouch.

Collecting

Kill not the Moth nor Butterfly,
For the Last Judgment draweth nigh.
—William Blake, "Auguries of
Innocence" (1803)

Some field observers use nonconsumptive techniques to record their experiences—notepads, sketch books, cam-

[65] This subject was moved a few feet for the photo, to obtain full visibility. The plant selected was of the same type as that on which she had originally been roosting.

Jeffrey Glassberg

I-35. First author photographing butterflies

eras, even GPS locators. Others do this by taking specimens. Certain objectives in butterfly study require collecting to achieve; others preclude collecting.[66] In most cases, though, the choice of collecting vs. noncollecting is a matter of preference rather than necessity.

The terms of this discussion have shifted dramatically in recent years. A decade or two ago, an essay of this type would have automatically been devoted to the techniques of collecting—how to use a net, how to "relax" butterflies in a jar in the field, how to mount specimens, how to label and arrange a collection in customized display cabinets, where to purchase supplies, etc. Traditionally, collectors defined virtually every aspect of butterfly study. Indeed, until recently they were virtually its sole practitioners.

An alternate paradigm has been introduced in recent years by a group of "study and observe" naturalists, often birders or other general natural history observers, who have turned their attention to butterflies (Pyle, 1984; Glassberg, 1993a). Many of the newcomers have expressed open reservations about recreational collecting, and some vigorously oppose it. This issue is not a new one (as suggested by the lines from Blake's early-19th-century poem, quoted above). But the arrival of noncollectors in the field has scaled up the debate.

A few key questions recur frequently:

Does collecting diminish populations of rare species?
It would take armies of collectors to make a dent in the population of common and widespread butterflies, such as Cabbage Whites or Orange Sulphurs—if a dent could be made at all. The more relevant question here is whether

small and isolated colonies of rare species, which are a primary target of both observers and collectors, can be damaged by overcollecting. No experiments have been conducted on this issue that take metapopulation dynamics fully into account, and the circumstances of historic cases-in-point have been hotly debated.[67] It does seem clear that active collecting in small, stressed populations does nothing to *enhance* their survival.

Is collecting necessary for identification?
Ardent collectors argue that critical identification requires dissection of genitalia, and that no identification is positive without a specimen. Even if we accept the premise that a positive identification is needed for all sightings (in the extreme, no living butterfly can properly be assigned to a species, since it has not as yet been dissected), we think this argument is overstated and has limited application in our region, in any case, as there are few species for which sight identification, authenticated where necessary by a photograph, is not sufficient. Certain azures and duskywings come to mind as occasional exceptions, although with further careful study—of both live individuals and specimens—many additional field marks will become known, as occurred in birding a generation ago. Also, the availability of high-quality, close-focusing binoculars will continue to enable greatly enhanced fieldwork.

Does restricting collectors thwart science?
Collections of biological specimens are absolutely necessary for scientific investigation, and many of the specimens in existing museum collections were originally taken by diligent amateurs, rather than by professional entomologists (of which there are relatively few in the butterfly world). The only problem with carte blanche approval based on this argument is that (1) for most species in our area, an abundance of specimens *already exists,* including geographic races and ecotypes; (2) a large number of specimens are collected each year at traditional sites where local populations of rare species are known to exist, and which are vastly overrepresented in collections from a scientific standpoint; and, finally, (3) specimens collected in bulk for commercial sale or barter may not fall within the scientific rationale for collection.

Is collecting the only valid scientific approach to butterfly study?
Another issue is whether collecting and scientific investigation are inseparable. We know of no field

---

[66] Specimens are necessary when detailed anatomical or genetic investigation is required (although some field marks disappear after death, such as the blue antennal tip of Great Southern Whites, or distinctive eye pigments in some species). At the opposite extreme, structured observations of uninterrupted field behavior are sometimes inconsistent with collecting. We know of several cases where lengthy observations have provided behavioral information not likely to have been acquired on a typical collecting trip (*see,* e.g., Cofaqui Giant-Skipper account).

[67] In the late 1980s, colonies of Mitchell's Satyr in small wetlands in western New

Jersey were reportedly visited on an almost daily basis by collectors, who captured every individual they could find. After a few seasons, the colonies vanished and have not returned (Gochfeld & Burger, 1997). Colonies of giant-skippers also have reportedly been lost after collectors dug up yucca plants where the larvae were burrowing so that they could hatch out emerging adults for mint specimens. *See also* Ford, 1977, p. 142 (extirpation of U.K. race of Large Copper). We are content in the belief that collecting has played a demonstrable role in some of these losses and frankly wonder at the impartiality of some who vehemently deny even the possibility of such a result.

of biology in which specimen-taking has a monopoly on the generation of scientific knowledge. The study of living organisms in the field, while challenging, is perfectly good science, often essential, and likewise a perfectly good source of hypotheses for further testing. It should not be overlooked or demeaned.

We find limited value in the "county dot maps" that are a mainstay of many traditionally oriented publications. In these compendiums, a dot is placed in a particular territory if a single "voucher specimen" is known from the region—no matter how old. This tells us nothing of current abundance, population trends over time, etc., and is often extremely misleading to users, both in terms of dots in areas where species no longer occur as well as gaps in areas where collectors do not happen to have visited "in season." If this approach is taken, it should be accompanied by observation-based phenograms. Successful examples of this "hybrid" approach include *Butterflies and Skippers of Ohio* (Iftner et al., 1992) and *Notes on the Butterflies of North Carolina* (LeGrand & Howard, 2003). Finally, we believe that reliable sight identifications and photographs should be accepted if "dot maps" are to be constructed.

In a different vein, many of the "unknowns" cited in the accompanying species accounts with respect to East Coast butterflies reflect gaps in our knowledge of the *behaviors* exhibited by living adult butterflies. These gaps can be filled only by observational studies, not by the accumulation of additional reference specimens.

In sites with limited populations, whose agenda should predominate? This is perhaps the greatest ultimate point of contention between the collector and observer communities. Even if we accept the premise that butterflies are a "renewable resource," the issue remains—in cases of scarce populations—of who has primary right of consumption. Individuals spend valuable time and money getting to sites where rare species can be found. Collecting may not permanently eliminate a local population, but it can certainly eliminate all individuals of a rare species flying at a particular time and place. It does an observer little good, after a long trip, to know that he or she might have better luck another year. Emotions will run high over such incidents until mutually accepted standards of conduct are achieved. Along these lines, The Lepidopterists' Society, whose membership includes many collectors, has adopted a constructive policy on collecting that addresses many sensitive issues.[68]

Is recreational collecting ethical? Another fundamental question is whether it is morally or ethically wrong to kill things without a specific need to do so. There is little general argument over killing for survival, to stop a pest intent on sucking one's vital fluids (or devouring crops), for scientific inquiry, etc. There are even special rationales based on situational factors (e.g., is it better for tropical forest residents to subsist by capturing and selling butterfly specimens, if the alternative is for the forest to be logged?). But absent such a specific rationale, the original question returns. This is an issue that each of us must face individually. Ethics inevitably play a role as we interact with nature. There is no universal resolution to such questions, but the issue is not without significance, and it occurs to us that small choices can be defining ones.

There will, to be sure, continue to be collectors and observers. Though neither of the authors collects personally, we have included photos of specimen from some of our friends' exceptionally well-curated collections in this book, to illustrate wing surfaces that are seldom displayed at rest, and thus difficult to photograph in the field. We believe that the need for collecting is likely to diminish over time, given the advent of compact digital cameras that allow routine observers with little photographic training to produce satisfactory, on-site documentation of field sightings. Still, much of our current knowledge of butterflies was generated by dedicated collectors, and the values of each tradition must be properly acknowledged, whatever the correct balance may be going forward.

CONSERVATION

The hard irony of butterfly conservation is twofold: (1) much of the loss that occurs in butterfly habitat is unnecessary to achieve any important human goals, and yet (2) once extirpated, butterfly populations are difficult to reestablish. The lesson is that we should take steps wherever possible to maintain the key populations and habitats that we have, since the costs of restoration greatly exceed those of conservation.

Most damage to threatened butterfly populations comes from loss of specialized habitat or excessive use of chemicals in the environment (chemicals almost never intended to control butterflies). Many of the species described in this book rely on rare or local habitats, and of these many are in serious decline along the East Coast. Maintaining such habitats is often the work of dedicated professional conservationists, since land acquisition, concerted political ac-

---

[68] The policy specifies that collecting should be limited to sampling that will not deplete populations, and only for valid purposes (mass collection for commercial sale is specifically excluded as a valid purpose). The policy specifies that where "the extent and/or the fragility of the population is unknown, caution and restraint should be exercised." Finally, it urges that "property rights and sensibilities of others must be respected."

One valid purpose of collecting, listed in the policy, is "introducing children and

adults to awareness and study of their natural environment." This we are prepared to debate. The first author has been an active naturalist virtually all his life, but kept away from butterflies most of that time after being obliged to collect and pin Eastern Tiger Swallowtails at a summer camp when young, an experience he found distasteful and upsetting. While collecting may encourage some young boys and girls to pursue nature, it is a serious deterrent to others.

tivism, conservation listing, and large-scale planning are often involved, all of which require dedicated resources. Even so, private individuals regularly play a critical role, by raising issues to public attention, by lobbying public officials and private landowners, and by supplying specific knowledge of individual threatened habitats or populations—knowledge that in many cases no on else possesses. The tireless efforts of John and Mary Yrizarry, residents of Orange County, New York, were instrumental in the recent effort to preserve Sterling Forest, a large wild area that includes numerous important butterfly habitats. But even when no immediate environmental Armageddon looms, butterfliers need to ensure continually that local and regional planners do not overlook butterfly habitat requirements for lack of adequate information or awareness.

At least as damaging as large-scale habitat destruction, in aggregate terms, is the threat from routine "prettification" of yards, parks, and roadsides. The most useful single message that butterfly conservationists can spread is that weedy fields and yard edges are alive and beautiful; while close-cropped, multiacre grass patches, doused with chemicals, are environmentally crippled and cost our wildlife dearly. It can take surprisingly little undisturbed space to support a population of butterflies—much less than to sustain large mammals or even birds. And only a small degree of forbearance may be required to preserve such habitats. Even simple mowing has been shown to reduce butterfly diversity—by killing larvae and removing nectar plants, and sometimes hostplants (Scoble, 1995, p. 180). Periodic mowing is necessary, of course, to keep habitats open and presentable, but all too often it is carried out routinely and indiscriminately, impairing large swatches of habitat without due regard for environmental management, and not to mention with a great attendant pollution and waste of fossil fuel.

On another front, the enormous recent increase in deer populations in our woodlands has led to overbrowsing of many low trees and wildflowers. To the extent that these plants are nectar sources or larval hosts, the damage to butterflies can be direct and significant. Deer browsing is known to deplete Shale Barrens Rock-Cress (*Arabis serotina*), a key hostplant for Olympia Marbles (Braunschweig et al., 1999). The East Coast population of this species has declined precipitously in recent decades.

On a much larger scale, evidence is quickly accumulating that global warming is causing a northward shift in the distribution of specialized plant and animal habitats. (*See,* e.g., Parmesan, 1999.) This poses a distinct problem for many East Coast butterflies: in a world with extensive habitat fragmentation and destruction, specialized communities are often trapped in small biological "island" zones from which they cannot relocate quickly. The more tightly integrated the community, the more difficult it becomes to accommodate changes, and the more dire the potential consequences.

I-36. Excessive or indiscriminate verge mowing eliminates critical butterfly resources.

I-37. Increasing habitat destruction in the East Coast region diminishes vital resources.

Discussion of these and related issues could be extended to great length. There is the collateral impact of biologically modified organisms released in the environment, the effects of aerial spraying, habitat degradation caused by acid rain, imbalances created by introduced species (such as fire ants, which eat caterpillars), etc. Species conservation objectives face many hurdles in a crowded world. But significant gains can be made in terms of butterflies if people simply reorient their approach to managing nature—starting in their individual backyards. It is encouraging to see significant butterfly habitats being preserved in some parts of the United States (we especially note the recently opened NABA Butterfly Park in the lower Rio Grande Valley of Texas, outside our area). Those who care for butterflies must act to ensure that the populations we currently enjoy will continue to persist into the future.

Most of the photos in this book were taken at protected sites, but many were not. In the accompanying footnote, we have listed 14 specific sites where approximately 60 of our photographs were taken, and which are no longer intact as a result of human activity.[69] This list represents only a small subset of the total habitat areas compromised or destroyed in the region during the course of this project. But it accentuates the imminent peril facing our open spaces generally.

USING THE SPECIES ACCOUNTS

The "framework" for the species accounts is derived from the introductory materials, in terms of habitat descriptions, characterizations of behavior, lifestyle traits, and the like.

Ranges: Range information is based on personal sightings, reports from active field observers, and a number of excellent published sources. Recent example include Struttman (USGS, on-line), Allen (1997), Glassberg (1999); Glassberg et al. (2000); Gochfeld & Burger (1996), Layberry et al. (1997), LeGrand & Howard (2003), Opler (1992), Opler & Krizek (1984).

Habitat: Terminology used to describe habitats are those defined in the Introduction. In certain cases, generic phrases such as "open spaces" may refer to a range of open, low-growth habitat types.

Hostplants: Principal hostplants in the East Coast region are noted. Plants used elsewhere may be excluded, and note is made (where known) of plants eaten by caterpillars in captivity, but not verified in the wild.

Occurrence: Estimates of occurrence can be misleading. Rare species can be locally common in colonies, and some widespread species are seldom seen in large numbers.

Flight period descriptions occasionally exclude unusually early or late records that may be misleading in a general overview.

ABBREVIATIONS

The following abbreviations are used in the text:

ARG	Argentina
BRA	Brazil
CAN	Canada
C. Amer.	Central America
Co. (Cos.)	County (Counties)
FW/HW	forewing/hindwing
Gr. Dismal Swamp	Great Dismal Swamp
Gr. Lakes	Great Lakes
Gr. Plains	Great Plains
L. Amer.	Latin America
LI	Long Island
MEX	Mexico
NF	National Forest
NP	National Park
N. Eng.	New England
No. Amer.	North America
ONT	Ontario
Pac.	Pacific
QUE	Quebec
So. Amer.	South America
SP	State Park
UK	United Kingdom/U.K.
US	United States/U.S.
W.I.	West Indies
WMA	Wildlife Management Area

[69] The sites include Boone Co, WV (mountaintop removal); Hickory, VA (residential construction); Carrol Co, NH (parking lot construction); Collier Co, FL (road construction); Daytona, FL (roadside "channelization"); Fairfield County, CT (site destruction); Ft. Indiantown Gap, PA (habitat reallocated to military use; ultimate effect to be determined); Gainesville, FL (industrial park development); Homestead, FL (residential construction); Lakehurst, NJ (golf course development); Miami-Dade Co, FL (racetrack development); Somers, NY (residential housing development); Suffolk Co, NY (residential housing development); Tavernier, FL (hospital construction).

Map Key

- Primary range
- Scarce/seasonal range
- Extirpated
- ○ Isolated colonies
- △ Isolated sightings

Butterfly Underwing

apex eyespots fringe
marginal band
submarginal band
FW costal (leading) margin
HW costal (leading) margin
FW cell
palps
outer angle
outer HW
postmedian line
basal (inner) HW inner (trailing) margin post basal line median (central) HW

Skipper Upperwing

club
palps antenna apiculus
costal (leading) margin basal FW
cell-end (mid-FW) dot median FW
shoulder thorax outer FW
subapical wrist band ("bracelet") apex
outer margin
lower post-median spots HW fringe
inner (trailing) margins abdomen
outer (anal) angle

"True" Butterfly Upperwing

FW costal (leading) margin FW cell club
apex antenna
palps
FW outer margin
thorax costal stripes
subapical band
FW inner (trailing) margin
eyespots
outer (anal) angle submarginal band
abdomen marginal band

The Swallowtails:
Family Papilionidae / Subfamily Papilioninae

Swallowtails are large, conspicuous butterflies. Only a few individual species from other families (such as the Monarch) can rival them in terms of public recognition. They are also well-recognized by the scientific community, which has generated a large and active research literature (e.g., Scriber et al., 1995).

The Swallowtail Family (Papilionidae) includes both the pierid-like parnassians (none of which occur in our area) and the "typical" swallowtails. The Papilionidae has between 475 and 563 species worldwide (Scriber et al., 1995). The broad range of this estimate reflects differing scientific views on taxonomy.

The "true swallowtails" (subfamily Papilioninae) number between 460 and 490 species, of which about 28 have occurred in North America. Twelve are known from the East Coast, including nine residents and two strays.

The swallowtails are an old family. They probably left the "main line" of butterfly evolution (along with the pierids) sometime after the skippers "split off" (Scott, 1986). Swallowtails subsequently evolved a distinctive physical and biochemical identity of their own, rooted in an acquired tolerance for certain hostplant defensive chemicals (specifically aristolochic acids, furnocoumarins, and acetogenins). Many species appear to be engaged in long-term, coevolutionary duels with their specialized hostplants, in which each continues to evolve increasingly refined mechanisms to outmaneuver the other (Ehrlich & Raven, 1964). Mimicry is common among our Papilioninae. It centers around the Pipevine Swallowtail, whose toxicity provides a shield for at least six other butterflies (including two nymphalids, the Red-spotted Purple and female Diana).

Male swallowtails generally patrol for females (an exception being the Black Swallowtail, which forms hilltop leks). They are strong fliers, and some can cross significant stretches of open water, such as between Caribbean islands. But few are regular migrants. Most males take minerals, often from moist dirt, and all species visit flowers, in many instances using long proboscises to probe deep blossoms. Adults are relatively long-lived. Females of many species are able to remate during their lifetimes. Except for the Pipevine Swallowtail Group (genus *Battus*) and Black Swallowtail, all our species feed on trees or other "apparent" plants, rather than on evanescent herbs.

The characteristic hindwing "tails" that give the Papilioninae their common name are significant aerodynamically: they create directed airflows over the wings that enable extended glides at high angles—angles greater than those at which a butterfly with such broadly cut wings normally would "stall" (Brodsky, 1994, pp. 176–81). Swallowtail eggs are plain and generally spherical. Most caterpillars feed at night. Some rest by day in silked-up leaf shelters. Early instars often have white "saddle patches" that cause them to resemble bird droppings. Later instars in some species sometimes develop thoracic "eyespots" that many believe make them resemble small snakes. And all larvae have eversible, Y-shaped organs (*osmeteria*) that are brandished in times of threat, chemically deterring many predators. All of our swallowtails diapause in the pupal stage. Pupae have silk "girdles" that secure them to their resting substrate (along with the usual cremaster), a trait they share with pierids.

The swallowtails are divided into a number of subgroups, but there is disagreement as to the proper generic arrangement in some cases (*see* Hancock, 1983; Miller, 1987, p. 386; Layberry et al., 1998, pp. 24–25). We continue to recognize a widely defined *Papilio* genus. Major swallowtail subgroups include:

Aristolochia Swallowtails (Tribe Troidini): Worldwide, includes the birdwings, largest of all butterflies. Two species in our area (in genus *Battus*). Both sequester aristolochic acids, making them distasteful to many predators. Females lay eggs in clusters and larvae feed gregariously in early instars.

Kite Swallowtails (Tribe Leptocircini [= Graphiini]): Over 140 species worldwide, mainly tropical; the lone species in our area feeds on pawpaws. Cuban Kite Swallowtail (*Eurytides celadon*) is a hypothetical stray from Cuba, similar to Zebra Swallowtail, but with a differing stripe pattern. No authenticated records exist. Leptocircini larvae are "humpbacked," without eyespots.

Fluted Swallowtails (Tribe Papilionini): Named for an obscure hindwing groove ("flute"). The tribe includes: Old World Swallowtail Group (mainly western, feeds on herbs in the Carrot/Umbellifer Family; excels at detoxifying furanocoumarins); Giant Swallowtail Group (rue/citrus-feeders; sometimes placed in genus *Heraclides*); and Tiger Swallowtail Group (sometimes placed in genus *Pterourus*; includes two subgroups, typical tiger swallowtails, often highly polyphagic, and the Palamedes/Spicebush subgroup, which feeds on laurels).

Mature larva of Polydamas Swallowtail with unhatched eggs

Zebra Swallowtail *Eurytides marcellus*

Dorsal male (spring form): 4/22/95, Elk Neck State Park, Cecil Co., MD

Ventral (summer form), on bear dung: 6/24/94, Gr. Dismal Swamp NWR, Nasemond Co., VA

Elegant in appearance, the Zebra Swallowtail is the lone member of its tribe to colonize temperate areas of North America. By comparison, more than 140 kite swallowtails inhabit the New World tropics. The Zebra's occurrence is closely linked to that of another tropical emigrant, the common Pawpaw—a member of the Custard-apple Family (Annonaceae). Pawpaws produce noxious chemicals (annonaceous acetogenins) that deter most herbivores. But Zebra Swallowtail caterpillars can ingest these toxins, which persist in adult wing and body tissues. Acetogenins are believed to confer protection from avian predators (Martin et al., 1999).

The flight has been described as bobbing, but the males' distinctive patrol flight is actually fast and direct, at a relatively constant height of about 2–6 feet. "Bobbing" better describes ovipositing females, which carefully inspect potential hostplants up and down before depositing a single egg. This segregation of egg sites is prudent, since Zebra Swallowtail caterpillars are "notoriously cannibalistic" (Klots, 1951, p. 179). When disturbed, young caterpillars brandish a yellow scent organ (*osmeterium*) that chemically deters small spiders and ants. Older larvae defend against predators by thrashing or by physically leaving the hostplant (Damman, 1986).

Males puddle frequently, and can be quite approachable when so engaged. The short adult proboscis cannot access long-tubed flowers—including many showy garden blooms. Rotting or putrid material is often used.

Identification	A comparatively small swallowtail, especially in spring. The graceful wing shape and long tails are distinctive. Summer-brood individuals have especially *long tails*, broadly *outlined in white*. Few species in our area are more easily identified than the Zebra Swallowtail.
Dorsal	Longitudinal black stripes on a white field; broader in summer. The dark outer FW stripe is especially prominent. A cherry-red crescent on the inner HW is flanked by a cloud of blue-green iridescence.
Ventral	Similar to dorsal surface, but a nearly continuous cherry-red line runs through the central HW.

Habitat	Rich, moist deciduous forests and river bottomlands, seldom far from pawpaws. Also in dry, open southern pine woods, savannas, dry prairie. Not in higher mountains (> 2,500 feet in NC).
Hostplants	About a dozen of the world's 2,300 custard-apples grow in our region. The main hostplant, Common Pawpaw (*Asimina triloba*), is regular in rich bottomland woods north to upper Chesapeake Bay. Several southern pineland/savanna pawpaws (e.g., *A. reticulata, tetramera incarna*) are also used. Fresh leaves are favored.
Occurrence	Regular from FL to MD and w. NY. May have once extended farther north. Extirpated in southernmost FL by habitat loss; accidental on Keys, where pawpaws are absent. Also west to Gr. Plains. Two flights in North (Apr-Aug), 3–4 in Deep South (Feb-Dec). First brood usually more numerous; large 2nd broods noted when pawpaws defoliated mid-season. Pupae overwinter.
Ecology	**Medium Specialist.** Limited to a single hostplant genus, but its principal member in the East (Common Pawpaw) is widespread. Tolerates development poorly; fares best in intact habitats.

Pawpaw (*Asimina triloba*). **Inset**: Leaves and fruit (also called "pawpaw")

av wingspan: 3.1"

Pipevine Swallowtail *Battus philenor*

Dorsal male: 9/5/93, Greenwood Sanctuary, Bergen Co., NJ

Ventral female: 7/8/95, Fork Creek Pub. Rec. Area, Boone Co., WV

We normally use the term "dazzling" as a compliment (albeit sometimes a slightly guarded one). But in nature, the open flaunting of bright colors is usually a warning, meant to elicit caution or avoidance. Creatures that produce such cautionary displays are termed "aposematic." The Pipevine Swallowtail is a prime example.

Pipevine caterpillars sequester toxic alkaloids (aristolochic acids) from their hostplants. These chemicals persist in the bodies and wings of adults, as well as in their eggs, larval shells, and pupae (Sime et al., 2000). This makes them one of the more consistently distasteful butterflies in our area. When snatched by an unwary avian predator, the adult's thick cuticle often resists the bite long enough for the attacker to taste an acrid liquid oozing from the body pores. This experience is usually sufficient to deter further assaults (Codella, 1986). Indeed, so potent is the Pipevine's chemistry that even parisitoid wasps are repelled, and their eggs will not survive if injected (Sime, 2002). The defense is not flawless, however, since anoles (*Anole carolinensis*) can reportedly prey on Pipevines (Odendaal et al., 1987). Still, Pipevines have numerous mimics, as discussed in individual species accounts.

Male Pipevines patrol likely habitats most of the day in search of receptive females. Their flight is rapid and choppy, with shallow wingbeats. Females spend most of their time searching for hostplants without egg clusters already present (Rausher, 1979). Males actively seek minerals, and may join other swallowtails at puddles. Pipevines show a native preference for yellow nectar flowers, but learn to accept different colors in nature—or in the lab, with proper reinforcement (Weiss, 1997). Communal roosting has been reported.

Identification	Large, but barely medium-sized for a swallowtail. Tails short and stubby. The outer HW is scalloped. Body *dark with white spots along the sides*. Individuals of spring generation small.
Dorsal	Black, with *no red or orange mark* at the HW anal angle. Males have *dazzling, blue-metallic iridescence* on the HW (females also, but less pronounced). A *row of cream spots* on the outer HW extends onto the FW (faintly in male, prominently in female).
Ventral	Black, with a submarginal row of *seven round, bright orange spots*, set in a *blue iridescent field*. White marks run along the HW margin. FW pattern mirrors the dorsal.

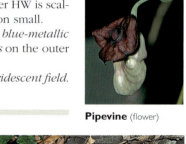

Pipevine (flower)

Habitat	Various open habitats, usually in or near deciduous woodlands; also gardens, scrub. Elsewhere, deserts, tropical forest, riparian woods, etc.
Hostplants	Pipevines (*Aristolochia*), including Dutchman's Pipe (*A. macrophylla* [= *durior*]), Virginia Snakeroot (*A. serpentaria*), others in West. Highly toxic exotics are not used. Early reports of Wild Ginger, knotweed, and morning glory seem incorrect. Eggs laid in batches beneath host leaf, larvae feed voraciously, consuming entire plants.
Occurrence	Common in central Appalachians and Deep South (not s. FL); less common north to central N. Eng. on ornamentals. Also west to CA, south to s. MEX. Two broods in North, Apr-Oct; 3+ in FL, Feb-early Nov. Pupa hibernates.
Ecology	**Medium Specialist**. Narrow hostplant range, but widely adaptable.

Pipevine (or Dutchman's Pipe) (*Aristolochia macrophylla*) showing caterpillar damage

av wingspan: 3.6″

Polydamas Swallowtail *Battus polydamas*

Dorsal: 9/24/95, Tradewinds Park, Ft. Lauderdale, Broward Co., FL

Ventral: 9/24/95, Tradewinds Park, Ft. Lauderdale, Broward Co., FL

Many dangers beset East Coast butterflies, from human development to adverse weather events, but the *Battus* swallowtails are unusual in being subject to fashion risk. The Polydamas Swallowtail, specifically, is a tropical species living at the edge of its natural range in Florida. It must search carefully in our area for native hosts that contain the desired, high concentration of aristolochic acid. Many non-native aristolochias meet the Polydamas' dietary needs, however, and the species has extended its local presence using cultivated plants.

The risk to newfound Polydamas colonies is that horticultural fancies may shift, leaving them suddenly without hosts. In 2001, temporary removal of a large stand of aristolochias at Butterfly World in Ft. Lauderdale reduced the incidence of wild-flying Polydamas Swallowtails in the vicinity. And the fortunes of the closely related Pipevine Swallowtail reportedly rose in the North along with the popularity of ornamental Dutchman's Pipe during the 1800s—then fell again when tastes later changed (Iftner et al., 1992; Gochfeld & Burger, 1997).

On a different note, researchers have long puzzled over the ecological rationale for cluster-feeding in *Battus* swallowtail caterpillars. In their early instars, Pipevine and Polydamas larvae attack a series of single leaves en masse, eventually consuming/defoliating entire plants. Several theoretical advantages have been proposed to explain this practice, from predator avoidance to thermoregulation. It now appears that cluster-feeding may allow 1st instar caterpillars to take advantage of a short "window" between their first attack on a leaf and the production of defensive chemicals that would slow larval growth; fast progress through the 1st instar is important, since this is the one phase in the life of a *Battus* swallowtail in which it is not well-protected against predators (Fordyce, 2003). Older caterpillars cease cluster-feeding and forage individually.

There are a dozen isolated Caribbean races of Polydamas Swallowtail, vs. just one widespread race in Central and South America (Smith et al., 1994). The race found in our area is *lucayus* (also occurs in the Bahamas). Adults nectar often, fluttering their wings continuously while feeding. Males take minerals from the ground.

Identification	Large, especially females. *Our only tailless swallowtail.* HW rounded with *scalloped margins*. Body dark with *red spots* along the sides.
Dorsal	Black with *satiny-green iridescence*. A band of *closely packed, elongate yellow spots* on the HW becomes *separate, triangular spots* on the FW.
Ventral	FW dark, mirroring dorsal pattern. HW paler *gray-brown*, with concentric rows of yellowish, bluish, and especially *reddish marks on outer HW*.
Habitat	Subtropical scrub and hammock edges, beaches, gardens, open woods.
Hostplants	Marsh's Dutchman's Pipe (*Aristolochia pentandra*), Virginia Snakeroot (*A. serpentaria*); also cultivated Giant Pipevine (*A. gigantea*), Calico Flower (*A. littoralis* [= *elegans*]), Pelican Flower (*A. ringens*), and other species. Eggs laid in clusters of 10–14. The osmeterium is yellow.
Occurrence	Generally uncommon, Keys and s. FL; rare n. FL. Also W.I. and s. TX south to ARG. Rarely strays north (GA, KY, MO). All year in South; 2–3 flights n. FL, Mar-Nov. Pupa overwinters.
Ecology	Specialist. Prefers highly potent aristolochias in subtropical surroundings.

Giant Pipevine (*Aristolochia gigantea*), an exotic

av wingspan: 3.65"

Eastern Tiger Swallowtail   *Papilio glaucus*

Dorsal female: 5/9/99, Northwest River Park, City of Chesapeake, VA

Ventral (taking minerals): 7/9/94, Orange Co., NY

A highly successful generalist, the Eastern Tiger Swallowtail is among the most widely recognized of all eastern butterflies. Its familiarity has early origins, as it was the subject of the first known drawing of any North American butterfly species (sketched by John White in 1587, while in Roanoke on Sir Walter Raleigh's third expedition). Linnaeus formally described the species in his original *Systema Naturae* (10th ed., 1758). Some taxonomists place this and other members of the "Tiger Swallowtail Group" in the genus *Pterourus*.

In 1991, single-brooded northern populations of Tiger Swallowtail were recognized as a separate species (the Canadian Tiger Swallowtail). And still more recently, a large, single-brooded spring population centered in the Appalachians, with reduced hindwing blue clouding, the Appalachian Tiger Swallowtail, has been reported as a sibling species (*P. appalachiensis*) (Pavulaan & Wright, 2002).

Adult Eastern Tiger Swallowtails and pupae are palatable to birds and small mammals (Brower, 1985). Dark-phase females gain protection by mimicking the appearance of the toxic Pipevine Swallowtail. Melanism is inherited through maternal genes (Clarke & Sheppard, 1962). The percentage of dark females is highest in areas where Pipevines are most common, reaching 90% in the southern Appalachians (Brower & Brower, 1962). Eastern Tigers eat a wide variety of tree leaves. This requires them to cope with diverse chemical defenses. And they are well-prepared, producing enzymes that neutralize furanocoumarins they infrequently encounter (Li et al., 2003).

Male Eastern Tigers patrol for females, which they entice with perfume-like pheromones. Both sexes nectar, preferring sturdy plants, and males visit puddles, drawn by sodium (Lederhouse et al., 1990).

Identification	Very large; southern females rank as our largest butterflies. Small, spring individuals can closely resemble Canadian Tiger Swallowtails.
Dorsal	Yellow to ochre-yellow with conspicuous *black stripes* and variable *blue clouding* on HW. *Melanistic female* similar but *all-dark*. Orange spots on HW apex, at anal angle and on innermost end of the HW marginal band.
Ventral	Striped, as on dorsal surface. Postmedian bands include orange spots and blue clouding. FW submarginal band *usually* divided into discrete spots. *Melanistic females show visible shadow of dark "tiger striping."*

Tuliptree, or Yellow Poplar (*Liriodendron tulipifera*)

Habitat	Nearly anywhere with deciduous trees.
Hostplants	Hosts in multiple families, especially Tuliptree (*Liriodendron tulipifera*), wild cherries (*Prunus*), ash (*Fraxinus*), Wafer Ash (*Ptelea trifoliata*), etc.; Sweet-Bay (*Magnolia virginiana*) in Deep South. Refuses laurels, aspens. Females often oviposit on hosts near nectar sources (Grosmueller & Lederhouse, 1987).
Occurrence	Common to abundant, from central N. Eng. to FL (scarce in far s. FL, absent on Keys). Also west to Gr. Plains, central TX. Two overlapping broods in North, May-Sep; three in South, Feb-Nov. Pupae from different generations may overwinter.
Ecology	Generalist. Extremely adaptive in hostplant and habitat use.

Sweet-Bay (*Magnolia virginiana*), southern host

av wingspan: 4.8"

Ventral female (dark form): 7/9/95, Big Ugly WMA, Boone Co., WV

Dorsal female (dark form): 4/26/96, Fork Creek Pub. Rec. Area, Boone Co., WV

Canadian Tiger Swallowtail   *Papilio canadensis*

Formerly a race of the Eastern Tiger Swallowtail, the single-brooded Canadian Tiger was recognized as a separate species just over a decade ago (Hagen et al., 1991). The two are closely related, nonetheless, and hybrids occur in central New England. The validity of the split has been reinforced by the discovery of biological differences between the two taxa. Canadian Tiger Swallowtail pupae, for example, can survive in colder winter temperatures than Easterns (Kukal et al., 1991). And they experience elevated mortality when exposed to heat (Scriber et al.. 2002).

Canadian Tigers adapt flexibly to cold habitats. It has been shown, for instance, that Canadian caterpillars from Alaska grow more rapidly in cold conditions than those from Michigan (Ayres & Scribner, 1994).

Ventral: 6/13/93, Errol, Coos Co., NH

Canadian Tiger Swallowtails rarely produce dark females. They have a low incidence of the gene that causes melanism, and a high incidence of an autosomal gene that suppresses it—even when the gene for melanism is not present (Scriber et al., 1990).

Identification	Smaller than Eastern Tiger. Spring Easterns can closely resemble the Canadian. Local brood timing is significant, and should be noted.
Dorsal	Normally paler (more *whitish-yellow*) than the Eastern Tiger. The dark border next to the body is *thicker*.
Ventral	Yellow submarginal band on the FW is *largely continuous* (not interrupted by wide, black lines at the veins).

Habitat	Northern deciduous and transitional forests and edges. Also nearby open areas.
Hostplants	Birch (*Betula*), aspen (*Populus*), and black cherry (*Prunus*). The Tuliptree (*Liriodendron tulipifera*), a common Eastern Tiger host, is toxic to Canadians. Conversely, Easterns cannot eat aspen (Lindroth et al., 1988). Caterpillars feed in silked-up leaf shelters.
Occurrence	Common to abundant, N. Eng. south in Appalachians to ne WV. Also west across CAN to central AK. One brood, mid-May to mid-Jul. Pupa hibernates.
Ecology	Generalist. Limited to northern climates, but otherwise wide-ranging and adaptable.

Dorsal: 6/13/93, Errol, Coos Co., NH

av wingspan: 2.9″

Spicebush Swallowtail *Papilio troilus*

Dorsal male: 7/8/95, Fork Creek Pub. Rec. Area, Boone Co., WV

Ventral male: 5/16/93, Mauricetown, Cumberland Co., NJ

Earlier called the Green-clouded Swallowtail, the Spicebush Swallowtail is southern-based, but still relatively cold-tolerant, and thus a familiar generalist throughout most of our region. Its nearest local relative appears to be the Palamedes Swallowtail (also a laurel-feeder), but it is placed on this page because field observers confuse it most often with the Black Swallowtail.

A palatable species, the Spicebush Swallowtail is one of several eastern swallowtails that mimics the toxic Pipevine Swallowtail for protection (Brower, 1985). It is a habitat generalist. In studies designed to determine the importance of corridors between natural habitats, the Spicebush Swallowtail stood out as a species that radiated widely, with no clear "home territory" or habitat, and that was not confined to narrow corridors (Haddad & Baum, 1999). A green-clouded race has recently been described from South Florida (Gatrelle, 2000).

Male Spicebush Swallowtails patrol in woods and woodland edges and roadsides, to locate receptive females. The low, fairly rapid flight (3–9 feet in elevation) is intermediate between the Black Swallowtail's direct flight and the more languid, gliding flight of the Eastern Tiger.

Spicebush Swallowtails court and mate in the afternoon. During courtship, both sexes engage in slow flight, with the male hovering above the female. Males puddle actively, and both sexes use long proboscises to nectar.

Identification	Large. Body dark with two rows of pale spots on the sides. Tails have dark centers.
Dorsal	Black. *Single row of* spots on outer margin, *ivory* on FW, *bluish-white* on HW (an additional, faint FW row is sometimes visible, especially in males). HW has iridescent *blue-green clouding* (*green* especially in s. FL); brighter in male.
Ventral	Dark with no "shadow" of wing stripes (*compare* dark female Tiger). *Two rows of postmedian orange spots* enclose a field of blue iridescence. The inner row is *missing one orange spot* (third in from the inner margin). Two rows of light spots on FW.

Sassafras (*Sassafras albidum*)

Habitat	Woods, fields, roadsides, pine barrens, swamps, parks.
Hostplants	Primarily Spicebush (*Lindera benzoin*), on which larvae fare better than on Sassafras (*Sassafras albidum*). Sassafras with extensive leaf pubescence deters larval growth (Haddad & Hicks, 2000). Also Redbay (*Persea borbonia*) in FL. Young caterpillars live in folded-over leaf shelters, older ones in silked-up leaf-tubes. Mature larvae have "snake head" eyespots.
Occurrence	Common, from central N. Eng. south to FL (declining in far south; gone from Keys). Also west to e. NE, central TX. Two overlapping broods in North, May-Sep; three in South, Apr-Oct; FL, Mar-Dec. Pupae overwinter; some from each generation. Short photoperiod produces brown, overwintering pupa.
Ecology	**Generalist**. Uses widespread hostplants in a variety of habitats.

Spicebush (*Lindera benzoin*) flower and leaves

av wingspan: 4.1"

Black Swallowtail   *Papilio polyxenes*

Dorsal male: 7/22/95, near Jefferson, Coos Co., NH

Dorsal female: 7/9/00, Orange Co., NY

Widespread and generally common, the Black Swallowtail is a well-known butterfly of open spaces throughout our region, although its distribution is sometimes patchy. This is the lone eastern member of the "Old World Swallowtail Group," most of whose members are western. The female is a Pipevine Swallowtail mimic.

Black Swallowtails have a distinctive, coevolutionary relationship with members of the Carrot Family (Apiaceae), which produce a diverse array of chemicals (furanocoumarins) that interfere with DNA transcription in herbivores—with toxic results. Black Swallowtails use specialized enzymes to denature these toxins, very effectively with most (especially "linear" furanocoumarins) but somewhat less so with certain more newly evolved types. In early instars, Black Swallowtail caterpillars have a white "saddle" composed of uric acid that makes them resemble bird droppings. In addition to camouflage, some suggest,

Ventral pair (mating): 7/11/99, Staten Island, NY

the uric acid "saddle" helps prevent sunlight-induced damage to DNA from ingested plant toxins (Timmermann, 1999).

Most eastern swallowtails patrol for mates, but the Black engages in a well-documented, lek-style perching courtship on hilltops (Lederhouse, 1982). Females fly close to the ground at midday, seeking suitable hosts. In cold weather, Black Swallowtails raise their abdomens above their hindwings for warmth (Rawlins, 1980).

Identification *Dorsal*	Large. But males/1st brood individuals smaller. Sexually and seasonally dimorphic. Body dark with *light spots.* Upper surface mostly black, with *two rows of yellow spots* (male) *or creamy yellow spots* (female) along outer HW, which *do not converge* at apex (*compare* Giant and Schaus'). Note isolated spot on FW costa, and (usually) yellow cell-end bar.
Ventral	Dark. *Two concentric orange spot rows* enclose a blue cloud. The postmedian spots are not aligned in a neat row, but there are *no gaps* (as in Spicebush).

Habitat   Varied open spaces: fields, suburbs, fresh and tidal marshes, deserts, roadsides, etc. Seldom forest interiors. Some populations are coastal.

Hostplants   Carrot/Parsley Family (Apiaceae). Queen Anne's Lace (*Daucus carota*), Fennel (*Foeniculum vulgare*) (preferred in Delmarva [Dombrowski & Mills, 1996]); also cultivated carrots, celery, caraway, dill, parsley. Caterpillars eat leaves or flowers (later instars prefer flowers). Osmeterium orange.

Occurrence   Variably common, n. N. Eng. south to s. FL (rare on Keys for lack of hostplants). Also most of e. US, south to Peru. Two or three broad, overlapping broods, Apr-Oct in North, Feb-Oct in South. Pupa overwinters; color varies from green to brown based on environmental triggers (West et al., 1972).

Ecology   Generalist. Widely adaptable.

Queen Anne's Lace (*Daucus carota*)

av wingspan: 3.2"

Giant Swallowtail   *Papilio cresphontes*

Dorsal female: 9/29/95, Key Largo, Monroe Co., FL

Ventral: 9/29/95, Key Largo, Monroe Co., FL

The Giant Swallowtail is not our largest butterfly (some female Eastern Tiger Swallowtails are larger). And it is much smaller than the endangered Giant Swallowtail (*Papilio homerus*) of Jamaica, whose average wingspan exceeds 5 inches. On the East Coast, however, our Giant Swallowtail remains noteworthy for its size.

This is the most familiar local member of the "Old World Swallowtail Group," a closely knit clan of large citrus-feeders. They are treated as a separate genus (*Heraclides*) by some taxonomists, but at present most still refer them to *Papilio*, their traditional genus.

Giant Swallowtail larvae are called "orange dogs" by citrus growers—not for their color, but rather for their shape (which resembles a dog's head at the anterior end) and for their occurrence on orange trees. The larvae of some other swallowtails have been given similar names: Androgeous Swallowtail caterpillars, for example, are referred to as "orange puppies" (Scott, 1986).

Male Giant Swallowtails patrol throughout the day seeking females. Flight is strong but often low and usually un-hurried, intermixed with frequent glides. Courtship occurs in the afternoon. Females oviposit around midday (Allen, 1997). Both sexes nectar frequently, and males seek minerals. Giant Swallowtails sleep with spread wings, thus pre-senting their less conspicuous surface.

Although they are not regular migrants, Giant Swallowtails have appeared with some consistency as vagrants in the North, and occasionally form colonies (*see* essay, following page).

Identification	Very large. Tails *spoon-shaped* with *yellow centers*. The body is dark on top with yellow sides (not spotted). Can be confused with several other dark swallowtails, including Schaus', Bahaman, and Black. Sexes are similar.
Dorsal	Blackish-brown; *two yellow wing bands* converge near the FW apex, with yellow spots extending from the meeting point to the FW costa. Only blue on HW is semi-circular "eyebrows" over orange-and-black eyespots at anal angle.
Ventral	Mostly yellow, with a *bluish postmedian band* and *single, irregularly shaped chestnut patch* (not an almost complete band, as in Schaus').

Habitat	Wide-ranging in natural and disturbed open areas, including parks, sub-urbs, and citrus groves. Open riparian slopes northward.
Hostplants	Rue (Rutaceae): Wild Lime (*Zanthoxylum fagara*), especially growing in the open, Hercules Club (*Zanthoxylum clavaherculis*), Common Rue (*Ruta graveolus*), Torchwood (*Amyris elemifera*), cultivated citrus (*Citrus*), Northern Prickly-Ash (*Zanthoxylum americanum*), Wafer Ash (*Ptelea trifoliata*). Eggs laid singly on twigs and leaves. Larvae maintain "birdlime" appearance in all instars.
Occurrence	Common in SE, progressively rarer northward. Also west to Rockies and SW deserts, south to COL. Two broods in North (May-Jun and Jul-early Sep), three or more in Deep South, all year in s. FL. Pupae overwinter.
Ecology	Medium Generalist. More adaptable (especially with respect to latitude) than other citrus-feeding Swallowtails.

Mature caterpillar ("orange dog")

av wingspan: 4.8"

Rueful Behavior: Dietary Themes in the Giant Swallowtail Group

On first impression, the dietary habits of the Giant Swallowtail may seem confusing. In Florida, this conspicuous and wide-ranging member of the "Giant Swallowtail Group" (assigned by some taxonomists to the genus *Heraclides*, as earlier noted) feeds exclusively on plants in the Rue, or Citrus, Family (Rutaceae). In commercial citrus groves, its larvae can become a minor pest.

But Giant Swallowtails occasionally form sustained, temporary colonies in the North, formerly to New Jersey (Gochfeld & Burger, 1997), well beyond the range of commercial citrus plants. Such northern colonies use highly specific local hosts, such as Hercules-Club, Wafer Ash, and Northern Prickly-Ash. These plants may seem an odd departure from the Giant's traditional diet. Yet they are all in fact members of the Rue/Citrus Family. Northern Prickly-Ash is not a true ash (ashes belong to the Olive Family, Oleaceae). Its common name is based on the ash-like shape and arrangement of its leaves, rather than its phylogeny. All of the Giant's observed, northern host plants are thus quite logical.

Worldwide, there are about 150–160 genera of rues, with 1,500–1,700 species. Most live in the tropics or temperate zone. They are chemically active, producing aromatic "essential oils" that give citrus fruits their distinctive flavor. Others are used in the production of perfumes. Many Rue/Citrus Family members also produce toxic chemicals (furanocoumarins) that protect them from general insect herbivores. Giant Swallowtails have the ability to detoxify these chemicals, and thus have an advantage in browsing on plants producing them. Black Swallowtails specialize in detoxifying furanocoumarins, which are highly diversified in the Carrot Family (Apiaceae). In Giant Swallowtails, this skill is more incidental.

The other members of the Giant Swallowtail Group also feed on rues. One, the Schaus' Swallowtail, is a noted resident in Florida. Others live in the tropics (nine in the nearby West Indies alone), and at least two have limited contact with our area:

The Bahaman Swallowtail (*Papilio andraemon*) occurs in very low densities on the Upper Keys, especially Key Largo and the islands of Biscayne National Park. Its hostplant and habitat use resembles the Schaus'. First noted in the United States in the 1970s, the *bonhotei* race of the Bahaman Swallowtail (from the Bahamas) may be a natural resident of the Keys, on either a permanent or transitory basis (Minno & Emmel, 1993). But its natural history here is not well understood, and populations appear to have declined in recent years. We are aware of no recent, verified sightings from upper Key Largo.

The Bahaman Swallowtail is best distinguished from the Giant and Schaus' Swallowtails by the yellow cell-end bar on the dorsal forewing and by a single long, rectangular chestnut-colored marking on the ventral hindwing.

The Androgeous Swallowtail (*Papilio androgeous*) was established in citrus groves near Miami from the late 1970s to the early 1980s. But it has not been seen recently and is presumed extirpated. The male has broad, yellow forewing bands. The female is all-dark, resembling a long-winged Pipevine Swallowtail, with two short hindwing "mini-tails," flanking a central, normal-length tail.

Citrus blossom (*Citrus* sp.)

Wild Lime (*Zanthoxylum fagara*)

Hercules-Club (*Zanthoxylum clava-herculis*)

Northern Prickly-Ash (*Zanthoxylum americanum*)

Schaus' Swallowtail   *Papilio aristodemus*

Jaret C. Daniels, Ph.D.

Dorsal male: 5/18/97, Key Largo, Monroe Co., FL

Ventral male (conservation-reared individual): 6/96, Eliot Key, Monroe Co., FL

The endemic Florida race of this handsome swallowtail (*P. a. ponceanus*) was named for Dr. William Schaus, a physician who discovered it in a hardwood hammock south of Miami in 1898 (now destroyed), while tending Spanish-American War patients. Described in 1911, the Schaus' Swallowtail was last seen on the mainland in 1924 (Minno & Emmel, 1993). It was later rediscovered on Lower Matecumbe Key, and became a collector's prize, fetching up to $150 a pair (Klots, 1951). It was believed extinct after this colony was destroyed by a hurricane in the 1930s. Hurricanes can affect Schaus' colonies by damaging foliage, including hostplants. Other threats include development, mosquito spraying, harsh droughts, and intense predation.

A different Schaus' population was later discovered on the Upper Keys (mainly Key Largo and Elliott Key). This small population numbered about 1,000. It was declared federally endangered in 1984, before Hurricane Andrew reduced its complement to less than 100. Schaus' rebounded in Biscayne National Park, and has been reintroduced on the Keys and near Miami (Deering Estate), following a captive breeding program. The recent arrival of Fire Ants in Schaus' habitats has raised fresh concerns for the butterfly's well-being (Forys et al., 2001).

Male Schaus' Swallowtails patrol hammock edges and trails for mates, most actively in the morning and late afternoon. Females inspect the hammock interior seeking suitable hostplants. Both sexes nectar infrequently. Schaus' are not migratory, but do "island hop" in the Upper Keys.

Identification	Large (but less so than the Giant, which is similar and occurs in the Schaus' habitat). *Male's antennal clubs yellow; female's black.* Two *straight, yellow-edged tails* (Giant's are yellow-centered and rounder); tails *angled away from the body.*
Dorsal	Dark brown to blackish. *Two yellow bands* cross wings, *converging near the FW apex;* yellow spots extend from upper line to FW costa (as in Giant). Red spot on inner HW, slight bluish clouding.
Ventral	Yellow. Note *large, chestnut-rust HW patch* (much smaller in Giant).
Habitat	Rarely leaves dry, tropical hardwood hammocks or surrounding scrub. Second-growth wooded habitat is utilized.
Hostplants	Rue Family (Rutaceae), primarily Torchwood (*Amyris elemifera*) and Wild Lime (*Zanthoxylum fagara*), in partially shaded settings. Torchwood has toxic, resinous sap that repels termites. Eggs laid on leaves of young plants, usually within 6 feet of the ground (Rutkowski, 1971). Caterpillars feed on leaves. Osmeterium is white.
Occurrence	Endangered, in extreme s. FL and Upper Keys. Also in Bahamas, Greater Antilles. Single-brooded, flies late Apr to mid-Jun. But an apparent partial 2nd brood is occasionally seen in Aug or early Sep. Pupae hibernate, can remain dormant for several years during droughts (Opler & Krizek, 1984).
Ecology	Specialist. Lives exclusively in a specialized, locally endangered habitat.

Hardwood hammock, Key Largo, Monroe Co., FL
Inset: Torchwood (*Amyris elemifera*)

av wingspan: 3.9"

Palamedes Swallowtail   *Papilio palamedes*

Dorsal male: 5/25/96, Gr. Dismal Swamp NWR, Nasemond Co., VA

Ventral: 5/25/96, Gr. Dismal Swamp NWR, Nasemond Co., VA

A signature butterfly of southern cypress swamps (Pyle, 1981), this large swallowtail creates a conspicuous flourish as it moves through rich, wooded wetlands. (It also inhabits several other, less dramatic environments.) Few lepidopteran behaviors can equal the Palamedes' graceful courtship flight, in which the male and female flutter along slowly in unison, hovering a foot or so apart, their broad wings beating quickly but shallowly. The male repeatedly lofts above her from behind to disperse pheromones.

The Palamedes belongs to the Tiger-Spicebush Group (which some place in the separate genus *Pterourus*). A laurel-feeder, it appears most closely related to the Spicebush. The nominate race is found throughout the Palamedes' U.S. range. An additional Mexican race (*leontis*) has smaller yellow markings.

Like others in this group, the Palamedes appears palatable to birds (Brower, 1985). It is not considered a Pipevine mimic, although its ventral markings are vaguely similar. And even limited similarity may be of some use at communal roosts, which are sometimes reported (Scott, 1986).

Male Palamedes patrol in wood openings and along edges for females. Flight is comparatively slow and deliberate. Newly emerged males actively seek minerals, and both sexes nectar regularly. Occasionally strays north, although this species is by no means a regular migrant.

Identification	Very large, with round, full-cut FWs (the leading edge appears almost circular). *FW tips bow downward* (ventrally) when perched. The HW tails are *thin and twisted*, so that their yellow centers may not be visible from above. The body is *striped* black-and-yellow lengthwise. Sexes are similar.
Dorsal	*Blackish-brown* (becoming browner with wear). *Yellow spot bands*, submarginal and postmedian, on both wings, plus a *yellow FW cell-end bar*. Note the single, blue spot on the inner HW, where the postmedian lines meet. Some variable, greenish-yellow dusting on outer HW.
Ventral	Dark, with two *pale spot bands* along the FW (mirroring the dorsal surface). HW has a series of bands—*orange, pale cream, and bluish*. Note the *thin, yellow line along inner edge of HW*.

Habitat	Southern swamplands (cypress/coastal), wet riparian woods, pocosins, bay forests, savannas. Also, less abundantly, in sandhills.
Hostplants	Red Bay (*Persea borbonia*). Accepts other laurels in the lab, but geographic range conforms closely to that of its main host. Sweet Bay (*Magnolia virginiana*), unrelated to Red Bay, is toxic to larvae. Caterpillars feed from a silk shelter.
Occurrence	Locally common, s. MD to FL (but not resident on the Keys). Old, isolated records from NY, NJ. Also along Gulf Coast to central MEX. Two or three long broods with few gaps, late May-Oct in VA, Mar-Dec in FL. Pupa overwinters.
Ecology	Medium Specialist. A hostplant and (to a lesser degree) habitat specialist.

Redbay (*Persea borbonia*), with characteristic leaf galls

av wingspan: 4.65″

The Pierids (Whites and Sulphurs): Family Pieridae

Typical larva (Cloudless Sulphur)

To many observers, the pierids are "typical" or "regular" butterflies. Their bright, simple colors and energetic flight style probably first inspired the English name "butterfly"—which some believe refers to the buttery-yellow color of a sulphur in flight; others suggest it is an early spoonerism on "flutter by." In any case, there are 1,000–1,200 species of pierid worldwide. Most are tropical, but quite a few live in the Arctic. About 60 species occur in the United States and Canada, 27 in our area. Most of our species are resident (22), the others are strays (5).

The popular image of pierids, as energetic and cheerful, is consistent with certain aspects of the family's biology. Males patrol unflaggingly for mates throughout the day. Their rounded wings produce a distinctive flight style that is active and lively without being overly choppy. Pierids frequent open habitats (where they often encounter humans). No other family in our area has such a high percentage of generalists, feeding on herbaceous hosts to achieve rapid growth. Most of our species are common, and several are abundant. A few pierids are single-brooded, but most have long, overlapping flights. It is a rare day in the field that one cannot find a Cabbage White or Orange Sulphur.

Pierids lay tall, spindly eggs on hostplant leaves, usually singly. Caterpillars are long and cylindrical, with short hairs that may exude chemicals. Voracious eaters, they can become agricultural pests. The pupa in most species has a pointed "head" and is connected to its substrate by a silk girdle, or "tie line," which supplements the cremaster.

Pierid wing pigments (called pterines—mainly yellow, orange, or white) are derived from stored uric acid. In many species, the dorsal wing surface reflects ultraviolet light. We note these reflectance patterns in individual species accounts: although invisible to humans, they are important in pierid courtship behavior and mate recognition. Seasonal and sexual dimorphism is well-developed among pierids. Many migrate or emigrate during the summer, and their wing pigments often serve thermoregulatory functions.

The Whites (subfamily Pierinae)

The whites (Pierinae) are a widespread subfamily, consisting mainly of medium-sized butterflies. There are about 700 species worldwide, including 22 in the United States and Canada and 8 in the East Coast region. True to their name, members of this subfamily are mostly white, often accented with black wing borders and various, isolated dark markings. The ventral hindwing sometimes has greenish "marbling" or a pale, yellowish wash. In one group, the forewing tips are bright orange. Most whites dorsal-bask, angling their wings so as to reflect solar warmth onto their bodies.

Larval whites feed mainly on mustards ("crucifers"), including some cultivated species such as cabbages and turnips. East Coast species consume plants in several related families (Cruciferae, Capparidaceae, etc.), but all contain glucosinolates, or "mustard oils" (Scott, 1986), which may render adults and caterpillars at least partially unpalatable in some species. Most whites overwinter in the pupal stage.

Two tribes (formerly considered subfamilies) occur in our region:

Typical Whites (Tribe Pierini): Our "everyday" whites belong to this tribe. As a group, they are multibrooded, open-area generalists, noted for their geographic dispersal and periodic, seasonal emigrations. (The Mustard and West Virginia Whites deviate somewhat from this general schema, as discussed separately.) Many female whites reflect ultraviolet light dorsally, allowing males to recognize them during courtship. Among sulphurs, it is the *males* that reflect UV (Silberglied & Taylor, 1978). The caterpillars of Cabbage Whites secrete noxious fluids, called mayolenes, that deter aggressive ants and probably other invertebrate predators (Smedley et al., 2002). This assists them in targeting nutritious hostplants that grow in exposed, weedy environments.

Marbles & Orangetips (Tribe Euchloini): Two species in our area. Both are single-brooded barrens specialists, in sharp contrast with the Pierini. Their ventral surfaces are marbled with an irregular, greenish cobweb pattern (produced by a mixture of black and yellow wing scales). Caterpillars feed on the flowers of small mustards, mainly cresses. Males rarely seek minerals.

Florida White *Appias drusilla*

Ventral male: 6/23/98, Matheson Hammock, Miami-Dade Co., FL

Ventral female: 3/24/93, Matheson Hammock, Miami-Dade Co., FL.

Widespread in the neotropics, this sleek butterfly illustrates the many versatile uses of the color white in butterfly wing decoration. The Florida White's wing scales range from lustrous white to pearly to glossy to off-white to plain, flat white. These subtleties are not always visible while the butterfly remains in the dim interior of a subtropical hammock (which it usually does). But on occasion a Florida White ventures into the open, usually to seek nectar, at which point its full, quietly stated splendor becomes evident in full light.

This is our only local member of the genus *Appias*, seemingly a very old clan (Smith et al., 1994). The Florida race (*neumoegenil*) is named for Berthold Neumogen, a mid-19th-century New York banker and amateur tropical lepidopterist (Opler & Krizek, 1984). Taxonomists distinguish between the nominate, continental race, found from Texas south to Brazil, and a series of interrelated West Indian subspecies (including *neumoegenil*).

Male Florida Whites patrol for females within tropical hammocks, moving quickly from clearing to clearing. When not patrolling, they may seek nectar in the canopy and canopy gaps. Females patrol more slowly within the hammock, looking for hostplants (Minno & Emmel, 1993). It has been stated that males do not seek minerals in Florida, but they reportedly do so quite regularly in Costa Rica (DeVries, 1987), and in Hispaniola they have been observed "like a shimmering white dome" atop piles of horse and cow dung (Smith et al., 1994, p. 133).

Florida Whites engage in dry-season emigrations in the tropics. Lacking cold tolerance, they have little opportunity to move northward from Florida. Yet summer strays do reach Gainesville, and historically (when the species was more common) there were scattered records from New York State, Maryland, and Massachusetts.

Identification	Medium-sized. FW is pointed, with a *slightly concave outer edge*, especially in male. *Antennal clubs white*. Light individuals resemble Great Southern Whites. Female has two variable seasonal forms, with many intermediates.
Dorsal	Lustrous/silky sheen at base of both wings. Male *all-white* except for *black FW costal edge*. *Dry-season* female similar, but usually has *darkened FW apex*, with *no inward extensions* along the wing veins. *Wet-season* female is *black along inner FW costal margin*; *HW* is *yellow-orange*.
Ventral	Note *yellow scaling* on *basal FW* (absent in Great Southern). Both sexes *glossy white*, with faint yellowish wash on HW.

Habitat	Interior of subtropical hammocks, also in canopy. Occasionally leaves to nectar or emigrate. Riparian second growth in tropics.
Hostplants	Capers and spurges, including Limber Caper (*Capparis flexuosa*), Milk-bark (*Drypetes diversifolia*), Guiana Plum (*D. lateriflora*). Crucifers are reported in Costa Rica (DeVries, 1987). Females lay in clusters of 2–3 eggs on young leaves. Caterpillars feed at night and on cloudy days, develop quickly, and are cannibalistic.
Occurrence	Sometimes common, but cyclical. Declining with habitat loss. S. FL and Keys, also W.I., s. TX, and south to BRA. Three or more broods, including a long, adult reproductive diapause generation during winter.
Ecology	Specialist. Climate- and habitat-limited in our area. Retreats with development, though can survive in some remnant, urban hammocks.

Roger Hammer

Limber Caper (*Capparis flexuosa*)

av wingspan: 2.1"

Checkered White *Pontia protodice*

Dorsal male: 5/3/96, Owaissa Bauer Park, Miami-Dade Co., FL

Dorsal female: 9/4/94, Jamaica Bay NWR, Queens, NY

Common in the West, the Checkered White is something of a puzzle in our region. Most commentators believe it has declined here during the last half-century. But some suggest it is an irruptive colonist, whose past historic occurrence was overstated as a result of occasional, periodic invasions. Both opinions may have some merit.

The Checkered White is a southern resident, but it forms sizeable, temporary colonies that occasionally reach Canada, and which can endure for several years (providing winters remain mild). To our knowledge, however, Checkereds are seldom regularly common almost anywhere in the East.

If the Checkered White has declined, it is likely from intense development or other causes, rather than from com-

Ventral male: 5/3/96, Owaissa Bauer Park, Miami-Dade Co., FL

petition with Cabbage Whites, as some have suggested. The two species often fly together. They use different primary hostplants and feed on different hostplant parts (mainly flowers vs. leaves). Direct competition seems unlikely.

Male Checkered Whites patrol quickly throughout the day, seeking receptive females. Mating typically ensues quickly once courtship is initiated (Rutowski, 1980). Females reflect UV light moderately; males key in on individuals of intermediate reflectiveness (Scott, 1986). Both sexes nectar on a wide variety of flowers. Young males seek minerals. Females disperse when courted too persistently by amorous males in dense colonies (Ehrlich, 1985).

Identification	Medium-sized. Individually variable and strongly sexually dimorphic. FW more pointed than in *Pieris* whites. Males resemble Cabbage Whites in flight, and should be allowed to settle before identifying them (females often can be identified on the wing).
Dorsal	Male *all-white* except for *dark FW spots* and *vein-end markings* at FW apex. Female *more heavily* marked, with *diffuse, dark HW markings* (white in male/other species).
Ventral	Male white with *vague HW pattern*; female has *yellowish-green* markings on HW and FW apex.
Habitat	Dry, weedy, waste areas: fields, beach edges, old railroad beds, etc. Will not enter woods even if frightened.
Hostplants	Almost any crucifer; often Shepherd's Purse (*Capsella bursa-pastoris*) and Virginia Peppergrass (*Lepidium virginicum*); also cultivated mustards. Caterpillars eat buds, flowers, fruits, sometimes leaves.
Occurrence	Usually rare and sporadic. N. Eng. south to FL; also w. US, s. CAN, and n. MEX. Multiple flights in the Deep South, Feb-Nov, variable northward. Pupa hibernates.
Ecology	Generalist. A nomadic opportunist.

Ventral female: 5/3/96, Owaissa Bauer Park, Miami-Dade Co., FL

av wingspan: 1.6"

"Waste" Areas:
An Overlooked Resource

Most people can readily appreciate the value of wilderness, including those who actively study nature as well as many who do not. But the same cannot be said of another local habitat type: the waste area. When exploring these bereft-seeming locations, a natural history enthusiast is generally on his or her own. Curious glances are likely from passers-by, as are puzzled questions, such as "Can I help you find something?"

By "waste area" we mean a habitat substantially impaired by human activity (more so than a field or pasture, which retains some natural character). A waste area must have been left unattended for long enough that feral vegetation has become established. As the only quasi-natural oasis in many heavily developed urban ecosystems, waste areas play an important role. They provide important emergency stopovers for varied strays and migrants. And, in addition, they host an interesting community of dedicated plants and animals that capitalize on disruption, forming characteristic local assemblages.

We are unaware of any reliable statistics concerning the extent of waste areas in our region. But everyday experience suggests that the total is large and rapidly growing. From a conservation standpoint, waste areas support active natural communities, sometimes including unusual species, although most are common. They take comparatively little effort to maintain—usually nothing more than restraint in mowing and the avoidance of pesticides.

Some of the generalist butterfly species that frequent these habitats, including Common Sootywing, Phaon Crescent, and Checkered White, may be considered "waste area specialists." Each occurs in various natural settings of its own, yet all show a strong affinity for disturbed or abandoned sites.

Waste areas have certain potential advantages. Newly disturbed sites, for instance, may produce a lush growth of herbaceous hostplants before being discovered by the usual cadre of predators or parasitoids. (And in urban settings certain predators may have difficulty becoming established at all.)

But "mature" waste areas are a challenging home for butterflies. Ants and other predators abound, providing incentive for the development of specialized chemical defenses (Smedley et al., 2002). And waste area plants themselves are often well-defended. Crucifers synthesize mustard oils (glucosinolates) to deter herbivores, often stepping up production when attacked (Agrawal et al., 2002). This does not deter Pierini larvae, however, which can detoxify glucosinolates. With this advantage, they actively seek out local, waste area mustards. Some plants "fight back," e.g., by acquiring concentrations of toxic, heavy metals: in a recent experimental study, Cabbage Whites avoided leaves of Mustard Greens (*Brassica juncea*) with elevated selenium concentrations—and died if such leaves were eaten (Hanson et al., 2003).

Among butterflies found in "waste areas," some are more at home there than others. Cabbage Whites feed on two open-space weeds (Wintercress and Garlic Mustard) that are not accepted by woodland pierids, such as the Mustard and West Virginia Whites. But Cabbage Whites in turn have difficulty with two hard-core waste area hostplants. Specifically, they develop slowly on Virginia Peppergrass (*Lepidium virginicum*) and refuse Shepherd's Purse (*Capsella bursa-pastoris*) (Scott, 1986). Meanwhile, the Checkered White, a consummate waste area survivalist, actively targets both of these plants.

Virginia Peppergrass (*Lepidium virginicum*), a common weedy crucifer of waste areas

Waste Area: Broad Channel, Queens Co., NY

Great Southern White   *Ascia monuste*

Dorsal female: 4/30/96, Mary Krome Sanctuary, Miami-Dade Co., FL

Ventral (light form): 6/26/93, Matheson Hammock, Miami-Dade Co., FL

Great Southern Whites are an everyday sight along the Florida coast. Yet their lifestyle is anything but ordinary. The species' restless disposition is marked by periodic mass shoreline emigrations in which millions of individuals fly in tight, churning clusters, some 10–20 feet high. They maintain a fixed bearing despite wind and physical obstacles (Nielson, 1961). The trigger for these dramatic irruptions is not specifically known; overcrowding seems to be an element in the process, but may not be its root cause. Emigrant females begin traveling soon after emergence and postpone egg-laying until arrival. Marked individuals have moved as far as 100 miles before recapture. Colonies become scarce north of Georgia for lack of coastal hostplants—though it is not clear why emigrants do not continue north inland, using common hostplants like Virginia Peppergrass. Some southbound emigration is also reported.

Ventral female (dusky "nigra" form): 3/23/02, Port Mayaca, Martin Co., FL

Identification	A bit larger than other East Coast whites, *FW apex rounder than Florida White. Antennal clubs electric blue* in life. Markings are geographically variable, but subspecies poorly defined, perhaps from recurrent gene mixing.
Dorsal	Male all-white, except for black FW margins. Black intrudes inward along the trailing FW veins to form a *"zigzag" pattern*. Female has prominent *black FW cell dot*, also a *variable suffusion of smoky-gray* in the dusky wet-season form (form "nigra," triggered by day length during larval stage).
Ventral	All-white, except *yellowish cast on HW*. Wet-season female has smoky-gray suffusion below, as on upper surface.

Habitat	Mainly coastal except in FL. Often near salt pans or mangroves, but also in weedy, disturbed habitats with nectar. Highly tolerant of human settlement. Inland strays and transient colonies occur in summer and during emigrations, especially in peninsular FL. A disturbed habitat species elsewhere in its wide range.
Hostplants	Uses Saltwort (*Batis maritima*) or Coastal Searocket (*Cakile lanceolata*) on FL coast. Inland, Virginia Peppergrass (*Lepidium virginicum*) and numerous other capers and crucifers. Sometimes a pest on cabbage, radishes, etc., especially in Caribbean. Hostplants contain glycosides that make adults semi-unpalatable. Young caterpillars prefer new leaves, but first devour their own eggshells (*chorions*), providing important nutrition (Barros-Bellanda & Zucoloto, 2001). Eggs often laid in clusters, can withstand saltwater.
Occurrence	Common to hyperabundant on FL coast, less regular inland. Intermittent summer colonies in coastal GA and SC. Strays north to e. MD. Also widespread in neotropics, north to TX and sw US. Permanent range expands northward in warm years, cut back by harsh winters. One relatively long adult generation overwinters in "dry season diapause."
Ecology	**Medium generalist.** Though populations in our area tend to specialize in coastal habitats with succulent crucifers such as Saltwort, the species overall has many generalist features, including a transient lifestyle and opportunistic use of hostplants.

Roger Hammer

Saltwort (*Batis maritima*)

av wingspan: 2.25"

Cabbage White *Pieris rapae*

Dorsal female: 6/24/00, Suffolk Co., NY

Ventral: 7/4/98, Morris Co., NJ

By nearly any standard, this lively pierid is among the world's most successful and prolific butterflies. Native to Eurasia and North Africa, it has spread with agriculture across temperate portions of the Northern Hemisphere, and also is established in Australia and New Zealand. Cabbage Whites first arrived in North America in the 1860s (in Quebec), and immediately spread. By 1889 they were "painfully common" in Philadelphia (Gochfeld & Burger, 1997, p. 133), and by 1900 had spanned the entire continent.

Ecologically, the Cabbage White is a model generalist, frequenting open, disturbed habitats, where it feeds on a large variety of fast-growing, nutritious crucifers. (Some of its hosts are agricultural crops, specifically bred for fast growth and nutrition, and in these cases the larvae can become serious pests.) When ovipositing, females detect plants with high levels of photosynthesis (Langan et al., 2001). If denied adequate dietary nitrogen, weakened larvae suffer increased mortality (Loader & Damman, 1991). Cabbage Whites favor "pioneer" habitats not yet targeted by predators or parasitoids (Ohsaki & Sato, 1994). They disperse instinctively, without being triggered by high density or starvation (Ehrlich, 1985; Emmel, 1973). Open-space devotees, Cabbages will ignore a favored hostplant when introduced experimentally in a woodlot (Cromartie, 1975). Larvae secrete specialized chemicals ("mayolenes") from body hairs that deter marauding ants, and probably other invertebrate predators found in open places (Smedley et al., 2002). They apparently do not sequester mustard oils (glucosinolates)—or at least not from certain hostplants—and thus can be vulnerable to vertebrate predation (Vlieger et al., 2004).

Cabbage Whites are "facultative diapausers," meaning they do not begin to overwinter until temperatures fall, regardless of day length (Scott, 1986, p. 28). Males patrol for females most actively in the morning (Hirota & Obara, 2000), with a fast, bounding flight. Anti-aphrodisiac pheromones are included in the spermatophore, which deter subsequent males from courting (this also allows the female to oviposit unmolested) (Andersson et al., 2000).

Identification	Medium-sized, *smaller in spring*. Some spring males are almost pure white, and must be distinguished with care from Mustard Whites (the latter has *dark ventral wing veins* in the spring brood).
Dorsal	White with *dark FW tips* and *central FW spots* (one in male, two in female).
Ventral	FW tips and HW *yellowish* or *grayish-green*; mostly unmarked, but may show scattered dark flecking.

Habitat	Almost any open, sunny habitat, including open woods in spring.
Hostplants	Crucifers, native and cultivated, including some (such as Garlic Mustard) that are toxic to Mustard and West Virginia Whites. But Cabbages fare poorly on peppergrass and Shepherd's Purse, favored by Checkered Whites. Some populations are genetically predisposed to certain hosts (Hovanitz & Chang, 1963). Single eggs are laid beneath host leaves.
Occurrence	Abundant, especially late in the season. N. Eng. south to FL (rare in far s. FL and Keys). Also west across US and MEX. Multiple broods, Feb or Mar-Dec, depending on latitude; earliest pupal diapauser to emerge in spring; flies in continuous, overlapping broods until first hard frost.
Ecology	Generalist. A generalist in virtually every way.

Cultivated cabbage (*Brassica oleracea*)

av wingspan: 1.7"

Mustard White *Pieris napi* [= *oleracea*]

Dorsal: 6/15/97, Wentworth Location, Coos Co., NH

Ventral (dark-veined spring form): 6/12/93, Wentworth Location, Coos Co., NH

The plain appearance of this northern white is deceptive. Its lifestyle, reproductive biology, and taxonomy are all actively studied by scientists. Experts disagree whether it is conspecific with the Green-veined White of Europe (*P. napi*), or a separate species (*P. oleracea*) (Geiger & Shapiro, 1992), and whether western populations constitute separate, sibling entities.

Like other native Pierini, the Mustard White has declined in our area since the mid-1800s—withdrawing in particular from open habitats, where it was formerly common (and where the European *P. napi* still is). Some have blamed the decline on competition from Cabbage Whites, which arrived here in the mid-1800s, or on loss of forests, the current primary habitat. We are inclined to associate the Mustard White's decline with the proliferation of two alien mustards, Wintercress (or Garden Yel-

Ventral (light-veined summer form): 7/19/99, Scott Bog, Coos Co., NH

lowrocket) and Garlic Mustard. Both attract ovipositing females, but neither supports the development of Mustard White caterpillars (Haribal & Renwick, 2001; Scott, 1986). Cabbage Whites, by comparison, can develop on both plants. Two chemicals in Garlic Mustard have been implicated in thwarting larval growth. Both vary seasonally in their production, with a combined ebb in mid-June that apparently allows limited use by Mustard Whites (Haribal & Renwick, 2001; Courant et al., 1994). Since the European *P. napi* feeds successfully on Garlic Mustard, our Mustard Whites may eventually adapt.

Males patrol throughout the day for females, mating most often around midday. Males produce an aphrodisiac to elicit female receptiveness, and also an *anti*aphrodisiac, chemically similar to aspirin, that deters future males from courting (Andersson et al., 2000). The male's "nuptial gift" to the female, transferred with the spermatophore, is enormous, up to a quarter of the female's body weight (Wiklund, 2003).

Identification *Dorsal* *Ventral*	Medium-sized. Wings thin and fragile-seeming, more angular than in West Virginia White. Body dark black above. Spring brood linked to day length. White with some darkening on FW apex in spring. Rarely a faint FW spot. Prominent, *dark vein lining* in spring; *faint vein lining, or none at all*, in summer brood.
Habitat	Mainly deciduous and transition wood/edges, bogs, streamsides, old fields.
Hostplants	Various mustards (*Arabis, Cardamine, Barbarea orthoceras*; also cabbage, turnips, radishes, etc.), differs with brood. Eggs laid singly beneath hostplant leaf, caterpillars bore into leaf.
Occurrence	Still locally fairly common in n. N. Eng. and n. NY. Also westward, where several sibling species may occur. Two or three flights, May to mid-Jun and early Jul to mid-Sep in our area. Pupa overwinters.
Ecology	Medium Generalist. Remaining populations confined to a limited habitat zone, but still more mobile and adaptable than West Virginia White.

Cut-leaved Toothwort (*Cardamine concatenata*)

av wingspan: 1.6″

West Virginia White *Pieris virginiensis*

Dorsal: 5/4/02, Clay Co., NC

Ventral (Appalachian, light): 4/23/94, Fork Creek WMA, Boon Co., WV

A fragile-seeming resident of Appalachian deciduous forests, the West Virginia White was formerly considered a race of Mustard White (Klots, 1951; Hovanitz, 1963). But it is now regarded as a single-brooded, woodland pierid that specializes on ephemeral, spring mustards (especially toothworts).

No sooner was the West Virginia White recognized as a separate species than it began a precipitous decline in the Northeast. It is now rare or absent at many former colony sites. Loss or fragmentation of woodland habitat is one factor in this decline (e.g., Stanton, 2001), since West Virginia Whites are poor colonizers. And the early die-back of hostplants in dry years can starve entire colonies, as in 1982 (Cappuccino & Kareiva, 1985). Viruses and Gypsy Moth spraying likewise have been implicated.

Ventral (northern): 6/8/97, Mt. Greylock, Berkshire Co., MA

But there is another, very specific culprit in this case: Garlic Mustard (*Alliaria petiolata* [= *officinalis*]), a European crucifer that has slowly but persistently invaded the East Coast and central-western states since introduced on Long Island in the 1860s. Female West Virginia Whites oviposit readily on this plant (Porter, 1994), but their caterpillars fail to survive eating it (Courant et al., 1994). Thick stands of Garlic Mustard grow in woods and along edges in Connecticut, where West Virginia Whites no longer exist (this is generally true in southwestern Connecticut and areas near New York City).

Male West Virginia Whites patrol slowly through open, spring woods in search of females, usually remaining within a few feet of the forest floor. Mating occurs from noon through mid-afternoon. Young males seek minerals.

Identification	Medium-sized. The rounded wings are papery thin. Body dark black above. A very rare, partial summer brood is difficult to distinguish from the summer Mustard White (Courant & Chew, 1995).
Dorsal	Nearly all-white, except for some basal and costal FW gray scaling.
Ventral	Whitish or cream-colored, with *no yellow wash*. HW veins are *faintly outlined with pale, smoky-gray scales* (in North) or almost pure white (southward). Never shows the very dark veining of Mustard White.

Habitat	Rich, moist deciduous woods, mixed woods, cove forests, shale barrens. Seldom in open habitats.
Hostplants	Mainly Crinkleroot (*Cardamine* [= *Dentaria*] *diphylla*), Cut-leaved Toothwort (*C. concatenata* [= *laciniata*]). Eggs laid singly on underside of host leaf.
Occurrence	Common only in central Appalachians. N. Eng. south to n. GA (well inland). Also Gr. Lakes. Normally one flight, late Mar to mid-May in South, mid-Apr to mid-Jun in North. Pupa formed in mid-summer, overwinters.
Ecology	Medium Specialist. Given its recent decline, this species requires careful monitoring and management.

Garlic Mustard (*Alliaria petiolata*), a toxic decoy

av wingspan: 1.55"

Falcate Orangetip   *Anthocharis midea*

Dorsal male: 4/18/98, Assunpink WMA, Monmouth Co., NJ

Dorsal female: 4/18/98, Assunpink WMA, Monmouth Co., NJ

Easy to identify, but difficult to observe at rest, the male Falcate Orangetip is a study in perpetual motion. It emerges as much as a week before the female, and patrols seemingly without pause during early spring, until a mating opportunity arises.

This is our only East Coast orangetip, yet its taxonomy is complicated. It has been assigned at times to four different genera and has borne two different species names. Dos Passos and Klots (1969) realigned the Falcate's races, designating one general eastern race (*annickae*), along with the southeastern *midea*; the latter is restricted to the immediate coast (Gatrelle, 1998a).

The Falcate Orangetip flies low and actively, but is fluttery and makes headway slowly. Males retrace fixed routes during the day, whereas females appear to wander more at random

Ventral: 5/8/92, Hook Mountain, Rockland Co., NY

(Opler & Krizek, 1984). Courting males may swarm around an available female. Mating and oviposition begin in the morning, and may persist well into the afternoon. Adults appear to be moderately unpalatable to avian predators.

Identification	Small-medium. Hooked (or "falcate") FW tip is conspicuous. FW edges checkered, especially in female. Female can be confused with other whites in flight, especially the Olympia Marble.
Dorsal	Male is unmistakable, white with *bright orange FW tips* (more extensive in SE *midea* race, where orange reaches FW cell-end dot). Female lacks orange FW tips, but often has some yellowish apical FW wash. The HW apex of male *midea* often has a yellowish wash.
Ventral	*Finely marbled*, in dark greenish pattern more diffuse than in Olympia.

Habitat	Bluffs, rocky outcrops, shale barrens; also sandhills, pine barren edges, sandy coastal woods; and swamps, open woods near waterways.
Hostplants	Various mustards (Brassicaceae), Hairy and Small-flowered Bittercress (*Cardamine hirsuta* & *parviflora*); Smooth, Lyre-leaved, Purple and Shale Barrens Rockcress (*Arabis laevigata, lyrata, divaricarpa,* & *serotina*); etc. Eggs laid singly. Caterpillars eat flowers, buds, seeds, mostly at night; pupate by June.
Occurrence	Locally fairly common, s. N. Eng. south to GA, isolated population in FL panhandle. Uncommon/absent in mountains. Also west to WI, NE, and e. TX. One extended brood (perhaps two rarely). Feb-Mar in FL, mid-Apr to mid-May in VA, late Apr to Jun in NY. Pupa overwinters, sometimes diapausing a full year before emerging (an adaptation to drought).
Ecology	**Medium Generalist**. Often local and "picky" in its requirements, but uses a substantial range of hosts in a varied set of habitats.

Hairy Bittercress (*Cardamine hirsuta*)

av wingspan: 1.45"

Olympia Marble *Euchloe olympia*

Dorsal: 5/17/98, Limerick Barrens, Jefferson Co., NY

Ventral (on hostplant): 5/13/00, Limerick Barrens, Jefferson Co., NY

In our region, the Olympia Marble is an early spring barrens specialist. It is found on shale barrens, in the central Appalachians, and on limestone barrens (or alvars) in western New York. It is rare and local even within its limited eastern range, however, and like other biological specialists it has come under increased pressure in recent years. Gypsy Moth spraying has been cited as a likely cause of its sharp recent decline in the Appalachians (Parshall, 2002). But we should not rule out the effects of acid rain, nor of deer browsing on key hostplants such as Shale Barren Rockcress (in West Virginia) or Purple Rockcress (in New York). Both of these plants are endangered, even more so than the butterfly itself. Fortunately, the Olympia's modest appetite poses no discernible threat to either of these hosts.

Mainly a butterfly of midwestern prairies, the Olympia Marble was described by William Henry Edwards in 1871 from the disjunct, West Virginia population. There are no defined subspecies, but Eastern Olympias tend to show less of a rosy blush at the base of their hindwings than western ones, and to have better-defined markings.

The Olympia's translucent wings are thin and delicate, almost papery. But you would not know this to see one fly. Males patrol for females close to the ground in a steady, very direct flight that "leaves one with the impression that this species knows where it is going" (Iftner et al., 1992, p. 76). Among the more strenuous exertions encountered in the production of this book was dashing at full speed after a patrolling Olympia Marble as it traversed a rutted alvar in western New York State—trying to remain nearby in order to catch a photo on one of the rare occasions when it paused to nectar (and also trying, all the while, not to break an ankle).

Olympias have not been observed puddling, but they nectar regularly, often on distinctive local flowers such as Birdsfoot Violet and Hoary Puccoon.

Identification	Small to medium. FW elongated (but not falcate). Wings thin and translucent. Appears all-white in flight; difficult to distinguish from female Falcate Orangetip in the air.
Dorsal	*Chalky white*, with black basal scaling. FW tip and subapical markings are gray (both fade with wear). Note *dark, mid-costal dash*.
Ventral	*Discrete greenish-yellow "marbling"* (more defined than in Falcate Orangetip). Living butterflies often have a *rosy wash* at HW base, both above and below (generally more pronounced in West).

Habitat	Mainly shale and limestone barrens in the East. Elsewhere dunes, prairies, foothills, open woodlands, meadows.
Hostplants	Small rockcresses (Brassicaceae), including Lyre-leaved (*Arabis lyrata*), Purple (*A. divaricarpa*), Smooth (*A. laevigata*), and Shale Barren (*A. serotina*). Eggs laid singly. Larvae eat buds, flowers, sometimes leaves.
Occurrence	Local and uncommon, central Appalachians and nw NY. Also Gr. Lakes and Gr. Plains, s. CAN. One flight, Apr-May in South, mid-May to Jun in North. Pupa hibernates, may not emerge for several years during droughts (Opler & Krizek, 1984).
Ecology	Specialist. A niche player, needs careful management.

Purple Rock-cress (*Arabis divaricarpa*)

av wingspan: 1.55″

The Sulphurs (subfamily Coliadinae)

The sulphurs (subfamily Coliadinae) were once reportedly called "red horns" because of their colorful antennae (Weed, 1917), even though the antennae are in fact usually pink rather than red. There are about 300 species world-wide, mostly found in the tropics. Approximately 37 species occur in North America and Canada, and 19 in our area (including five that have been recorded only as strays, some very rarely).

A number of our most abundant and conspicuous butterflies belong to this subfamily. They are distinguishable from the whites (subfamily Pierinae) in several basic features, such as color and diet. In fact, they have a number of diametrically opposite traits that act to reinforce the groups' evolutionary separation, by allowing subfamily members to coexist alongside each other without competing for hostplants or mistakenly interbreeding.

With some exceptions, sulphurs are medium-sized. Most are yellow, orange, or cream-white, with black wing edges and other dark markings. They have characteristic wing-vein patterns and shorter antennae (on average) than the whites—though neither of these points is a reliable mark in the field. Most sulphurs feed on legumes, which support rapid growth. Their hosts provide few protective chemicals, however, so most are palatable.

Four subgroups of sulphurs are recognized in our area:

Common Sulphurs (*Colias/Zerene*): These medium-sized sulphurs include many wide-ranging generalists, such as Clouded and Orange Sulphurs. But there are also specialist subgroups within the genus, whose lifestyles are adapted to "niche" environments. The Pink-edged Sulphur is our only species in this category.

Giant Sulphurs (*Aphrissa/Phoebis*): This New World group is strongly migratory/emigratory, often flying over large stretches of open water. Tropical in origin, they have little cold tolerance.

Small Sulphurs (*Eurema*/etc.): Several of these tropical species occur in southern Florida only. Others range well north along the East Coast in summer, but none can successfully overwinter in the North.

Dainty Sulphur (*Nathalis*): One species. The Dainty is often lumped with other small sulphurs, but is distinctive and arguably could be assigned to a separate subfamily.

Most sulphurs produce distinctive seasonal and sexual forms, which play an important adaptive role in courtship, interspecific recognition, and temperature management. But some variations in appearance also arise because *Colias* sulphurs (particularly Clouded and Orange) regularly hybridize with their close relatives. At times, it can be quite difficult to sort out the various dimorphisms, diphenisms, and hybridizations.

Many male sulphurs reflect ultraviolet light, a trait linked to individual recognition and courtship behavior (Silberglied & Taylor, 1978). Females are non-reflective but have more pronounced dorsal wing markings in the visible range.

Almost all northern-dwelling sulphurs have dark scales on their basal hindwings (the Pink-edged is a peculiar exception, *see* species account). This "melanization" aids in thermoregulation, by allowing the hindwing to absorb heat more quickly while ventral basking (i.e., with wings closed). Melanization extends a dark-form individual's critical flight period during the day, enhancing fecundity (Kingsolver & Watt, 1983). More southerly species often produce dark-form phenotypes early and late in the season, triggered mainly by day length in the late larval stage. In mid-summer, these same species produce a lighter-scaled, and more heat-tolerant, phenotype.

Most female sulphurs have white ("alba") forms, which are usually most prevalent either northward or in the spring and fall. The northern bias of this phenotype runs contrary to what we might expect from a purely thermoregulatory standpoint. But it has been shown that white female sulphurs divert nitrogen from yellow pigment production into more rapid growth and the creation of extra body fat and larger eggs (Scott, 1986; Opler & Krizek, 1984). It also has been suggested that female white sulphurs may be mimics of "true" whites (Pierinae), whose mustard oil diet confers some level of unpalatability, but if this is so, it is not clear why albinism varies by latitude and season.

Seasonal phenotype: Dorsal female Orange Sulphur (fall "semialba" form)

Seasonal phenotype: Dorsal male Orange Sulphur (spring)

Orange Sulphur *Colias eurytheme*

Dorsal male (left), **female** (right); **specimen photos**

Ventral: 8/21/92, Blackwater NWR, Somerset Co., MD

Among our most familiar butterflies, the Orange Sulphur was a rare stray on the East Coast until around 1920, when it radiated eastward with the cultivation of its preferred hostplant, Alfalfa (Klots, 1951). It reached Virginia in 1925, New York around 1930, and Maine in the 1940s (Roble et al., 2000; Gochfeld & Burger, 1997).

The Orange Sulphur has a small, spring phenotype with dark basal hindwings and reduced orange patches on the dorsal wings (Hoffman, 1973). The main importance of this phenotype appears to be thermoregulatory (Watt, 1969). On the other hand, the white-form female ("alba") appears to be selected on other adaptive criteria that outweigh thermoregulation (*see* essay, facing page; Scott, 1986).

Ventral (white form): 5/23/99, Macon Co., NC

Male Orange Sulphurs patrol in low flight throughout the day for females—who recognize them by the UV reflections from their forewings. Pheromone exchange is required to finalize the arrangement, however. Consummate generalists, Orange Sulphurs use a series of nectar plants and seek out disturbed sites to colonize.

Identification	Medium-sized (slightly larger than Clouded). Highly variable. Individuals in Southeast are often reportedly less orange, causing potential confusion with the Clouded. Wings pink-edged.
Dorsal	Males *bright yellowish-orange* (very rarely plain yellow) with *well-defined dark borders on both wings*. Female orange (or white in "alba" form), with *broader, rough-edged wing borders* and *light oval patches* within. Small, dark FW central spot in both sexes, also red HW central spot, more prominent in female.
Ventral	Yellow to yellow-orange (whitish in "alba" females). Well-defined, red-rimmed, silvery central HW spot, with smaller "satellite" spot above and post-median row of small, dark spots on HW. Small, apical FW spot is within the shadow of the dorsal FW apex border, visible from below (Gatrelle, 2001d).

Habitat	Nearly any open, weedy site. Can be a pest in Alfalfa fields.
Hostplants	Various legumes (Fabaceae), favors Alfalfa (*Medicago sativa*); also White Clover (*Trifolium repens*), White Sweet Clover (*Melilotus alba*), etc. Female lays single eggs on host leaves. Larvae feed mostly at night, first chewing through leaves, then eating down from the tip.
Occurrence	Throughout, but rare in far s. FL/Keys. Also entire continental US, much of CAN, south into MEX. Numbers grow during season. Multiple overlapping broods, continuing as long as warmth permits, Mar-Nov. Pupa overwinters.
Ecology	Generalist. A paradigm of the generalist lifestyle.

Alfalfa (Medicago sativa)

av wingspan: 1.9″

Clouded Sulphur   *Colias philodice*

Dorsal male (left), **female** (right); **specimen photos**

Ventral: 8/3/94, Pondicherry Wildlife Refuge, Coos Co., NH

Like its close relative, the Orange Sulphur, this species was formerly a regional butterfly, based in the North and East (vs. South and West for the Orange). The Clouded Sulphur originally was common on the East Coast, "rising in swarms from muddy roads, or dancing by hundreds over clover fields" (Klots, 1951, p. 185). But with the advent of widespread forage plant cultivation across the country (including many legumes that both species savor), the two sulphurs' ranges have effectively merged, creating a turbid, hybridizing complex that delights geneticists—while confounding field observers. Clouded Sulphurs have lost ground steadily since the Orange Sulphur's arrival, probably in substantial part due to genetic competition.

Cloudeds have an early-/late-season phenotype, similar to that of the Orange Sulphur, with increased hindwing melanization. But in the West, the Clouded also has season-long dark-hindwing forms, with melanization varying by altitude, that is genetically controlled (Ellers & Boggs, 2002). High-altitude western males favor yellow females when mating, or else these populations might well break off to form a separate species. For our purposes, it is significant to note that melanization in western Cloudeds is not triggered by seasonal cues, since a day length trigger would not produce stable altitude stratification.

Clouded Sulphurs have a more northerly distribution than Orange Sulphurs. In keeping with this preference, their night-feeding larvae prefer slightly cooler temperatures and their average development cycle is a few days shorter (Scott, 1986). Males do not reflect ultraviolet light, so females must rely entirely on pheromones or other cues during courtship. While many Orange × Clouded hybrids occur, the two generally "breed true" when in contact and are well-accepted as full species. Behavior is generally similar to that of the Orange Sulphur.

Identification	Medium-sized, slightly smaller than Orange. Spring phenotype smaller, has narrower borders, is darker below. White form females ("alba") may be visually indistinguishable ventrally from white Orange females.
Dorsal	*Bright, clear yellow*, without orange. Thin, black wing edges (thicker in female, but encloses *smaller light spots* than in Orange). Note dark central FW spot and reddish central HW spot (more prominent in female).
Ventral	Yellow with *well-defined central HW spot, rimmed in pinkish-red, usually doubled* (but "satellite" spot may be tiny, as in photo above). Note *postmedian median row of dark spots on HW.* Subapical FW spot at edge of the shadow of the dorsal FW apex border, visible from below.

Habitat	Open, weedy sites; favors natural-type habitats more than Orange.
Hostplant	Various legumes (Fabaceae), especially White Clover (*Trifolium repens*); also Red Clover (*T. pratense*), Alfalfa (*Medicago sativa*), many others. Eggs laid singly on leaves, which larvae eat.
Occurrence	Common, but not as abundant as formerly. Throughout our area, except SE coastal plain and FL. Also west across US and CAN to AK, south to C. Amer. Multiple, overlapping broods, Mar-late Nov (rarely Dec). Pupae (possibly larvae) overwinter.
Ecology	Generalist. Very wide-ranging and adaptable.

White Clover (*Trifolium repens*)

av wingspan: 1.8"

Pink-edged Sulphur *Colias interior*

Dorsal male (left), **female** (right); **specimen photos**

Ventral: 7/20/97. Moose River Plains, Hamilton Co., NY

The boreal woodlands of northern New England have a deep, rugged character. In mid-summer, the Pink-edged sulphur adds a spark of color to the sober surroundings. They are often common in boreal scrub and bog habitats (both of which support large heath thickets). This species was first discovered by the eminent 19th-century naturalist Louis Agassiz, on a celebrated natural history expedition to Lake Superior in 1848 (Weed, 1917).

The Pink-edged Sulphur stands apart from the other sulphurs in our region. It is our only local representative of the "northern sulphur" complex, a set of specialized groups within the genus *Colias* whose limited tastes and well-defined habitat associations stand in stark contrast to the cosmopolitan lifestyle and broad palate of "main line" sulphurs (like the Orange and Clouded). Some are known to produce digestive enzymes in the larval stage that are less diverse than those of its southern relatives, consistent with their narrower hostplant range (Burns, 1975).

As our sole northern sulphur, however, the Pink-edged is peculiar in lacking the dark, melanized ventral hindwing scaling that characterizes most other northern-dwelling pierids. Instead, its ventral surface is distinctively "clean." We have no ready explanation for this departure from "standard ecological practice."

Male Pink-edged Sulphurs patrol rather slowly in search of females. Both sexes nectar regularly on a variety of flowers and rest quietly on low foliage much of the time. They are mainly sedentary. A disjunct population occurs in the Appalachian highlands of eastern West Virginia and in nearby parts of Virginia and Maryland. Given its limited range, this population should be monitored carefully.

Identification	Small to medium in North; larger in Appalachians. Both sexes have *pink wing edges*—but so do many other sulphurs, including many Cloudeds, and we do not find this to be a consistently helpful character in the field. Easiest to identify from above, but here (as with most sulphurs) the dorsal surface is seldom displayed.
Dorsal	Both sexes yellow, with weakly defined central HW spots and a *greatly reduced or absent FW spot* (*compare* Clouded). Male has comparatively thin, dark wing borders, not continuous on HW. Female has little bordering on the FW, and little or none on the HW. Males do not reflect UV light.
Ventral	Clear yellow, *without dark scaling*. HW has a *single, round, silvery central spot, rimmed in pinkish-red*, less well-defined than in Clouded. *Lacks small brown dots on median HW* that characterize the Clouded; also *usually lacks small, dark mark on leading HW edge*.

Habitat	Boreal scrub and bogs with hostplants; also balds and barrens in the Appalachians; flies along woodland edges and roadsides. May visit ashes at campfire sites.
Hostplants	Various heaths (Ericaceae), Lowbush and Highbush Blueberry (*Vaccinium angustifolium* & *corymboisum*), Velvet-Leaf Huckleberry (*V. myrtilloides*), etc. Eggs laid on leaves.
Occurrence	Can be locally common. Occurs from n. NY/N. Eng. to the central Appalachians; also west along 50th latitude across CAN. One protracted brood, mid-Jun to early Sep. Third instar caterpillars diapause (Allen, 1997).
Ecology	Specialist. Has adopted a narrowly focused lifestyle.

Lowbush Blueberry (*Vaccinium* sp.)

av wingspan: 1.65"

Southern Dogface   *Colias* [= *Zerene*] *cesonia*

Dorsal male: 4/19/96, Ocala NF, Marion Co., FL

Ventral: 4/29/96, Ocala NF, Marion Co., FL

Observers on the East Coast seldom encounter this species, since it has few long-term strongholds in the region. Furthermore, it strays northward only rarely—most often during widely separated "irruption" episodes. Temporary colonies have been documented quite far north, on occasion, but none that has endured the next harsh winter. Southern Dogfaces can be plentiful farther west. In southern Texas, for instance, they can be among the most common butterfly species on the wing at times.

Although it is still widely included in the genus *Colias*, there is a substantial case for reassigning to *Zerene*—a longtime *Colias* subgenus (*see*, e.g., Pollock et al., 1997). As one commentator has put it, "with *Zerene* excluded *Colias* is structurally a homogeneous genus" (Layberry et al., 1998, p. 25).

Dogfaces patrol rapidly, with little indirection, usually flying at eye level or a bit below. Both sexes pause often for nectar, and both reportedly puddle (males do so actively). Individuals may cover a significant amount of territory in the course of a single day. In the Midwest, they stray northward regularly, sometimes reaching southern Canada. It is not surprising, therefore, that colonies may appear suddenly in formerly remote locations throughout their range. A weedy, abandoned housing development in the Big Cypress region of South Florida attracted one colony, in a setting that, if undisturbed, might have been too overgrown to be suitable (Smith et al., 1994). In 2001, a colony was discovered in southeastern Virginia (LeGrand & Howard, 2004).

As with many other sulphurs, Southern Dogfaces have a long, winter generation in which sexual maturity is delayed until spring. This generation is largely inactive, spending much of its time perched on low foliage (Opler & Krizek, 1984).

Identification	Medium-large (the biggest of our *Colias*-type sulphurs). *FW tip is pointed*, nearly falcate. Female has a rare, unmarked whitish form.
Dorsal	Prominent *"poodle head" shape on FW*, complete with *dark eye*. Similar pattern, but less clear-cut on female (occasionally absent). Male reflects UV from front part of "poodle face," and has a reddish sex-scale patch near the leading HW base (visible in photo).
Ventral	Yellow with dark HW eyespot in summer; mottled with black and rosy-pink color in winter ("rosa" form). Dorsal pattern visible from below.

Habitat	In Florida, mainly xeric scrub and sandhills, often more or less near water (B. & L. Cooper, pers. comm.). But may wander extensively.
Hostplants	Small-leaved peas, including Feay's Prairieclover (*Dalea feayi*), Summer Farewell (*D. pinnata*), Indigobush (*Amorpha fruticosa*), etc. Eggs are laid beneath terminal leaves of hostplant.
Occurrence	Locally uncommon in FL, n. GA; rare and episodic elsewhere. Very rarely strays up the East Coast (mostly historic). Also on US southern rim, south to ARG. All year in s. FL; 3+ broods in n. FL, Mar-Nov. Breeding erratic northward. Adults overwinter in reproductive diapause.
Ecology	**Medium specialist.** Confined in our area to a fairly limited habitat zone; less of a specialist elsewhere.

Feay's Prairieclover (*Dalea feayi*), a Florida hostplant in xeric scrub habitats

av wingspan: 2.25″

Occasional Pierid Strays to Southern Florida: Infrequent Visitors to a Tropical Oasis

Butterfly enthusiasts interested in migration often visit southern Texas in the fall, to experience the sometimes enormous lepidopteran movements that occur there, as myriads of butterflies pour across the Rio Grande River out of Mexico. The list of potential species is extensive (though even in "peak" years only a sampling of the full gamut actually is seen). Many butterfly gardens have been established recently across "the Valley." As vagrants stop to nectar, they can easily be observed. The North American Butterfly Association's new International Butterfly Park near McAllen seems well-situated to allow systematic canvassing of rarities.

By comparison, observers in South Florida need to wait more patiently for their strays. Given the vast overwater distances that separate landfalls in the Caribbean, lepidopteran traffic across the region is sometimes measured in terms of decades, as opposed to single seasons.

Still, regular movements do occur in the Caribbean, as evidenced by several recent arrivals noted in these species accounts. Hurricanes may occasionally accelerate distributions through the region, although debate persists as to whether creatures as frail as butterflies are actually assisted by hurricanes, as opposed to being torn apart by them. (We suspect that at least in some instances "hurricane dispersal" can occur.)

A number of Caribbean vagrants have colonized South Florida. Established residents have been given individual species accounts in this book. But other pierids occur in our region only as strays, and seem unable to establish long-term colonies. A few examples warrant consideration.

The Lyside Sulphur (*Kricogoia lyside*) verges at times on colonial status. Smaller than giant sulphurs (around 1.8"), it is an abundant open-space generalist in the Caribbean and northern neotropics. Lysides sometimes undertake "huge emigrations" in late summer or fall (Opler & Krizek, 1984) that carry them to the Keys. Sporadic reports exist from July to October. (A large influx was observed in 2004, reaching the southern mainland.) Lyside's Caribbean hostplant, Holywood Lignum-vitae (*Guajacum sanctum*), grows sparingly on the Keys.

Lysides have pointed forewings with bright yellow basal patches. (Florida individuals lack a black bar near the apex seen in some other populations.) Below, the Lyside is greenish-yellow with a faint, pearly sheen and a bit of yellow at the forewing base. A white-scaled vein runs through the central hindwing.

A less regular emigrant to the Keys is the Yellow Angled-Sulphur (*Anteos maerula*), a large, dramatic-looking pierid with falcate forewings and a small hindwing tail. Yellow above and greenish-yellow below, it has white-scaled wing veins and central cell-end spots on both wings. The Yellow Angled-Sulphur's known hostplants are introduced in Florida, including Quickstick (*Gliricidia sepium*) and Hairy Sensitive Pea (*Chamaecrista pilosa*), but it might subsist on other cassias. So far, only rare, late summer strays are known. Resident on Cuba.

The rarest transient pierid on the Keys is the Cuban endemic race of the Orbed Sulphur (*Aphrissa orbis orbis*), which was recorded once, on Big Pine Key in April 1973 (Minno & Emmel, 1993). The male is white with a bright orange circular patch on the basal forewing. Its host is Dwarf Poinciana (*Caesalpinia pulcherrima*), an ornamental plant in southeastern Florida and the Keys. Perhaps the Orbed Sulphur will find its way to our region again, in an area with suitable conditions for breeding.

Yellow Angled-Sulphur (*Anteos maerula*): 10/27/99, Santa Ana NWR, Hildago Co., TX

Lyside Sulphur (*Kricogonia lyside*): 10/29/99, Santa Ana NWR, Hildago Co., TX

Holywood Lignumvitae (*Guajacum sanctum*), tree and flowers

Cloudless Sulphur   *Phoebis sennae*

Dorsal male: specimen photo

Ventral male: 8/22/92, Gr. Dismal Swamp NWR, Nasemond Co., VA

This active, sun-loving butterfly epitomizes the genus *Phoebis* (giant sulphurs). It is probably the most widespread and common of all New World sulphurs. And although it is little more cold-tolerant than others in its genus (if at all), it ranges in summer from southern Canada to Patagonia.

The East Coast race of Cloudless Sulphur is *eubule*, but with the species' constant migratory gene mixing there is some doubt as to the validity of stated subspecies designations, which are largely statistical and "imprecise at best" (Smith et al., 1994, p. 157).

Males patrol for females throughout the day in a fast, usually low flight (to about 10 feet). Migrants are especially fast and direct, seldom stopping to nectar. A male begins courtship by touching the female with his wings or legs; she signals receptiveness with a quick wing flap. If she flattens her wings and lifts her abdomen, to signal rejection, the male may land on top of her, presumably depositing

Ventral female: 8/19/95, Gr. Dismal Swamp NWR, Nasemond Co., VA

pheromones to reiterate his case (Rutowski, 1983). Famous for their seasonal movements, Cloudless Sulphurs fly both north and south in fall, covering a dozen miles a day (Scott, 1986). Migrants navigate using a "sun compass" (Srygley & Oliveira, 2001). In compensating for wind drift, females act to conserve energy needed for egg-laying, while males seek to minimize the time to their destination, expending extra energy if necessary.

Identification	Medium-large. FW pointed and angular (especially in males).
Dorsal	Male *clear, unmarked yellow*, somewhat paler toward margins; does not reflect UV. Female *orange-yellow to white*; dark, with dark *"perforation" marks along both wing edges*. Note *dark-ringed central FW spot*.
Ventral	Lemon- to greenish-yellow. Male with sparse reddish marks, female with reddish blotches ("grafitti"), more so in winter ("rosa" form). Two white central HW spots (in both sexes; larger in female).

Habitat	Parks, gardens, wood edges, beaches, etc. Nearly any open disturbed area.
Hostplants	Various sennas, including Partridge Pea (*Chamaecrista fasciculata*), Sicklepod (*Senna obtusifolia*), Maryland Senna (*Senna marilandica*), etc. Eggs laid singly on host; larvae eat young leaves or buds, suffer mortality from ants (Barton, 1986).
Occurrence	Common to abundant, resident in warm years to NC, may breed north to ne MA along the coast in fall (variable; not every year). Many flights with overlapping broods; all year central FL and south, Feb-Nov in n. FL. Overwinters as adult in reproductive diapause.
Ecology	**Generalist.** Mobile and adaptive, a model of the generalist lifestyle. Why large numbers fly far north in fall, never to return, is a mystery.

Partridge Pea (Chamaecrista fasciculata)

av wingspan: 2.5″

Statira Sulphur *Phoebis* [= *Aphrissa*] *statira*

Dorsal male: specimen photo

Ventral male: 12/29/98, Easterlin Park, Ft. Lauderdale, Broward Co., FL

The Statira is an established resident of South Florida. Its endemic local race (*floridensis*) resides mainly along the southern coast. It has never been common, although its numbers appear to have increased in recent times, with the growing availability of open, disturbed habitats created by rapid development.

Widespread in the neotropics, the Statira is noted for long migratory flights—e.g., over the open sea from Colombia to Panama. Overwater migrants concentrate in corridors just 2 kilometers wide (Srygley & Oliveira, 2001). They use solar and landscape cues, and perhaps magnetic field information, to navigate. Back on land, they change direction to skirt coastal Panama's mountains, then resume a southerly flight across the Canal Zone.

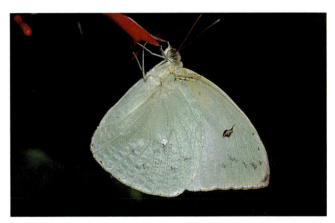

Ventral female: 12/28/96, Hugh Birch Taylor Park, Ft. Lauderdale, Broward Co., FL

These migratory feats are not emulated by the Florida race, however, which is comparatively sedentary. Almost no northward ventures have been recorded up the East Coast. Active puddlers, males patrol during the day for females. Both sexes nectar often. The Statira's behavior generally is similar to that of other giant sulphurs.

Identification	Our smallest giant sulphur (often placed in *Aphrissa*, a small Antillean genus, vs. *Phoebis*).
Dorsal	Male is *bright yellow near the body*, but *whitish ("mealy") toward the margins* (yellow areas of the FW reflect UV light). Otherwise unmarked (but some races have darkening at apex). Female is *creamy white* with some yellow near body, and *small all-dark spots* near the central FW; also *thin but continuous* dark FW borders and apex.
Ventral	Male *unmarked pale yellow* or *greenish-yellow* (no central HW dot). Female *yellow at FW base* with a *dark FW spot* and a *white HW dot*.

Habitat	Salt marsh and mangrove edges, and disturbed sites with hostplant.
Hostplant	Coinvine (*Dalbergia ecastophyllum*), Powderpuff Tree (*Calliandra haematocephala*), perhaps others. Eggs laid singly, larvae eat fresh leaves.
Occurrence	Local in s. FL, mainly near coast; strayed once to GA. Does not breed on Keys. Also s. TX, south to ARG. Three or more broods, Mar-Oct. Dry-season reproductive diapause only.
Ecology	**Specialist.** Mainly uses a single, restricted host.

Coinvine (*Dalbergia ecastophyllum*)

av wingspan: 2.4"

Orange-barred Sulphur   *Phoebis philea*

Dorsal male: specimen photo

Dorsal female (damaged): 11/24/00, Stock Island, Monroe Co., FL

The Orange-barred Sulphur is among our most impressive-looking butterflies. Yet its flight is high and fast, so that it can be difficult to get a good look at its colorful upper surface (and, when nectaring, it seldom opens its wings).

Orange-barred Sulphurs arrived in Florida in the late 1920s or early 1930s. The nominate mainland race occurs here, vs. members of any of several nearby Caribbean populations (Smith et al., 1994). Orange-barreds currently are less migratory than our other *Phoebis* sulphurs.

DeVries describes an abrupt and forceful courtship style from Costa Rica. A female "may be intercepted by the male, bashed around in the air, forced to the ground, and coupled to the male, all within fifteen to thirty seconds" (De Vries, 1987, p. 103). This approach has not been reported here to our knowledge. Orange-barreds are partial to nectaring on red flowers. Fresh males seek minerals, often puddling.

Ventral male: 6/29/93, Ft. Lauderdale, Broward Co., FL

Identification	Our largest *Phoebis* sulphur. Wings rounded, full-cut. Sexually and seasonally dimorphic.
Dorsal	Male *bright yellow* with a conspicuous *red-orange FW bar* (the bar reflects UV light); HW yellow with a *marginal red-orange patch*. Female varies from *gold-yellow to creamy/white*, and has conspicuous (but not continuous) *dark marginal spots*.
Ventral	Male yellow with FW cell-end spots, has *small, red basal spots* (more in winter). Female as above, with pinkish tinges at margins, and conspicuous, white cell-end spots on both wings; much *red grafitti* in winter form. Vague postmedian FW line is *offset at midwing* (*not* continuous, as in Large Orange).

Habitat	Open habitats near hostplants or nectar. Adaptable to urban settings. Edges, parks, gardens, roadsides, etc.
Hostplants	Sennas, including Chapman's Wild Sensitive Plant (*Senna mexicana* var. *chapmanii*), Privet Senna (*Senna ligustrina*), and others. Larvae that feed on leaves are green and produce white female adults; flower-feeders are yellow and produce yellow adults (Minno & Emmel, 1993).
Occurrence	Resident s. FL and Keys. Breeds in summer to GA/SC. Rare vagrant to NJ, NY, PA, RI (many in 1930s, just after becoming established in FL). Also s. TX south to BRA. Three or more broods, Mar-Dec in n. FL, all year in south and Keys. Adult overwinters in reproductive diapause.
Ecology	**Medium Generalist.** Somewhat less wide-ranging in natural habitats than other *Phoebis* sulphurs

Ventral female: 10/24/99, Santa Ana NWR, Hildago Co., TX

av wingspan: 3.0"

Large Orange Sulphur   *Phoebis agarithe*

Dorsal male: **specimen photo**

Ventral male: 5/1/96, Everglades NP, Miami-Dade Co., FL

If there is an "average" *Phoebis* in our area, it would be the Large Orange, which is middle-of-the-road in size, markings, population, and even extent of seasonal movement. It is common in southern Florida, but its numbers decrease rapidly with latitude. In the North, it is an accidental stray.

Like the Statira (but unlike more recent arrivals), the Large Orange has been in Florida long enough to have evolved an endemic race (*maxima*). Still, the distinctions between *maxima* and races in the West Indies are "not impressive," since regular gene mixing occurs in this species during mass emigrations within the Caribbean Basin (Smith et al., 1994, p. 156).

Fast-flying, like others in its genus. Large Oranges pause frequently for nectar, but remain alert and difficult to approach. Chances are better with puddling males, which often gather in sizeable groups.

Ventral female: 3/23/94, Key West, Monroe Co., FL

Identification Medium-sized, both generally and in relation to other *Phoebis*. Sexually and seasonally dimorphic, this species can be confusing, and needs to be distinguished carefully from other giant sulphurs.

Dorsal Male is *bright golden-orange*, especially the FW (which reflects UV light; HW does not), largely unmarked. The female ranges from orange to whitish ("albarithe" form), sometimes with salmon tinges. Note *discontinuous dark edging along both wing margins* (not very sharply delineated), and the *small, all-dark* central FW spot.

Ventral Male *yellow-orange* in summer, female as above, with central white FW spot; both sexes with *rusty scaling in winter* (particularly female). Both sexes have a *straight, continuous postmedian FW* line ("broken" in other species; most prominent in female), also central white HW spot.

Habitat Open subtropical scrub, disturbed areas, gardens, fields, roadsides, parks, etc.

Hostplants Mimosas, including Blackbead (*Pithecellobium keyense*), Cat's-Claw (*P. unguis-cati*), Wild Tamarind (*Lysiloma latisiliquum*). Caterpillars eat hostplant leaves.

Occurrence Common to abundant in peninsular FL, especially southward. Strays rarely north to ME, NJ, NY (Jun 1995), SC (Aug 1997); but cannot survive winter cold. Also s. TX to So. Amer. Three or more broods, all year in s. FL; Mar-Dec to north. Adults overwinter in reproductive diapause.

Ecology Generalist. Tolerant and adaptable (except to cold).

Blackbead (*Pithecellobium keyense*)

av wingspan: 2.5"

Sleepy Orange   *Eurema* [= *Abaeis*] *nicippe*

Dorsal male (left) and **female** (right): **specimen photos**

Ventral (light summer form): 6/15/96, Edisto Island, Charleston Co., SC

This exceptionally pretty butterfly is not a typical *Eurema*. This is in part because it *is* exceptionally pretty—a trait that not all *Eurema*s share—and partly because some taxonomists now assign it to the genus *Abaeis*, formerly a sub-genus under *Eurema* (Opler & Warren, 2003). *Abaeis* originally was described by Hübner in 1819, so the proposed reclassification does not break new ground in taxonomy. Still, decisions on lumping and splitting regularly shift, and there is long-term value in regularly revisiting the current arrangement.

The Sleepy Orange has a misleading common name that commentators have long delighted in criticizing (*see* Tveten & Tveten, 1996). Some commentators now refer to this lively, fast-flying pierid as the "Rambling Orange." This describes the butterfly better than the official name, which refers to *nicippe*'s small, forewing dots, whose crescent shape reminded Comstock of closed eyes—hence "sleepy" (Iftner et al., 1992).

The Sleepy Orange is primarily a tropical butterfly. It now resides year-round in the Deep South, but like other *Eurema* it must colonize the northern parts of its range anew each year. It appears hardier than most of its cold-intolerant relatives, on the other hand, as it may be able to survive at least some winters as far north as North Carolina or even West Virginia (LeGrand & Howard, 2004; Allen, 1997). Much remains to be learned about the life history of this interesting species, including its exact northern permanent residency limits.

Male Sleepy Oranges patrol throughout the day for females. Fresh males spend much time puddling or taking minerals, vs. nectaring (males are exceptionally prone to puddling).

Identification	Medium-sized. Antennae are black (vs. "hot pink" in *Colias* sulphurs). Wings rounded, but male's FW is somewhat pointed. Sometimes confused with Orange Sulphur, but shape and pattern are very different. Alternating bright orange and yellow "flashes" in flight are distinctive.
Dorsal	Male *bright golden-orange*, with *wide, dark margins* on both wings; no UV reflectance. Female is paler orange or yellow. *Dark HW border fades out* before outer angle. Both sexes have *cell-end "sleepy eyes"* on the FW.
Ventral	Orange-yellow (*summer*); brick-red/tan (*winter*); note *long mark on outer HW*.

Habitat	Various open, disturbed habitats: old fields, roadsides, wood edges, swamps, wet meadows, etc. Spreading in Caribbean with slash-and-burn agriculture (Smith et al., 1994). Open woods in early spring (LeGrand & Howard, 2004).
Hostplants	Sennas and chamaecristas, including wild sennas (*Senna hebecarpa, marilandica, obtusifolia, mexicana*, etc.), Wild Sensitive Plant (*Chamaecrista nictitans*), Partridge Pea (*C. fasciculata*). The literature lists several plants (e.g., clovers) accepted in captivity but not recorded in the wild. Eggs are laid singly under host leaves, which the caterpillars eat.
Occurrence	Common resident in Deep South, but rare in s. FL/Keys; breeds north annually to latitude 40, strays as far as north as s. N. Eng.; also west to SD and south to Costa Rica. Multiple regular-season flights during the summer (depending on local arrival date). A single long, adult reproductive-diapause generation in the South. Diapause stage farther north, if any, is not described.
Ecology	Generalist. Mobile and flexible in use of habitats and hostplants.

Privet Senna (*Senna ligustrina*)

av wingspan: 1.65"

Dina Yellow *Eurema* [= *Pyrisitia*] *dina*

Dorsal female: specimen photo

Ventral male: 6/6/99, Castellow Hammock, Miami-Dade Co., FL

Also called Bush Sulphur. Klots thought the Cuban race of Dina Yellow might someday "stray to Florida" (Klots, 1951, p. 284). But it actually was the Bahaman race (*helios*) that colonized the state in 1962. Population levels have varied since, but the Dina's numbers in Florida are effectively limited, given the rarity of its two known hostplants—both state endangered—and the loss of successional habitats needed to support them. Careful management will be required for this attractive pierid to persist.

Some taxonomists have reclassified the Dina (along with several other *Eurema* sulphurs) into the genus *Pyrisitia*, formerly a *Eurema* subgenus (Opler & Warren, 2003).

Dina Yellow males patrol for mates. Their flight is lively, but not as fast (nor as frenetic-seeming) as that of some other pierids. Both sexes actively nectar on hammock flowers and *Bidens*.

Ventral female: 6/25/93, Owaissa Bauer Park, Miami-Dade Co., FL

Identification	Medium-small; our largest *Eurema* other than Sleepy Orange. Wings rounded but with points near the apices. Dry-season individuals larger, darker (Opler & Krizek, 1984).
Dorsal	Male *bright orange-yellow* with *very narrow black FW border* (wings reflect UV). Female yellow with *darker, wider borders.*
Ventral	Male *lemon-yellow* with tiny black dots. Female paler, *with dark HW smudges* and *reddish apical patches on both wings.* Note *single, dark postbasal mark* on HW.
Habitat	Scrubby, second-growth areas in or near subtropical hardwood hammocks.
Hostplants	Two members of the Quassia Family (Simaroubaceae) in Florida: Mexican Alvaradoa (*Alvaradoa amorphoides*) and Bitterbush (*Picramnia pentandra*) (at Deering Estate) (Cech, 2002). Females lay single eggs on new growth (but potentially many eggs at different locations on the same plant). We have found the pupa suspended from a horizontal hostplant branch.
Occurrence	Extremely local, but can be common at colony sites. Originally found at Matheson Hammock in Miami, now known principally from the Deering Estate and various hammocks near Homestead, where hostplants occur. Another race occurs in TX, AZ, and C. Amer. Multiple wet-season generations (3-4), plus one long, winter adult generation in reproductive diapause.
Ecology	Specialist. The Dina's hosts belong to an ecologically specialized family of tropical plants, in contrast to relatives that feed on very common pea-family hosts.

Mexican Alvaradoa (*Alvaradoa amorphoides*)

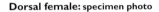
av wingspan: 1.4"

Little Yellow   *Eurema* [= *Pyrisitia*] *lisa*

Dorsal male: specimen photo

Ventral (summer form): 8/15/93, Gerritsen Beach, Queens, NY

Small sulphurs (genus *Eurema*) generally cannot withstand winter cold, and the Little Yellow is no exception. Yet it pushes farther north in summer than any of its relatives. And it leaves time for a modest return migration in autumn. In earlier times, enormous migrating swarms of Little Yellows were reported at sea, as far offshore as Bermuda (Weed, 1917).

It is a mystery how Little Yellows can return with such consistent precision to a particular site each year (such as one small patch of sandy scrub, less than a quarter acre in extent, at Gerritsen Beach, in New York City). This causes some observers to suspect the existence of a resident population, or sibling species, with fewer annual broods than in the main, southern population. But no evidence for any such resident populations has yet been forthcoming.

Ventral female (fall albinistic form): 9/28/97, Gainesville, FL

Male Little Yellows patrol throughout the day for females. When encountering a female, the male flies above his intended mate, causing her to land. Through ritualized signals, the female indicates her acquiescence (or lack thereof). The Little Yellow's flight is low and active, but fairly slow. Both sexes nectar and males take minerals.

Identification	Medium-small. Apex tips elongated. Some females albinistic (especially in fall). Dry-season individuals have reduced dark markings. Usually easy to identify, but several species must be ruled out in southern Florida (*see* Mimosa Yellow account). We have observed Little Yellows flying alongside Mimosa Yellows in Homestead, FL.
Dorsal	FW yellow with *wide, black apex*, and a small but distinctive *black spot* at the end of the FW cell (seldom displayed in the field). HW has black borders; less defined on female. Male reflects UV light.
Ventral	Yellow with *scattered, black HW markings*. Female has *large, rusty spot at HW apex*. Note *two tiny black dots at HW base*.

Habitat	Various low, open sites, fields, sandy waste areas. Tolerates "even the driest habitats" (Smith et al., 1994).
Hostplants	Wild Sensitive Plant (*Chamaecrista nictitans*), Partridge Pea (*C. fasciculata*), Key Cassia (*C. lineata*). Larvae eat leaves, hide on host stems and leaf ribs.
Occurrence	Resident in relatively frost-free areas of SE, numbers in North vary annually. Reaches NYC area by Jun/Jul, remains till late Oct. Also west to Gr. Plains, south to C. Amer. Multiple broods, all year in South. May overwinter solely in adult stage, in reproductive diapause.
Ecology	Generalist. Mobile, widespread, and adaptive.

Wild Sensitive-Plant (*Chamaecrista nictitans*)

av wingspan: 1.3″

Mimosa Yellow *Eurema* [= *Pyrisitia*] *nise*

Dorsal male: specimen photo

Ventral male: 12/24/96, Owaissa Bauer Park, Miami-Dade Co., FL

Also called Jamaican Sulphur. Widespread in tropical lowlands, this species is marginal in our area, confined to just a few subtropical hammock edges south of Miami (it is also found rarely on the northern Keys).

First discovered in Florida in 1933, the Mimosa has been common in some years (1947), but virtually absent in others (Klots, 1951). Its vagaries are well-illustrated by the fact that it was first described from Jamaica in 1775, but then not seen there again until 1951 (Smith et al., 1994). It colonized Cuba around 1900.

Behaviorally, the Mimosa Yellow is noted for flying near the edge of hardwood hammocks, darting inside for cover if disturbed (Smith et al., 1994). Little Yellows, by comparison, fly more in the open, and normally would not flee into dense woods, even if alarmed.

Ventral female: 12/24/96, Owaissa Bauer Park, Miami-Dade Co., FL

Identification   Medium-small (the "Little" Yellow is slightly larger). Wings rounder than Little Yellow's. Observers should keep in mind the possibility of Boisduval's Sulphur (*E. boisduvaliana*), found once on the lower Keys (9/20/73) (fairly large, with wide, irregular FW borders and a slight tail-point). Also Shy Yellow (*E. messalina*), found once near Orlando; the latter is white above and has a dark spot just below the FW apex.

Dorsal   Both sexes yellow (slightly more lemon-yellow than Little; a difference that can be seen in flight). Male FW reflects UV light (Scott, 1986). FW has *narrow darkening at apex*, little or no HW band (*compare* male Little).

Ventral   Both sexes yellow, with two small, black dots on the central HW. *Lacks two small basal dots* at wing base, characteristic of Little Yellow. Female has a large, dark-rusty patch at the HW apex. The narrow, apical FW border on the dorsal surface can be seen from below, through the Mimosa's thin wings.

Habitat   Various open, brushy habitats in the tropics, but in our area found mainly along the edges of subtropical hardwood hammocks.

Hostplants   Wild Tamarind (*Lysiloma latisiliquum*); Sensitive Plant (*Mimosa pudica*) also reported. No details on life history, but with such common hosts, it is curious that this species is so scarce (*compare* Dina).

Occurrence   Rare and sporadic, Homestead to northern Keys. Also s. TX throughout lowland tropics to ARG. Rare stray to AZ, CO, KS (once). Three or more broods, at least May-Dec in FL. Adult passes winter in reproductive diapause.

Ecology   Specialist. More generalist elsewhere, but not here.

Wild Tamarind (*Lysiloma latisiliquum*)

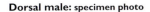

av wingspan: 1.2"

Barred Yellow   *Eurema daira*

Dorsal male: specimen photo

Ventral (white summer form): 6/25/93, Fruit & Spice Park, Miami-Dade Co., FL

A better name for this complex, wide-ranging pierid might be "Variable Yellow." Its two dissimilar races in southern Florida interbreed confusingly on the Keys, and each has distinct seasonal forms, which can be either "pure types" or intermediates. The long winter generation (4-5 months) lives in dry-season diapause, with reduced daily activity, its sexual maturation postponed until spring. Adults in other generations have a more typical life span of 1-10 days (Opler & Krizek, 1984).

Males patrol throughout the day in search of females. When one is encountered, the male lands alongside her and waves a single forewing in her direction. If she is impressed, by either his pheromones or his ultraviolet wing reflections, copulation will proceed. Both sexes nectar actively, and males puddle. In fall, a southward migration occurs on the Florida peninsula (Walker, 2001).

Ventral ("red" winter form): 9/21/96, Gainesville, FL

Identification   Medium-small (larger than Dainty Sulphur). Highly variable by season and race (the resident *daira* race is widespread, whereas the Caribbean *palmira* occurs mainly on the Keys). *Summer/wet-season* form is smaller than *winter/dry-season* form, and has darker markings. Intermediate phenotypes and hybrids occur.

Dorsal   Male: Yellow with *thick black bar along trailing FW*; strongly reflects UV light (Opler & Krizek, 1984). HW has dark outer border in summer, partial in winter. Male *palmira* is *white on upper HW*. Female: Similar to male, but FW bar is *faint* in summer; *very faint or absent* in winter. Female *palmira* is *all-white above.*

Ventral   Summer phenotype ("fairy white") has *satiny white HW*. Very different winter form is *brick red to tan*, with two small dots in HW cell. *Palmira* race similar to summer *daira* (with some brown dusting in winter.)

Habitat   Low-growing subtropical scrub, waste areas, roadsides, dunes, sparse open woods, etc.

Hostplants   Small, weedy peas, especially Pencil Flowers (*Stylosanthes hamata*), joint vetches (*Aeschynmene*), probably others. Female lays single eggs on tip of plant growth. Caterpillars eat leaves.

Occurrence   Usually common, SE coast and peninsular FL; strays north late in season, rarely to NC/VA. Also west to TX, south to ARG. All year in Deep South, three or more broods (Feb-Oct in GA). Long-lived, nonreproductive adults overwinter.

Ecology   Generalist. Typical generalist lifestyle.

Pencil Flower (*Stylosanthes hamata*)

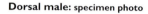

av wingspan: 1.2″

Dainty Sulphur   *Nathalis iole*

Dorsal male: **specimen photo**

Ventral (summer form): 9/27/95, Okeechobee Co., FL

The Dainty Sulphur is distinctive among our pierids, in shape and other traits, and may warrant its own subfamily.

This small butterfly colonized Florida in the early 1900s, probably from the Caribbean (Opler & Krizek, 1984). Our population is far less inclined to wander northward during the summer than those in the Midwest. (There, they follow river courses quite far north.) Yet there *is* evidence of limited movement up the southeastern coastal plain (LeGrand & Howard, 2004). Certain "outlier" sightings, like that on southern Long Island, October 25, 2000, may be of western strays.

This southern species is unable to survive winter cold. If day length is short during its larval period, the Dainty produces a winter phenotype with three times the usual number of dark ("melanic") scales; this allows it to absorb solar warmth more efficiently (Douglas & Grula, 1977).

Dainty Sulphur males patrol just inches from the ground

Ventral (winter form): 12/25/96, Ft. Lauderdale, Broward Co., FL

in pursuit of females. When faced with rejection, males are likely to engage in a distinctive, open-winged display, showing off their handsome dorsal bars and androconial spots. This last-ditch effort can often induce a female to reconsider her unfavorable decision (Rutowski, 1981). Both sexes nectar avidly, often on the hostplant.

Identification	Our smallest pierid. Elongated FW shape is distinctive. A rare population with mostly white individuals is known from Homestead (Smith et al., 1994). Appearance highly variable, but identification is seldom a problem.
Dorsal	FW yellow with dark tips. *Black bars* extend along *leading HW and trailing FW* edges. Male has oval scent (androconial) patch within dark HW bar (near body); this is reddish-orange in life, pale yellow in specimens. Female's HW orangish above.
Ventral	*Dingy yellow* (summer) or *greenish-gray* (winter). FW has a yellow or orange patch near base, prominent black spots toward the margin.

Habitat	Nearly any low-growth open space: waste areas, roadsides, coastal flats.
Hostplants	Aster Family (Asteraceae). In our area, mainly Spanish Needles (*Bidens alba* var. *radiata*). Female lays eggs singly on emerging leaves.
Occurrence	Fairly common to abundant, peninsular FL; strays/immigrants in SE, rarely to PA/NY. Also US SW to n. So. Amer.; northward to MN, CAN in summer. All year in s. FL; Mar-Dec in SE; northward movements Jun-Nov. Overwintering stage not reported.
Ecology	Generalist. Adaptable and mobile, except in terms of cold tolerance. One main hostplant used here, but an extremely common one.

Spanish Needles (*Bidens alba* var. *radiata*)

av wingspan: 0.975″

The Gossamer-Wings: Family Lycaenidae

Typical lycaenid caterpillar, retiring and inconspicuous (Appalachian Azure, on Black Cohosh buds)

The small size and delicate-seeming construction of lycaenid wings combine to create an impression of sheerness and delicacy—hence the family name "gossamer-wing." But while this epithet does apply well to many lycaenids, it should not be taken too literally. Cassius Blues *do* have wings that are thin enough to be translucent. But so do some nymphalids, such as Melissa Arctic, and in tropical "clearwings" the wings are fully transparent!

Lycaenids are among the largest of the world's butterfly families (if not the largest), with some 4,000 to 6,000 species. Yet they are not necessarily the most numerous in terms of population, since the family has a significant number of specialists, more than 55 percent on the East Coast, by our estimation. East Coast lycaenids are modest in size, including our smallest butterfly (Eastern Pygmy-Blue, about 0.6"), and extending up only to the medium-sized Atala (about 1.75"). Most lycaenids feature flower parts in their diet. Indeed, an early shift to flower-feeding probably helped to define the group taxonomically (Scott, 1986, p. 99 *et seq.*). Most species overwinter as eggs or pupae, which allows them to produce caterpillars in the spring, when fresh flower buds are common. Many larvae also eat leaves, and a small number feed on aphids or ant pupae.

We follow Layberry et al. (1998) in not referring to lycaenid caterpillars as "slug-like," an epithet that has long appeared in the literature. Slugs are more elongate than lycaenid caterpillars, and lack their characteristic wrinkles at segment joints. "Sowbug-like" often would be closer to the truth, even "trilobite-like." But whatever the shape, lycaenid larvae tend to be secretive and inconspicuous, often burrowing invisibly into flowerheads. Lycaenid eggs are shaped like sea urchins ("echonoid") and may be quite beautifully detailed on close view. Most eggs are deposited singly, but Atalas lay in clusters.

Many lycaenids associate with ants in one fashion or another, and many are ant-attended (especially blues), sharing sweet "honeydew" secretions in exchange for protection from enemies. Many pupae emit chirp-like squeaks ("stridulation"), which may appease predatory ants.

Four lycaenid subfamilies occur in our area:

Harvester (Miletinae): Fifty or so species worldwide, mostly in Asia and Africa. One species in North America. Larva feeds on aphids. *See* Harvester account.

Coppers (Lycaeninae): A north-temperate/holarctic group with scattered outliers in New Zealand and the Afro-Asian tropics. Seventeen species in North America, four in our area, one of which is common and widespread; the others are more local and specialized. Caterpillars feed on docks (Polygonaceae). The name "copper" refers to the bright, copperish iridescence seen on many males' dorsal surfaces. Males lack scent patches and signal females using bright forewing iridescence (reflective of UV light). Most bask and display with open wings. Coppers frequently occur in low, open habitats. Flight is strong but usually not very direct. Most species hibernate in the egg stage.

Hairstreaks (Theclinae): The largest lycaenid subfamily, with about 2,000 species (1,000 in the New World tropics). There are about 70 species in the United States and Canada, around 30 in our area. They are named for their thin, hindwing tails and bold, streaky ventral patterns (also perhaps for their rapid flight?). Larvae tend to feed on the flowers and young leaves of "apparent" plants such as trees (which do not die back in winter); this provides a place where their eggs can diapause in relative security, poised for the following spring's "leaf out." In most genera, hairstreaks keep their wings closed when not flying. Males have forewing scent patches used in courtship, though many display bright iridescence (usually blue-toned) as well. Many hairstreaks have a "thecla" spot near the base of the tail. This and the tails themselves are thought to resemble a "false head," luring predators to the wrong end when attacking (Robbins, 1981). With their tall, relatively pointed wings, hairstreaks have a characteristically rapid, jerky flight style. Most males perch to await females.

Blues (Polyommatinae): Found mostly in north-temperate habitats, like coppers, but blues are fairly common worldwide. They are most diverse in Eurasia. About 34 species live in North America, 13–17 on the East Coast (depending on taxonomic viewpoint). The dorsal surface of most males is indeed blue (less so with females). The ventral surface is usually light-colored with dark spots. Wings are rounded, and flight seems comparatively weak or fluttery. Most males patrol for females in open habitats, basking with wings partially open. Blues typically overwinter as pupae. Females emerge in spring to oviposit on emerging flower buds (mainly of herbs). Many species are ant-attended and most are sedentary, remaining near hostplant concentrations.

Harvester *Feniseca tarquinius*

Dorsal: 8/6/97, Lancaster, Coos Co., NH

Ventral: 6/12/93, Wentworth Location, Coos Co., NH

Our lone representative of its small subfamily (others occur in Asia and Africa), the Harvester is an obligate insectivore as a caterpillar. It is in fact our only dedicated butterfly predator, except for certain blues whose older larvae feed on ant pupae in their nests, and some species whose larvae engage in cannibalism.

The Harvester's main "hosts" are aphids. On hatching, the caterpillar spins a protective silken web around itself. It then uses its specialized mouthparts to grasp, pierce, and suck dry the aphids it encounters. Their desiccated shells are stuck to its back, gradually forming a "sheep's clothing" that further cloaks the marauder. Perhaps because of this comparatively rich food source, the Harvester has the fastest "cycle time" of any of our butterflies, taking just around three weeks to progress from egg to adult (Scott, 1986).

Earlier called "The Wanderer" (Weed, 1917), the Harvester is attuned to the shifting whereabouts of the aphid hosts. Aphids gain their livelihood by sucking sugary juices from a variety of trees (mainly alders in the Harvester's case, but also others; *see* below). Aphids typically utilize two different host trees each year. The Woolly Alder Aphid (*Prociphilus tessellatus*) shifts between Silver Maples and alders, and may overwinter on either (Kyhl & Hahn, 2002). Grazing aphid colonies usually consist of flightless females, who reproduce by parthenogenesis. But at certain times (either seasonally, or when stressed), a winged generation with both sexes appears. This generation reproduces sexually and relocates, typically to the alternate host. It is these errant aphid colonies that "The Wanderer" somehow must pursue, much to the confusion of its human observers, who would rather it stayed put.

The Harvester's pupa has a characteristic "monkey face" shape. It is capable of making scraping noises ("stridulation"), perhaps as a signal to deter ant attacks (Douglas, 1989, p. 29). Adults often perch quite high (to 20 feet), flying out to engage rivals like a hairstreak (though they are less rapid in flight). With a very short proboscis, adults do not nectar, but rather feed on aphid "honeydew," and also at times on mud, sap, dung, and carrion.

Identification	Medium-small. Dark individuals in e. CAN formerly considered a separate race, *novascotiae* (Layberry et al., 1998).
Dorsal	*Irregular yellow-orange or orange-brown patches.* Row of *dark submarginal spots* on trailing HW.
Ventral	Unique. Orange-brown to purplish-brown; many *dark spots circled with white.*

Habitat	Alder swamps, creeksides, ditches, along streams in deciduous or mixed woodlands.
Hosts	Woolly aphids (several genera), sometimes other Homoptera; mainly on alders (*Alnus*), but also beech, ash, hawthorn, etc. Female lays eggs alongside aphid colony.
Occurrence	Local, uncommon to rare. N. Eng. south to central FL; rare/absent on SE coastal plain. Broods unclear; may be six-plus (Allen, 1997). Late Apr-Sep in North, Mar-Oct in FL. Diapause stage unsure; probably mature larva or pupa.
Ecology	Medium Specialist. Its lifestyle is obviously unique, but the Harvester's prey uses a range of hosts and habitats.

av wingspan: 1.2"

Speckled Alder (*Alnus incana*) thicket, Washington Co., ME, with fruits (above left). Woolly Aphids (lower right) on branch (caterpillar mostly concealed beneath aphids).

American Copper   *Lycaena phlaeas*

Dorsal: 5/22/93, Pound Ridge Reservation, Westchester Co., NY

Ventral: 5/22/94, Albany Pine Bush, Albany Co., NY

Also known as Small Copper (mainly in Europe). Pyle (1981) lists "Flame Copper" as an alternate name in the United States. We have not heard this name in use, but it does accentuate the fact that the American Copper is (overall) the most brightly colored of eastern *Lycaena*s. (On the other hand, the name "Flame Copper" would better describe one of several colorful western species, such as the Ruddy or Lusterous Copper.)

Though widespread and often abundant, American Coppers are actually somewhat local, absent in many likely locations. The East Coast race of this holarctic species (*americana*) was originally described as a separate species, and looks quite different from the disjunct, western races of North America. The nominate race (*phlaeas*) occurs in Eurasia. Some commentators have suggested that the American Coppers may have traveled to this country with early European colonists (Opler & Krizek, 1984). This view is based on the fact that American Coppers dwell mainly in human-disturbed habitats and feed on introduced European buckwheats. Also, they resemble the *polaris* race of *Lycaena phlaeas* from Scandinavia (more ostensibly than any U.S. race). Others, however, note differences between American Copper and *polaris*, and also point to several other butterflies, such as Hayhurst's Scallopwing and Wild Indigo Duskywing, that were formerly niche specialists, but have now shifted their feeding and extended their ranges on alien weeds (Glassberg, 1999; Layberry et al., 1998).

Sedentary and colonial, adult American Coppers usually move little during their life (Scott, 1986). Males are quite pugnacious within their small, well-defined territories. They perch for females throughout the day, but also seemingly patrol at times. Both sexes are active nectarers, feeding with their wings partially cocked.

Identification	Small, but still larger than Bog Copper. Quite variable, seasonally, geographically, and individually. Several named aberrations have been documented, including "fasciata," where dark FW spots fuse into a single, dark patch; "obliterata," in which FW dots are very small or missing, etc. Spring individuals tend to be small and brightly colored, with comparatively small dots.
Dorsal	*Iridescent red-orange*, with varying yellow overtones and *variable dark spotting* on FW. HW dark with *orange border* along outer margin.
Ventral	FW *bright orange basally* with *large, dark spots*. HW gray with *smaller spots* and a *weakly defined, red-orange submarginal border*.
Habitat	In the East, disturbed places, pastures, vacant lots, road edges, pine-oak barrens, etc. Other habitats throughout holarctic range.
Hostplants	Alien buckwheats (docks), including Sheep-Sorrel (*Rumex acetosella*) and Curled Dock (*Rumex crispus*). Mountain Sorrel (*Oxyria digyna*) to our North. Eggs laid singly on hostplant leaves. Young larvae chew holes in leaf, older ones "mine" channels along leaf surface.
Occurrence	Common in NE, rarer in South and West; *americana* race from Maritimes west to ND, south to MO. Other races in West and Arctic, widely across Eurasia. Several overlapping broods, Mar-early Oct in e. NC; Apr-Oct in NYC; Jun-Sep in ME. Disagreement over diapause stage, probably pupa (*but see* Allen, 1997 [egg]; Thomas & Lewington, 1991 [larva]).
Ecology	Medium Generalist. Though sedentary, has mainly generalist traits.

Sheep-Sorrel (*Rumex acetosella*)

av wingspan: 1.0"

Bronze Copper *Lycaena hyllus*

Dorsal male: 8/21/92, Blackwater NWR, Somerset Co., MD

Dorsal female: 8/21/92, Blackwater NWR, Somerset Co., MD

The moist, weedy patches in which Bronze Coppers fly seem lifted from the pages of *Watership Down*—except that the novel, of course, was set in England. Still, many British meadow plants are now established in our area. And rank, weedy meadows are in decline in both countries, threatening resident butterflies.

Bronze Coppers have declined sharply in much of their northeastern range in recent decades. Yet despite their large size and dramatic appearance, they have attracted little attention among conservationists. Strongholds remain in our area, but this is a troubled species that needs stewardship in our increasingly urbanized region. During the decade we spent preparing this book, we encountered Bronze Coppers just twice, and we found them absent at many formerly reliable sites, especially near large urban centers.

Male Bronze Coppers perch on low foliage with wings

Ventral: 8/21/92, Blackwater NWR, Somerset Co., MD

widespread to await females. Mating often occurs in the afternoon. We agree with Iftner et al. (1992) that Bronze Coppers are active nectarers, on smartweeds, milkweed, and other flowers. Flight is low and "floppy," usually in the immediate vicinity of a hostplant.

Identification	Our largest copper, striking in appearance in the field. The female's dorsal markings superficially resemble those of the American Copper, but in life the size difference between the two greatly simplifies the comparison.
Dorsal	Sexually dimorphic. Male is *coppery-brown with purple iridescence* and dark spotting; female has *yellow-orange FW patches* and bold, slightly smudgy spots. Both have *orange subterminal band on HW.*
Ventral	Numerous dark spots on both wings. *FW bright orange* (except near margin, where gray); *HW with a broad orange band along the margin only;* otherwise light gray.

Habitat	In the East, low, wet areas, such as damp meadows, poorly drained bottomlands, sedge meadows, ditches, brackish salt marsh edges, pond edges, etc.
Hostplants	Mainly docks, including the native Water Dock (*Rumex orbiculatus*) and sometimes the non-native Curly Dock (*R. crispus*); knotweeds (*Polygonum*) also reported. Eggs are laid singly on hostplant leaves, larvae emerge in spring to feed.
Occurrence	Widespread but very local, uncommon to rare, usually seen in small numbers. ME south to Delmarva (where still relatively common), WV. Also in US and CAN west to Rockies. One to three broods, Jun-Sep in North, May-Nov in South. Eggs hibernate.
Ecology	**Medium Specialist.** Requires special conditions, not fully understood.

Curled Dock (*Rumex crispus*)

av wingspan: 1.35"

Bog Copper *Lycaena epixanthe*

Dorsal male: 6/16/95, Klot's Bog, Lakehurst, Ocean Co., NJ

Dorsal female: 6/22/94, Klot's Bog, Lakehurst, Ocean Co., NJ

Bog Coppers live almost exclusively in northern acid bogs, many of which are relicts of the Pleistocene Ice Age. They nectar almost exclusively on their hostplant as adults (Wright, 1983), and "puddle" on dew drops clinging to bog foliage. Thus, generations of Bog Copper can spend their entire life-cycle within the confines of a single acid bog. And they apparently often do just that, for while this species can reach exceptionally high concentrations—a condition often associated with dispersal—they are feeble fliers and apparently colonize poorly (Ehrlich, 1985, p. 32; Allen, 1997).

Bog Coppers spend nearly 85% of their life span in the egg, glued to the underside of a cranberry leaf only a few inches above the bog surface. This subjects them to regular immersions. They survive because their eggs can exchange respiratory gasses underwater (Wright, 1983).

Males perch on low foliage throughout the day. Females mate shortly after emergence, usually in the afternoon.

Ventral: 6/22/94, Klot's Bog, Lakehurst, Ocean Co., NJ

Identification	Our smallest copper. Variable across populations. Also within populations: 15% of individuals in the southern nominate race (*epixanthe*) have whitish undersides, resembling the northern *phaedra* race (Wright, 1983).
Dorsal	Brown to gray-brown with *dark spotting* and *orange marks* along trailing HW. Male has few FW spots, mostly basal, and *blue-purple iridescence* (glows strongly under UV light).
Ventral	*Creamy yellow to grayish-white*, with *small, mostly round spots* and a vaguely defined *red-orange zigzag submarginal HW band*.

Habitat	Confined almost entirely to acid bogs with cranberries.
Hostplants	Large Cranberry (*Vaccinium microcarpa*) in southern portion of range; Small Cranberry (*V. oxycoccos*) north of *microcarpa*'s range. Female lays on plants concealed by sedges or in hummocks, lands and walks down to oviposit. Young larvae eat old leaves, older ones feed on fresh growth. Early stages described in astonishing detail by Wright (1983).
Occurrence	Highly colonial, annual abundance varies. N. Eng. south to s. NJ (many in abandoned cranberry bogs). Disjunct population in w. MD/WV. Also west through Gr. Lakes region. One flight, mainly mid-Jun-early Jul in NJ, occasionally later; mid-Jul-mid-Aug in ME. Eggs overwinter (with 1st stage larvae already formed inside).
Ecology	**Specialist.** Active conservation of fragile acid bog habitats essential, for this and many other rare, bog-dependent species.

Wild Cranberry (*Vaccinium macrocarpon*)

av wingspan: 0.9"

Dorcas Copper *Lycaena dorcas claytoni*

Dorsal male: 8/4/00, Aroonstock Co., ME

Dorsal female: 8/5/00, Aroonstock Co., ME

The Dorcas Copper is a widespread northern butterfly. It ranges east across boreal Canada from Alaska, and south into the Great Lakes region. But it occurs in our area only as an isolated race (*claytoni*) in northern Maine and New Brunswick. Fewer than a dozen global sites are known for this entity (called Clayton's Copper). It is on the endangered species list in Maine. Some of its known sites are protected, especially those in fens or acid bogs, but the hostplant also occurs in dry, rocky uplands, and here the colonies are far less secure. Even if undisturbed, such colonies eventually will be crowded out, as abandoned pastures revert to forest.

But another danger can be far more immediate: the *claytoni* race's "type location" in Springfield, Maine, where it was described in 1940, was cleared for development in 2000. Former stands of hostplant were destroyed

Ventral: 8/4/00, Aroonstock Co., ME

and the primary habitat site leveled, very likely eliminating the colony (we have not monitored the site since).

The Dorcas Copper is a sedentary butterfly that seldom wanders far from its local colony. (Some dispersal inevitably occurs among local metapopulations, but the dominant mode is one of clustering.) Males perch and patrol near hostplants to locate females. Both sexes nectar actively, and may stray short distances from bog habitats to feed. We find their flight fairly fast and direct, at least in comparison with Bog and American Coppers.

Identification	Medium-small (but larger than Bog). Must be separated from the similar Purplish Copper (*L. helloides*) in many places, but in our area the only similar, sympatric species is the Bog Copper.
Dorsal	Brown with scattered dark markings. Fresh males show bright, blue-purple iridescence. Faint orange spots appear at HW outer angle.
Ventral	FW orange, HW rosy brown (fades to pale when worn; never yellow); note indistinct, inward-pointed orange scallops along outer HW margin.
Habitat	Acid bogs, fens, streamsides, old brushy fields, in all cases with hostplant.
Hostplants	Shrubby cinquefoil (*Potentilla fruticosa*), possibly other cinquefoils. Eggs laid under leaves near top of hostplant; caterpillars eat leaves. Though unrelated, docks and smartweeds both contain flavenoids that induce ovipositing in female Dorcas Coppers (Douglas, 1989, pp. 178–79).
Occurrence	Can be locally common in colonies. One flight, mid-Jul-late Aug in ME. Eggs overwinter; young caterpillars crawl to hostplant in spring.
Ecology	Medium specialist. Closely tied to a single hostplant and limited habitats.

Shrubby Cinquefoil (*Potentilla fruticosa*)

av wingspan: 1.05″

Atala   *Eumaeus atala*

Dorsal male (left), **female** (right): **specimen photos**

Ventral: 6/24/98, Palm Beach Co., FL

The fate of this spectacular butterfly was uncertain during much of the 20th century, as its specialized lifestyle was overmatched by hostplant and habitat loss, fires and hurricanes, biological competitors—and on occasion by undisciplined collection (Klots, 1951; Kimball, 1965). Florida Atalas were believed extirpated by 1940. One remnant colony was found unexpectedly in 1959, but then disappeared in 1965. The entire current population seems to have descended from a single colony on Key Biscayne, newly discovered in 1979 (R. Hammer, pers. comm.). Its progeny were "seeded" to other likely habitats and are now propagating steadily (sometimes traveling on nursery plants). We may never know whether these last individuals were holdovers from the original population or fresh Caribbean emigrants.

The Atala's unhurried, mothlike flight reflects the presence of deadly cycasins in its wings and body, derived from its hostplant. All life stages may be protected (the eggs perhaps indirectly, by scales pasted on from the female's abdomen). Cycasins may double as a component of the male's sex attractant. Courting males hover over females, wafting air past the "hair pencil" extended from their abdomens to deliver pheromones; cycasins have been detected in this exudate (Dietrich, 2001).

Identification	Boldly marked, with bright, aposematic warning colors. The largest hairstreak in our region; tailless with rounded wings. *Abdomen bright, red-orange.* No range overlap at present with the roughly similar Great Purple Hairstreak.
Dorsal	Deep blackish base. Bright *iridescent patches on FW*, extensive metallic *green* (male), or less extensive metallic *blue* (female). Wings almost always closed at rest.
Ventral	FW unmarked black. Conspicuous *red-orange spot on HW near abdomen*; also three concentric rows of gaudy, *neon-blue metallic spots.*

Habitat	Originally dry pine rocklands and nearby hammock edges (where adults perch at midday). Commoner now in settled areas with cultivated cycads. Wanders to nearby scrubby areas and gardens for nectar (lantanas preferred).
Hostplant	Coontie (*Zamia pumila*), a "fossil" gymnosperm of Paleozoic lineage; endemic to dry, semitropical pinelands. A starchy food (when detoxified), harvested commercially in the 1800s. Exotic cycads also used, making Atalas unwelcome at some botanic gardens. Eggs laid in clusters (usually 5–15) on fresh foliage. Unlike species that may select fresh leaves to avoid toxins, Atala larvae seek high cycasin concentrations in new growth.
Occurrence	Once common in s. FL—abundant where downtown Miami now stands. Original northern limit unclear. Nearly continuous broods. Also Cuba and Bahamas. Intolerant of cold (Coontie grows north of Atala's range). Communal caterpillars can wipe out entire hostplant stands, especially if weakened by fire (Hall & Butler, 2000), thus undermining colonies. Females are disposed to periodic dispersal, lessening local concentrations.
Ecology	Specialist. All aspects of the Atala's life are highly specialized.

Coontie (*Zamia pumila*), also young larvae and pupas on hostplant.

av wingspan: 1.75″

Great Purple Hairstreak   *Atlides halesus*

Dorsal male (left) and **female** (right): **specimen photos**

Ventral female: 8/10/97, Gr. Dismal Swamp NWR, Nasemond Co., VA

The Great Purple Hairstreak is not purple above, as often noted, but rather blue. Still, from almost any vantage point it is strikingly beautiful. Observers marvel at how its upper surface (which is often briefly visible in the field) can be so intensely brilliant. The answer may lie in the physics of its black wing borders: some butterflies use specialized, refractive scales to create "ultrablack" wing borders that absorb nearly all light, thereby enhancing the brilliance of adjoining iridescent display patches (Vukusic et al., 2004).

The Great Purple Hairstreak is the only butterfly whose larvae depend entirely on American Mistletoe (two unrelated western hairstreaks feed on Dwarf Mistletoes [*Arceuthobium*]). Commentators often assume that the Great Purple Hairstreak gains protection from predators by sequestering toxins from its larval host (Smith et al., 1994; Whittaker, 1984). But adult Great Purples engage in active "false head" displays, usually associated with predator diversion and palatability. "Wing rubbing" may be merely an ingrained, ancestral "habit" in this case. Still, ant-tending may be an alternate means of larval protection—although this has not been observed in the field, and parasitism of larvae is reportedly quite high.

Male Great Purples perch in the treetops during the afternoon, awaiting females. Flight is active and erratic. Many nectar plants are used, but Devil's Walkingstick (*Aralia spinosa*) is a favorite in late summer.

Identification	Our second-largest hairstreak (next to Atala). Two tails on each HW, longer on female. Abdomen is *bright, orange-red* below, body has *white spots*. Difficult to confuse with any other species in its range (does not occur with Atala).
Dorsal	Male wing is *jet black* with extensive *blue iridescence* and dark stigmas near FW costa. Female lacks stigma and has *less blue*.
Ventral	*Dark black* initially; fades to brownish-purple with wear (maybe this is the Great Purple's "purple"); *blue and gold iridescence* near HW tails, blue iridescence on inner FW. *Three red spots* at base of wings.

Habitat   Moist hardwood forests, bottomland woods and swamps, pocosins, savannas, or other sites with mistletoe-infested trees and nectar. Other habitats in West.

Hostplants   Oak or American Mistletoe (*Phoradendron leucarpum*), a hemi-parasite on several tree species (it photosynthesizes some nutrients, but also draws on host trees, gradually weakening them). Female lays up to 22 eggs on the host, but not in a cluster. Young larvae eat male flowers, new leaves, and the outer layer of old leaves; mature larvae eat mature leaves, pupate under loose bark at base of tree.

Occurrence   Fairly common in proper habitat. Delmarva south to s.-central FL (formerly to Miami); also west to CA, south to C. Amer. NJ before 1900, but mistletoe collecting led to extirpation (Gochfeld & Burger, 1997). Three or more broods, Mar-Nov in VA; all year in s. FL. Pupae overwinter.

Ecology   Medium Specialist. Adaptable in climate and habitat, but host-specific.

Oak, or American, Mistletoe (*Phoradendron leucarpum*)

av wingspan: 1.45″

Coral Hairstreak   *Satyrium titus*

Ventral: 7/2/95, Sterling Forest, Orange Co., NY

Orange Milkweed (*Asclepia tuberosa*), a habitual nectar plant

The sight of a Coral Hairstreak nectaring on Orange Milkweed, while highly familiar, is still a welcome experience in early summer. Displayed against the milkweed's exceptionally bright flowerhead, the butterfly's simple but attractive pattern creates an appealing study in orange.

The Coral Hairstreak is the first *Satyrium* hairstreak listed on our regional checklist. But it is not an entirely typical member of the group. Indeed, it was long placed in a separate genus (*Harkenclenus*), named for the 20th-century lepidopterist Harry K. Clench: i.e., *Har*[ry]+*Ken*[lion]+*Clen*[ch]+*us*. Commentators cite the Coral's lack of tails and uncharacteristic habitat associations as points of distinction. The lack of tails is certainly a valid difference, but the second point is less certain. True, the Coral does not fit the standard "*Satyrium* image," in that it is not a woodland hairstreak, like the Banded or the Hickory. Rather, it inhabits open, scrubby, and second-growth areas. But it is not alone in this, even within its genus. We prefer to think of the Coral Hairstreak as one of an ant-attended subgroup within the genus *Satyrium* that favors lower-growth habitats (fields, barrens, etc.). Others in this ecological subgroup are Acadian, Edward's, and (to some extent) Striped.

Male Coral Hairstreaks perch from mid-morning on to await females. They may hilltop (when possible) on perches 10-12 feet off the ground. From this vantage point, they aggressively confront rival males and other intruders. Mating has been observed in the afternoon. Many plants are used for nectar, but Orange Milkweed is clearly a favorite.

Identification	Medium-small (but fairly large for its genus). Our only *Satyrium* with *no tails*. Male's FW is *tall and pointed*, female's is more rounded. Variable geographically. The weakly differentiated northern race (*titus*) is darker below than the pale, southern race (*mopsus*). Usually not difficult to identify, though can resemble Acadian, or to some extent Edward's, especially if either of these is missing its blue "tail spot." This is not an unlikely possibility, since the tailed hairstreaks often perch with their heads slanted downward and actively rub their hindwings, giving the impression of a "false head," complete with threadlike "antennae." An approaching predator might very well strike this part of the wing. *See generally* Robbins (1993; 1981).
Dorsal	Dark brown. Males have a pale FW scent patch, females a variable row of orange spots along the trailing HW. Dorsal surface seldom displayed.
Ventral	Light brown with a row of *red-orange ("coral") spots along the HW, with no blue thecla spot*. Also note postmedian row of black dots encircled with white (the white may be faint or missing in northern individuals).

Habitat	Shrubby open areas: old pastures, second growth, clearings, etc.; sandhills.
Hostplants	Rose Family (Rosaceae): Wild Cherry (*Prunus serotina*); Chokecherry (*P. virginiana*); American Plum (*P. americana*), Common Plum (*P. domestica*), Chickasaw Plum (*P. angustifolia*), etc. Eggs laid at base of hostplant on trunk; larvae are ant-attended, feed at night on flowers, young fruit (Allen, 1997).
Occurrence	Common in center of range; rarer to the North and South. N. Eng. south to nw FL, rare on coastal plain south of VA. Also west to BC, n. CA, south to n. TX. One brood, mid-May-Jun (FL), mid-Jun-Jul (NY), mid-Jul-Aug (north). Eggs overwinter.
Ecology	Medium Generalist. Uses common and widespread hostplants and habitats.

Chokecherry (*Prunus virginiana*)

av wingspan: 1.2"

Oak Hairstreak   *Satyrium favonius*

'Northern' Oak Hairstreak (*ontario race*): 6/23/94, Queens Co., NY

'Southern' Oak Hairstreak (*favonius race*): 3/25/94, Hendry Co., FL

'Northern'

'Southern'

Many *Satyrium* hairstreaks have large, brightly colored populations in the South, but none differs as dramatically from its northern counterpart as the 'Southern' Oak Hairstreak, long considered a separate species. The 'Northern' Oak Hairstreak (*ontario*) is the rarest *Satyrium* in its range—so much so that Holland (1931) wondered if it might not be an aberrant form of some other species. The 'Southern' Oak (*favonius*) is locally more prevalent. In Florida, *favonius* is the most common *Satyrium.*

The Oak Hairstreak has been the center of much taxonomic uncertainty. Systematists have placed it in several different genera—including *Strymon* (with the scrub-hairstreaks), *Euristrymon*, and *Fixenia*—before it settled in *Satyrium.* At the species level, it was not until the last decade that evidence of clinal interbreeding in the Southeast caused the northern and southern populations to be regarded as a single entity (Robbins, 1994).

The Oak Hairstreak's common name also has fluctuated. Traditionally called Northern and Southern Hairstreak (as two separate species), the name "Southern Hairstreak" was at first applied to the combined entity. But this left *ontario* as the 'Northern' Southern Hairstreak. And since the north-south distinction was no longer meaningful, Oak Hairstreak was adopted. (Many other *Satyrium* hairstreaks feed on oaks, but only this and the Edward's are oak specialists in the East, so the current name is at least an improvement.)

Little behavioral information has been published on this species. The 'Northern' Oak Hairstreak usually is seen nectaring in the vicinity of an oak (or scrub oak) forest, frequently on Common Milkweed or dogbane, and often in the company of other hairstreaks. The 'Southern' is seen most often in or near Live Oak mottes.

Identification	Medium-small (average for a *Satyrium*; largest in South). Two tails on each HW, one short (both much longer in the South; *see favonius* photo, above). Lack of cell-end bars makes patterning look "clean" (Glassberg, 1999).
Dorsal	Gray-brown (Northern) to blackish-brown (Southern). Pale yellowish-orange patches on all wings, pronounced in Southern but reduced or absent in Northern. Male has gray, oval stigma. Dorsal surface seldom displayed.
Ventral	Northern is light gray-brown with a *thin black-and-white postmedian band forming a "W" near the tails.* Orange marks on HW margin are *small and separate.* Southern has *browner base color,* conspicuous *wedge-shaped red-orange mark on outer HW,* and a *single, white dot* near the leading HW edge, toward the body.

Habitat	Oak forests (coastal and upland), oak mottes, pine-oak barrens, nearby areas with nectar.
Hostplants	Oaks. 'Northern' uses White (*Quercus alba*), Post (*Q. stellata*), Laurel (*Q. laurifolia*), etc. 'Southern' uses Live (*Q. virginiana*), Blackjack (*Q. marilandica*), etc. Female lays on twigs that will bear male catkins (Tveten & Tveten, 1996). Larvae eat these, later young leaves. Pupates in leaf litter.
Occurrence	Mainly uncommon to rare, s. N. Eng. to s. FL (formerly Miami); other populations in West. One flight, earlier than other *Satyrium*, late Mar-May in FL; May-mid-Jun (NC); mid-Jun-early Jul (NY). Eggs hibernate.
Ecology	Medium generalist. Widespread but usually rare.

Virginia Live Oak (*Quercus virginiana*)

av wingspan: 1.1"

Acadian Hairstreak   *Satyrium acadica*

Ventral: 7/3/98, Wyandanch, Suffolk Co., NY

Willow (*Salix* sp.) in wet meadow

With morning dew still clinging to the vegetation of a dank, northern meadow, it is memorable to watch adult Acadian Hairstreaks walk quietly from their concealment, one by one, to bask on flat leaves before taking flight. Scores may be seen this way in the center of an active colony. But most sightings of this species are not so dramatic, as we usually see scattered individuals nectaring alone.

Aptly named, the Acadian is the most northerly of our *Satyrium* hairstreaks. (The Coral Hairstreak does range farther north in western Canada, but not so on the East Coast. And Corals occur much farther south throughout their range, reaching northwestern Florida, and thus are not as distinctively northern as the Acadian.)

The nominate race (*acadica*) occurs throughout our region. There are other races in Canada and to our west.

The lifestyle of this large, rather handsome *Satyrium* is not well-reported. Nearly a century ago, Clarence Wood remarked, "This is one of the numerous butterflies that offer some young student an opportunity to make real contributions to science" (Wood, 1917, p. 248). Some 87 years later, the Acadian's early stages are now documented, but Wood's statement still retains some validity. The Acadian is known to be ant-attended (Scott, 1986), for example, yet few details have been reported. It may be that this species' life history is as distinctive as the Edward Hairstreak's (*see* next account). Both the Edward's and the Acadian live in compact, often densely populated colonies.

Male Acadians perch on low vegetation to await females, especially in the afternoon, and sometimes patrol (Scott, 1986). Both sexes regularly seek nectar.

Identification   Somewhat larger than most *Satyrium* hairstreaks. One tail on each HW. Resembles Edwards, Banded, Hickory, even Coral Hairstreaks to some extent, but positive identification is rarely problematical.

Dorsal   Uniform brownish-gray with a well-defined *orange spot* at base of the HW tail; but dorsal surface rarely visible.

Ventral   *Pale gray*, even bluish-gray (others noticeably browner). Conspicuous *orange markings* along outer HW (includes an *orange cap over the blue lunule*, which is lacking in Banded, and is less prominent in others); orange markings continue on FW, but are less conspicuous. Postmedian band is expressed as *round, well-separated spots, circled in white* (vs. oval and less well-separated in Edward's).

Habitat   Damp fields, meadows, weedy streamsides, sometimes sandy plains, but always with willows. Other habitats in the West.

Hostplants   Various willows, including Black (*Salix nigra*), Silky (*S. sericea*), Sandbar (*S. exigua*), Pussy (*S. discolor*), Slender (*S. petiolaris*), and Beaked (*S. bebbiana*). Caterpillars eat hostplant leaves.

Willow catkins

Occurrence   Locally uncommon to rare, but can be abundant at colony locations (125+ individuals). N. Eng. south to central PA, also Nova Scotia west to w. CAN, south to Rockies. One flight, generally late Jun-Jul; Jul-early Aug in far North. Eggs hibernate.

Ecology   **Medium Specialist.** Focused lifestyle, but some variations.

av wingspan: 1.2"

Edwards' Hairstreak *Satyrium edwardsii*

Ventral: 7/6/96, Pound Ridge Reservation, Westchester Co., NY

Scrub (Bear) Oak (*Quercus ilicifolia*) growing in sandy barrens

Named for William Henry Edwards, prominent 19th-century butterfly authority. Few of the lepidopteran life histories chronicled in these accounts is any more intricate nor unexpected than that of the Edwards' Hairstreak. Edwards' live in densely concentrated colonies, with a single, highly synchronized annual brood. Their isolated occurrence reflects a specific suite of biological requirements. Namely, Edwards' are confined to low-growth scrub oak habitats with nearby ant colonies (principally, if not exclusively, *Formica integra*). The ants tend and protect Edwards' larvae in exchange for sweet secretions. The ant colonies are located not just where Edwards' occur, but also where they can obtain an uninterrupted supply of "honeydew"—both during and after the Edwards' brief larval period. The presence of larval treehoppers and scale insects—which the ants also attend—is thus important (Webster & Nielsen, 1984). Attendant ants construct conical, ant hill-like shelters (called "byres") at the base of host trees, where older Edwards' caterpillars rest during the day. At night, the caterpillars emerge to feed, escorted by protector ants. (Younger larvae do not produce honeydew, and feed alone on oak flowers; still, the ants seem to recognize and shield them—or at least do not attack them as intruders.)

The Edwards' is mainly a northern butterfly, but it occurs rarely in the Appalachians and sandhills of the Southeast, where a larger, longer-tailed population occurs (Gatrelle, 2001a; reported subspecies *meridionale*).

Adult males are aggressively territorial. They perch on scrub oak leaves and other relatively low roosts. From there, they engage in frequent aerial dogfights, with very rapid flight. Nectaring is comparatively infrequent, and may draw Edwards' temporarily away from their dry, primary habitat, where nectar sources can be scarce.

Identification	Medium-small (average for a *Satyrium*; larger in the South). One small HW tail (much longer in South). Closely resembles several other *Satyrium*, including Banded, Hickory, Acadian, and possibly King's; but positive field identification usually possible.
Dorsal	Light gray-brown. Usually a *small orange spot* on trailing HW.
Ventral	Pale gray-brown. Postmedian band consists of *disconnected ovals, circled in white* (but closer to Banded in Deep South). An orange "spear" extends along the inner HW margin, more extensive than in Banded (but less than in Acadian). Blue lunule has *very slight* orange cap, which is lacking entirely in the Banded.

Habitat	Scrub oak thickets in rocky open habitats, including shale barrens and limestone outcrops. Also oak savannas and sandhills.
Hostplants	Scrub oaks, principally Scrub (or Bear) (*Quercus ilicifolia*); Blackjack (*Q. marilandica*) in South, where Bear Oak rare/absent; also Black (*Q. velutina*), Scarlet (*Q. coccinea*), Turkey (*Q. laevis*) in NC. Eggs laid in tree wounds or under dead wood or bark.
Occurrence	In colonies from s. ME to n. GA; rare in South; also west to ND, ne TX. One short flight (as little as a week, varying with weather), May-Jul in South; Jun to Jul or early Aug in North. Eggs hibernate.
Ecology	Specialist. We have relatively good information on this butterfly's highly specialized lifestyle.

Oak scrub along utility easement, Westchester Co., NY

av wingspan: 1.1"

Banded Hairstreak   *Satyrium calanus*

Ventral (darker variant): 6/22/96, Pound Ridge Reservation, Westchester Co., NY   **Ventral (lighter variant)**: 6/28/92, Wyandanch, Suffolk Co., NY

This is the most common and widespread of our *Satyrium* hairstreaks. Bandeds can be hyperabundant during peak flight years, mainly in the northern states, but in other years few may be seen.

The Banded Hairstreak epitomizes the *Satyrium* lifestyle. There is a single annual brood in early to mid-summer. Females lay eggs on hardwood twig ends, near buds that will develop the following spring, providing food for emerging caterpillars. (By comparison, Lycaenids that feed on the flowers of herbaceous plants have no such secure repository for their eggs. They overwinter instead as pupae. In the spring, newly emerged adult females seek out the fresh buds of sprouting herbs on which to place their eggs.)

The Florida race of Banded Hairstreak (*calanus*), once considered a separate species, was thought to be rare, or even possibly extinct (Klots, 1951). It is now recognized as a large, locally uncommon race, with long tails and enhanced hindwing markings. The subspecies found in most of our area is *falacer*, which is more plainly marked. Other subspecies occur in the West.

Male Banded Hairstreaks perch at various levels in host trees (commentators indicate low perches, from 4 to 12 feet, but we often see individuals perched much higher in host trees as well). Banded are renowned for their swirling aerial dogfights, with jousting duos (or clusters) ascending high into the air. The victor usually returns to his original perch afterward. Both sexes actively nectar, often on milkweeds or dogbanes in the North. With short proboscises, they are limited to short-tubed flowers (Opler & Krizek, 1984).

Identification	Medium-small (average for a *Satyrium*; larger in the South). Two HW tails, one shorter. The tree-feeding *Satyrium* hairstreaks present significant identification problems, particularly as regards the Hickory (*see* discussion in next account). Wing wear often complicates already difficult identifications.
Dorsal	Dark, blackish brown; male has FW scent patches; female has orange patches in *calanus* race. Dorsal surface rarely displayed in the field.
Ventral	*Slate gray to blackish-gray* (vs. brownish-gray). Postmedian bands narrow, usually *white-edged only outwardly* (sometimes both sides). Blue marginal patch *not capped with orange*, darker blue than in Hickory, *extends little (if at all) beyond line of orange marginal spots.*

Habitat	Mixed and deciduous forests; most often seen at forest edges, fields, and disturbed areas with nectar. "Its only requirements are the presence of suitable hosts and nearby nectar sources" (Opler & Krizek, 1984, p. 92).
Hostplants	Many oaks, walnuts, hickories, walnuts. Oaks including White (*Quercus alba*), Chestnut (*Q. prinus*), Bluejack (*Q. incana*), Turkey (*Q. laevis*); also Butternut Hickory (*C. cinerea*), etc. Larvae eat catkins, then young leaves.
Occurrence	Variably common, s. ME to central FL. Also west to central CAN, south in Rockies, Gulf Coast to e. TX. One brood: Apr-May in FL; May-Jul in NC; mid-Jun to mid-Jul in North. Eggs overwinter
Ecology	**Medium Generalist.** Tied to forests, but flexible and adaptive.

Black Oak (*Quercus velutina*), with basking hairstreak

av wingspan: 1.125″

Hickory Hairstreak *Satyrium caryaevorum*

Ventral: 7/4/95, Bedford, Westchester Co., NY

Ventral (variant): 7/8/01, Orange Co., NY

Much remains to be learned about the lifestyle and distribution of this mainly northern hairstreak. The lack of available details might be attributed to the recentness of its split from the Banded Hairstreak—but in fact the Hickory was described some time ago, in 1942. The issue is more the visual similarity of the Hickory and Banded, which complicates accurate observation and recordkeeping, and also the Hickory's secretive nature. It seems to spend a significant portion of its time in the treetops, out of sight.

Thus, disagreement persists as to whether the Hickory is rare or merely local and cyclical. All agree it can be abundant in occasional boom years. In 1992, for example, Hickories outnumbered all other hairstreaks in southern Canada combined, with "thousands" nectaring on a single Basswood tree (Layberry et al., 1998, p. 134). Between such peaks, however, it may virtually disappear in many locations, leaving observers to wonder whether they have overlooked it in the interim, or whether it undergoes broad, geographic extension during peak years.

Whatever the case, Hickories *do* seem rare in the southern portion of their range. They have been recorded in just three counties in Virginia, three in North Carolina, and four in West Virginia (LeGrand & Howard, 2004; Allen, 1997). And a number of these sightings represent isolated records.

Rather little is known of the Hickory's behavior. Individuals often descend to nectar in the morning and evening, probably perching in the canopy at other times. Yet they may be observed at any time of day, nectaring or engaged in aerial dogfights. They often associate with other hairstreaks (especially Banded).

Identification	Medium-small. One HW tail. Highly similar to Banded—and both show individual variation. Many individuals present confusing "mixed signals" regarding "diagnostic" marks. Identification always should be based on multiple traits, and even then not all individuals can be safely identified in the field, especially if worn. (Genitalic shape is diagnostic, if required.)
Dorsal	Uniform dark brown; male has dull, FW scent patches.
Ventral	Light gray-brown (banded, usually slate gray, even blackish). *Sky-blue patch* at outer angle is *pale and shiny* (darker in Banded), extends *farther inward on wing* than Banded's (breaks through the imaginary line formed by the inner edge of the surrounding orange marginal dots). The postmedian HW band is usually *white-edged on both sides*. On the FW, it *widens* near leading edge (at the costa; *compare* Banded). The incomplete, median HW band is *wider* than the Banded's.

Habitat	Rich deciduous forest edges and nearby nectar sites (fields, etc.).
Hostplants	Hickories: Shagbark (*Carya ovata*); Bitternut (*C. cordiformis*). Also Black Ash (*Fraxinus nigra*); accepts other hardwoods (including oak) in captivity. Larvae feed on lower surface of leaves.
Occurrence	Usually uncommon or rare, from s. N. Eng. south to n. GA; absent from coastal plain below central NJ. Also west to MI, south to AR. One brood, Jun-late Jul (occasionally early Aug). Eggs hibernate at twig ends.
Ecology	**Medium Generalist.** Apparently flexible in terms of hostplant and habitat, yet normally much rarer than Banded in most locations.

Shagbark Hickory (*Carya ovata*); eggs overwinter on exposed branch tips

av wingspan: 1.1″

King's Hairstreak   *Satyrium kingi*

Ventral: 6/7/98, Weymouth Woods Sandhill Nature Preserve, Moore Co., NC

Sourwood (*Oxydendrum arboreum*), frequent nectar source in NC

Our only southern-based *Satyrium*, the King's Hairstreak was not formally described until 1952. It was originally found by Harold L. "Verne" King, a Latin American hairstreak and metalmark specialist (Opler & Krizek, 1984). The King's Hairstreak was not included in Klots's 1951 field guide, although Klots was the lead author of the article describing it a year later (Klots & Clench, 1952).

Initially, the King's Hairstreak was described by commentators as exceptionally rare and local. In retrospect, this was not fully accurate. While King's is a "good find" in most locations, Hickory and Edwards' are rarer in the Southeast. Indeed, few *Satyrium* of any kind are common in the South (H. LeGrande, pers. comm.). King's is known from many counties in North Carolina (LeGrand & Howard, 2004). But in Delaware and Maryland, at the northern edge of its range, King's is legitimately conservation-listed.

The distribution of King's Hairstreak closely mirrors that of its principal—and probably sole—hostplant, Sweetleaf (sometimes called Horse Sugar). Sweetleaf is the only member of a widespread tropical family (Symplocaceae) found in North America. It is a small, understory tree that grows along forest and stream edges, and is a component of the "shrub layer" in many pocosins. It flourishes after disturbance, when overstory trees are temporarily unable to suppress its growth. Such young hosts are favored by the butterfly. Sweetleaf is absent on much of the Piedmont, which accounts for the King Hairstreak's odd distribution: it is often found in the mountains and on the coastal plain, but rarely on the Piedmont (LeGrand & Howard, 2004).

Male King's perch at midday, sometimes on fairly low foliage. They engage in territorial jousting, but are otherwise rather staid (for a *Satyrium* Hairstreak), in both flight style and approachability. Nectaring is opportunistic: few flowering plants are available during flight season in many of the King's habitats, and in such cases adults will concentrate on whatever is locally available (examples include Sparkleberry, Sourwood, and New Jersey Tea).

Identification	Medium-small. Wings rounder than in typical *Satyrium*. Two HW tails, one very short. Can be confused with Edwards', Hickory, and Striped Hairstreak where ranges overlap.
Dorsal	Uniform dark brown, with bronze-green iridescence when fresh. Dorsal surface seldom exposed.
Ventral	Light gray-brown. The *pale blue marginal patch* is *prominently orange-capped* (the cap is often more pointed and inwardly convex in Striped). Postmedian HW bands resemble those of Banded or Hickory (they are much wider in Striped); trailing end of the band *points toward blue marginal patch* (vs. toward nowhere in particular in the Striped). FW cell-end bars are well separated from the postmedian line (*compare* Striped).

Habitat	On the coastal plain, pocosin edges (especially where they meet sandhills), brushy forest edges, streamsides, swamp edges, forest clearings. Mountain habitats less well-reported.
Hostplants	Sweetleaf (*Symplocos tinctoria*); has accepted Flame Azalea (*Rhododendron calendulaceum*) in captivity. Eggs laid on twig ends; larvae eat leaves only, refuse flowers or fruit.
Occurrence	Uncommon to rare, occasionally coastal DE south to nw FL, also west to e. TX. One flight, May-Jun in South; mid-Jun-Jul in se VA; mid-Jul-mid-Aug in mountains. Eggs overwinter.
Ecology	**Specialist.** Closely linked to the ecology of a single hostplant.

Sweetleaf (*Symplocos tinctoria*)

av wingspan: 1.2″

Striped Hairstreak   *Satyrium liparops*

Ventral (northern *strigosum*): 6/27/92, Van Cortlandt Park, Bronx Co., NY

Ventral (southern *liparops*): 5/29/95, Gr. Dismal Swamp NWR, Nasemond Co., VA

Taxonomically, Striped Hairstreaks belong to the genus *Satyrium*, but they behave in some respects more like scrub-hairstreaks (genus *Strymon*). This is due partly to their hostplants, which include small trees and second-growth shrubs that thrive in comparatively scruffy habitats. But they also lack the strong territorial tendencies of other *Satyriums*: Striped Hairstreaks are not "macho dogfighters" like some of their relatives, who constantly jostle in tight clusters. Indeed, they are most often seen as singletons, perched in isolation or quietly nectaring.

Many commentators have noted that Striped Hairstreaks are low-density butterflies, "never recorded in abundance like some other hairstreaks" (Layberry et al., 1998, p. 135). There is a least one East Coast "hot spot," however, Great Dismal Swamp, where in the mid-1990s Strideds were seen in large numbers, in one case outnumbering all other butterfly species (LeGrand & Howard, 2004; R. Cech, pers. obs.; *compare* Roble et al., 2000).

There are several races of Striped Hairstreak, which meet in broad mixing zones (or "clinal intergrades"). The northern *strigosum* race is short-tailed and cleanly striped, with reduced outer hindwing markings. The southern *liparops* race is larger and longer-tailed, and has more pronounced hindwing marks. A newly reported Florida race (*floridensis*) is a subset of *S. l. liparops* (Gatrelle, 2001b). Other, differing races occur to our west.

Male Striped Hairstreaks perch throughout the day, rarely more than 20 feet high, and nectar frequently. Some colonies are transient, but we are aware of a number that have remained stable for many years (taking into account the annual fluctuations typical in this genus).

Identification	Small to medium (larger in the South). One or two HW tails: shorter in North (males sometimes have just one), much longer in southern races. Outer margin of HW indents inward, just above the base of the second tail, but this also occurs in King's Hairstreak and to some degree in some of the other species as well.
Dorsal	Dark brown with variable reddish or yellow-orange patches; patches rare in northern race (*strigosum*), variable in southern (*liparops*), always present in proposed Florida race (*floridensis*) (Gatrelle, 2001b). Dorsal surface seldom seen in the field.
Ventral	Base color varies, gray-brown to blackish-brown. *HW bars wide, thinly outlined in white*, giving impression of stripes. Blue spot at base of HW tail *capped with orange* (also in King's).

Habitat	Openings/edges of second-growth mixed and deciduous woods; also shaded swamps, bogs.
Hostplants	Numerous: mainly roses (Rosaceae), cherry/plum (*Prunus*), hawthorn (*Crataegus*), etc., and heaths (Ericaceae); also American Hornbeam (*Carpinus caroliniana*). Larvae feed on flowers, seeds, and young leaves in spring.
Occurrence	Most common in North. N. Eng. south to central FL. Also west to BC, Rockies, e. TX. One flight: Apr-May (FL); May-early Jun (e. NC); mid-Jun-Jul (NY); Jul-early Aug (ME). Eggs overwinter.
Ecology	**Medium Generalist.** Uses a wide variety of hosts and habitats in our area.

Black Cherry (*Prunus serotina*)

av wingspan: 1.1"

Brown Elfin   *Callophrys* [= *Deciduphagus*] *augustinus*

Ventral (nominate race): 5/25/97, Carroll Co., NH

Ventral (*croesioides* race): 5/6/93, Warren Grove WMA, Ocean Co., NJ

In early spring, this plainly marked butterfly can be found perched on low twigs in barrens and other dry, wooded habitats (also, curiously, in bogs). Like other elfins, it is single-brooded, with a flight period coinciding with new spring growth—in this case the leaves and flowers of various heaths, which are the caterpillar's main food.

This is often described as our commonest elfin—although none of our elfins is common in the true sense, i.e., compared with species like Pearl Crescent or Cabbage White. Still, within its close-knit group, the Brown Elfin is generally the most often seen, if only because it utilizes a comparatively broad range of hostplants and habitats. (Generalizations tend to fare poorly with elfins, on the other hand, and even here a caveat is required: in Ohio, where suitable habitat is rare, the Brown Elfin is considered the rarest of the group [Iftner et al., 1992].)

Despite ample-seeming opportunities for gene exchange among colonies, the Brown Elfin has evolved a number of geographic races. The northeastern race (*augustinus*), found in northern New England and Canada, is relatively dark, with heavily checked wing fringes. The southern race (*croesioides*), more common and widespread in our area, is usually lighter brown and has less pronounced fringe markings. Western Brown Elfins vary greatly in appearance and lifestyle from their eastern relatives; some may be separate species.

Male Brown Elfins perch throughout the day to await females. Individuals in flight usually cover only short distances, returning to a central perch or one nearby. Adults may live up to 23 days in the wild (Opler & Krizek, 1984). Such a long life span could have adaptive significance, to the extent it allows elfins to "ride out" adverse early season weather events and mate successfully.

Identification	Small and plain. *No tails or frosting*, and little obvious patterning. As noted above, wing fringes are checkered, especially in northern race.
Dorsal	Brown to gray-brown, female more reddish. But this surface is seldom exposed to view.
Ventral	*Rich, often reddish brown* when fresh, darker near base (fades to a uniform pale brown with age). Our only elfin with *no marginal frosting* on either wing. When fresh, HW shows a green or purplish iridescence.

Habitat	A variety of habitats with acidic, usually low-nutrient soils, including barrens (pine, shale, and serpentine), dry pine-oak woods, sandhills (in NC), outcrops, acid bogs, boreal scrub, etc.
Hostplants	Numerous heaths (Ericaceae), including blueberries (*Vaccinium*), huckleberries (*Gaylussacia*), Labrador-Tea (*Ledum groenlandicum*), Bearberry (*Arctostaphylos uvaursi*), Leatherleaf (*Chamaedaphne calyculata*); also Sheep Laurel (*Kalmia angustifolia*). Caterpillars eat flowers and fruits.
Occurrence	Fairly common, in colonies. N. Eng. south to n. GA; rare or absent on se coastal plain. One flight, Mar-Apr in GA, early Apr-May in NY; mid-May-Jul in North. Pupa overwinters in litter at base of hostplant.
Ecology	Medium Generalist. Reasonably adaptable—for an elfin.

Dwarf Huckleberry (*Gaylussacia dumosa*)

av wingspan: 1.0"

Hoary Elfin *Callophrys* [= *Deciduphagus*] *polios*

Ventral (on hostplant): 5/9/95, Warren Grove WMA, Ocean Co., NJ

Bearberry flats in pine barrens, Ocean Co., NJ

It is difficult to say which elfin appears earliest in the spring, since most live in isolated habitat pockets that are subject to local climate differences; such variations can advance or retard the date of emergence at any given location. Nonetheless, the Hoary Elfin is definitely on the wing very early, and is well adapted to living on the seasonal "edge." In low, exposed Bearberry flats, where Hoaries customarily fly, it is possible to see scores of them darting actively in the bright sun. But moments later, when a dark cloud passes over, the entire assemblage will vanish in seconds, and locating their hiding perches afterward is surprisingly difficult. This fast response allows Hoaries to avoid warm-blooded predators, who, unlike themselves, do not become sluggish when cold. Ducking into a sheltered niche thus can help preserve life and limb, as well as maintain critical body warmth.

As a sedentary species, we might expect the Hoary Elfin to have many local races, but there is in fact just one race in the East (*polios*). Perhaps a continuous supply of heathlands in the Hoary's northern range has allowed populations to maintain close genetic contact. But even in our region, where colonies are more isolated (and recently declining), we have seen them flying in open pine-oak woods, in apparent dispersal flights.

Males generally perch throughout warm periods of the day, on or very close to their ground-hugging hostplants. They remain low both in flight and when perched. They nectar in their primary habitat, often pollinating their own host (perhaps some of the very flowers their caterpillars will soon devour!).

Identification	Very small (among elfins, only the Bog is smaller). Lacks HW tails. Habitat is a useful cue in identification, especially in relation to the similar-looking Henry's and Frosted Elfins, since the Hoary infrequently occurs in close association with either of these species.
Dorsal	Upper surface rufous-brown in eastern race. Male has a dark FW scent patch. But wings are seldom held open, so these features are rarely visible.
Ventral	*Dark brown* (darker basally). Outer *HW broadly frosted*, across entire outer margin; *no dark thecla spot*, as in Frosted. *FW margin narrowly frosted* (not so in other elfins). Thin, *postmedian FW line is white*. Brown Elfin entirely lacks ventral frosting.

Habitat	In our region, a variety of low, open habitats with hostplant ("heathlands"), including barrens (shale and pine), alvars; also forest and bog edges, rocky ridges, lakeside dunes. High-altitude meadows in the Rockies.
Hostplants	Primarily Bearberry (*Arctostaphylos uva-ursi*), a low, trailing heath. Also at times Trailing Arbutus (*Epigaea repens*) (Layberry et al., 1998; Allen, 1997). But while Trailing Arbutus often grows alongside Bearberry, it is not limited to such sites and is much more widespread than the butterfly. Larvae eat flowers and leaf buds.
Occurrence	Can be locally common, but usually rare; gone from L.I., declining in Appalachians. Also west across CAN to AK, south in Rockies. One brood, Apr-May in South, mid-May-Jun in North. Pupa overwinters.
Ecology	**Specialist.** Rarely out of sight of a hostplant. Inhabits scarce and threatened barrens habitats. Eastern colony sites should be protected.

Bearberry (*Arctostaphylos uva-ursi*)

av wingspan: 0.9"

Frosted Elfin   *Callophrys* [= *Deciduphagus*] *irus*

Ventral (typical): 5/4/91, Assunpink WMA, Monmouth Co., NJ

Ventral (variant): 4/24/94, Brattons Run Shale Barrens, Rockbridge Co., VA

The Frosted Elfin's broad, geographic distribution—as shown on a typical range map—significantly overstates the butterfly's occurrence. Rare and local, it lives almost exclusively in small, widely scattered colonies. Frosteds are conservation-listed throughout their northern range, as well as in the Great Lakes states and in Florida. And they appear little more abundant (if at all) in the Southeast, where they are not specifically listed.

The fragile status of this butterfly relates to its selective habitat associations. Frosteds have adapted to more than one habitat type, and to at least two major hostplant groups (lupines and Wild Indigo), but all settings are rather specialized. Most colonies occur in barrens or in dry, disturbed habitats, most which usually are maintained by periodic surface fires that prevent succession (*see*, e.g., Wagner et al., 2003). Lupine-feeding Frosteds have developed the ability—unique as far as we know among our butterflies—to pupate below the surface (D. Schweitzer, pers. com.). This allows them to survive seasonal burns that can kill other species.

This commitment to early successional habitats and fire-maintained environments (favored by butterfly and host-plant alike) is now a disadvantage, however, as these sites are increasingly subject to development and fire-suppression (both of which destroy hostplants). Excessive deer browsing is also a growing threat. It can quickly eliminate entire populations of barrens-specialist lepidoptera (Wagner et al., 2003).

Some observers note differences between lupine-feeding and *Baptisia*-feeding Frosteds, in larval markings and other specifics, and suggest the possibility of sibling species (D. Schweitzer, pers. comm.).

Frosted Elfins are low-flying and low-perching. Efficient colonizers, they quickly adopt and abandon sites as conditions dictate (at least where fragmentation is not so extensive as to interfere with dispersal).

Identification	Slightly larger than other elfins, with a lobed HW outer angle and short HW tails. Easily confused with other elfins, especially when worn. Habitat provides some clue; Frosteds infrequently mix with Browns or Hoaries.
Dorsal	Male is brown or gray-brown, with a long, thin, dark *FW stigma* (lacking in Henry's). Female may be somewhat more reddish-brown. Dorsal surface is rarely exposed to view in the field, however.
Ventral	Usually *less contrast* between basal and outer area of wings than in Henry's or Hoary. Broad HW frosting extends toward body past midline, but not along the entire outer margin (*compare* Hoary). Note *dark thecla spot near HW margin* (diagnostic). The postmedian FW line is surrounded by variable white clouding, usually vague, that is absent in other elfins (Glassberg, 1999).

Habitat	Clearings in dry oak woods, shale and pine barrens, sandhills, coastal scrub; sometimes more mesic habitats.
Hostplants	Mainly Wild Lupine (*Lupinus perennis*) and Wild Indigo (*Baptisia tinctoria*); but also Blue False Indigo (*B. australis*), Rattle-box (*Crotalaria sagittalis*). Larva bores into inflated seedpods.
Occurrence	Widespread but globally rare. ME south to n. FL, west to WI and e. TX. One flight, emerging slightly later than other elfins. Late Feb-early Apr in FL, May-Jun in North. Pupa overwinters.
Ecology	Medium Specialist. Often locally specialized, but adapts to a significant variety of different settings at the species level.

Wild Indigo (*Baptisia tinctoria*), with Frosted Elfin

av wingspan: 1.1"

Henry's Elfin   *Callophrys [= Deciduphagus] henrici*

Ventral (holly feeder): 4/14/96, Belleplain SF, Cumberland Co., NJ

Ventral (Redbud feeder): 4/23/95, Flintstone, Allegany Co., MD

The Henry's Elfin emerges early in the spring, though exact dates vary with location and weather. The species' lifestyle is difficult to summarize in general terms, since regional populations have developed a wide variety of differing habitat and hostplant associations.

Observers and taxonomists long confused the Henry's with other elfins, especially Frosted. And even after Henry's was correctly classified, a perplexing array of local forms remained to be addressed.

Henry's has two major population groups: first, a set of obligate holly-feeders on the Atlantic coastal plain; and, second, a number of polyphagous populations found mainly in the North and Appalachians, many of which feed on flowering Redbuds (Gatrelle, 1999b).

Male Henry's perch above eye level throughout much of the day to await females. Females also perch, but usually lower (Allen, 1997). In cool weather, either sex may roost on sun-warmed rocks. Both sexes also nectar and males seek minerals with some frequency. A sedentary butterfly, the Henry's is seldom found far from the immediate vicinity of a hostplant.

American Holly (*Ilex opaca*), Atlantic coastal hostplant, with Henry's Elfin

Identification	Small (but average for an elfin). Shares certain traits with the Frosted, such as short, nubbin-like HW tails (other elfins are tailless), and an enlarged lobe at the HW outer angle. Males lack a stigma, but this is seldom visible.
Dorsal	Mostly dark brown, sometimes orange-toned outwardly. Wings remain closed at rest.
Ventral	*Strongly two-toned*, darker at base, with *white outlines at either end* of the HW median separation line. Outer HW *frosted*. Outer FW often *green-toned* in Atlantic coastal populations (Pavulaan, 1998).

Habitat	Highly variable: barrens; coastal holly forests; dry upland clearings/edges; pocosins; etc.
Hostplants	Also varied. Most populations use a single host (or a few related ones). Redbud (*Cercis canadensis*), hollies (*Ilex*, especially *opaca* and *vomitoria*), blueberries/huckleberries (*Vaccinium*), Black Buckthorn (*Rhamnus frangula*). Some larvae eat flowers/seeds, others leaves.
Occurrence	Once considered rare, but in fact often the most common elfin locally. Coastal and s. N. Eng., south to central FL; also west to WI and e. TX. Increasing in N. Eng. with use of alien buckthorn. One brood, Mar-Apr in FL to Apr-May/early Jun in North. Pupa overwinters in leaf litter.
Ecology	**Medium Generalist.** Highly adaptable; even specialized populations utilize widespread hosts.

Redbud (*Cercis canadensis*), Appalachian host

av wingspan: 1.0"

Eastern Pine Elfin   *Callophrys* [= *Incisalia*] *niphon*

Ventral (lighter variant): 4/18/98, Chatsworth, Ocean Co., NJ

Ventral (rubbing wings): 5/14/00, Albany Pine Bush, Albany Co., NY

Pine Elfins are exceptionally attractive butterflies, but their most distinctive biological feature may be their diet. Specifically, Pine Elfin caterpillars (along with those of the next three species) are the East Coast's only dedicated conifer feeders. This specialization allows them to occupy a broad and little-exploited niche, shared mainly with moth larvae. Some taxonomists assign these species to their own genera based on food preference, pine-feeders being assigned to *Incisalia* and cedar-feeders to *Mitoura* (these taxonomists place other elfins in *Deciduphagus*).

The Western Pine Elfin (*C. eryphon*) occurs—as its name suggests—mostly in western North America. But its range also extends into eastern Canada. An isolated population is known in western Maine/northeastern New Hampshire (*see* purple mark on range map). Western Pine Elfins probably feed on White Pine (*Pinus alba*) in our area, yet they commonly visit spruce bogs for nectar—a habit that has created some confusion (Layberry et al., 1998). Observers should watch for Western Pine Elfins across the East Coast's "northern rim," while taking care not to confuse them with the Eastern Pine's northern *clarki* race, whose variable patterns and coloration may be unfamiliar to observers used to seeing the nominate race (found in most of our region).

Male Pine Elfins perch at the top of host trees to await females. Many authors state that Pine Elfins prefer young, low-growth pine stands, but they may simply be more difficult to detect when perched in the upper canopies of 60-foot pine trees. As with other lycaenids, Pine Elfins angle their wings perpendicular to the sun's rays when basking and rub their hindwings almost constantly. Adults nectar readily on pinewoods flowers, but also wander to fields or gardens to feed. Puddling is fairly common. Not an exceptionally rapid flier, unless disturbed.

Identification	Small (but average for an elfin; the similar-looking Bog Elfin is some 20% smaller). Fringe checkered, no tails.
Dorsal	Male is *dark brown* above, female *tawnier* with dark borders. Dorsal surface shows when HWs are rubbed up and down (as in photo).
Ventral	HW has *complex, variably colored bands and chevrons*, better defined than in Bog, but outer HW marginal chevrons are *less jagged and angular* than in W. Pine. Outer margin is usually *frosted* (not in W. Pine). A dark bar or patch usually shows in the *center* of the FW cell that is absent in Western (both have a single FW cell-end bar).

Habitat	Various pine or pine-oak habitats, including dry and sandy areas such as barrens. Also edges, glades, roadsides, old fields.
Hostplants	Various "hard" pines: Scrub (*Pinus virginiana*), Jack (*P. banksiana*), Loblolly (*P. taedo*), Sand (*P. clausa*); also "soft" White Pine (*P. alba*). Eggs laid singly on fresh needles, young caterpillars bore through the sheath to feed, older ones eat needles to the base.
Occurrence	Widespread but usually seen alone or in low densities, N. Eng. south to n.-central FL; also e. CAN west to Alberta, south to n. TX. One long brood (up to 3 months in some locations), late Feb-Apr in South, Apr-Jun in North. Pupae overwinter.
Ecology	Medium Generalist. Uses a variety of common hosts, adapting to a wide range of habitat types.

Pitch Pine (*Pinus rigida*)

av wingspan: 1.0"

Bog Elfin *Callophrys* [= *Incisalia*] *lanoraieensis*

Ventral: 5/26/97, Carroll Co., NH

Black Spruce (*Picea mariana*) **bog**

If the ground moves beneath your feet on first observing this diminutive elfin, it may be because you are standing on a quaking peat bog. The interlaced mesh of tangled roots and vegetation under your feet is likely the only thing separating you from the depths of a dark, glacial "kettle hole" pool lying beneath.

This exotic habitat is only part of the allure of this tiny, hard-to-see butterfly. For those who enjoy demanding quests, the pursuit of a Bog Elfin colony should provide an ideal challenge. The Bog Elfin was first described in 1934, from a Black Spruce bog located in the town of Lanoraie, Quebec (east of Montreal). To our knowledge, southern Quebec remains this specie's global stronghold (Layberry et al., 1998). There are only four known sites in Nova Scotia and four in New Brunswick, plus a handful in the United States (in Maine, New Hampshire, Massachusetts, and New York), not all of which currently appear active (R. Dirig, pers. comm.). We suspect, however, that with continued effort additional populations will be discovered.

Location of a promising site can be difficult, since available descriptions of preferred habitat vary significantly. Most accounts place colonies in the open, central portions of bogs, where tree growth is lowest (Layberry et al., 1998). But in New Hampshire, where we rediscovered a formerly documented colony in 1997, the elfin was found in a spot where the spruce growth was *tallest*. And in New York State, Bog Elfins found in the mid-1980s were reported from "a small area of *mostly closed-canopy* Black Spruce-Tamarack (*Larix laricina*)-Red Maple (*Acer rubrum*) peatland"; other elfins, meanwhile, flew in more open bog areas nearby (Miller, 1995, p. 121, emphasis added). Finally, individuals have been found in dry woodlands in Nova Scotia and Massachusetts— although both times in the vicinity of spruce bogs (Layberry et al., 1998; C. Kamp, pers. comm.). Plainly, much remains to be learned about the lifestyle of this intriguing butterfly.

Male Bog Elfins perch to await females, often on the highest accessible branch tip. But individuals regularly descend to nectar (or simply to walk on the bog surface, perhaps to drink or obtain minerals). Even in active colonies individuals can be hard to find, probably because they remain quietly perched during much of the day. Flight is reportedly slow and weak, but we have found Bog Elfins quite capable of quick evasion tactics when startled.

Identification	Our smallest elfin and smallest northern hairstreak. Indeed, the Bog's size is distinctive enough to be useful in field identification. Habitat is also a clue in many cases; their most common look-alike (the Eastern Pine Elfin) seldom occurs in open spruce bogs, where the species most likely to be encountered is the much different-looking Brown Elfin. But note that Western Pine Elfin also can occur in spruce bogs. Fringes checkered, no tails.
Dorsal	Male's dorsal surface is dark with *orange scaling*. Female's is plain brown. We have observed dorsal basking on a cool morning in New Hampshire.
Ventral	Like a dark, miniature pine elfin, but patterns vaguer and more obscure (especially toward HW margin, which can be heavily frosted). Fresh individuals show a purplish sheen basally and green FW overtones.

Habitat	Open, acid Black Spruce-Tamarack bogs, and additional habitats as discussed above.
Hostplants	Black Spruce (*Picea mariana*). Young caterpillars mine into pine needles and eat inside; older ones feed on the needles from the outside.
Occurrence	Rare and extremely local in e. CAN and n. N. Eng./NY. One brood, May-early Jun. Pupae overwinter.
Ecology	**Specialist.** Among the most specialized and restricted of East Coast butterflies.

av wingspan: 0.8"

Juniper Hairstreak   *Callophrys* [= *Mitoura*] *gryneus*

'Sweadner's' Juniper Hairstreak (*sweadneri* race): 10/7/01, Yankeetown, Levy Co., FL

'Olive' Juniper Hairstreak (*gryneus* race): 5/6/93, Sandy Hook, Monmouth Co., NJ

This bright, green jewel of a butterfly is one of three greenish hairstreaks found in its range. (The others are Hessel's and Early; there is no overlap with two additional green East Coast species, the Silver-banded and Amethyst, both found only in South Florida.) Though widespread and fairly common, the Juniper Hairstreak is also extremely local, inhabiting only a fraction of suitable-looking hostplant sites. It has undoubtedly benefited from the abandonment of agricultural lands, since its primary hostplant is a successional habitat specialist. But this also leaves it vulnerable to habitat loss as fields mature into forests.

Until recently, the eastern race of Juniper Hairstreak was considered a separate species (Olive Hairstreak). But based on evidence of interbreeding in isolated contact zones, it was lumped together with western races during the 1980s to form a large, highly variable "megaspecies," with representatives across the continent. (The present discussion addresses only the original, eastern forms.)

A sedentary butterfly, the Juniper Hairstreak has a number of local varieties, including more than one in our area. The nominate *gryneus* race occurs most widely. But the 'Sweadner's' Juniper Hairstreak of northern Florida is well-differentiated, and it appears that as many as two additional subspecies may dwell in Florida and along the southeastern coast as well (Gatrelle, 2001c).

Male Juniper Hairstreaks are highly territorial on redcedar trees, where they perch throughout the day in search of females. They can be set in flight by jiggling the trunk, and if this is done gently they will return to their original post after darting around with great energy for a few moments. Both sexes nectar, often near the host. The Sweadner's race seems especially apt to visit flowers (Glassberg et al., 2000).

Identification	Small. Two HW tails of unequal length; the *lower tail is especially long in Sweadner's*. Easily confused with Hessel's; *see* that account.
Dorsal	Varies by sex, season, and subspecies. Male tawny orange or bronzy above (brown in Sweadner's) with small FW stigma. Female tawny in spring, dark, blackish-brown in summer (Allen, 1997).
Ventral	Bright green with variable brown scaling. White postmedian band *edged inward* with brown. Two white spots near HW base (variable in shape). Outer HW usually frosted. *Compare* Hessel's.

Habitat	Open, dry, alkaline areas, often on slopes, including bluffs, old fields, barrens, etc. Also in coastal redcedar stands.
Hostplants	Eastern and Southern Redcedar (*Juniperus virginiana* & *silicicola*). Others in West. Eggs laid on branch tip, and the well-camouflaged green larvae feed there. References to Catbriar (*Smilax rotundifolia*) in early literature are in error (Gatrelle, 2001c).
Occurrence	Widespread but local; closely tied to hostplant range; eastern forms in w. ME south to n. FL, also west to e. TX. One brood in far North (Mar-Jul), but usually two (Mar/Apr-May and Jul-Aug or Sep-Oct, depending on latitude). Pupae overwinter.
Ecology	Medium Generalist. Few hostplants, but widely available.

Eastern Redcedar (*Juniperus virginiana*)

av wingspan: 1.0″

Hessel's Hairstreak *Callophrys* [= *Mitoura*] *hesseli*

Ventral (on Sand Myrtle): 5/9/95, Chatsworth, Burlington Co., NJ

Atlantic White Cedar (*Chamaecyparis thyoides*)

Rarity and beauty are an alluring combination, and this exceptional hairstreak possesses both. Beginners should be aware that this is no backyard butterfly. A dedicated trip to an Atlantic White Cedar swamp in season is required to see this East Coast endemic, and even then adults tend to congregate high in the canopy where they are difficult (sometimes impossible) to observe.

This species was described only in 1950, based on the observations of Sidney Hessel, a banker and amateur lepidopterist from Connecticut. It was earlier confused with the Juniper Hairstreak, which it closely resembles. Southern individuals are darker-toned and have longer tails. Based on this and certain other traits, subspecies status has been proposed (Scott, 1986; Gatrelle, 2001c).

A true specialist, the Hessel's Hairstreak is threatened by the diminishing footprint of Atlantic White Cedar swamps along the East Coast. The Atlantic White Cedar (actually a cypress) is rot-resistant, which has made it a valuable timber tree since colonial times. Few efforts have been made to replant clear-cut forests, however (this is difficult), and its acreage in North Carolina alone has declined by 90%. Regular fires are required for perpetuation and regrowth, moreover, and fire suppression has been an additional source of habitat degradation.

Males perch high in the cedar canopy to await females. On rare, unpredictable occasions puddle parties convene. Individuals also descend sporadically for nectar (sometimes traveling significant distances, since nectar sources can be rare in their prime habitat). Often seen near dusk (4:00 p.m. or later), but not exclusively so: the photo above was taken at half past noon, and we have had many other midday sightings.

Identification	Small. Two HW tails, one stubby. Can be difficult to separate from Juniper when seen away from primary habitat. Variable geographically and individually; best to use a suite of field marks for identification.
Dorsal	Male dark brown above (darker in summer brood). Female with some reddishness. Upper surface seldom displayed in the wild.
Ventral	*Green or blue-green*, with variable brown scaling. White postmedian line is *brown-edged on both sides*. Note *white subcostal mark* on inner FW. The *white spot* near the trailing HW edge is *convex to the body* (concave in Juniper); the mark near the median HW costa is *displaced toward the body*; that on FW costa is *displaced toward the apex* (vs. both straight in Juniper).

Habitat	Altantic White Cedar swamps, pocosins, bay forests, sandhill seeps, and nearby areas with nectar.
Hostplants	Atlantic White Cedar (*Chamaecyparis thyoides*). Some larvae accept Eastern Redcedar (*Juniperus virginiana*) in captivity (Opler & Krizek, 1984). Green caterpillars eat new growth at branch tips.
Occurrence	Extremely local, s. ME to nw FL, along coast and in sandhills; suitable habitats are more local and disjunct than indicated on the range map. One flight in N. Eng. (May-Jun); two southward (Apr-May, Jul-Aug); three in FL (Mar-May, Jun-Jul, Sep-Oct). Pupae hibernate.
Ecology	Specialist. Depends on a single rare and declining East Coast habitat.

Road through White Cedar swamp, where Hessel's may descend to take minerals

av wingspan: 0.95"

White M Hairstreak   *Parrhasius m-album*

Dorsal male: specimen photo

Ventral: 9/28/97, Gainesville, Alachua Co., FL

This large, handsomely appointed hairstreak projects an air of apparent serenity as it works its way along a flowerhead. It is an arboreal oak feeder whose visits to the ground are irregular, especially so when adults can obtain nectar from flowering trees (Gochfeld & Burger, 1997). Ground-level sightings are comparatively reliable in the fall at Higbee Beach, in Cape May, New Jersey—perhaps because White Ms are drawn to fields of Slender Fragrant Goldenrod then in bloom, while tree nectar is rare in September.

Though it can be difficult to distinguish population shifts from variations in adult nectaring patterns, it is nonetheless clear that White M Hairstreaks experience significant population swings from year to year. In 2002, for example, regular sightings occurred throughout the season, in both established locations and novel ones, testifying to a successful breeding year by any measure.

This mainly southern butterfly has extended its range northward gradually in recent decades. Shapiro (1974) knew of only two historic records for the species in New York State, but there are now many more. Today it is at least a temporary colonist as far north as Massachusetts. Isolated fall emigrants are also seen with some regularity as far north as Michigan, Wisconsin, and Pt. Pelee in Ontario.

Males perch, occasionally near eye level, to watch for females. Flight is fast and direct. Individuals may drop almost straight down from a tree to land upon a flower to feed (Iftner et al., 1992).

Identification	Fairly large. Two tails of very uneven length. HW lobe at outer angle is rounded, giving wing a distinctive shape. Can be confused with 'Northern' Oak and Gray Hairstreaks.
Dorsal	*Bright, iridescent blue*, easily noted in flight, with dark borders (wider in female).
Ventral	Grayish-brown. White postmedian line forms a clear "M" pattern near the tails. A squarish, red-orange submarginal patch is recessed from the margin. Note *isolated white dot along inner HW costa* (otherwise only seen in 'Southern' Oak Hairstreak).

Habitat	Deciduous wood edges and nearby open areas with nectar.
Hostplants	Primarily oaks, especially Live Oak (*Quercus virginiana*) in the South, White Oak (*Q. alba*) and others northward. Basswood (*Tilia*) reported from Ohio (Iftner et al., 1992).
Occurrence	Commonest in SE and in Appalachians; not found on Keys. Also west to IA and MO, south to e. TX. Three broods in North (late Apr-Oct), four in South (Feb-Oct). Pupae overwinter in litter beneath host; some in 2nd brood overwinter. Pupae "squeak."
Ecology	**Medium Generalist.** Hosts and habitat are widely available; shows tendency to disperse during the season.

White Oak (*Quercus alba*)

av wingspan: 1.25"

Gray Hairstreak   *Strymon melinus*

Dorsal male: 9/28/97, Daytona, Volusa Co., FL

Ventral male: 6/24/93, Jonathan Dickenson SP, Martin Co., FL

In deference to long-established custom, the NABA English Names Committee has elected not to call this butterfly a "scrub-hairstreak"—the group name it applies to other members of the genus *Strymon*. Nevertheless, the Gray Hairstreak is by far the most common and widespread scrub-hairstreak in the United States, and arguably the most common hairstreak of any kind. Grays seldom occur in dense, local aggregations, as do a number of their kin. A half-dozen sightings a day would constitute a respectable field total. Yet the species occurs with great persistence, both geographically and throughout the year, and this justifies their classification as prolific.

The Gray Hairstreak is a celebrated multiphage—it reportedly will eat nearly anything. Grays feed on the seeds and flowers of plants in at least 20 families. But they show a decided preference for mallows and legumes. Many incidental hostplants cited in the literature are plants that have been accepted by larvae in captivity. Typical food selection in the wild—while still highly diverse—seems to be less eclectic. (To be cautious, we excluded such secondary hostplant associations in the "hostplant by plant family" chart in the Introduction, p. 3.)

Male Gray Hairstreaks perch from mid-afternoon until dusk, awaiting females. Territorial dogfighting is common. Mating reportedly occurs after dark (Opler & Krizek, 1984). The Gray's flight is rapid and jerky, interrupted by frequent nectar visits. Both sexes are given to frequent dorsal basking, more so than most other East Coast hairstreaks (the Early is an exception; and all our scrub-hairstreaks will dorsal bask at least occasionally).

Identification   A small butterfly, but medium-sized among East Coast hairstreaks. Spring-brood individuals are smaller and comparatively dark. The *male abdomen is orange above*; the *female's is gray*. Two HW tails, one quite stubby. Usually not difficult to identify, but compare 'Northern' Oak and White M Hairstreak.

Dorsal   Uniform dark, bluish-gray, with a prominent *orange-and-black spot* on the trailing HW. Often dorsal basks.

Ventral   Varying shades of gray, but paler than the dorsal surface. A thin, fairly straight *postmedian dash-line* is usually tri-colored (white, black, and reddish-orange). A large, rather shapeless *orange patch* and smaller, *pale blue field* appear near the base of the tails.

Habitat   A wide variety of open, weedy, and disturbed sites.

Hostplants   Extremely varied. Primarily mallows and legumes, but also many others (*see* above). Still, uses a single plant, Sweetfern (*Comptonia peregrina*), at the edge of its range in Canada (Layberry et al., 1998). Sometimes uses balloon vine (*Cardiospermum*) along with other hairstreaks. Can be a pest in cotton and bean fields. Caterpillars eat flowers and fruits/seeds, and sometimes young leaves. Reportedly ant-tended (Allen, 1997).

Occurrence   Widespread and regular but usually not abundant. Northern N. Eng. south to the Keys, also west across continent and south to n. So. Amer. Not in W.I. (but apparently introduced on Great Bahama). Two often overlapping broods in North (Apr-Oct), three in South (Feb-Nov), all year in s. FL. Pupa hibernates.

Ecology   Generalist. We know of few better examples of the generalist lifestyle.

Musk Mallow (*Malva moschata*)

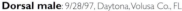
av wingspan: 1.1″

Martial Scrub-Hairstreak   *Strymon martialis*

Dorsal male: 12/31/96, Big Pine Key, Monroe Co., FL   **Ventral**: 12/23/94, Big Pine Key, Monroe Co., FL

Also called Cuban Gray Hairstreak, although it is reportedly rare in Cuba (Smith et al., 1994). And in the Caribbean it is sometimes called Grey Hairstreak, confusingly enough (Riley, 1975).

South Florida's two most specialized scrub-hairstreaks, Bartram's and Martial, share overlapping ranges and are fairly similar in appearance. But there the similarity ends. As indicated in their respective species accounts, the two species' hostplants, habitat selections, and general lifestyle are all very different. The Bartram's, for one thing, has endemic races on many Caribbean islands, while the Martial Hairstreak has a single race throughout its range, suggesting a more vagrant lifestyle.

The Martial Hairstreak's footloose ways are consistent with its penchant for living in "at-risk" coastal habitats, which are exceptionally vulnerable to tidal surges and other storm damage. In September 1998, Hurricane Georges devastated Bay-Cedar stands on several Lower Keys where Martial colonies had been thriving. And it has been vulnerable even in seemingly more sheltered environments on the mainland. We are aware of comparatively few mainland sightings of Martial since the early 1990s. And not having seen the species north of the Keys ourselves, we cannot confirm its reported association with Nettletree (*Trema micranthum*) as a mainland hostplant (Minno & Emmel, 1993).

Male Martial Hairstreaks perch much of the day on or near hostplants, often near eye level. They sometimes display or dorsal bask. On several occasions, we have seen marauding anoles lurch at perching males, and many may be predated in this fashion. Both sexes sip nectar from Bay-Cedar and other available flowers.

Identification	Small, about the size of Bartram's (which it resembles). *Two tails on each HW*, one short. Top of male abdomen is gray, not orange (*compare* Gray Hairstreak). Habitat may provide a useful first clue as to identity.
Dorsal	Base color dark, almost blackish. *Bright iridescent blue* on trailing FW and most of HW (darker in male). Male has dark, *smudgy sex scales* near the costal FW edge (female has similar marking, but paler). Note *red spot on outer lobe*—similar to Disguised (*see* p. 126), but that species lacks the Martial's blue dorsal scaling.
Ventral	Soft, light gray with a *continuous white band on both wings* (intermediate in thickness between dash-line on Gray Hairstreak and the thick bands of Bartram's; the latter also has two HW bands). *Lacks two basal HW spots of Bartram's.*

Habitat	Open sunny, coastal areas; pinelands and hammock edges on mainland. May wander for nectar.
Hostplants	Bay-Cedar (*Suriana maritima*); Nettletree (*Trema micranthum*) on the mainland. Young larvae eat new leaves, flowers, and fruit; older ones eat leaves.
Occurrence	Local in s. FL and Keys, also Bahamas and w. Gr. Antilles. Flies all year, but usually most common in the winter dry season. Many broods.
Ecology	Specialist. Has greater mobility than Bartram's, but is still a hostplant and habitat specialist.

Bay-Cedar (*Suriana maritima*)

av wingspan: 1.0"

Bartram's Scrub-Hairstreak   *Strymon acis*

Dorsal male: 12/24/94, Long Pine Key, Everglades NP, Miami-Dade Co., FL

Ventral (on Narrow-leaved Croton): 11/27/00, Big Pine Key, Monroe Co., FL

Bartram's Hairstreak (*Strymon acis bartrami*) is the endemic Florida race of a complex and regionally variable Caribbean species, the Acis Hairstreak (Pyle, 1981)—sometimes called Drury's Hairstreak (Riley, 1975). Captive to their own specialized habitat requirements, Bartram's populations have dwindled steadily in recent decades with the loss of pine rocklands in southern Florida—at the hands of human developers, as with other species, but also from the ravages of Hurricane Andrew in 1992, which obliterated many acres of tropical pinelands in Miami-Dade and Monroe counties.

The name of this handsome butterfly honors John and William Bartram, pioneering father and son naturalists. Beginning in the colonial era, they became America's first eminent, native-born natural historians. John Bartram used a £50 annual stipend he received from King George III as Royal Botanist to conduct traveling botanical surveys. In 1765–66 he and his son journeyed through the Carolinas, Georgia, and northern Florida studying natural history. This general contact with Florida is seemingly honored in the hairstreak's nomenclature, since Bartram's did not reach South Florida, where the species lives.

The Acis Hairstreak is a sedentary species, prone to develop local forms. It is found on numerous Caribbean islands, where additional endemic races occur (Smith et al., 1994). Adults can be seen throughout much of the day in close association with their hostplant, either perching (often low), nectaring, or ovipositing. A number of nectar sources are utilized. Fast-flying, as with most hairstreaks.

Identification	Slightly larger than some of the tropical hairstreaks with which it occurs, and notably larger than most tropical blues flying nearby. Usually not difficult to identify, by field marks or habitat, but it is often fast-moving and may be hard to see until it settles to nectar. *Two pairs of recurved tails* on each HW, one short. *Male's abdomen is white, the female's is gray* (Minno & Emmel, 1993).
Dorsal	Plain, dark gray (*compare* Martial). Male has dark patch of sex scales near FW costa. Dark spot on trailing HW is *orange-capped*.
Ventral	Base color gray. *Bold, distinctive white lines*, with inner black edging, cross both wings. Note *two white spots on the basal HW*; also a large and relatively shapeless red-orange outer HW patch.

Habitat	Pine rocklands, especially Big Pine Key and Long Pine Key (Everglades).
Hostplants	Narrow-leaved Croton (*Croton linearis*). Eggs laid singly on flower stalks; young caterpillars eat fruit, flowers, and top of leaves, older ones leaves.
Occurrence	Highly local, but can be common occasionally in colonies. S. FL and Keys. Other races in W.I. Three or more flights all year, but gaps occur between flights, so individuals are not always present.
Ecology	**Specialist.** Highly specialized and sedentary. Requires careful monitoring and protection.

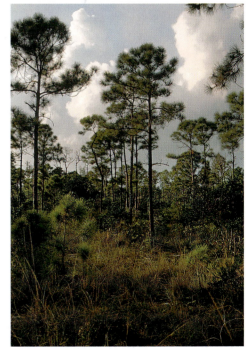

Pine rockland, Big Pine Key, Monroe Co., FL

av wingspan: 1.0″

Mallow Scrub-Hairstreak   *Strymon istapa* [= *columella*]

Dorsal male: 11/24/00, West Summerland Key, Monroe Co., FL

Ventral: 12/23/94, Big Pine Key, Monroe Co., FL

Also known as Modest or Columella Hairstreak. Several times we have noted this imperturbable little butterfly nectaring on short-cropped lawn weeds as we sat in traffic along busy Florida streets. Open and disturbed habitats are the Mallow's principal home, both in Florida and throughout its wide geographic range.

The Mallow Scrub-Hairstreak is the most common and widespread of a close-knit group of very similar-looking small hairstreaks, based mainly in the Caribbean. The Florida race of Mallow was classified until recently as *Strymon columella modesta*. But it has now been split from the Caribbean *columella* and realigned under a separate species, *Strymon istapa* (Robbins & Nicolay, 1998).

This taxonomic shift is no great moment to the average Florida observer, since apart from the Gray this is the only scrub-hairstreak they will commonly see. But there is another possibility: the Disguised Hairstreak (*S. limenia*). First seen on the lower Keys in 1972 (Minno & Emmel, 1993), the Disguised was absent for many years, until three to five individuals were observed on Big Pine Key in March 2004. This may herald the onset of a local rebound.

We have seen male Mallows perching at different times of day, often on hostplants. But, as others have noted, they perch quite consistently in late afternoon. At other times, they can be seen actively seeking nectar from a variety of flowers.

Identification	Small in absolute terms, but medium-sized among South Florida's hairstreaks. Each HW has a *single tail*. The Disguised is browner and has a more pointed FW. Many authors describe this species as "plain" or "modest," but we think these terms are better applied to western Mallows, which are grayer and more blandly marked.
Dorsal	Male *dingy brown or gray-brown*, with *smudgy black sex scales* near the leading edge of the FW and smaller dark spots near the base of the HW tails. Female plain dark with variable blue scaling (but male can also have blue tones, as in photo). Disguised has a *red spot at the outer HW angle* (visible when rubbing wings).
Ventral	Gray to brown, with a *curved, postmedian row of black marks* edged in white (straighter in Disguised), and a *large, orange-capped thecla spot*, near base of tail. The two spots at the leading end of the HW postmedian row are *nearly equal-sized* (vs. very different-sized in Disguised).

Habitat	Varying open and disturbed habitats, usually weedy.
Hostplants	A variety of plants in the Mallow Family (Malvaceae), including Sleepy Morning (*Waltheria indica*), Indian Mallow (*Abutilon permolle*), Wireweed (*Sida acuta*) (Jamaica), and Bay-Cedar (*Suriana maritima*). Many others noted in Cuba (Fernández, 2001). Larvae eat stems and young leaves of hostplants.
Occurrence	Common and widespread, but seldom in large numbers at any single location. *S. istapa*, as newly defined, occurs in s. FL and across the southern rim of the US to s. CA; also into So. Amer. to at least BRA; but with recent split, details are uncertain. Many broods, throughout the year, in s. FL.
Ecology	**Medium Generalist**. Not cold-tolerant, but adopts a generalist lifestyle.

Roger Hammer

Sleepy Morning (*Waltheria indica*)

av wingspan: 0.95"

Fulvous Hairstreak *Electrostrymon angelia*

Dorsal male: specimen photo

Ventral: 5/1/96, Mary Krome Sanctuary, Miami-Dade Co., FL

Also known as Angelic Hairstreak. This small, Caribbean butterfly was first reported from Florida in 1973–74, during a biologically active decade when a number of other butterflies also took up residence in the state (*see* Gray Ministreak account). The Fulvous' only verified U.S. hostplant is Brazilian Pepper, which is notable as one of Florida's most aggressive and unwelcome invasive plants—in a state noted for harmful invaders. Native to South America, Brazilian Pepper dwells in shady woodlands until a disturbance occurs, then spreads quickly to dominate the landscape, forming dense, snarly thickets of foliage that crowd out native plants and animals. They produce chemicals that inhibit the growth of competing plants nearby. Peppers resprout rapidly after fires and are in general highly difficult to eradicate. Their leaves cause a rash in certain individuals (they are related to Poison Ivy). Despite intensive control efforts, nearly three-quarters of a million acres in Florida currently are overrun by this hellacious plant (Ferriter, 1997).

Given the abundance of Brazilian Peppers in Florida, it is surprising that the Fulvous Hairstreak is not one of the state's most prevalent butterflies. But while it does often occur in swarming local colonies, the Fulvous is not an everyday sight throughout its range. It poses no visible threat to its hostplant's viability, based on its apparently modest appetite as an herbivore.

A close relative, the Ruddy Hairstreak (*Electrostrymon sangala*), was reported once from the Keys, in February 1982 (Minno & Emmel, 1993). Otherwise, it has been seen in the United States only in southern Texas. Distinguishable in appearance from the Fulvous, the Ruddy is unlikely to be seen in our area.

Fulvous Hairstreaks spend much of the day in treetops, where they are reportedly active into the evening, sometimes even after sunset (Smith et al., 1994). They descend regularly during the day to nectar or perch on low foliage. Aerial "dogfights" occur frequently in active colonies.

Identification	Small. Two pairs of HW tails (one stubby). The Ruddy, while congeneric, is much paler and is visually distinguishable. FW pointed.
Dorsal	Blackish-brown with *tawny, red-orange patches* (reduced or absent in female), sometimes visible in flight. (Ruddy male is pale orange above, female mostly dark.)
Ventral	Dark brown with an *incomplete white HW line*, including an *isolated white dot* near the HW leading edge. (Ruddy is pale/tannish with a red-and-white HW dash-line.)

Habitat	Suburbs, hammock edges, near hostplant.
Hostplants	Brazilian Pepper (*Schinus terebinthifolius*) (*see* above). Interestingly, *Salvia* recently reported in Cuba (Fernández, 2001). Some observers suspect additional hosts in FL. Larvae eat leaves.
Occurrence	Locally common, s. FL. Also W.I. Three or more broods, essentially all year.
Ecology	**Medium specialist.** Single hostplant, but very common; quite tolerant of human settlement.

Brazilian Pepper (*Schinus terebinthifolius*)

av wingspan: 0.9″

Red-banded Hairstreak   *Calycopis cecrops*

Ventral: 7/25/94, Allaire SP, Monmouth Co., NJ

Winged Sumac (*Rhus copallina*)

With its bold, op art wing patterns, this handsome little butterfly should draw immediate attention, even from nonenthusiasts. But its markings are produced on such a small scale as to be virtually invisible to casual observers—who seldom carry close-focusing binoculars and are rarely inclined to crouch down for close looks at Lepidoptera. It is a common sight, nonetheless, along the southeastern coastal plain.

Like many small East Coast hairstreaks, the Red-banded belongs to a tropical genus (quite a large one in this case, with some 70 species). It is unique among *Calycopis* in having adapted to north-temperate climates, with significant exposure to winter cold. Yet despite a marked northward expansion in recent years, the Red-banded betrays its tropical origins, as population densities remain low at the northern edge of its range.

Considerable uncertainty persists as to the specifics of this familiar butterfly's early stages. Klots (1951) listed no specific information concerning its life history. Both he and Pyle (1981) recognized several hostplant associations, but the caterpillar's means of utilizing its hostplant was uncertain. Opler and Krizek (1984) knew that females oviposit on the underside of fallen hostplant leaves, but assumed that larvae crawled onto a nearby hostplant to feed. Scott (1986) suggested that the caterpillars "probably eat mainly detritus on the ground" (p. 367). Allen (1997) stated that larvae select detritus or living plants depending on where their eggs were laid; in captivity, he noted a preference for moist, decaying material. Detritus-feeding appears unique among East Coast butterflies, and deserves further careful study. For example, are variations in the hindwing color band linked to larval food?

Male Red-bandeds perch low on foliage to await females. They remain active throughout much of the day, but are especially lively in late afternoon. Both sexes nectar actively, often on their sumac host, and take minerals from damp soil (though care should be taken to distinguish puddlers from ovipositing females).

Identification	Small, but less so than Gray Ministreak, the only species with which it is likely to be confused in our area. Two tails on each HW, one stubby.
Dorsal	Dark, blackish-brown, with varying degrees of *iridescent blue*, mainly on HW and basal FW.
Ventral	Grayish-brown, with a *conspicuous, wide orange or red-orange post-median band*, edged outwardly in white. Large, red-capped marginal eyespot is *mostly black*. Light blue patch at the HW outer angle.

Habitat	Extremely variable, mainly in open, scrubby fields and edges; also savanna, swamp, and pocosin edges, sandy fields, damp woods, etc.
Hostplants	Associates with sumacs, including Winged (*Rhus copallina*), Wax Myrtle (*Myrica cerifera*), Brazilian Pepper (*Schinus terebinthifolius*), oaks, possibly many others. Early-stage behaviors described above.
Occurrence	Common in SE, less so in mountains and northward; s. N. Eng. south to FL; also west to e. TX, OK. Range extending northward. All year in s. FL; three broods in South (Apr-May; Jun-Jul; Aug-Oct); two in North (May-Jun; Aug-Oct). 4th-stage caterpillars overwinter.
Ecology	Medium generalist. Though larval lifestyle is specialized, Red-bandeds use common hosts that occur in numerous habitat types.

Wax Myrtle (*Myrica cerifera*)

av wingspan: 1.0″

Gray Ministreak *Ministrymon* [= *Tmolus*] *azia*

Ventral: 9/8/01, Homestead, Miami-Dade Co., FL

Ventral, ovipositing on hostplant bud, 9/8/01, Homestead, Miami-Dade Co., FL

Also known as Azia Hairstreak. This miniature butterfly was first seen in southern Florida in 1973, and has since established a modest toehold. Yet it remains a transient colonist, usually uncommon. It is one of a half-dozen or so tropical butterflies that took up residence in Florida during the 1970s (or late 1960s). Others include Silver-banded and Fulvous Hairstreaks, Bahaman Swallowtail, Malachite, Dorantes Longtail, Tropical Buckeye, and Dingy Purplewing.

The Gray Ministreak's arrival in Florida was by no means a case of isolated island hopping (which also does occur in the region). It was entirely unknown in the West Indies before 1975, when Riley published. Multiple reports since seem to reflect a broad, Caribbean-basin range extension, of which Florida is just a part (Smith et al., 1994; we must question the authors' suggestion that Gray Ministreaks simply may have been overlooked in all of their current island homes simultaneously prior to the 1970s).

As a colonist, the Gray Ministreak has a reputation for abandoning colony sites as abruptly as it populates them. Few reliable, long-term sites are known. This may be a function of the ephemeral habitats the Ministreak selects, where its main hostplant grows as an opportunistic weed, or of the fact that this hostplant is an invasive alien, subject to periodic eradication campaigns.

Lead Tree (*Leucaena leucocephala*)

Adult Gray Ministreaks perch a moderate distance off the ground, often on or near a hostplant. They are most conspicuous when taking nectar (often from *Bidens*). In flight, they dart very rapidly from perch to perch, in a rapid, hairstreak fashion, making them extremely difficult to follow.

Identification	Tiny. Our smallest hairstreak; indeed, among our smallest butterflies of any kind. *One HW tail* on each wing. Might be confused with Red-banded, but that species is larger, browner, and has a thicker red HW "sash."
Dorsal	Upper surface rarely displayed. Usually gray or tannish-gray, sometimes with white suffusion on HW.
Ventral	Variable; shiny, *silvery gray to flat, slate gray*, depending in part on wear. *Bright red-orange postmedian line, jagged on HW.* Reddish eyespots on outer margin.

Habitat	In our area, subtropical scrub, disturbed sites with hostplant. Sometimes pine rocklands.
Hostplants	In Florida, Lead Tree (*Leucaena leucocephala*), possibly others. Eggs laid on unopened buds.
Occurrence	Currently local and uncommon, s. FL. Also W.I., south to ARG. Colonies north to Pasco Co., but winter cold limits permanent residency. All year.
Ecology	**Medium specialist.** Locally restricted lifestyle.

Scrubby field habitat, Homestead, Miami-Dade Co., FL

av wingspan: 0.7"

Silver-banded Hairstreak   *Chlorostrymon simaethis*

Ventral: 3/24/94, Tavernier, Monroe Co., FL

Balloon Vine (*Cardiospermum corindum*), with flower and "balloons"

Also known as St. Christopher's Hairstreak. (Another choice might be Key Lime Hairstreak.) Before the early 1970s, there was only one small, greenish hairstreak in South Florida—the Amethyst Hairstreak (*Chlorostrymon maesites*), not the closely related Silver-banded. In 1973, as Amethysts were declining in South Florida (for reasons unknown), Silver-bandeds arrived, just as inexplicably. But after an initial population spike, Silver-banded populations also languished, and this species is now rare and local as well.

The "little green hairstreak" saga has at least one more chapter, as the Amethyst has lately reappeared. For the first time since the late 1970s, it was seen on the Keys and southern peninsula in 2003–2004. Some ecological dynamic (such as differing response to drought cycles) may explain the seesaw-like changes in these butterflies' relative abundance, but their differing occurrence patterns also may be simply a matter of chance.

The Silver-banded is a widespread neotropical species. The western *sarita* race occurs in southern Texas and across Mexico to Baja. The nominate *simaethis* race is found in South Florida, as well as in much of the Caribbean and south as far as Argentina.

Silver-bandeds are most active in the morning and evening. They often perch on foliage and seldom stray far from a tangle of hostplants. When disturbed, they dart rapidly and can be difficult to track. (Amethysts may remain unseen for long periods, emerging occasionally to nectar, but they can be active at any hour.)

Identification Very small (Amethyst is even smaller). *One pair of HW tails* (Amethyst has a second, shorter "stub tail" on each HW). Male's FW is usually more pointed, with a less convex trailing edge, than the female's.

Dorsal Wings usually closed, but occasionally displays. Base color is brown. Male has considerable *violet-blue iridescence*. Female shows less iridescence (or none), and it is confined when present to the wing bases. The Amethyst Hairstreak is a *vivid, purplish-blue* above (male) or *bright, iridescent blue* with a dark FW apex (female).

Ventral *Iridescent, yellowish-green* in both sexes. A *conspicuous, silvery-white postmedian line* crosses *both wings*, forming a "v" near the tails. In the Amethyst, this line *does not extend across the entire HW* and is *absent from the FW* (replaced by dark dash-marks, sometimes outlined in white). Beyond the postmedian line, both species have a *dark, chocolate-brown patch* with variable *white frosting*. In Silver-banded this patch is usually thinner, but extends along *the entire outer HW edge*.

Habitat Silver-banded favors subtropical scrub or hammock edges with hostplant; often in seasonally dry habitats. Amethyst less predictable; usually seen nectaring along hammock edges or in weedy areas or gardens.

Hostplants Silver-banded lays on buds of balloon vine (*Cardiospermum*). Larvae feed on seeds within the lantern-like pods, one larva per pod (Brown, 1981). Amethyst early stages poorly known. Reportedly feeds on flowers, possibly of Woman's Tongue (*Albizia lebbeck*); Creeping Wild Bean (*Calopogonium coeruleum*) reported in Cuba; probably others.

Occurrence Currently rare and local, mainly on upper Keys. Three or more broods, all year. Mature larvae chew out of balloon vine pod and pupate in debris at base of plant (Brown, 1981).

Ecology Specialist. Silver-banded has an exceptionally specialized lifestyle. Amethyst's appears somewhat more varied, but poorly known.

Amethyst Hairstreak (*C. maesites*), 8/7/04, Castellow Hammock, Miami-Dade Co., FL

av wingspan: 0.875"

Early Hairstreak   *Erora laeta*

Dorsal male: 6/11/94, Mt. Greylock Reservation, Berkshire Co., MA

Dorsal female: 6/11/94, Mt. Greylock Reservation, Berkshire Co., MA

Ventral female: 6/11/94, Mt. Greylock Reservation, Berkshire Co., MA

In 1995, a chance encounter with an Early Hairstreak in western North Carolina, by an observer named Scott Hartley, confirmed a long-held assumption. Specifically, as he climbed a tall oak tree to find a *Satyrium* Hairstreak he had missed during an annual summer count, Hartley unexpectedly came "face to palp" with an Early Hairstreak, perched high in the forest canopy (LeGrand, 1995). Canopy perching has been offered as an explanation for why this small, elusive butterfly is so seldom observed. But whatever the reason, the mystique surrounding Early Hairstreaks is such that an episode of this type is likely to be long-savored by the participants.

The exact occurrence of Early Hairstreaks is not clear, given their cryptic appearance and secretive lifestyle, but they are definitely among the most prized and sought after of northeastern butterflies. Their maddening lack of predictability has challenged generations of devoted pursuers. Yet we think their distinctive character is best appreciated from a different perspective. In the 1990s, specifically, some longstanding mysteries were resolved concerning the Early Hairstreak's early stages. Newfound knowledge may diminish a species' original appeal. But rather than deflating the Early Hairstreak's cachet, these new insights have instead created a deeper appreciation.

Early Hairstreaks are closely associated with American Beech (*Fagus grandifolia*), a stately, slow-growing tree of shady, moist forests that may live up to 400 years. Captive female Early Hairstreaks will not oviposit on beech leaves, however. Only by mistake was it discovered that females *would* readily oviposit on the developing beechnuts themselves—and these are in fact a principal larval food (Layberry et al., 1998). Since the American Beech takes 40 to 60 years to begin producing fruit, Early Hairstreaks generally are confined to forests with mature trees. And abundant beechnut crops occur only in 2- to 8-year intervals (Burns & Honkala, 1990). Annual flowering is highly sensitive to late frosts and mature trees are vulnerable to blights, harsh weather, and environmental stresses. We can see, therefore, that Early Hairstreak populations are closely tied to deep rhythms within ancient beech forests. They are a sparse indicator, however, since apart from one fairly reliable site in northwestern Massachusetts, encounters seem to be a matter of chance.

Identification	Very small. Shape, coloration, and behavior (especially their tendency to remain still when drinking at damp soil, even when disturbed) all conspire to make them inconspicuous. No HW tails. Fades with wear.
Dorsal	Male *dingy gray* with slight blue patches on trailing HW and sometimes blue flecking. Female *iridescent blue* with wide, dark FW borders.
Ventral	*Iridescent blue-green* with an irregular *red-orange band*, lined in white.
Habitat	Paths, dirt roads, and outcrops in moist but well-drained beech forests.
Hostplants	American Beech (*Fagus grandiflora*); probably Beakel Hazel (*Corylus cornuta*) in some places. Young 1st-brood larvae feed on seed husks, older ones burrow into the nut, eating still-soft seeds. 2nd brood said to shift to galls or catkins; accepts willows in captivity.
Occurrence	Rare and local, n. N. Eng. south through Appalachians. Also se CAN to MI and WI. Two-plus broods in South, Apr-May and late Jun-Aug; one to two in North, late-May-Jun, maybe partial second in mid-summer. Larvae drop to ground to pupate (R. Webster, pers. comm.). Pupae overwinter.
Ecology	Medium specialist. Tightly linked to a fairly widespread host.

American Beech (*Fagus grandifolia*)

av wingspan: 0.85"

Eastern Pygmy-Blue   *Brephidium isophthalma*

Dorsal: 12/31/98, Big Pine Key, Monroe Co., FL

Ventral: 9/25/95, Coastal Prairie Trail, Everglades NP, Monroe Co., FL

The coastal salt pans of the southeastern states are a unique and extremely specialized habitat, formed in the wake of tidal overwashes and storm surges. Evaporating seawater leaves salty concentrations that can be tolerated by only tough and specialized plants, such as glassworts (*Salicornia*) and saltworts (*Batis*). Salt pans can be extremely local vest pocket habitats, or thin ribbons surrounding subtropical shorelines, or broad coastal prairies, as found near Flamingo in the Everglades. There, on occasion, one can stand on a wide, open expanse, beset by hordes of tiny, salt marsh mosquitoes, and watch the fluttering of many thousands of eastern Pygmy-Blues (perhaps millions), extending off toward the horizon. This experience is worthwhile despite the torment.

Pygmy-blues (genus *Brephidium*) are for the most part brown rather than blue. Often considered our smallest butterflies, they have a volatile taxonomic history. Originally described as *B. pseudofea*, from the Keys (Morrison, 1873), they were considered separate from both their Antillean relatives and the closely allied Western Pygmy-Blue (*B. exilis*). Scott (1986) regarded all three as a single species (*B. exilis*), but the current prevailing view (not unanimous) combines the Caribbean and Florida races as *B. isophthalma*, while the Western Pygmy-Blue is regarded as a separate species (*exilis*). Distinctive physical traits have recently been reported in northern coastal populations (northern Florida to South Carolina) that are ascribed taxonomic significance (Pavulaan & Gatrelle, 1999).

Male Eastern Pygmy-Blues patrol merely inches over coastal prairie hostplants seeking females. Flight is slow and fluttery, but individuals are nonetheless wary and difficult to approach closely. Nectaring is relatively sparse. Wide local population swings cause some to suspect mass emigration, but other factors may be involved (such as patchy and shifting distribution of ant attendants, subtle habitat fluctuations, etc.).

Identification	Tiny; considered our smallest butterfly (along with Southern Skipperling and Acacia Blue—both of which also can be extremely small). Nothing on the East Coast closely resembles this species, especially in its isolated coastal haunts. *Brown fringes* distinguish it from Western-Pygmy Blue. Antennae of southern individuals are tipped orange/red, vs. slightly colored or black to the north (Pavulaan & Gatrelle, 1999). Underside of abdomen is white, extending up along sides (in south), or grayish and not extending upward (in north).
Dorsal	Brown or reddish-brown, little or no blue dusting; note *dark trailing HW spots*.
Ventral	Yellow-brown (south) or gray-brown (north); not whitish basally (*compare* Western). *Metallic blue-black HW marginal spots*. White FW dashes weak (south) or defined (north).

Habitat	Coastal salt marshes, prairies, and salt pans.
Hostplants	Glassworts (*Salicornia perennis & bigelovii*), probably saltworts (*Batis*). Sea Blite (*Suaeda linearis*) reported in Cuba (Fernández, 2001). Larvae eat flowers, fruits, stems; ant-attended.
Occurrence	Disjunct coastal colonies, SC to FL. Also W.I. Three or more broods, all year s. FL. Pupae overwinter in north.
Ecology	Specialist. A quintessential specialist.

Coastal Prairie, Everglades NP, Monroe Co., FL, **Inset**: Glasswort (*Salicornia perennis*)

av wingspan: 0.6"

Blue Palate Special: Varied Food Selections of Neotropical Blues

The small, neotropical blues in our area belong to a number of different genera (*Leptotes*, *Hemiargus*, and, according to some taxonomists, *Cyclargus*). But all share eclectic and often overlapping tastes for the flowers and seeds of various tropical legumes. Ant-tending figures in some of these lifestyles, but often not as prominently as with other blues (subfamily Polymmatinae). Perhaps because of earlier confusion over species identity within the neotropical group, exact patterns of hostplant use are still fraught with questions. Or, perhaps the confusion arises from the complex and malleable hostplant relationships that vary with time and location (*see* Fernández, 2001).

A classic type of hostplant for small blues is the Pineland Acacia (*Acacia pinetorum*), a Florida endemic plant with a "star burst" blossom whose buds can be fed on individually by larval blues without preventing the plant from reproducing (other buds may be left undamaged). This is the primary hostplant of the Acacia Blue in Florida. Other legumes with "star burst" flowers have been recorded as hosts for neotropical blues, including Blackbead (*Pithecellobium keyense*) and Catclaw Blackbead (*P. unguis-cati*).

Several of our small blues also feed on the flowers of nickerbeans, or nickers (*Caesalpinia*), a group of shrublike, often thorny legumes. Indeed, one species has been assigned a common name based (incorrectly, we think) on its presumed use of *Caesalpinia* as a host. As noted in the species account, the last known colony of Miami Blue feeds exclusively on Gray Nicker (*C. bonduc*) in the Keys.

The oddest hostplant of small blues in Florida is Balloon Vine (*Cardiospermum* sp.). Larvae chew their way into its hollow, lantern-like "balloon pods" to feed inside on developing seeds. Historically, Miami Blues reportedly used this strategy (Minno & Emmel, 1993). Earlier commentators believed that Miamis fed on an exotic balloon vine (*C. halicacabum*), as native species succumbed to development. But this seems to be in error; *C. halicacabum* is rare and does not occur at all on the Keys (R. Hammer, pers. comm.). Two native balloon vines with which *C. halicacabaum* may have been confused include *Cardiospermum microcarpum* and *C. corindum*. The latter grows wild on the Keys.

Balloon Vine (*Cardiospermum corindum*) "balloons"

Gray Nicker (*Caesalpinia bonduc*), seed pods above and leaves below

Pineland Acacia (*Acacia pinetorum*)

Cassius Blue *Leptotes cassius*

Dorsal male: 11/28/00, Port Mayaca, Martin Co., FL

Dorsal female: 3/22/94, Valhalla, Monroe Co., FL

Ventral: 6/27/93, Palm Beach Co., FL

This sun-loving butterfly is a familiar sight throughout the Florida peninsula. It extends its presence into Georgia and South Carolina on occasion. Like the Ceraunus, with which it often associates, the Cassius is a dedicated habitat generalist. But neither species can tolerate hard winters.

A number of Cassius races occur in the Caribbean and South America; our sole race is *theonus*.

Both sexes bask early in the day with wings partly cocked. Males are particularly active in flight, gyrating feverishly about for some minutes in a small area without alighting. While often found near the ground, this species also flies high into the canopy, to either nectar or oviposit.

Leadwort (*Plumbago* sp.), a cultivated hostplant

A closely related and quite similar-looking species, the Marine Blue (*L. marina*) is not an established East Coast butterfly, but a single stray was found on August 29, 1993, in Brooklyn, New York. Severe floods in the Midwest that summer may have pushed this (and other stray Marine Blues) eastward.

Identification

Dorsal — Usually quite small. Thin wings are translucent, allowing underside patterns to be seen faintly, especially in females. Male *theonus* is generally *pale blue* above (other races generally more violet). Often some white streaks on the HW (*compare* male Marine Blue). *No trailing HW eyespots.* Female bluish-white to almost pure white basally, with wide, dark borders and a conspicuous black dot on the trailing HW (along with other, smaller ones).

Ventral — White with a series of *irregular bands and spots* on both wings. (Similar bands on Marine Blue are longer and straighter, leaving fewer white patches, especially on the outer-trailing FW.) *Lacks black basal HW spots.*

Habitat — Highly variable. Weedy fields and waste areas, roadsides, parks, forest edges, pine rockland, etc.

Hostplants — Again highly variable. Leadwort (*Plumbago*), Rattlebox (*Crotalaria incana*), Wild Mimosa/Mimosa Tree (*Lysiloma latisquum*), milk peas (*Galactia*), beans (*Phaseolus*), etc. Larvae eat flowers and seedpods; ant-tended.

Occurrence — Common, FL north to se SC; strayed to se VA in 2001. Also s. TX south to ARG. Multiple broods all year in s. FL. Mar-Nov in mid-FL; strays mainly in fall.

Ecology — Generalist.

Marine Blue (*Leptotes marina*), both surfaces, for comparison: 8/2/96, Ramsey Canyon, AZ. A very rare eastern stray.

Ceraunus Blue *Hemiargus ceraunus*

Dorsal male: 11/23/00, Eco Pond, Everglades NP, Monroe Co., FL

Dorsal female: 11/24/00, Stock Island, Monroe Co., FL

Widespread in the neotropics, the Ceraunus Blue is probably the most common of any blue (Polyommatinae) in the Americas (Opler & Krizek, 1984). But it does not range very far north on the East Coast: it is a rare stray beyond southeastern South Carolina, and the northernmost records in our area are from southern North Carolina.

The Ceraunus has a number of geographic races. Some are quite distinctive and may be sibling species. Some regard our southeastern U.S. race (*antibubastus*), plus Caribbean forms, as a separate species (*H. hanno*).

Ceraunus Blues generally fly low to the ground. Males patrol throughout the day for females, and females seek hosts on which to oviposit. Both visit a variety of flowers, including several ground-hugging species such as Mexican Clover (*Richardia grandiflora*), a common lawn weed. Known to emigrate in some regions (migrates altitudinally in the neotropics).

Ventral male: 12/28/96, Hugh Birch Taylor Park, Ft. Lauderdale, Broward Co., FL

Identification	Small, but averages somewhat larger than the Cassius and Acacia Blues. Variable in appearance.
Dorsal	Male is blue above (sometimes purplish), with a narrow black border and a *single dark spot* on the trailing HW. Female is dark, with variable blue patches basally on both wings. In neither sex is the dark trailing HW spot capped with orange.
Ventral	Drab gray-brown in both sexes, with a median row of dark bars, edged in white (reminiscent of a *Satyrium* hairstreak). *One outer HW eyespot* (two in TX race). Dark *basal HW spots*, similar to Miami Blue. Variable line of *white, postmedian marks*, usually vague or absent on FW.

Habitat	Highly varied. Low, open habitats: lawns, scrub, dunes, roadsides, vacant lots, tropical pinelands.
Hostplants	A wide variety of legumes, including Partridge Pea (*Chamaecrista brachiata*), Crab's Eye (*Abrus precatorius*), Sesbans (*Sesbania herbacea*). Caterpillars feed on buds and flowers, sometimes young leaves. Sometimes ant-tended (Calhoun, 2000).
Occurrence	Common in FL, less so s. GA/se SC. Also W.I., TX, sw US, south to ARG. Strays north in Midwest. All year in s. FL, late summer in North.
Ecology	**Generalist.** Highly tolerant and adaptable (except to cold).

Crab's-Eye (*Abrus precatorius*)

av wingspan: 0.85″

Miami Blue   *Hemiargus* [= *Cyclargus*] *thomasi*

Dorsal male: 3/25/00, Bahia Honda Key, Monroe Co., FL

Dorsal female: 3/25/00, Bahia Honda Key, Monroe Co., FL

It is difficult to pinpoint why the Miami Blue has declined so severely in recent decades, to the point of being federally endangered. Unlike some tropical butterflies, which exist in the United States only in isolated colonies in far south Florida and the Keys, the Miami Blue was an endemic race, once fairly common and widespread, with temporary colonies reaching as far north as St. Petersburg and Daytona (Calhoun et al., 2000). Development of coastal habitats is regularly cited as the cause of decline (e.g., Opler & Krizek, 1984). But this does not account for all cases of disappearance, especially at colonies in well-protected areas. Additional factors, such as drought, pesticide spraying, hurricane damage, and probably others not stated, have also played a role.

But whatever the reason, Miami Blues disappeared from all but a handful of former colony sites by the early 1990s, after a long but steady decline. Hurricane Andrew ravaged one of these (on Key Biscayne) in August 1992, and there were no verified sightings after that date for approximately seven years. Then, on November 29, 1999, an experienced field observer discovered a colony of about 50 individuals at Bahia Honda State Park in the Florida Keys (Ruffin & Glassberg, 2000).

A petition to list the Miami Blue as a federally endangered species was filed in June 2001 by the North American Butterfly Association (NABA). The Miami Blue has now been declared endangered by the State of Florida as well as the U.S. Fish and Wildlife Service. A captive breeding program is being undertaken at the McGuire Center for Lepidoptera, University of Florida, with controlled releases anticipated in the near future.

The Miami Blue was described (in the 1940s) as an endemic Florida race of a Caribbean butterfly. Its scientific name "is really something to make even a hardened entomologist wince a bit: 'Hemiargus (Cyclargus) thomasi bethune-bakeri Comstock and Huntington'" (Klots, 1951, p. 162).

The flight of the Miami Blue is variously described as strong, rapid, or directional. Some of this is undoubtedly attributable to its being relatively large. Lands often to nectar.

Identification	A small butterfly, but larger and stronger-flying than similar s. FL relatives, often visibly so. The Florida race (*bethunebakeri*) is relatively large and has well-defined ventral white bars. *Compare* other small blues.
Dorsal	Males *light blue* (variously described as turquoise, powdery, or cornflower), with thin dark borders and usually *two plain, dark spots on the trailing HW*. Female has *wide dark borders* and an *orange cap on one HW spot*.
Ventral	Two trailing HW eyespots, one orange-capped. White postmedian band on both wings. Four dark postbasal dots.

Habitat	Edges of subtropical hammocks, thorny subtropical scrub; sometimes dry pinewoods.
Hostplants	*See* discussion p. 133. Remaining colony uses Gray Nicker (*Caesalpinia bonduc*), apparently exclusively. Larvae feed on flower parts.
Occurrence	Bahia Honda SP. Also W.I.; recently Cuba. Three-plus broods, especially Mar-May, Sep-Oct. Longer-lived winter adults enter reproductive diapause.
Ecology	**Medium specialist**. Prior to recent decline, was fairly wide-ranging, appears to have used various hosts.

Ventral female: 3/25/00, Bahia Honda Key, Monroe Co., FL

av wingspan: 0.9"

Acacia Blue   *Hemiargus* [= *Cyclargus*] *ammon*

Dorsal male: 11/25/00, Big Pine Key, Monroe Co., FL

Dorsal female: 11/26/00, Big Pine Key, Monroe Co., FL

Blues are often considered to be a fairly sedentary group, so it may seem a bit surprising that in the last decade the only new additions to the East Coast species list have come from this subfamily (the "newcomers list" includes this species and the Northern Blue, recently noted in coastal bogs of southeastern Maine).

Lucas' Blues (as this species was formerly known) were reported with some frequency from southern Florida and the Keys around the turn of the 20th century. But at the time the Miami and Acacia Blues were considered a single species, and the Florida race of Miami Blue (*buthenbakeri*) had not yet been recognized as a separate, endemic form (Calhoun et al., 2000). None of the early Florida records appears to have referred to what we now consider the *true* Acacia Blue (*H. ammon*), a common butterfly on Cuba and elsewhere in the West Indies.

At some point in the late 1990s, however, by means unknown, the *true* Acacia Blue became established in the United States, colonizing the pine rocklands of Big Pine Key (also found very recently on nearby Bahia Honda Key). When a photo taken in 1998 was first published, it was labeled as Miami Blue (Salvato, 1998). Shortly afterward, two collectors found and correctly identified the population. (*Note:* Several photos taken in 1997 on Big Pine Key were subsequently studied and identified as Acacias [Calhoun et al., 2000].)

This tiny addition to the U.S. butterfly list flies throughout the day, low to the ground, in a manner reminiscent of Ceraunus. It pauses frequently to perch on tall blades of grass, budding acacias, and other low foliage. Flight is generally slow and fluttery. Nectars quite often.

We have elected to call this species Acacia Blue (rather than Nickerbean Blue, as adopted by NABA), since (1) there is little evidence that the species specializes on nickerbean in the Caribbean, (2) the original and predominant hostplant in the United States is Pineland Acacia (*Acacia pinetorum*), (3) nickerbean is the apparently exclusive host of the lone remaining *Miami* Blue colony in the Keys, and (4) this name has been recommended by a number of commentators, including the species' earliest discoverer in the Keys (M. Salvato, pers. comm.).

Identification	Very small. We have seen some truly tiny individuals, especially those emerging in drought periods, that rival the Eastern Pygmy-Blue in size. Fringes are white, with some checkering (especially on FW).
Dorsal	Male mostly *bright blue*, with *two dark spots* on the trailing HW, one usually capped with a *pinkish-orange dot*. Female with dark borders and two prominent trailing HW dots, one capped with a thin, *elongate* orange spot.
Ventral	Similar to Miami, but *three rather than four postbasal dots*; two outer HW spots, one with a long, prominent, *bullet-shaped orange cap*.

Habitat	Pine rocklands, Big Pine Key; also Bahia Honda.
Hostplants	Flowers of Pineland and Sweet Acacia (*Acacia pinetorum & farnesiana*); once reported laying on a rare, native Fewflower Holdback (*Caesalpinia pauciflora*). Other hosts in W.I. (Fernández, 2001).
Occurrence	Florida Keys; also W.I. Multiple broods throughout the year. No diapause.
Ecology	Specialist. Highly restricted ecology in our area.

Ventral female: 8/30/99, Bahia Honda Key, Monroe Co., FL

av wingspan: 0.7"

Eastern Tailed-Blue   *Everes comyntas*

Dorsal male: 6/24/00, Suffolk Co., NY

Dorsal female: 6/24/00, Suffolk Co., NY

This diminutive butterfly is a common sight in short-cropped fields and waste areas throughout our region, except in northern Maine and the lower Florida peninsula. Unlike most types of Spring Azure (which it resembles), the Eastern Tailed-Blue is given to regular dorsal basking. Thus, the male's brilliant upper surface is a familiar sight to field observers.

The Eastern Tailed-Blue belongs to a small, holarctic genus with two representatives in North America. A single race (*comyntas*) occurs throughout the East.

Males patrol near hostplants during the day, seeking females. Mated pairs can be seen from midday until mid-afternoon. When dorsal basking the wings are usually held partially open, at about a 45 degree angle. At rest, individuals rub their wings up and down, in a manner reminiscent of hairstreaks, creating a distraction effect with the hindwing tails. Adults nectar regularly, but their short proboscises limit the range of flowers they can visit.

Ventral: 8/3/94, Pondicherry Wildlife Refuge, Coos Co., NH

Identification	The smallest of our northern blues, especially in spring. Note the small, *thread-like tail* (may become lost or damaged) and white wing fringes. Patterns somewhat crisper in spring. Female resembles Gray Hairstreak from above.
Dorsal	Male is *deep, iridescent blue* with dark wing fringes; note *small orange HW dots* near base of tails. Female is a *dark, brownish gray* with better-defined HW black-and-orange markings and sometimes blue basal flecking.
Ventral	Pale grayish-white with distinct black dots and *several orange marks* on outer HW near base of tails.

Habitat	Nearly any kind of open or disturbed space with low growth and hostplants.
Hostplants	Legumes, including vetch (*Vicia*), clovers (*Trifolium*), bush clovers (*Lespedeza*), many others. Females oviposit on flower buds. Larvae eat flowers and seeds, sometimes leaves. Early stages meticulously described by Lawrence and Downey (1966).
Occurrence	Common except in the far north and south. Also sporadically across the West and south to Costa Rica. Three or more overlapping broods, Apr-Oct. Mature caterpillars overwinter.
Ecology	Generalist. Adapts to a wide variety of habitats and hostplants. Ubiquitous.

Cow Vetch (*Vicia cracca*)

Hop Clover (*Trifolium agrarium*)

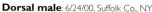

av wingspan: 0.9"

The *Celastrina ladon* Complex— Complex Indeed

For a wee bit of a gossamer-winged creature that expands scarcely an inch across its outstretched wings, the Spring Azure has caused American scientists an immense amount of patient labor.
—Butterflies Worth Knowing, Weed (1917, p. 258)

The genus *Celastrina* (commonly known as azures) is thought to have originated in the Orient (Eliot & Kawazoe, 1983). But peak biological diversity occurs in the eastern United States. Here, the genus exhibits enormous phenotypic variation, along nearly every conceivable line—seasonal, sexual, ecological, individual, host-plant-linked—and, at times it would seem, utterly at random.

The broad diversity of the azures is evident in a single eastern "species," the Spring Azure, *C. ladon*. The scientific species name *ladon* refers to a mythical, 100-headed dragon (Opler & Krizek, 1984). This is fitting, since *C. ladon* presents numerous, inscrutable faces. The modern study of eastern azures is credited to West Virginian William Henry Edwards, who recognized the butterfly's tantalizing complexity, but considered all its forms to be a single, polymorphic entity (Edwards, 1884). Edwards correctly differentiated the Spring Azure from a close European relative, the Holly Blue (*C. argiolus*), and identified several characteristic "forms" that correspond to later taxonomic splits. But he believed that the complex consisted of several discrete populations, connected via loose interbreeding, which presented a variety of phenotypes over the course of a season.

Today, most investigators consider the eastern Spring Azure to be a biological complex consisting of five or more "sibling species," all similar in physical appearance but each biologically distinct (e.g., Pratt et al., 1994). (And this is with two of Edwards's original "forms," the Appalachian and Dusky Azures, already split). *C. ladon* races/sibling species are similar genitalically (Wright & Pavulaan, 1999), but consistent physical and lifestyle differences exist—including segregated breeding populations—that defy traditional concepts of a unitary "species."

Flowering Dogwood (*Cornus florida*)

Wild Cherry (*Prunus serotina*) showing leaf galls

Why has all this odd complexity been visited on the azures? Several explanations are possible. We may, simply enough, be witnessing the fragmentation of an ancestral species, much as astronomers occasionally observe a supernova erupting somewhere far off in the universe. But this does not explain how a number of very similar entities are able to live together (in "sympatry") without interbreeding. Perhaps, during earlier glacial epochs, certain azure populations became isolated, enough so that when rejoining they were already specifically distinct. But there is a second possibility, intriguing yet controversial. Specifically, the larval diet of azures consists mainly of flower parts (or leaf galls in some groups). Since flowers may contain fewer protective chemicals than leaves, caterpillars can "jump" from one plant group to another with comparative impunity. Ovipositing females might specialize on the chemical signature of various locally prevalent flowering plants. Since many target plants have specific flowering periods (especially in spring), this eventually could lead to brood separation among populations, and then ultimately to speciation. Such a segregating mechanism could be ongoing, moreover, not just a singular "supernova" event.

An issue with this explanation is that it implies "sympatric speciation," i.e., species formation without physical isolation. Some lepidopterists believe that the azures may pose a challenge to prevailing orthodoxy on this point.

We have chosen to remain cautious in dealing with the *Celastrina* complex, showing candidate species' names in quotes, without opining as to final taxonomy. A core of dedicated investigators have made significant progress recently in sorting out this difficult group, but the study is an evolving one and should be given time to solidify. Moreover, field observers note certain continuing, unresolved local situations.

Spring Azures   *Celastrina ladon / spp.*

Dorsal female: 4/18/98, Assunpink WMA, Monmouth Co., NJ

Ventral ("marginata" morph): 4/9/95, Cumberland Co., NJ

A traditional harbinger of spring at northern latitudes, the Spring Azure is among the earliest butterflies on wing each year (especially when compared with other species that do not overwinter as adults).

But it is now apparent that only certain members of the Spring Azure species complex are first-line emergers. The early flyers are among the single-brooded segregates now being proposed as separate species. The lifestyles of some segregates are highly specialized, but there are also wide-ranging generalists, such as the pale, multibrooded Summer Azure (*C. 'neglecta'*).

As a group, Spring Azure males patrol for females and are active puddlers. Females prefer nectar and sometimes fly quite high in search of tree buds on which to oviposit. Caterpillars are ant-tended and all members of the group overwinter as pupae.

Ventral ("lucia" morph): 4/9/95, Cumberland Co., NJ

Separation of azures by sight (not always possible) is complicated by the fact that each proposed species presents similar-looking phenotypic forms (or morphs). In the North, well-marked "lucia" and "marginata" morphs are prevalent. "Lucia" is distinguished by a large, amorphous "blotch" on the central hindwing; "marginata" has pronounced dark ventral borders on both wings. (There is also an intermediate "lucimargina" morph that combines both traits [Scott, 1986].) The dark pigments of northern morphs are useful in absorbing solar radiation, thereby maintaining body heat in cool spring or northern weather. The paler "violaceae" form, by comparison, has small, often indistinct markings that are better-adapted to warm surroundings, as found either in the south or later in the season. The currently proposed forms of Spring Azure are as follows (beginning with spring forms, most of which overlap geographically):

Spring (or Edwards') Azure (*Celastrina ladon*): The most southerly of the sympatric early spring forms, well-known to Edwards. It occurs in deciduous woods from central N. Eng. to n. FL, but is rare on the coastal plain. The "violaceae" morph predominates, but "lucia" or "marginata" sometimes occurs in the North. Feeds mainly on buds of Flowering Dogwood (*Cornus florida*), but is omnivorous; e.g., uses viburnums (H. Pavulaan, pers. comm.). Single-brooded, although a few emerge in captivity without diapausing. Difficult to identify by sight, but the dorsal male has distinctive, matted FW scales, visible in the hand with a strong magnifier (Wright & Pavulaan, 1999; Wright, 1995).

Northern Azure (*Celastrina 'lucia'*): Not systematically described; ranges south to about NJ. Flies early (Mar or Apr to May). Omnivorous, often feeds on blueberries, but also cherry and viburnum buds. Varying wooded habitats and scrubby areas, including NJ Pine Barrens.

Ventral ("violaceae" morph): 5/22/95, Albany Pinebush, Albany Co., NY

av wingspan: 1.0″

Dorsal male Summer Azure (C. 'neglecta'): specimen photo

Dorsal female Summer Azure (C. 'neglecta'): 5/18/02, Clay Co., NC

Ventral Summer Azure (C. 'neglecta'): 7/26/97, Weymouth Woods Sandhill Nature Preserve, Moore Co., NC

Atlantic Azure known range (Wright & Pavulaan, 1999).

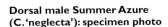

Usually appears as "lucia" or "marginata" morphs, but also rarely as "violaceae." (Confusingly, "lucia" is the name of both a phenotypic morph and this proposed species.) Wing fringes are often more checkered than in Spring Azure (*ladon*) or Atlantic ('*idella*'), but this is not diagnostic.

Cherry Gall Azure (*Celastrina* sp.): A confusing segregate, not yet systematically described; earlier combined with the next form (Wright, 1995). Best noted for the fact that in many populations caterpillars feed on leaf galls, especially those of cherries. Habitat varies with hostplant, may include open woods and abandoned fields with cherry trees. This butterfly's unusual "hostplant" selection may allow it to persist at times and in places where suitable blooming plants are scarce. Single-brooded, the Cherry Gall flies slightly later than Northern or Spring Azure at any given location (overall mid-May to early Jun). Comparatively pale above, usually appears in "violaceae" form. Occurs mainly in s. N. Eng. and NY, including L.I., but not in NJ Pine Barrens; also west into central PA (H. Pavulaan, pers. comm.). Populations in CAN, the central Appalachians, and GA may be this form—or something else.

Atlantic (or Holly) Azure (*Celestrina* '*idella*'): It may not be premature to recognize this distinctive segregate as a distinct species, based on privately published evidence (Wright & Pavulaan, 1999). Others have already begun to acknowledge the split (LeGrand & Howard, 2004). This fluttery, low-flying azure is relatively pale above, sometimes with a purplish cast. Unlike most other azure types, Atlantic Azures "show their colors" regularly, dorsal basking with some frequency (others tend to remain closed at rest, except when warming themselves in exceptionally cool weather). About 95% of Atlantics surveyed are of the "violaceae" morph. Limited to coastal plain woodlands and pine barrens, often occurring near acid bogs or cedar swamps. One brood, late Feb in the South to early Jun in NJ. This butterfly appears genetically programmed to emerge from its pupa slightly later than the Northern; indeed, Atlantics can be seen newly emerged in the NJ pine barrens when "lucia" and "marginata" form azures are declining or have already "gone by." The Atlantic Azure feeds primarily on male holly flowers (*Ilex*), including those of Inkberry (*I. glabra*), American Holly (*I. opaca*), and at least two others. These same hosts have proven toxic to other azure types (Wright & Pavulaan, 1999). Accepts willows in captivity. The originally suggested name of Holly Azure is not used here, as it is likely to cause confusion with a close relative, the Old World Holly Blue (*Celastrina argiolus*).

Summer Azure (*Celastrina* 'neglecta'): Another azure that is almost certainly a distinct species. It begins flying in most locations just as others are waning (as early as Feb in FL and as late as Jul in ME, but usually in May-early Jun). Additional broods fly throughout the summer (single-brooded in the far North, though); flies until Aug or Sep. Summer Azures feed on a wide variety of flowers at different seasons and locations, including New Jersey Tea (*Ceanothus americana*) and Meadowsweet (*Spirea*), but they reportedly will die on a diet of holly flowers or Black Cohosh leaves. Pale dorsally, often with whitish HW rays, light fringe; nearly always "violaceae." The thin, black submarginal line on the dorsal HW is shadowed inwardly by a *thin white line*, rarely seen in the Appalachian Azure. Early flying individuals can easily be confused with "violaceae" morphs of other azures, but during much of the summer this is the only *Celastrina* on the wing. Less confined to woodlands (and better adapted to disturbed habitats) than some of its relatives.

Ventral Atlantic Azure (C. 'idella'): 5/9/95, Cumberland Co., NJ

Appalachian Azure   *Celastrina neglectamajor*

Dorsal male (left), **female** (right): **specimen photos**

Ventral: 5/23/98, Bucks County, PA

The common name assigned to this butterfly describes it adequately, given its mainly upland distribution along the Appalachian Mountain chain. But Eastern Mountain Azure might have been better, since it also lives in the Ozarks, and the next species (Dusky) is at least as closely associated with the Appalachians.

The Appalachian Azure has been recognized as a full species only since the 1980s. Edwards (1884) considered it a large form of Spring Azure specializing on Black Cohosh. He named it *pseudoargiolus*, for its similarity to the European *argiolus*, and believed it was connected to the rest of the Spring Azure complex via sparse summer broods that interbred with Summer Azures (*C. 'neglecta'*). But no such interbreeding is apparent, and the Appalachian Azure is now considered a discrete entity, based on its average size, specialized ecology, early larval features, and other traits (*see* Wright & Pavulaan, 2000).

Male Appalachian Azures emerge before females. They seek minerals at puddles, often not in direct association with hostplants. Once females are in flight, both congregate near Black Cohosh.

Larva being tended by ant

Identification	Our largest eastern blue (on average), often visibly larger than other *Celastrina*. But there is significant size overlap, with both Spring Azure (*C. ladon*) in late spring and Summer Azure (*C. 'neglecta'*) in early summer (Wright & Pavulaan, 2000). Easily confused with the Spring, Summer, or Cherry Gall Azure. It is advisable to study populations rather than individuals, and to carefully note habitats and local flight sequences. HW fringes normally pale (vs. more checkered in many *ladon*).
Dorsal	Wings usually closed. Males blue to violet-blue (lighter than *ladon*) with narrow black margins. Male lacks microscopic matted FW overscaling of *ladon* (Wright & Pavulaan, 1999). Females usually blue on HW (40% all-blue; a few are extensively white, as in Summer).
Ventral	Usually *chalky white*, vs. more grayish in other forms (but color differences are minor vs. *'neglecta'*). Dark HW markings are pale, much reduced, and in some cases virtually absent.
Habitat	Rich, moist deciduous woods, cove forests, often near streams. But also on ridge tops with hostplants, at times in comparatively dry areas.
Hostplant	Black Cohosh (*Cimicifuga racemosa*). Female lays on "corn cob" budspikes, ant-tended. Caterpillars eat buds or leaves if buds are exhausted. Summer Azures also feed on buds, but die eating leaves. Larval color highly variable; has brown "saddle-patch" near head.
Occurrence	Local, usually scarce, but can irrupt. In Appalachians, se NY to n. GA; also populations in Ohio River Valley, to sw MO. One brood, Apr/May-Jul (usually between *ladon* and *'neglecta'*). Pupa overwinters.
Ecology	Specialist. Highly specialized in terms of hostplant use and habitat.

Black Cohosh (*Cimicifuga racemosa*)

av wingspan: 1.25"

Dusky Azure   *Celastrina nigra* [= *ebenina*]

Dorsal male: 4/26/96, Fork Creek Pub. Rec. Area, Boone Co., WV

Ventral (worn): 4/26/98, southeastern WV

We are inclined to think of this species as the *true* Appalachian Azure. It flies in early spring, mostly in the central Appalachian highlands (exact dates depend on latitude and elevation). In these timeless haunts, it can be found along with such other, distinctive early emergers as the 'Appalachian' Silvery Blue and, if one is truly fortunate, the 'Appalachian' Grizzled Skipper.

Considering its memorable surroundings and notable flight companions, the Dusky Azure itself is rather modest in appearance. Yet it has a distinctive, highly specialized lifestyle, also an interesting taxonomic history.

This was the earliest and most obvious of the "splits" from Edwards's original Spring Azure complex (Clench, 1972). Edwards knew of *nigra* in the 1880s, but considered it a variety of his *pseudoargiolus* "megaspecies" (specifically *Lycaena pseudoargiolus*, form *nigra*). His failure to realize the Dusky's distinctiveness is undoubtedly due in part to the fact that he incorrectly thought the dark-winged individuals he saw were females (in keeping with the normal sexual color pattern among azures). Other observers perpetuated this mistake until the 1970s, when Clench finally discovered that *male* Dusky Azures have dark forewings and subsequently described the species (Iftner et al., 1992). Clench christened his new butterfly *ebenina*, but it was later determined that *nigra* was a prior available name under governing taxonomic rules, so that name is used today (Scott & Wright, 1991).

Male Dusky Azures patrol for females, but also may be seen perched on dried hostplant stalks from the prior year (Iftner et al., 1992). As in other azures, males actively seek minerals from wet soil. Indeed, at "puddle parties" Dusky males can be spotted quite easily on the wing as "dark shadows" among other Spring Azure forms. Females remain near hostplants, except when nectaring on a variety of spring flowers (Allen, 1997).

Identification	Small. Males are distinctive as one of the few *Celastrinas* that can be clearly identified in flight in the field. Females are far less determinable. A certain amount of care must be taken not to mistake a male Dusky for a female Eastern Tailed-Blue, also dark above. The two species can fly in close proximity.
Dorsal	Male is a *uniform, dark gray-brown*, often with flecks of blue scaling. No orange dots near HW base (as in Eastern Tailed-Blue). The female is *light, whitish-blue* with rather broad, dark margins and darkened veins.
Ventral	Pale, ash-gray to bluish white, *hard to distinguish from other "violaceae" morph azures*. Well-defined dark dots along HW margin sometimes cited (Scott, 1986).

Habitat	Moist deciduous woods, cove forests, ravines; especially shaded northern slopes where hostplant abounds. Often near streams.
Hostplants	Goatsbeard (*Aruncus dioicus*). Eggs laid on new, unfurled foliage, later on buds. Early larvae eat leaves, later ones leaves or buds. Ant-tended.
Occurrence	Local and restricted, nw MD to w. NC, also OH, west to nw AK. One brood, late Mar-May. Pupa overwinters.
Ecology	Specialist. Range limited by that of single hostplant; larval feeding strategy and distinctive color pattern.

Goatsbeard (*Aruncus dioicus*)

av wingspan: 1.0"

Silvery Blue   *Glaucopsyche lygdamus*

Dorsal male: 4/25/95, Romney, Hampshire Co., WV

Dorsal female: 7/15/00, Pittsburgh, Coos Co., NH

Ventral (northern *couperi* race): 6/12/93, Wilsons Mills, Oxford Co., ME

Bright and active, the Silvery Blue emerges in early spring in the central Appalachians and in early summer in the North.

Two physically isolated races occur in our area. They differ significantly in appearance and lifestyle, and may well be separate species (Dirig & Cryan, 1991). The southern nominate race (*lygdamus*) has withdrawn recently from the northern edge of its historic range, while the northern *couperi* race is pushing south along highway corridors, feeding on introduced Cow Vetch. It may soon overtake retreating southern populations. Several isolated races of Silvery Blue in California are either threatened (*paloverdensis* and *pseudoxerces*) or extinct (*xerces*).

Male Silverys patrol near the ground seeking females, and often take minerals. Females fly near hostplants, and actively visit flowers for nectar.

Ventral (southern *lygdamus* race): 4/25/95, Romney, Hampshire Co., WV

Identification	Small-medium. Northern race statistically larger than southern (Dirig & Cryan, 1991), an unusual condition within a single butterfly species. Northern's FW also more pointed. Fringes white.
Dorsal	Male is *shimmering, silvery-blue* above, narrowly outlined in black. Female is *blue toward the wing base*, with broad, brownish-gray edges (broader in northern)—but sometimes all-brown, as in Dusky Azure male.
Ventral	Gray (northern) to brownish-gray (southern). Note the *single row* of *black spots, outlined in white*, on both wings; dots notably larger and more conspicuous on southern race.

Habitat	Openings in damp woods, but also brushy areas, shale barrens (southern race). Uses additional habitats in the West and CAN.
Hostplants	Southern race specializes on Carolina Vetch (*Vicia caroliniana*); northern uses various legumes, many non-native, including Cow Vetch (*V. cracca*), White Sweet-Clover (*Melilotus alba*), Veiny Peavine (*Lathyrus venosus*), Crown Vetch (*Coronilla varia*). Those caterpillars that feed on flowers or seedpods are reportedly white to purple, while those eating leaves are green (Layberry et al., 1998). Ant-attended, but in one recent study only one of four attending ant species significantly reduced parasitism; two others reduced it somewhat, and one seemed to prey on the larvae (Fraser et al., 2001). Larger ant species reportedly tend older larvae (Spomer & Hoback, 1998).
Occurrence	Southern race fairly common. Northern race arrived in N. Eng. in the 1960s, n. NY in the 1980s. Also found in central Appalachians, across CAN, and in western states. Southern race flew on shale barrens in central NY until the 1960s (Dirig & Cryan, 1991). One brood, early Apr-late Jun, synchronized with hostplant flowering; 15-30 days in any location. Pupae overwinter in leaf litter.
Ecology	Medium generalist/specialist. Southern race specialized, northern very adaptable.

Carolina Vetch (*Vicia caroliniana*)

av wingspan: 1.05"

Transient Blues of Northern Maine: Road Trippers on Curious Journeys

When thinking of sudden, unpredictable shifts in butterfly ranges, we are inclined to envision fragmented tropical island chains, such as the Florida Keys, where new species regularly become established while other, more familiar ones vanish just as rapidly. In the lycaenid subfamily Polyommatinae, for example, the Floridian race of Miami Blue has collapsed in recent decades while a new colonist, the Acacia Blue, has developed a seemingly secure foothold. And a similar switch was seen in the 1990s with the Amethyst and Silver-banded Hairstreaks.

It might not occur to us to look for rapid distributional change in the staid and timeless-seeming woodland expanses of Maine. Yet biological systems are everywhere in flux, and the effects of human activity regularly create subtle new opportunities. Three recent incursions, all involving blues, illustrate this point.

At some time during the early 1900s, the Greenish Blue (*Plebejus saepiolus*) appeared quietly in southeastern Canada and northern Maine. The species was unknown in Quebec as of 1912 (although an early, isolated Nova Scotia record does exist [Scott, 1986]). Yet by 1929, Klots found the Greenish Blue "well-established" in the Gaspé, and by 1950 he declared it "quite common" in Maine as well (Klots, 1951, p. 167). Klots imagined that the species would continue to extend its range southward into the United States.

Instead, the Greenish Blue has lost its foothold in northern Maine. To our knowledge, it has not been reported in our area for more than a decade. And even if remnant colonies remain in this large and incompletely surveyed state, the butterfly is by no means any longer "quite common."

The Greenish Blue is readily identifiable in New England. It has dark cell-end spots on its dorsal forewings and usually a single rusty-orange submarginal mark on its dusky, ventral hindwing. It still should be sought in northern New England between late May and early July, in areas where it may still persist.

It is thought that the Greenish Blue spread into Maine along highway corridors, feeding on roadside plants such as Alsike Clover (*Trifolium hybridum*), an Old World legume first established in North America in the 1830s. We have less of a clear vision as to why the species' southward extension suddenly failed. And this is all the more perplexing when we compare the current range expansion of the northern Silvery Blue (*couperi* race), a second recent invader, which seems to have arrived in Maine during the 1960s and northern New York in the 1980s (Dirig & Cryan, 1991). Like its predecessor, the Silvery Blue has spread along highway corridors, in this case those lined with Cow Vetch (*Vicia cracca*). And it has recently been observed using Crown Vetch (*Coronilla varia*) as well, like another butterfly, the Wild Indigo Duskywing (Shapiro, 1979; R. Dirig, pers. comm.).

Invasions such as that of the Greenish Blue can fail for many reasons, including predation, disease, or a period of unfavorable weather. We often lack specific information to draw well-founded conclusions when such reverses occur.

Observers in Maine should also keep a lookout for another invading polyommatine, the Western Tailed-Blue (*Everes amyntula*). This butterfly has isolated populations in New Brunswick and the Gaspé. It feeds on a variety of legumes, and is quite a good candidate to extend its range eventually into our region (R. Webster, pers. comm.). Unfortunately, it is highly similar to the Eastern Tailed-Blue, a common resident as far north as central Maine, and this could mask an incursion. (The Western is generally paler, with less distinct orange markings.)

Greenish Blue (*Plebejus saepiolus*), failed early 20th-century colonist.

Western Tailed-Blue (*Everes amyntula*), occurs in New Brunswick: colonist of the future?

'Crowberry' Northern Blue   *Lycaeides idas empetri*

Dorsal male: 7/13/01, Washington Co., ME

Dorsal female: 7/13/01, Washington Co., ME

Traditionally, the Karner Blue was northern New England's only medium-large blue. But Silvery and Greenish Blues eventually appeared, and now we are aware of still another: the 'Crowberry' Northern Blue (*L. i. empetri*), a very distinct race that inhabits coastal "raised bogs" in far eastern Maine—where it can be locally common.

The Northern Blue separated recently from the Melissa. Its genetic isolation appears to rely on behavioral factors—such as hostplant differences and the male's ability to distinguish female wing patterns—more than on genetic incompatibility per se (Fordyce et al., 2002; Nice & Shapiro, 1999).

Flies early in day. Males patrol for females; attracted to closed-wing individuals holding their heads down so that submarginal patterns show. Adults bask often and nectar actively.

Ventral (differently marked individuals): 7/13/01, Washington Co., ME

Identification	Medium-small, but a bit larger on average than most eastern blues. Geographically variable (this account considers only the 'Crowberry' Northern Blue, which has characteristics of a sibling species).
Dorsal	Male is *bright, iridescent blue* with a thin, dark wing border. Female grayish-brown with variable blue flecking and a *submarginal row of orange-capped black dots* (variable, most pronounced on HW).
Ventral	Pale, whitish-gray with black dots. Usually an *orange-and-black submarginal dot row*, but *may be absent*. Patterning more pronounced in most western races.

Habitat	Confined to coastal "raised bogs" of eastern ME (see Introduction, p. 15).
Hostplants	Local race feeds on Black Crowberry (*Empetrum nigrum*), a small, heathlike plant. Western races use hosts in other families. Eggs laid singly on or near host. We have seen females carefully examine plants before ovipositing. Larvae are ant-attended; later instars may live in ant nests, feeding on pupae (Tolman, 1997).
Occurrence	Holarctic. *L. i empetri* in CAN Maritimes. Should be sought in N. Eng. bogs (we may have seen one in a central NH bog). One brood, Jun-Aug. Eggs overwinter.
Ecology	Specialist. Our race highly specialized.

"Raised bog" habitat, inset: **Black Crowberry** (*Empetrum nigrum*)

av wingspan: 1.05"

'Karner' Melissa Blue   *Lycaeides melissa samuelis*

Dorsal male: 5/17/98, Albany Pine Bush, Albany Co., NY

Dorsal female: 5/17/98, Albany Pine Bush, Albany Co., NY

This imperiled butterfly was listed as federally endangered in 1992, less than 50 years after it was first described by novelist Vladimir Nabokov (who was an accomplished student of lycaenids; Johnson & Coates, 1999). Originally found in 12 northeastern states, the Karner Blue rapidly vanished during the 1990s, and is today known from just a few. In our area, isolated remaining colonies at Concord, New Hampshire, and the Albany Pine Bush teeter on the brink of disappearance, despite dedicated habitat conservation programs and captive breeding efforts.

Karner Blues are specialized to the point of fragility. They require healthy stands of Wild Lupine, which are beset by developers, habitat succession, deer browsing, and drought. Diapausing eggs also appear to need consistent snow cover, which has been lacking in several recent warm winters.

Ventral: 7//93, Albany Pine Bush, Albany Co., NY

The Karner Blue may be specifically distinct from the Melissa Blue, given its distinctive habitat, hostplants, and number of broods. It is at least a cohesive and well-differentiated race.

Male Karners patrol small territories around lupine patches seeking mates. Mating has been seen in the afternoon. Weak fliers, Karners are mostly sedentary, seldom moving far from colony sites.

Identification *Dorsal* *Ventral*	Medium-small. Can be confused in our area only with the prior species. Males are *bright, metallic pale-blue*. Females are dark, often with much *deep blue* scaling; note *orange-and-black markings* on HW submargin. Conspicuous row of *separated, submarginal orange markings* on HW, next to a series of blue-centered black dots. FW pattern similar but less pronounced. HW flecked with dark spots outlined in white.
Habitat	Sandy pine-oak barrens and savannas, lakeside dunes (IN), disturbed sites. Western *melissa* race uses prairies, varied open weedy habitats.
Hostplants	Wild Lupine (*Lupinus perennis*). Eggs laid on or near host, larvae are tended by as many as 20 ant species; most offer protection but some feed on the eggs (Dirig, 1997). Western race uses additional hosts.
Occurrence	In sharp decline. Currently limited to Albany Pine Bush and Concord, NH. Formerly ME, PA, w. NY. Captive breeding releases being undertaken. Also persists (or is being reintroduced) in WI, MN, MI, IN, OH, ONT. Two broods, late Jun and mid-Aug. Eggs overwinter.
Ecology	Specialist. A specialist in early successional habitats, A number of ephemeral conditions must coincide for Karner Blues to survive. Effective management techniques not yet definitely clear. Loss of contiguous regional sites for metapopulational dispersal is clearly a threat.

Wild Lupine (*Lupinus perennis*)

av wingspan: 1.1"

Northern Metalmark   *Calephelis borealis*

Dorsal: 7/1/94, Fairfield Co., CT

Ventral: 7/3/99, Springdale, Sussex,Co. NJ

At first glance, the scattered locations in which Northern Metalmark colonies exist in our region do not seem exotic, especially when compared with the visually distinct habitats of other, more remote isolates, such as the alpine satyrs or northern bog dwellers. Yet on close inspection the metalmark's needs are equally specialized. In Connecticut, for instance, Northerns require southern-facing calcareous outcrops, usually in redcedar glades, not more than 50 meters from suitable nectar plants (David Norris, pers. comm.). Hostplants tend to grow in or near the edge of associated rocky woods. Given these very selective habitat preferences, it is not surprising that this is a local and isolated butterfly.

The metalmarks are a diverse family, closely related to the lycaenids. Most live in the New World. Of about 20 species in the United States, three occur in the East and two in our region. Metalmarks differ from the lycaenids in their longer antennae, distinctive wing venation, and other, specialized anatomical features—including, in many species, the distinctive metallic wing markings that give the family its name. Male forelegs are half the length of the remaining four and are nonlocomotive. Female forelegs are only slightly reduced and are used for walking.

Male Northern Metalmarks perch near outcrops awaiting females. In the Northeast at least, they are distinctly attracted to Black-eyed Susans for nectaring (though other plants are used as well). When disturbed, individuals perch upside-down beneath a leaf with wings spread. We have seen mating in early afternoon in Connecticut.

Identification	Medium-small; larger than Little Metalmark. Abdomen above *mostly dark*. Note light areas in wing fringe, especially near FW apex. If seen well, there is little likelihood of misidentifying this butterfly, except in parts of the upper Midwest, where Swamp Metalmark also occurs. The latter is most easily separated by habitat and range.
Dorsal	Wings orange with dusky basal scaling on both wings (or in a vague median band). Note *concentric silver "metal marks."*
Ventral	Yellowish-orange, notably brighter than upper surface (more so than in Little). Underside "metal marks" silvery, mirror dorsal pattern.

Habitat   As described above, limestone outcrops, including old quarries, near hostplants and nectar. Shale barrens in WV.

Hostplants   Roundleaf Ragwort (*Senecio obovatus*) is the main (perhaps exclusive) host. Golden Ragwort (*Senecio aureus*) and Common Fleabane (*Erigeron philadelphicus*) require substantiation. The main host (*obovatus*) is far commoner than the butterfly (Gochfeld & Burger, 1997). Eggs are laid on the underside of hostplant leaves.

Occurrence   Seldom found far from a local colony site. Separate population clusters exist in nw CT to w. NJ; central PA through central Appalachians; also upper Midwest and e. KY; MO to e. OK. One brood, mid-Jun in South to early Jul in North (late May in MO). 6th instar caterpillars overwinter in leaf litter; 8th instar pupates in the spring (Allen, 1997).

Ecology   Specialist. Careful observations have provided detailed insights into this butterfly's specialized habitat requirements. They are not easily met, and scarce colony sites are being lost due to succession and disturbance.

Round-leaved Ragwort (*Senecio obovatus*)

av wingspan: 1.1″

Little Metalmark *Calephelis virginiensis*

Dorsal: 9/25/95, Everglades NP, Miami-Dade Co., FL

Ventral: 3/20/95, Everglades NP, Miami-Dade Co., FL

This small, rather mothlike butterfly is worth the effort to study closely. Its wing and body patterns create an intricate study in miniature, and the results can be striking. Little Metalmarks reside in the Southeast, where—oddly enough—they are the only resident metalmark. Between 1,000 and 1,500 species of metalmark occur in the neotropics, in a wide variety of mainly wooded and scrubby habitats (see, e.g., DeVries, 1997). Given this wide family presence, it is somewhat surprising that it is barely found at all in southern Florida and the West Indies. Perhaps the sedentary nature of the group as a whole, and their generally weak flight style, help explain their lack of dispersal, especially over water in the Caribbean basin.

Metalmarks in the genus *Calephelis* look much alike. Eastern species (including the Swamp, not in our area) tend to be identifiable by geographical location. But where several species occur together, as in the southwestern United States, field identification can be difficult. Many *Calephelis* metalmarks also occur in the tropics, where they are as confusing as ever, but there the family also assumes "nearly every imaginable combination of colors, patterns, and wing shapes" (Klots, 1951, p. 122). Metalmarks often mimic moths and butterflies, including swallowtails. Many are ant-attended. The metalmarks are a truly fascinating family, deserving of much study.

Little Metalmarks fly low to the ground, resting frequently in grassy tangles or on low leaves. Small and inconspicuous, they often go unnoticed even in flight, and are most visible when perched open-winged on light-colored flowers. They use a variety of nectar plants, but seem to favor flat-flowered composites. Males reportedly patrol for females during the day, but the patrol route is often quite confined.

Identification	Very small, especially males. This is the only metalmark to be expected in the se US. FW rounded, fringes dark. Abdomen mostly orange (vs. dark in Northern).
Dorsal	Variably rusty orange to bright, brownish-orange, with a series of roughly concentric dark bands, dots, and silvery metallic lines (like solder lines). Basal area of wings is not darkened, as in Northern.
Ventral	Bright orange with a series of lead-gray or silvery-gray markings, approximating the pattern of the dorsal surface.

Habitat	Pine flatwoods, savannas, salt meadows, grassy fields, and roadsides (common in Everglades).
Hostplants	Not fully studied. Yellow thistle (*Cirsium horridulum*) is usually cited as the sole host, but in NC Bo Sullivan recently reported apparently exclusive use of a local composite, Vanilla-Plant (*Trilisa odoratissima*) (LeGrand & Howard, 2004). Caterpillars feed mostly at night, rest beneath leaves.
Occurrence	Locally uncommon to common, resident se VA and south; but rare stray on the Keys. Also along Gulf coast to se TX. Three broods, Mar-Oct in North, all year s. FL. Caterpillars probably hibernate.
Ecology	**Medium Specialist.** Usually sedentary, in local colonies, with limited habitat use and few known hostplants.

Yellow Thistle, yellow form (*Cirsium horridulum*)

av wingspan: 0.75"

The Nymphalids (Brushfoots): Family Nymphalidae

The nymphalids are an unlikely assemblage of butterflies. They encompass, in one family, much of the biological diversity of all of the other five families combined—in shape, appearance, ecological lifestyle, early stages, hostplant use, and seasonal movements. Specialized associations with key hostplants (such as violets, milkweeds, and nettles) define discrete "species clusters" within the family. And in terms of sheer aesthetics, this family includes many of the world's most commandingly beautiful butterflies. The most brilliant among them (morphos, heliconians, agrias, etc.) are mainly tropical, but a number of our own species are also breathtaking to behold.

Generally medium-sized to large, nymphalids inhabit nearly every region in the world. The Monarchs and Painted Ladies have the most extensive ranges of any butterflies. Yet the family also includes many local species, a number restricted to single Caribbean islands (Smith et al., 1994). Worldwide, nymphalids comprise more than 5,000 species, of which just over 200 occur on the North American continent north of Mexico. About 70 species reside on the East Coast. The exact number depends on whether one accepts particular "lumps" and "splits."

The enormous diversity of the Nymphalid Family has long challenged taxonomists, who have struggled to produce a definitive taxonomy for the group. Even with careful study of lifestyles, anatomy, and the like, it has proven difficult to identify a single set of broad traits and behaviors that characterize all nymphalids above the subfamily level. (Indeed, most of the present nymphalid subfamilies were earlier classified, at one time or another, as separate families. This includes snouts, heliconians, emperors/leafwings, browns, and milkweed butterflies.)

The nearest thing to a universal field mark for nymphalids is the use of *only the four rear legs* for walking. In most species and sexes (female snouts being the principal exception) the front two legs are atrophied and nonfunctional, covered with brush-like "fur" and dedicated to sensory functions. (The only non-nymphalids to walk on their "rear fours" are male metalmarks.) Nymphalids have certain characteristic wing vein patterns and many have hairy palps (most obvious in the snouts). Wing colors are complex and varied, drawing on a broad palette of colors. Orange, black, yellow, white, silver, and brown are emphasized, with occasional use of blues, reds, and greens. The electrifying iridescence of some tropical species, such as morphos, is based on the mechanical structure of specialized, light-refracting scales (Scoble, 1995, pp. 75–76). Several groups engage in mimicry.

The longest-lived of our adult butterflies are all nymphalids. In a number of species, the last generation of the season emerges sometime during the summer and survives until the following spring, overwintering in a state of inactivity, or diapause. At our northern latitudes, only nymphalids overwinter as adults. (In the Deep South, butterflies from other families pass the cooler months as adults, in a state of reduced activity, but not full diapause.) The longest life span of any continuously active adult butterfly in our area is that of the Zebra Heliconian, which can survive for six months or more. All of our nymphalids overwinter as adults or larvae.

Long adult generations have allowed nymphalids to develop comparatively complex seasonal movement patterns. Monarchs and Painted Ladies travel great distances each year, with at least some "return migration." And even irregular emigrants, such as American Snouts, occasionally stage enormous mass flights.

Even the more sedentary nymphalids often show strong spatial orientation, being diligently territorial (especially commas, emperors/leafwings, etc.). Still other groups patrol to find mates (the fritillaries and crescents). When mated pairs are disturbed, females usually carry the pair—but either may do so in fritillaries, and only males carry among heliconians.

Nymphalid lifestyles differ greatly during the early stages, as described in the individual accounts. Communal feeding is relatively common. In most species, the pupa hangs suspended upside-down from a single point of attachment (the *cremaster*), with no silk "girdle." Arctics pupate at ground level, however, often in silk-lined "nests."

Nymphalids utilize a broad range of hostplants, generally dicotyledons (except for the grass-feeding satyrs). Adults have short proboscises and frequently feed on mud, dung, carrion, and sap, as well as on nectar. Certain heliconians are distinctive in feeding on pollen.

Typical nymphalid caterpillar (Harris' Checkerspot)

As already noted, the lifestyle and ecology of nymphalids are best described at the subfamily level. Beginners are best advised to gain familiarity with the major "lifestyle groups" as an entree to this broad assortment of species.

Snouts (Subfamily Libytheinae). An ancient and widespread group, but comparatively small, with only about a dozen species, and just one in our region. Earlier considered a separate family, the snouts are an "outlier" in the nymphalid complex. *See* species account.

Heliconians (Tribe Heliconiini). Also once considered a separate family, now listed as a tribe in the subfamily Heliconiinae (along with fritillaries). Many members of this group derive strong toxins from their larval hosts (e.g., passionvines). In the tropics, they form highly complex mimicry rings, often including moths. The wings are long and thin. Relatively sharp-eyed, for butterflies, even "intelligent" (especially genus *Heliconius*). Mostly sedentary, but Gulf Fritillaries regularly colonize northward in summer.

Fritillaries (Tribe Argynnini). Northern counterparts of the heliconians, fritillaries include two different subgroups. *Greater Fritillaries* (mainly genus *Speyeria*) are familiar generalists of open habitats. All use violets as hostplants. First-generation caterpillars overwinter near hostplant roots, feeding in the spring on fresh shoots. Males recognize females in part by scent. This is useful, since many of the Greater Fritillaries are extremely difficult to distinguish by sight (even, it seems, for male fritillaries). *Lesser Fritillaries* (genus *Boloria*) are a northern, often holarctic group, including many boreal niche specialists. Lesser Fritillaries feed on a variety of hosts, not always violets. Many species are declining.

Crescents and Checkerspots (Tribe Melitaeini). Mainly small, orange-and-black butterflies; includes some extremely common species. Most feed on composites or acanthuses, and many are noxious to predators. Cluster-laying and communal feeding are frequent in this group. Most species are sedentary.

Typical Brushfoots (Tribe Nymphalini). Anglewings, tortoiseshells, and ladies are a diverse and widespread group. Hostplants vary widely, including elms, nettles, willows, birches, asters, currants, vervains, and figworts. Adults overwinter in diapausing species. Sap and rotting material is the main adult food for many. Adults are often "territorial" and pugnacious.

Admirals and Relatives (Subfamily Limenitidinae). Many species in this interesting group engage in mimicry. Most are sedentary. They overwinter as mid-stage caterpillars. The subfamily includes a number of miscellaneous neotropical species that reach the southern edge of our region.

Discussed separately below: Leafwings and Emperors (Subfamily Charaxinae), Satyrs (Subfamily Satyrinae), and Milkweed Butterflies (Subfamily Danainae).

American Snout   *Libytheana carinenta*

Dorsal: 7/7/96, Jamaica Bay NWR, Queens, NY

Ventral (lighter variant): 7/26/97, Weymouth Woods Sandhill Nature Preserve, Moore Co., NC

This unusual-looking butterfly resides permanently in the South, but it is a marginal colonist/stray in the North. It belongs to a small, worldwide clan whose lineage can be traced back in the fossil record some 35 million years (Emmel et al., 1992). And fossil leaf imprints suggest that even the snout ancestors were associated with hackberries (Pyle, 1981). Today, the Snout is perhaps best noted for its enormous, mass irruptions out of the Southwest, which can at times darken the sky and halt traffic.

Early fossils have not eliminated debate over Snout taxonomy. Authorities disagree as to whether the group is a separate family or a subfamily within Nymphalidae, and most (but not all) recognize a single North American species. Assuming this is correct, it is interesting that several regional races persist despite regular emigrations across territorial boundary lines. The East Coast race is *bachmanii*.

Perched on twigs, Snouts resemble dead leaves, their feltlike "snouts" seeming to mimic a petiole. This bit of camouflage may be the snout's main adaptive feature. Adults congregate at patches of damp soil and actively nectar. Flight is bounding and satyr-like, taking a zigzag course (Kawahara, 2003). Puddling adults keep their wings shut; they are easily disturbed. Dorsal basking occurs briefly to achieve warmth. Adult males walk on four legs, females on six. We cannot contradict the interesting comment that mating has been seen only at night (Opler & Krizek, 1984).

Ventral (darker variant): 7/26/97, Weymouth Woods Sandhill Nature Preserve, Moore Co., NC

Identification	Medium-sized. Instantly recognized when its exaggerated labial palps ("snout") are seen. The FW shape is distinctive, *squared-off* at the apex and curved along the leading edge. Trailing HW edge is *scalloped.*
Dorsal	Orange on FW base and central HW (extent varying by race), dark brown outward with variable *white dots,* usually conspicuous (especially on females).
Ventral	Two forms: *plain gray* (often with violet sheen), and *mottled light gray-and-black.*

Habitat	In our area, thickets, river bottoms, coastal scrub, open wood edges. Usually near hostplants, but can be anywhere when emigrating.
Hostplants	Various hackberries (*Celtis*), including American (*C. occidentalis*), Dwarf (Georgia) (*C. tenuifolia*), Sugarberry (*C. laevigata*). Eggs laid in small clusters, caterpillars eat young leaves.
Occurrence	Seems to overwinter as far north as NYC. Less common s. FL/Keys. Also W.I., US southern rim, south to ARG. Mass emigrations mainly in SW. Apparently two broods, May-Aug. Adults overwinter in South.
Ecology	Medium Generalist. Roving migrant, but limited in host plant utilization.

American Hackberry (*Celtis occidentalis*)

av wingspan: 1.75"

Passionflowers and Butterflies: A Tangled Web

There are between 400 and 500 plants in the genus *Passiflora* (passionflowers), comprised mainly of vines and climbing herbs. As a group, they synthesize a fearsome array of deterrent chemicals, honed to a high level of sophistication in the evolutionary forge of the neotropics. Passionflowers have practical importance in areas as diverse as human medicine and agriculture, at one extreme, and to the livelihood of longwing butterflies in the Tribe Heliconiini, at the other.

Three native passionflowers are important hostplants in our region: Maypops (*Passiflora incarnata*), Corkystem Passionflower (*P. suberosa*), and Yellow Passionflower (*P. lutea*). *Passiflora* usually cannot withstand winter temperatures, but Maypops are hardier, ranging north to Virginia.

Passionflowers have strong chemical defenses against browsing herbivores, but a few kinds of insects (notably the Heliconiini) have learned to bypass them, and now utilize the antibrowsing chemicals for their own protection (Gilbert, 1982). But not content to surrender so easily, passionflowers themselves have evolved additional defenses, seemingly directed at their new larval tormentors. This is widely cited as a classic example of coevolution, in which the adaptive path taken by a particular group of organisms is shaped by another group of organisms, and vice versa. The end product is an "arms race" in which each combatant continually seeks to outmatch the other's latest advance (Ehrlich & Raven, 1964). Coevolution is a theoretical model, but it is difficult otherwise to account for some of the customized adaptations uncovered by heliconian researchers, such as Lawrence Gilbert, at the University of Texas, and his students.

Principal among the "secondary chemicals" utilized by passionflowers are cyanogenic glycosides, which induce vomiting in vertebrates (Brower, 1985) and, when eaten, can release deadly levels of hydrogen cyanide (HCN).

Longwings use secondary chemicals obtained from passionflowers for protection in the larval stage. They are generally not available to adults—who either must produce their own (as in the Zebra Longwing) or go unprotected. *Note:* An exception is a South American longwing (*Heliconius sara*), which, as recently learned, *does* have the ability to sequester cyanogenic glycogens obtained in the larval stage for use as an adult (Engler et al., 2000).

Among the specific counterdefenses evolved by passionflowers are: changes in leaf shape (to confuse ovipositing females, who use shape as an initial cue in locating hosts); growing yellow nodules that resemble heliconian eggs near oviposition points (to make females think another female has already visited this plant, thus deterring her); shedding new growth to periodically rid the plant of eggs and caterpillars; and growing tiny hooks along the stem that pierce a caterpillar's prolegs, thus immobilizing it. The plants also may produce a "cocktail" of several secondary chemicals, thus forcing a hungry herbivore to manage multiple detoxifications simultaneously. If these adaptations had been designed by human intelligence, they would be hailed as ingenious.

The term "passionflower," we should note, does not refer to emotions, but rather to the fact that the flowerhead reminded early observers of the crucifixion—with its three stigmas representing nails, a ring of floral filaments representing the crown of thorns, etc. Passionflowers were introduced to Europe from Peru by a 16th-century Spanish doctor. Initially used in herbal tea, they were later employed in traditional medicine as pain-relievers and sedatives. Some passionflowers produce pulpy fruits that are sold commercially. Also, chemistry of hostplants is a key to their importance to us and butterflies mutually.

Some passionflowers are highly invasive. It is not advisable for South Florida gardeners to plant showy, non-native varieties (in favor of plainer-looking native ones) in order to provide hostplants for local butterflies.

Maypops (*Passiflora incarnata*)

Corkystem Passionflower (*Passiflora suberosa*)

Passionflower with "false eggs"

Zebra Heliconian   *Heliconius charitonius*

Dorsal: 12/22/94, Long Pine Key, Everglades NP, Miami-Dade Co., FL

Ventral (in roosting crèche): 11/22/00, Homestead, Miami-Dade Co., FL

This neotropical butterfly is Florida's state butterfly. Its unique lifestyle is not immediately apparent as it meanders through the state's subtropical woodlands. Yet the Zebra's biology is intricate and intriguing. Many of its subtleties have been revealed in the research of Lawrence Gilbert and his students at the University of Texas. As a result of such efforts, *Heliconius* is among the best-studied of all butterfly genera (Douglas, 1989, p. 107).

The cornerstone of the Zebra's lifestyle is pollen-feeding, first described in the 1970s (Gilbert, 1972). Adults secrete enzyme-laden fluids to dissolve pollen wads collected on their proboscises. The released amino acids and other nutrients are absorbed into the body. These nutrients allow Zebras to extend their adult lives to as long as six months. Females also use them to produce additional eggs, laid in small clusters throughout their extended life spans. This creates (among other things) a regular influx of fresh, young adults into the population. Both sexes use pollen-derived chemicals to synthesize aliphatic cyanogens, which render them toxic to vertebrate predators (Wade, 2000)

Zebras capitalize superbly on the opportunities provided by pollen-feeding. They have evolved sharp distance vision (for a butterfly), which allows them to recognize environmental landmarks, and also have learning-enabled brains that allow them to retain landmark information. Taken together, these abilities permit Zebras to regularly "trapline," visiting a prescribed series of nectar/pollen sites each day. Young adults seem to learn the route from their elders, with whom they share communal roosts. Zebra sleeping aggregations, called *crèches*, include up to 25–30 individuals. A single roost may be used for several months at a time. Female roost occupants depart first each morning to forage (Douglas, 1989, pp. 192–93), and all occupants return each night at dusk.

Males Zebras patrol during the day for females, and are attracted to female chrysalids, with whom they seek to mate at or near the time of emergence; after mating the male deposits an "antiaphrodisiac" chemical onto the female's abdomen to deter further couplings. Zebras are basically sedentary, but some regular dispersal is apparent, given the low genetic "drift" observed among supposedly isolated colonies (Kronforst & Fleming, 2001). Individuals also stray north in the summer, beyond their survivable winter range.

Identification	Unmistakable in pattern, shape, and behavior. Long, rounded wings, slender body. Flies with fast, shallow wingbeats that produce a languid, floating effect, although more rapid flight is possible when alarmed.
Dorsal	*Pitch black with bold, yellow stripes* (sometimes infused with rust tones). Small red dots near body.
Ventral	Mirrors dorsal, but red basal marks more pronounced.

Habitat	Subtropical hammocks and pine-oak woods, also suburbs, parks, open fields with nectar.
Hostplants	Passionflowers (*Passiflora*), including Maypops (*P. incarnata*), Yellow Passionflower (*P. lutea*), Corkystem (*P. suberosa*), Many-flowered (*P. multiflora*). Eggs laid on new growth.
Occurrence	All year in s. FL; long adult life span may increase apparent population levels. Also along southern rim of US and in So. Amer. Strays as far north as NY may be human-assisted.
Ecology	**Medium generalist.** Adapts to disturbed sites, uses a variety of common larval hosts.

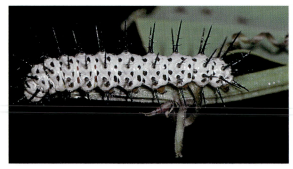

Larva: 5/22/00, Mary Krome Sanctuary, Miami-Dade Co., FL

av wingspan: 3.1″

Julia Heliconian   *Dryas iulia*

Dorsal male: 3/25/94, Fakahatchee Strand State Preserve, Collier Co., FL

Dorsal female: 3/25/94, Fakahatchee Strand State Preserve, Collier Co., FL

This attractive, fast-flying butterfly exemplifies the colorful dash of the neotropics. It spares no energy in making its daily rounds in the heat of an equatorial afternoon, following a regular itinerary in which it visits a series of chosen nectaring sites ("traplining"). In our area, Julias are restricted to southern Florida, where they are well-adapted to the scrubby habitats that have become increasingly common amid the state's rapid development.

Biologically, the Julia occupies an intermediate niche between the Zebra, a "true" heliconian, and the Gulf Fritillary, a more primitive member of the heliconian subfamily. Like the Zebra (and unlike Gulf Fritillaries), Julias are conspicuously long-winged and "trapline" during the day. But unlike the Zebra they do not ingest pollen, are palatable to predators, and have an adult life span of only a few weeks (Scott, 1986). Julias favor habitats that are neither as thickly shaded as the Zebra's nor as open and sunny as the Gulf Fritillary's.

Although they move around a good deal during the day, Julias are sedentary geographically, straying north occasionally only in the Midwest. Perhaps because of this, they have many local forms, up to 12 in the Caribbean alone (Smith et al., 1994). The southern Florida race is *largo* (formerly lumped with the Cuban *cillene*). The odd scientific species name ("*iulia*") is a variant of "Julia" used by Fabricius in his original 1775 description. Although Fabricius himself later called the butterfly "Julia," the scientific die already had been cast (Tveten & Tveten, 1996).

Julias fly at mid-level in tropical hammocks, descending to make brief nectar forays, often in open settings. Males patrol during the day for females, which can mate up to four times. A "scent club" at the end of the female's abdomen is extruded during the courtship (Scott, 1986), although its exact function is not verified. Both sexes nectar actively, and sometimes take minerals from cayman or turtle eyes (Tveten & Tveten, 1996).

Identification	A large butterfly by virtue of its *elongated FW*. While its shape is distinctive when clearly seen, in flight it can be confused with Ruddy Daggerwing, Florida Leafwing, or even Gulf Fritillary. Some believe that groups of butterflies with similar appearances can collectively dilute the attention of predators (Joron, 2003).
Dorsal	Male bright, satiny orange when fresh (much paler orange when faded). Note *two dark spots* near the FW cell ends, faint dark wing borders, and some dark wing veining. Females darker orange-brown with *transverse black markings*, one crossing the entire FW.
Ventral	Pale with indistinct smudges and mottling; male slightly darker. *No silver spots*.

Habitat	Subtropical woodlands and nearby open areas with nectar, including settlements, parks, gardens, abandoned areas.
Hostplants	Passionflowers (*Passiflora*). Females oviposit on new growth, larvae eat leaves.
Occurrence	Generally common in s. FL; once abundant on upper Keys ("a swarm . . . lined the roadsides . . . for many miles," Klots, 1951, p. 83). Also s. TX, south to ARG. Multiple broods, all year, with some "gap" periods. No diapause.
Ecology	**Medium generalist.** Mostly sedentary, but employs multiple locally common hostplants, adapts well to varied surroundings.

Ventral female: 3/25/94, Fakahatchee Strand State Preserve, Collier Co., FL

av wingspan: 3.3″

Gulf Fritillary   *Agraulis vanillae*

Dorsal: 12/24/94, Long Pine Key, Everglades NP, Miami-Dade Co., FL

Ventral: 6/24/93, Loxahatchee NWR, Palm Beach Co., FL

This bright, lively butterfly frequents open, scrubby habitats in the Deep South and Florida, often occurring near settlements. It is also common throughout the West Indies and neotropical South America. The Gulf's widespread familiarity owes as much to an active and conspicuous lifestyle as to its sheer biological abundance.

This species and the next are evolutionary "lynchpins" that straddle a blurry taxonomic line between the heliconians and fritillaries. The Gulf Fritillary was named (or misnamed, actually) on the basis of its appearance more than its ancestry, since it falls on the heliconian side of the divide, whereas the Variegated Fritillary is more properly labeled as a "fritillary." The Gulf Fritillary is the sole member of its genus. It has a number of geographic races, but regular migratory intermixing obscures clean delineations. The accepted race in our area is *nigrior*.

Gulf Fritillaries belong to a group of "straight-line migrants" that undertake defined seasonal movements on an annual basis (Walker, 1978; 2001; Scott, 1986, pp. 44–46). In Florida, they move northward between February and June, then southward from August to November. The species' common name reportedly refers to individuals seen flying over the Gulf of Mexico during seasonal irruptions (Pyle, 1981). Northern emigrants form temporary breeding colonies, occasionally reaching as far north as southeastern Virginia (Roble et al., 2000), but individual strays also occur farther north than the range of the species' native hostplants.

Male Gulf Fritillaries patrol throughout the day for females (but do not trapline). As with Zebras, they seek freshly emerging individuals, including those not yet able to fly (Tveten & Tveten, 1996). Both sexes nectar actively on a variety of flowers. Experiments show an ability to learn flower color associations (Weiss, 1991), and recent evidence shows that the butterfly's olfactory lobes grow with sensory stimulation (Kroutov et al., 2002). Despite these adaptations to a vagrant lifestyle, however, they lack the behavioral sophistications of Zebras.

Identification	Fairly large. FW long and pointed (especially males) with a concave trailing margin. Body striped orange-and-white. Not likely to occur with any of the "true" fritillaries in our area outside of the southern Appalachians.
Dorsal	Bright, intense orange with variable black markings (females darker, markings heavier). Note three white-centered dots in FW cell (especially in eastern *nigrior* race). Black veins thicken toward trailing FW edge.
Ventral	*Elongated silver-white HW markings* distinctive. Inner FW cell is bright pinkish.

Habitat   Prefers open, scrubby habitats with hostplants and available nectar: fields, waste areas, gardens, roadsides, open woodlands. Nearly anywhere during migration.

Hostplants   Maypops (*Passiflora incarnata*), Corkystem Passionflower (*P. suberosa*), other passionvines, native and cultivated. Female lays on or near the hostplant. Caterpillars hatching near the host crawl to it using unknown cues, having avoided predators in the egg stage (Gannon, 1986). Larvae are aposematic, protected by plant toxins, but adults are apparently palatable (Brower, 1985).

Mature larva

Occurrence   Common in Deep South, migratory northward, straying rarely to NY, PA. Continual, nonsynchronous broods, all year in South. Adults overwinter, in South only.

Ecology   **Generalist.** Adaptive generalist lifestyle, despite limited hostplant selection, lack of cold tolerance.

av wingspan: 2.8"

Variegated Fritillary   *Euptoieta claudia*

Dorsal: 8/4/96, Garden Canyon, Cochise Co., AZ

Ventral: 6/30/94, Virginia Beach City, VA

Permanent residents of the South, Variegated Fritillaries are fairly common on the outer coastal plain and upper Florida peninsula. They are less regular northward, often only sporadic immigrants, and in subtropical lowlands farther south, where they seem less at home (Smith et al., 1994). In New England, Variegated Fritillaries are always a good find—though not completely unexpected—usually seen in mid- to late summer during strong flight years.

Variegateds are closely related to *Speyeria* Fritillaries, but with several obvious differences: Variegateds have multiple yearly broods, vs. one in *Speyeria*; they are nomadic, vs. sedentary/metapopulational; and they use multiple hostplants, vs. just violets. Variegateds also have taxonomic links to the heliconians, as evidenced by their use of passionflowers as a host. With species such as this and the Gulf Fritillary blurring definitional boundaries, it is no wonder that the taxonomy of heliconians "seems to be in a perpetual state of refinement" (Penz & Peggie, 2003).

Early observers thought Variegated Fritillaries overwintered in the North, possibly in several different life stages (Weed, 1917). Most commentators now agree that adults overwinter mainly in southern, frost-free areas, where they likely undergo reproductive diapause. But a multiyear colony near Quebec City in Canada (overwintering stage not reported) suggests the species is capable of surprising adaptive feats (Layberry et al., 1998).

The Variegated is a wide-ranging generalist. It thrives where low-growth corridors link patches of suitable habitat (Haddad & Baum, 1999). Flight is low and rapid, but even at rest the species is exceptionally difficult to approach. Indeed, the genus name is from the Greek *euptoietos*, for "easily scared" (Opler & Krizek, 1984). Males patrol actively for mates, females more slowly, seeking egg-laying sites (Allen, 1997). Both sexes nectar actively on a wide variety of flowers. Males sometimes puddle.

Identification	Fairly small for a Greater Fritillary, with which it is sometimes classed—especially males, which may be only half the size of females (Iftner et al., 1992). Only somewhat "variegated." The outer FW margin is concave, creating a shape that reminds us of a painted lady (*Vanessa*).
Dorsal	Tawny orange (brighter in male, more duller/brownish in female), with a *pale, yellowish central band on both wings* and *pale, dark-rimmed spots in the FW cells*. Black veins and submarginal dots.
Ventral	*No silver spots.* Base pattern mottled and cryptic with smudgy submarginal HW dots, light outer border. Inner FW rusty orange.

Habitat	Open, disturbed habitats and barrens: fields, roadsides, mountain meadows, scrub. Not dense forests, or even savannas (LeGrand & Howard, 2004).
Hostplants	Widely variable: passionflowers (*Passiflora*), violets and pansies (*Viola*), cultivated flax (*Linum*) (sometimes an agricultural pest). Many others reported, though some older claims are dubious. Larvae eat seed pods, leaves, flowers.
Occurrence	Variably common in SE, emigrates to central N. Eng. Also US southern rim, south in highlands to ARG. One to four broods, depending on local arrival date and latitude; all year in South. Overwinters usually (always?) as an adult.
Ecology	**Generalist.** Mainly limited to "pioneer" habitats, but generalist traits include vagrancy, varied food plants and feeding styles, opportunistic breeding pattern.

Common Blue Violet (*Viola sororia*)

av wingspan: 2.1"

Diana Fritillary   *Speyeria diana*

Dorsal male: 7/8/95, Fork Creek Pub. Rec. Area, Boone Co., WV

Dorsal male: 7/8/95, Fork Creek Pub. Rec. Area, Boone Co., WV

The steep-sided valleys of the central Appalachians have a subdued and timeless feel. But they spark to life with the appearance of a Diana Fritillary, a genuine lepidopteran marvel, as it glides along a streamside tangle to perch on a tall milkweed.

Few of our other butterflies exhibit such complete sexual dimorphism. Indeed, only male Dianas were recognized when the species was originally described in 1775. It was not until 1864 that William Henry Edwards first described the female, from southern West Virginia. Edwards initially believed he had found a new species of admiral, but quickly realized his discovery was in fact a fritillary (Weed, 1917).

For all its grace and beauty, the Diana has fared poorly when faced with human disturbance. Of four originally known regional population centers, only two remain (in the Appalachian and Ozark Mountains). Populations in the Ohio River Valley faltered as early as the 1800s, with the removal of forests for iron smelting and agriculture (Iftner et al., 1992). And the earliest-known populations, from tidewater Virginia, were largely extirpated by around 1960, with the elimination of critical, old-growth pine woods (Opler & Krizek, 1984). Recent, unverified sightings from eastern Virginia raise hopes that some colonies may persist (Roble et al., 2000).

Males patrol along woodland edges and in openings, sometimes flying quite high to clear forest foliage. Their active lifestyle causes them to wear quickly. Females, by comparison, save themselves for their long adult life by resting quietly most of the day (active mostly in the morning and late afternoon, unless ovipositing). Dianas nectar on tall, sturdy plants such as milkweeds or large fall composites. But like other woodland butterflies they are also attracted to dung and decaying matter. Difficult to approach, females hide in brush when disturbed.

Identification	The largest of our Greater Fritillaries. Both sexes are instantly recognizable, although a female might be confused momentarily with a Red-spotted Purple (both mimic the distasteful Pipevine Swallowtail).
Dorsal	Male: Inner three-fifths of both wings *blackish-brown*, outer two-fifths *bright orange* with small dot rows. Female: Inner wings *deep blue-black*. Outer HW *pale, iridescent blue* or blue-green; outer FW with *blue-and-white* marks.
Ventral	*Lacks silver HW spots* of other *Speyeria*. FW spangled, similar to dorsal in both sexes. Male: HW two-toned tan and brown, *nearly unmarked*. Female's two-toned blue-black and brown, with *rust-purple overtones*.

Habitat	Mid-elevation wooded valleys (usually below 3,500 feet) with thick undergrowth. Formerly tidewater pine woods with surface water.
Hostplants	Violets (*Viola*). Females walk along the ground, scattering up to 1,000 eggs (Allen 1997; Harris, 1972).
Occurrence	Locally common in mountain colonies, s. WV to n. GA; also e. AL/KY, Ozarks. One brood, mid-Jun–mid-Aug (males), early Jul–Sep (females). 1st stage larvae overwinter.
Ecology	Medium Specialist. Never far from woods.

Fork Creek bottomland (July 1995), Boone Co., WV

av wingspan: 3.7″

Ventral male: 7/8/95, Fork Creek Pub. Rec. Area, Boone Co., WV

Ventral female: 7/8/95, Fork Creek Pub. Rec. Area, Boone Co., WV

Mountaintop Removal:
The Price of Coal?

The secluded valleys of the central Appalachians, favored by Dianas, leave an indelible impression. They are ancient, misty places, hewn slowly from the slopes of surrounding highland peaks—among the world's oldest mountains.

The forests of the central Appalachians are also a stronghold of biological diversity, a meeting zone of northern and southern life forms, with many endemics. Their species diversity ranks among the highest of any temperate region in the world. Especially notable are a rich and distinctive plant community, a large cadre of woodland songbirds (including such declining species as Cerulean and Kentucky Warblers), a host of endemic salamanders, a variety of small mammals, endangered aquatic life, and much else.

Butterflies of the central Appalachians are similarly distinctive. Their sheer number and variety is surprising (*see*, e.g., Cech, 1995). And these wooded valleys host populations of specific importance: the Diana Fritillary, the Dusky Azure, the southern race of Silvery Blue, and the eastern race of Golden-banded Skipper (quite possibly a distinct species). All of these have been reliable at sites such as Fork Creek Public Recreation Area in Boone County, West Virginia (southwest of Charleston), where many of the photos in this book were taken.

The Achilles' heel of these habitats is low-sulphur coal, which exists in significant quantity beneath the surface of the Appalachian Plateau. Narrow coal seams sometimes occur near the surface in areas of steep topography, making them inaccessible using traditional mining methods. Coalsburg, West Virginia, for example, lies at the center of the "Coalsburg seam," a primary example of this configuration (United States Environmental Protection Agency, 2003). Located within miles of Fork Creek, Coalsburg was the home, interestingly enough, of prominent 19th-century lepidopterist William Henry Edwards.

The method used in recent decades to extract coal from these seams is *mountaintop removal* (MTR), in which forests are cleared and mountain peaks then blasted away with explosives. Leftover rock and mining by-products are dumped into adjoining valleys and pits, leaving a flattened topography, lingering pollution, and buried headwater streams. Between 300 and 600 permits were granted annually for MTR operations in 1985–98 (United States Environmental Protection Agency, 2003). Some 500 square miles of land reportedly have been affected by MTR in West Virginia (OHVEC, 2003). An MTR operation began at Fork Creek in 1999.

None of the several studies undertaken to assess the effects of MTR has addressed butterflies. To date, potential adverse effects on certain specialized wildlife have been documented.

Current federal law calls for MTR sites to be recontoured. But even assuming that true ecosystem reconstruction is possible here (a proposition that verges on ludicrous), many MTR projects have been chronically underfunded, ultimately abandoned by their sponsors after completion for local communities to remediate.

Ironically, the coal extracted at Fork Creek was of poor quality, bankrupting the project before completion (T. Allen, pers. comm.). And thus a flickering, defiant remnant of Appalachian life still remains there.

Mountaintop Removal, 4/23/03, Boone Co., WV

Regal Fritillary   *Speyeria idalia*

Dorsal Male: 7/5/95, Ft. Indiantown Gap, Lebanon Co., PA

Dorsal female: 7/5/95, Ft. Indiantown Gap, Lebanon Co., PA

The Regal Fritillary is among our handsomest butterflies, but also among the most severely imperiled. Its sudden, poorly understood decline in the East has been startling. Locally common until the 1970s, it simply vanished by about 1988. At present, only a handful of known colonies remain, in Pennsylvania and western Virginia (the latter found only in 1997).

The disappearance of the eastern Regal Fritillary is important biologically. Recent genetic studies suggest that eastern populations may be well enough differentiated to constitute a separate race (Williams et al., 2003). Until recently, our Regals probably maintained genetic contact with western prairie populations via intervening colonies (assisted by the species' innate tendency to disperse over substantial distances), but differences nonetheless emerged, perhaps during Ice Age isolation.

Males patrol in a low, regular flight, pausing often for nectar or to perch. Females mate once in early summer, but do not produce eggs until August (Kopper et al., 2001a).

Ventral: 7/5/95, Ft. Indiantown Gap, Lebanon Co., PA

Identification	Very large, second in size only to Diana among our fritillaries. Difficult to confuse with any other butterfly in our region, with the possible exception of the *alcestis* race of Aphrodite, which is also dark on the ventral HW.
Dorsal	FW bold orange with intermittent dark markings. Female FW dark near apex with white dots. HW dark with velvety iridescent sheen. Submarginal HW dot row is *white* in female, *yellow-orange* in male.
Ventral	HW dark *olivaceous* with *bold white spots*.

Habitat	Dry, tallgrass prairies in West. But in the East, moister large, open grasslands, often with rolling topography and wet swales. Formerly boggy wetlands.
Hostplants	Arrow-leaved Violet (*V. sagittata*) in PA. Birdsfoot Violet (*V. pedata*) and others elsewhere. Female strews nearly 2,500 eggs over a wide area, usually near violets (Wagner, 1995). Caterpillars recently found to feed diurnally (Kopper et al., 2001b).
Occurrence	Local/colonial. Only at Ft. Indiantown Gap (PA) and two w. VA sites. One brood, mainly Jun-Sep (May-Oct southward). Caterpillars overwinter unfed.
Ecology	Medium specialist. Significant habitat dependencies.

Eastern "prairie" grassland: Ft. Indiantown Gap, Lebanon Co., PA. A successional ecosystem.

av wingspan: 3.2"

Eastern "Prairies" in Decline: Implications for Grassland Butterflies

Regal Fritillaries favor large tracts of open, low-growth habitat (Swengel & Swengel, 2001). And since they relocate from time to time to fresh colony sites, significant tracts of prairie-like habitat are required on a regional basis for the species' long-term well-being. Many suspect that the loss (or fragmentation) of suitable eastern grasslands has played a substantial role in the eastern Regal's dramatic decline.

The Regal Fritillary is often considered a foremost grassland specialist in our region. But this depends to some extent on how we define "grassland." If the term is applied to *any* low-growth habitat dominated by grass, permanent or otherwise, then it would include savannas, alpine tundra, balds, barrens, dunes, and sedge marshes—to say nothing of lawns, meadows, roadsides, and sod farms. Numerous competing formal definitions are currently in use (FAO, 2002), each stressing a particular set of factors, including soil type, level and variability of annual rainfall, dominance of grasses, limited occurrence of woody plants, and support of feral grazing animals. Under a broad definition, many of our butterfly species would qualify as "grassland specialists," including numerous skippers and even a number of barrens dwellers, such as the "Karner" Melissa Blue.

Definitional issues seem a bit less tangled in the West. America's prairies are quintessential grasslands, and they have well-defined communities of obligate, prairie-specialist flora and fauna—species that will cease to exist if their native grassland habitats are eliminated. Some of our eastern grassland butterflies in fact are disjunct prairie specialists (the Regal Fritillary is a good example).

But there are no true prairies today in the East—based on a standard definition of "prairie" as a seasonally dry, treeless habitat formed on level or rolling terrain, with coarse grass and usually fertile soil. Even the "dry prairie" habitat of southern Florida is generally considered a low-growth form of pine flatwood (Abrahamson & Hartnett, 1990). Given the high average levels of rainfall on the East Coast, forests are the "climax" habitat type nearly everywhere, as opposed to grasslands. Some have gone so far as to quip that salt marshes are the only "true" climax grasslands in the East. It is tempting to try to define the "natural" occurrence of eastern grasslands by imagining the landscape before human disturbance began. But this is unfortunately not practical, since native tribes altered habitats in many unrecorded ways long before European settlers arrived. And thus it is difficult to establish a creditable "baseline."

But none of this matters if we accept *all* existing, feral grasslands as important to safeguard, even if they are ephemeral, and even if they did originate with human activity or disturbance. These habitats presently exist and without them the diversity of eastern butterflies would be greatly diminished.

Having accepted this view, the underlying problem comes into sharper focus. Even without exact statistics, we can see that the supply of open, undisturbed grasslands in our region is in sharp decline. A major cause is succession of fields to forests (over 20 million acres have reverted to woodland in the Northeast since 1900, according to the U.S. Forest Service). Also at fault is direct, extensive grassland fragmentation and development.

Grassland butterflies are being displaced by the loss of available habitats. The importance of grassland preservation grows as a result. But this is highly challenging, since flat, open land is at a premium for development.

And even the maintenance of preserved low-growth habitats is complicated. Ill-timed or excessive prescribed burns can eliminate rare butterflies on "protected" lands. Debate continues as to the optimal approach, which may include alternatives to burning such as mowing, thatching, or grazing (Panzer, 2002; Swengel, 1996). Much remains to be learned on this important topic.

Florida "dry prairie": Technically neither a prairie nor a grassland, this unusual habitat illustrates the difficulties of classifying habitats. Kissimmee Prairie SP, Okeechobee Co., FL

Great Spangled Fritillary   *Speyeria cybele*

Dorsal male: 6/18/98, Pound Ridge Reservation, Westchester Co., NY

Dorsal female: 7/10/99, Pound Ridge Reservation, Westchester Co., NY

The most widespread and familiar of our Greater Fritillaries, these grand butterflies are a fixture in upland fields and milkweed patches during the summer.

This is one of three "orange-dappled" *Speyeria* fritillaries in our region. All share a specialized and unusual lifestyle (*see* box essay on facing page). They can be troublesome to identify, although the Great Spangled is usually the easiest among them. Males exhibit *protandry*, appearing some weeks before females to initiate the flight season. Like many *Speyeria*, this species is geographically variable. Our principal race is *cybele*.

Flight is strong and conspicuous. Males patrol in wide, meandering circuits, a few feet above the ground, in locations where they are likely to encounter females. But dispersing individuals fly quite directly. Avid nectarers, Great Spangleds often linger on or in the vicinity of a single flowerhead, once for eight consecutive hours (Gochfeld & Burger, 1997).

Ventral male: 7/5/93, Pound Ridge Reservation, Westchester Co., NY

Identification	Largest of our "orange-dappled" Greater Fritillaries, with full, rounded wings. Named for the silvery "spangles" on the undersurface, but this generic marking is not diagnostic. *Eyes are amber-brown* in life (Glassberg, 2000).
Dorsal	Variably bright orange. *No black dot near the trailing edge of the basal FW* (*compare* Aphrodite and Atlantis). Females conspicuously dark on basal wing surfaces.
Ventral	HW disk pale brown (male) or dark, orange-brown (female). Normally less reddish than Aphrodite. Note *broad, pale submarginal band*.

Habitat	Various open, often moist habitats with nectar or hostplants.
Hostplants	Many species of violets (*Viola*). Female oviposits on or near host in late summer. Caterpillars reportedly feed at night (*but see* Kopper et al., 2001b), then crawl away when mature to find a pupation site.
Occurrence	Usually common, N. Eng. south to n. GA, most common inland. Also widely in central and e. CAN/US. One brood, Jun-Sep in North, May-Oct in South; females seem to estivate in mid-summer. Newly hatched caterpillars overwinter unfed.
Ecology	**Medium Generalist.** Common and adaptable, but regional; needs intact habitats.

Open meadow nectary with Common Milkweed (*Asclepias syriaca*), Westchester Co., NY

av wingspan: 3.0"

Violets & Fritillaries:
A Close Association

Where, like a pillow on a bed,
A pregnant bank swelled up, to rest
The violet's reclining head,
Sat we two, one another's best
—*The Ecstasy*, John Donne (1633)

Violets (genus *Viola*, c. 500 species) have diverse connections with human culture. In poetry, they are used to symbolize a contrasting set of emotive themes, from young love and frail beauty to untimely death. Ancient and traditional cultures valued them for their medicinal and aromatic properties, appealing colors, and presumed mystical powers. And their leaves—both then and now—make a nutritious supplement to a green salad.

But for one group of temperate zone butterflies, the fritillaries, violets are the cornerstone of a distinctive lifestyle, a host genus with which they have evolved a dedicated and finely tailored relationship.

The fritillaries are a northern group, customarily subdivided into two subtypes, Greater and Lesser. Greater Fritillaries (genus *Speyeria*) are large, single-brooded butterflies, all obligate violet-feeders. Lesser Fritillaries (genus *Boloria* in our area) are, as their name indicates, smaller. They tend to be multibrooded and feed on a diverse group of hostplants, including but not limited to violets.

Male Greater Fritillaries emerge in late spring or early summer, well ahead of the females (*protandry*). They patrol actively for mates during the day. Once females emerge, active courtship ensues. Females apparently mate just once (e.g., Kopper et al., 2001a), after which males die off. But females persist, in a temporary state of reproductive diapause. Most egg-laying is deferred until the late summer or early fall.

Many observers suspect that female Greater Fritillaries estivate, at least sporadically, during July/early August. Some phenograms show an activity "trough" in mid-summer (LeGrand & Howard, 2004; Cech, 1993), but from another (Iftner et al., 1992) we might conclude that numbers simply dwindle over time, and that the midsummer "trough" may be a function more of data than of behavior. We suspect, however, that at least some estivation does occur.

In any case, ovipositing females disperse eggs near the ground, on twigs, foliage, or the like, customarily near a stand of violets—but often not upon them. Newly hatched caterpillars crawl to the nearby hostplant or hostplant rootstock. Once there, they exhibit an unexpected behavior: rather than feeding actively while good weather persists, to accumulate a store of nutrients to help pass the winter, they nestle into the leaf litter and enter diapause *without feeding*. For northern butterflies, this peculiar behavior is actually quite adaptive, since undigested food particles in the gut of a diapausing caterpillar can become "nucleators" on which ice crystals form in a cold winter, killing the larva (Leather et al., 1993, p. 77).

Canada Violet (*Viola canadensis*)

Greater Fritillary (*Speyeria*) larva feeding on violet leaf

In spring, the awakening caterpillar feeds on fresh violet shoots. What is the attraction of this particular food for a young fritillary? The answer is not clear, but violets do contain a variety of powerful chemicals, including glycosides, saponins, and flavenoids. They have been utilized by humans for products as varied as fine French perfume, ceremonial emetics, expectorants to treat catarrh, analgesics for labor pain, dyes, insecticides, and tobacco additives (*see*, e.g., Moerman, 1998, pp. 596–98). It is interesting that violets' rootstocks, the underground "bulbs" from which early leaves emerge, are inedible. Thus spring growth may draw up a specific concoction of stored secondary/deterrent chemicals whose properties the fritillaries prize.

Aphrodite Fritillary   *Speyeria aphrodite*

Dorsal: 7/5/95, Ft. Indiantown Gap, Lebanon Co., PA

Ventral: 7/5/95, Ft. Indiantown Gap, Lebanon Co., PA

The inspiration for this butterfly's name was Aphrodite, Greek goddess of beauty, fertility, and physical love. The fritillary acquits itself nicely as far as beauty is concerned, but is a bit suspect in terms of fertility, since Aphrodites are seldom the most common of the three "orange-dappled" *Speyeria* fritillaries at East Coast locations. They do outnumber other *Speyeria* in parts of the central and southern Appalachians, but more often we find them in the minority, mingling with aggregations of other *Speyeria* (whichever species is locally prevalent). One experienced field observer in northwestern New Jersey estimates that the typical ratio of Great Spangleds to Aphrodites there is approximately 200:1 (Wade Wander, cited in Gochfeld & Burger, 1997).

The geographic distribution of the Aphrodite is highly similar to that of the Great Spangled, and the two can be found alongside each other in summer meadows. The Aphrodite shows a small but consistent preference for northern-style habitats throughout its range, however, and is more likely to be seen at high altitudes (e.g., Kiel, 2003; LeGrand & Howard, 2004).

Aphrodites are variable geographically. As many as nine subspecies have been described, but some of these are weakly differentiated. The principal race in our area is *aphrodite*. A Great Lakes race with little or no submarginal band (*alcestis*) reaches western Pennsylvania. Individuals in the southern Appalachians (described as the race *callasaja*) lack the species' usually distinctive black forewing spot (Gatrelle, 1998d).

As with other Greater Fritillaries, males emerge as many as three weeks earlier than females. They patrol throughout the day seeking mates. Active nectaring occurs in the early morning and late afternoon (Allen, 1997).

Identification	A large species, intermediate in size between Atlantis and Great Spangled. In shape, it resembles the Great Spangled, with round, full-cut wings (*compare* Atlantis). *Eyes are amber-brown* in life (Glassberg, 2000). A dark aberrant form ("bakeri") has been reported (Iftner et al., 1992). Given the significant individual variation in all our orange-dappled *Speyeria*, it is best to use a combination of field marks for identification.
Dorsal	Bright, reddish- or yellowish-orange with a series of fritillary-style black dots, chevrons, and dark stripes and bars. The FW margin usually consists of *two thin, parallel black lines with variable amounts of orange in between*. The margins usually are *not entirely black* (*compare* Atlantis). Most individuals also have a *black dot near the trailing edge of the basal FW* (variable, often more conspicuous than in the individual illustrated).
Ventral	Inner two-thirds of HW is redder than in Atlantis, varying from a light, cinnamon-brown (males) to a deep reddish-brown (females) (Allen, 1997). HW spots always silvered in our region. The disc color bleeds (diffuses) into the pale, submarginal band on the HW, especially at the "anal angle" near the tail, making it look narrow (*compare* Great Spangled).

Habitat	Varied upland habitats with acid soils. Moist grasslands (meadows, fields, prairies farther west), shale barrens, open woods, mountain meadows, bogs. Additional sites when seeking nectar.
Hostplants	Violets, including Northern Downy (*Viola fimbriatula*), Lance-leaved (*V. lanceolata*), Primrose-leaved (*V. primulifolia*), Common Blue (*V. sororia*), etc. Female lays eggs singly near violets, often under shrubs (Scott, 1986), in Jul-Sep. Caterpillars eat leaves, make no nests.
Occurrence	Common in Appalachians, but otherwise scarcer than other *Speyeria*. N. Eng. and south in highlands to n. GA. Declining many places. Also boreal CAN, Rockies, northern plains, and Gr. Lakes states. One brood, Jun-Sep (May in NC). Newly hatched caterpillars overwinter unfed near hostplant rootstock.
Ecology	**Medium Generalist.** Exploits a significant range of habitats and hostplants, though currently losing ground.

av wingspan: 2.7"

Atlantis Fritillary *Speyeria atlantis*

Dorsal: 7/24/94, Moose River Plains, Hamilton Co., NY

Ventral: 7/23/94, Moose River Plains, Hamilton Co., NY

Trim and crisply marked, the Atlantis Fritillary is the most boreal of our three large orange-dappled fritillaries. Beginning observers often have difficulty distinguishing this species from the preceding two, even though the Atlantis averages 10 to 20% smaller and has several distinctive field marks. This may have something to do with the Greater Fritillaries' restless movements and bounding flight, which can make them difficult to study. But more important, undoubtedly, is the fact that the Greater Fritillaries are complexly marked, are highly similar in overall pattern, and exhibit wide individual variation. Taken together, these factors make them challenging to identify. But eastern observers should take comfort, at least, from not having to cope with the truly bewildering array of local forms that characterize *Speyeria* in the West, where a myriad of ecologically isolated populations create enormous perplexities for both observers and taxonomists. More than 100 species were originally described in this genus, but taxonomists then carefully reduced the number to 14 (Scott, 1986, p. 317).

The nominate race (*atlantis*) is found in our area. It was first described from the Catskill Mountains of eastern New York, but is rare in that area today, as well as in much of its formerly described southern New England range. Still common northward, it is a boreal specialist that cannot easily adapt to modern land use patterns and warming climate conditions (Robert Dirig, pers. comm.). Some authorities now classify many western populations of the Atlantis Fritillary as a separate species, the Northwestern Fritillary (*S. hesperis*).

Male Atlantis Fritillaries patrol throughout the day in search of females. On locating a potential mate, the male induces her to land, then faces her and flicks his wings to waft pheromones toward her antennae. Females also emit pheromones in this group, perhaps for mutual identification in a visually similar group (Scott 1986). Adults nectar actively, often on Joe-Pye Weed in our area, and sometimes take mud or decaying matter.

Identification	Smaller, on average, than our other *Speyeria*, a characteristic useful in side-by-side comparison. Wings comparatively pointed and angular. Easily confused with Aphrodite. Fringes black-and-white. *Eyes are blue-gray* in life (Glassberg, 2000). Description here refers to eastern *atlantis* race.
Dorsal	Ranges from fairly dark orange-brown to lighter or more yellowish-orange. *FW border solid black* (vs. usually with at least some orange in Aphrodite and Great Spangled). As with Aphrodite, most individuals also have a *black dot near the trailing edge of the basal FW.*
Ventral	Basal two-thirds of HW *purplish- to slightly greenish-brown* (vs. more reddish in Aphrodite). Pale submarginal band is *more distinct* than in Aphrodite, with less intrusion of the dark, basal color (or none). This makes the dark "follow-spots" associated with the postmedian silver dots more obvious (Glassberg, 2000).

Habitat	Boreal-style habitats, including open meadows, bogs, woodland openings, and roadsides. Fairly wide-ranging.
Hostplants	Various violets, including Northern Blue (*V. septentrionalis*) in NY (Shapiro, 1974). Life cycle typical of Greater Fritillaries. Female strews eggs near violet host late in season after mid-Aug.
Occurrence	Often common in suitable habitats. ME and n. NY south to mid-N. Eng. Currently rare south of this range. Separately in w. NY and PA near Gr. Lakes. An isolated population occurs in the WV highlands. Also e. BC across CAN to Newfoundland, Ohio River Valley, n. Gr. Plains south to CO, AZ, and NM. One brood, mid-Jun-Aug/early Sep; in N. Eng., mostly mid-Jul-mid-Aug. Newly hatched caterpillars overwinter unfed.
Ecology	**Medium Generalist.** Notable in our area as a boreal zone specialist, but within those limits it has a fairly robust distribution and is often very common.

av wingspan: 2.4"

Meadow Fritillary   *Boloria bellona*

Dorsal: 6/22/96, Pound Ridge Reservation, Westchester Co., NY

Ventral: 7/5/93, Pound Ridge Reservation, Westchester Co., NY

In specialized groups, such as the Lesser Fritillaries, it is not unusual for common and widespread species to differ in certain key respects from their more ecologically restricted relatives. This is certainly true of the Meadow Fritillary, a familiar resident of moist northern habitats. Unlike most *Boloria*, it has significantly loosened its evolutionary ties with intact, natural habitats, and now thrives in various disturbed settings such as hayfields, ditches, and damp meadows. It is also distinctive—in this mainly sedentary genus—for having extended its range boundary southward in the last century, invading qualitatively novel habitats in the Midwest (e.g., Iftner et al., 1992).

This is not to say that Meadow Fritillaries are immune to habitat degradation, nor that their recent history has been an unmixed success. In southern New England, many open farmlands that once hosted the species are reverting to woodlands, or have been "denatured" for suburban use, and Meadow Fritillaries have declined accordingly (Gochfeld & Burger, 1997). And in the western Carolina mountains, where the Meadow's range extends farthest south, farmland acreage has declined steadily in recent decades (LeGrand & Howard, 2004).

The eastern race of Meadow Fritillary (*bellona*) is relatively light, compared with the northern/western races *toddi* and *jenistae*. Some authorities place many of our Lesser Fritillaries in the genus *Clossiana*, but others retain the traditional classification (Layberry et al., 1998).

Male Meadow Fritillaries patrol for females, in a low, crisp flight. Both sexes pause frequently for nectar, often utilizing composites. Males also extract minerals from damp soil and putrefying matter.

Identification	Much smaller than the Greater Fritillaries (*Speyeria*). About the same size as the Silver-bordered, its only look-alike in most of our region. Size and appearance variable. Beginners should take care not to confuse the *Boloria* with checkerspots or crescents. Note the *clipped-looking appearance of the FW just below the apex*. Long palps.
Dorsal	Moderate orange-red with varied dark spots, lines, and zigzagged bands. Inner (basal) area of wings dusky. Most individuals *lack a dark border on the trailing HW*.
Ventral	Mottled orange and purplish-brown, with little clear pattern except for a vague, yellowish band across the central HW. *Lacks silver spots or markings*, but does have a *whitish-violet wash* on the outer HW when fresh and a small but distinctive white mark on the leading HW. FW orange brown-smudged near apex.

Habitat	Typically in wet, open places, including fields, pastures, streamsides. May nectar in drier sites.
Hostplants	Various violets, Northern White (*V. pallens*) and Common Blue (*V. sororia*) often mentioned. Female lays eggs on twigs or foliage near the host.
Occurrence	Widespread, fairly common, ME south and inland to s. Appalachians in w. Carolinas. Also across most of CAN and n. US, n. Rockies. Usually two to three broods, last may be partial, Apr to late Oct/Nov. 3rd or 4th instar caterpillars overwinter.
Ecology	Medium Generalist. Flexible (within limits).

Damp meadow in spring (during first flight), Westchester Co., NY

av wingspan: 1.6"

Silver-bordered Fritillary   *Boloria selene*

Dorsal: 6/19/94, Errol, Coos Co., NH

Ventral: 6/12/93, Wilsons Mills, Oxford Co., ME

The Meadow Fritillary is our most familiar *Boloria*, but the Silver-bordered's overall range is more extensive. It occurs widely in north-central Eurasia, from Portugal to Korea, and on this continent from western Canada to Labrador. Silver-bordereds also fly quite far south in the United States, in the Appalachian and Rocky Mountains.

Yet as with many Lesser Fritillaries, the Silver-bordered's wide distribution masks complex and specialized population dynamics. Metapopulations are often limited in both size and territory, especially southward, and are subject to extensive fluctuation. In Britain (where the species is called Small Pearl-bordered Fritillary), there has been "an extraordinary decline" since 1950 (Thomas & Lewington, 1991, p. 138). And commentators in our region have likewise noted a sharp, recent retrenchment (Gochfeld & Burger, 1997; Iftner et al., 1992). The single-brooded Maryland race (*marilandica*) now appears to be extinct (Klots, 1951). Silver-bordereds share the Meadow Fritillary's ability to inhabit wet fields, but do not seem as durable away from natural settings. Changing land use patterns (including fragmentation and extensive loss of wetlands) may help explain recent losses. Not very long ago, Silver-bordereds were reportedly common to abundant in the Northeast (Weed, 1917; Beutenmuller, 1902).

The race of Silver-bordered Fritillary presently found in our region is *myrina*. Darker subspecies (*terranova* and *atrocostalis*) occur in Canada, and a large race (*nebraskensis*) lives in the West. The nominate race is Eurasian. Males patrol in wet habitats much of the day for females, pausing regularly for nectar.

Identification	Medium-small, same size as Meadow Fritillary but wings are rounder and more full-cut, with *no flattening below the FW apex*. Fringes are checkered. The Meadow is the only other *Boloria* found in most of the Silver-bordered's East Coast range, but in the North it also might be confused with Bog Fritillary, or conceivably with Purplish.
Dorsal	Orange with a variety of black dots, scallop markings, and marginal chevrons (dark HW border usually indistinct or lacking in Meadow). Comparatively large, open orange areas on the outer wings give Silver-bordereds a bright, attractive appearance.
Ventral	The *"silver border" along both outer wing margins* distinguishes this species from the Meadow and Purplish. The Bog also has marginal HW spots, but they are white, not silver. *Postmedian dot row is black, not white-centered* (as in Bog). Also *compare* Harris' Checkerspot.

Habitat	Some overlap with Meadow (especially when nectaring), but favors wetter and more overgrown habitats: marshes, bogs, sedge meadows, etc.
Hostplants	Various wetland violets (*see* Allen, 1997; Iftner et al., 1992). Lance-leaved (*V. lanceolata*) in NJ. Females lay eggs singly on or near the host.
Occurrence	Local, declining. ME south to s. VA/WV. Now rare or absent in much of mapped se range. Also CAN, Eurasia. Two or three broods in our area, Jun-Sep. Last may be partial. 2nd–4th stage caterpillars overwinter.
Ecology	**Medium Specialist.** Widespread but ecologically fragile, struggling in much of our area.

Weedy pond habitat, Ocean Co., NJ

av wingspan: 1.6"

Bog Fritillary   *Boloria eunomia*

Dorsal: 6/19/94, Wilsons Mills, Oxford Co., ME

Ventral: 6/19/94, Wilsons Mills, Oxford Co., ME

A boreal remnant of the glacial past, this circumpolar butterfly is widespread yet extremely local. It is known in our area from only a few bogs in northern and western Maine (it also probably occurs in other, less well-studied locations). Bog Fritillaries are currently being researched, along with other small fritillaries and checkerspots, as examples of metapopulational species in fragmented habitats (e.g., Mennechez et al., 2003, Nekola & Kraft, 2002).

As noted earlier, taxonomists are not of one mind as to the correct generic arrangement of the Lesser Fritillaries. Some place the Bog Fritillary in its own genus (*Proclossiana*), citing a small but characteristic feature of the male genitalia. But others do not accept this classification (*see* Layberry et al., 1998).

A number of local forms of Bog Fritillary have been described, only some of which are differentiable as full races. The *dawsoni* race in our area is generally darker and more richly colored than the northern (*tridaris*) or western U.S. (*caelestis*) races. The nominate race (*eunomia*) is Eurasian.

In warm weather, male Bog Fritillaries patrol all day over the sodden vegetation of their boggy homes. They launch into active, aerial dogfights on encountering other males (Opler & Krizek, 1984). But in cold weather, they remain inactive in low vegetation. Bog Fritillaries are often found nectaring on Labrador Tea. Both sexes disperse through woods and other adjoining habitats to find nearby bogs. Usually jumpy and difficult to approach.

Identification	A smallish *Boloria*. Resembles Silver-bordered Fritillary, which is often found close by. A ventral view may be needed for confident identification. Fringes black-and-white checked. Wing base is hairy.
Dorsal	Reddish-orange with complex dark lines and dot rows. HW margin is dark with variable light spots (well-defined in the individual illustrated above).
Ventral	HW has white marginal spots and a series of bands and irregular patches (rust-red, white, or creamy). A concentric row of *postmedian black-rimmed white dots* is distinctive (visible on both wings, but especially the HW). FW pattern resembles dorsal.

Northern acid bog, Oxford Co., ME

Habitat	A strict acid bog specialist in our region. Elsewhere moist tundra, subalpine forests and meadows, willow thickets, etc. Traverses other habitats during dispersal.
Hostplants	Multiple hosts, including willows (*Salix*), bistorts (*Polygonum*), violets (*Viola*); in e. CAN, Small Cranberry (*Vaccinium oxycoccos*) and Creeping Snowberry (*Gaultheria hispidula*). Cranberries in ME (J. Ingraham, pers. comm.). Female lays two to four eggs under hostplant leaf.
Occurrence	Rare and local in nw ME (e.g., Wilsons Mills Bog, Mt. Katahdin). Also from AK east across most of CAN; isolated populations in central Rockies; n. Eurasia, and in isolated bogs from n. Spain to Balkans. One brood, mid-Jun–mid-Jul. (Flight period is *not* limited to one to two days, as earlier reported.) 3rd or 4th stage larvae overwinter.
Ecology	**Specialist.** Even among our bog specialists, the Bog Fritillary has a confined occurrence.

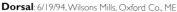

av wingspan: 1.5″

Purplish Fritillary *Boloria montinus* [= *Chariclea montina*]

Dorsal female: 8/2/94, Mt. Washington, Coos Co., NH

Ventral: 8/2/94, Mt. Washington, Coos Co., NH

This wide-ranging northern butterfly is holarctic—or not—depending on one's taxonomic views. Formerly called Titania Fritillary (*B. titania*), it was lumped with several similar taxa in North America and Eurasia. But two splits have recently occurred. First, the European Titania Fritillary was split off from the Arctic Fritillary (*B. chariclea*), a widespread, mostly North American butterfly with two general forms (tundra and boreal). Second, some authorities now recognize the more southern, boreal form of Arctic Fritillary in North America as a separate species, the Purplish Fritillary (*B. montinus*) (NABA Checklist, 2001; *but see* Layberry et al., 1998).

Our local race, the White Mountain Fritillary (*B. m. montinus*), seems unperturbed by these taxonomic contentions. It is cloistered in an isolated, alpine realm of less than 5 square miles, located at or above treeline on the Presidential Range of New Hampshire's White Mountains. Note that the eastern Canadian race of Purplish Fritillary (*grandis*) also could be a possibility in our area, however, along the northern Maine border.

Male White Mountain Fritillaries are described as patrolling all day in moist, alpine meadows and stunted woody scrub ("krummholz"). But at 4,500 feet on Mt. Washington, "all day" can be a relatively short and unpredictable time period, given chilly mornings, frequent strong winds, and thick cloud cover. Females perch to await patrolling males, and may issue a pheromone from their abdomens to suitors passing nearby (Scott, 1986). Both sexes actively nectar on Alpine Goldenrod and other available mountain flowers.

Identification	Slightly smaller than Meadow or Silver-bordered. There is little practical likelihood of misidentifying a White Mountain Butterfly, given its geographical isolation. And even the larger, generally darker *grandis* race has few regional look-alikes.
Dorsal	Similar to Silver-bordered, but with generally heavier dark markings. Inward facing triangles are distinct from checkered wing margins in *montinus* race (less well-separated in *grandis*).
Ventral	Deep, rusty red on HW (*no* purplish highlights in *montinus*). Light markings, including broken white band across mid-HW. Inward facing "toothy" chevrons, capping white spots on HW margin (less conspicuous in White Mountain race).

Habitat	Moist alpine meadows, especially near springs, mostly 4,000–5,500 feet in White Mountains (e.g., at Cragway Springs on auto road).
Hostplants	Multiple hosts: willows, violets, blueberries, bistorts. Surprisingly, the exact hosts are not reported in White Mountains (Kiel, 2003; McFarland, 2002). Eggs laid singly, young feed without nests.
Occurrence	Very local. Also occurs across boreal CAN to e. AK, south into Pac. NW; also n. Gr. Lakes and in Rocky Mountains. One brood, late Jul-mid-Aug. Newly hatched caterpillars evidently overwinter in our area, though other populations of the *montinus/chariclea* group have two-year brood cycle, diapausing again in the 4th instar.
Ecology	Specialist. A strong habitat specialist, vulnerable in our area to acid rain, global warming, and ozone depletion. Needs monitoring.

White Mountain ridge line, Coos Co., NH

av wingspan: 1.5"

Pearl Crescent   *Phyciodes tharos*

Dorsal male: 6/12/94, Pound Ridge Reservation, Westchester Co., NY   **Dorsal female**: 5/23/99, Macon Co., NC

Bright and energetic, the Pearl Crescent is among our commonest and most ecologically successful butterflies, well-known even to casual observers. It is a dedicated generalist whose pliable lifestyle enables it to exploit a wide range of open habitats, both natural and disturbed. Look for it almost anywhere that asters grow.

Despite its prevalence, we cannot easily explain why this one crescent, among many others, is so unusually successful. Its aster foodplants contain no glycosides, which confer toxicity on some other melitaeines (Wahlberg, 2000). A recent study found that Pearl caterpillars do seek out a common, aromatic hydrocarbon (germacrene D) found in preferred hostplants (Hauptfleisch & Kinnel, unpubl.), but this chemical has not been shown to be toxic to predators. Pearls also avoid certain chemicals found in nonhost asters, such as White Wood Aster (*A. divericatus*), New York Aster (*A. novi-belgii*), and Flat-topped Aster (*A. umbellatus*). The latter, interestingly enough, is the sole foodplant of the Harris' Checkerspot. And so, while a mosaic of food preferences is beginning to be deciphered, the question of ecological strategy remains. Perhaps adults synthesize their own protective chemicals.

The Pearl Crescent is the "lead species" in a cluster of closely related melitaeines, sometimes called the "*tharos* complex." In our area, the group also includes Northern and Tawny Crescents. This variable assemblage poses a significant challenge for taxonomists as well as for observers (*see*, e.g., next account).

Male Pearl Crescents patrol most of the day seeking mates. Their flight is low and rapid. Butterfly enthusiasts seeking unusual skippers may be disappointed when a whirling orange form alights and turns out to be "just another Pearl." Relatively sedentary, Pearls nectar on many flowers, pumping their wings slowly while sipping.

Identification	Medium-small, females larger. This is the only crescent in much of our area, yet wide variations in appearance (seasonal, sexual, and individual) can create confusion. Antennal clubs are orange-tipped in the North, vs. black (or black-and-white) in the South. Two seasonal forms, discussed below, are triggered by day length.
Dorsal	Complex spots and mottling on an orange base. Central HW is *not as broadly or continuously orange* as in the Northern. Female mid-FW band often *slightly yellow-toned* (but less so than in Phaon).
Ventral	Note *pearlescent crescent* set in a *dark, submarginal HW patch*. In summer form ("morpheus"), light HW is complexly and variably marked. In fall-spring form ("marcia"), HW is brown with a pale median area.

Habitat	Varied open habitats, fields, vacant lots, open woods, roadsides, etc.
Hostplants	Many aster species, including Panicled (*Aster lanceolatum*), Calico (*A. lateriflorus*), Heath (*A. pilosus*), Bushy (*A. dumosus*), etc. Eggs laid in layered clusters of 20–200 (avg. mid-60s). Gregarious early instars feed on leaves without making nests.
Occurrence	Common to abundant, central NE s. to Keys. Also w. to Rockies, s. to se CA and MEX. Three-plus broods, Mar-Nov; overlapping in South. 3rd stage caterpillars overwinter, some from early broods.

Heath (Frost) Aster (*Aster pilosus*)

Ecology	**Generalist**. A generalist in nearly all respects.

av wingspan: 1.3″

Ventral male ("marcia" form): 6/12/94, Pound Ridge Reservation, Westchester Co., NY

Ventral female ("morpheus" form): 7/11/99, Sterling Forest, Orange Co., NY

Northern Crescent   *Phyciodes selenis* [= *cocyta*]

This attractive woodland butterfly was recognized as a separate species just 20 years ago. Yet new evidence already indicates that it may not be a full species after all, but rather a well-defined segregate (as originally thought). It is largely—but not entirely—isolated from the main body of Pearl Crescents by geography and brood timing (Porter & Mueller, 1998). Northern × Pearl crosses have been found to be fully viable.

We treat the Northern separately here because it *is* a well-defined segregate, often differentiable in the field. Its flap-and-glide flight is slow and deliberate, as compared with the Pearl's rapid, darting movements. Northerns may fly hundreds of yards along a dry stream bed or woods road without stopping. They are frequently seen taking minerals at damp soil or puddles.

Dorsal male: 6/12/93, Errol, Coos Co., NH

Identification	Slightly larger than the Pearl Crescent in our region. Males are more easily identified than females (the latter may closely resemble female Pearls or Tawnys). Northerns tend to fly between Pearl Crescent broods, producing significant brood isolation. But some overlap does occur. Antennal clubs are orange-tipped in both sexes.
Dorsal	In males, the large, central area of the dorsal HW is *bright, unmarked orange*. Male Pearls tend to have some dark markings or reticulation in this area. Females are darker, like other northern female crescents.
Ventral	Dark patch surrounding HW crescent is more tan than blackish (as in Pearl). HW lineation is *orange rather than black*.

Habitat — Moist, partially open woods, stream or marsh edges; shale barrens in WV/VA.

Hostplants — Asters. Panicled (*A. simplex*) and Smooth (*A. laevis*) reported. Also, surprisingly, Flat-topped (*A. umbellatus*) (Scott, 1986) (Pearls reject it).

Occurrence — Often common, ME and n. NE, south in mountains to WV/VA and NC (once). Also Rockies, w. CAN. One brood, mid-Jun-Jul, partial 2nd in South (Aug). 3rd instar caterpillars overwinter.

Ventral: 6/13/93, Pondicherry Wildlife Refuge, Coos Co., NH

Ecology — **Medium Generalist**. Uses many habitats, but more consistently associated with woodlands than Pearl Crescent.

av wingspan: 1.4″

Tawny Crescent *Phyciodes batesii*

Dorsal male (*maconensis*): 5/23/99, Clay Co., NC

Dorsal female (*maconensis*): 5/23/99, Clay Co., NC

A butterfly of subtle habitat associations, the Tawny Crescent selects low-growth environments in much of its eastern range, often at sites with scattered bare patches. For unknown reasons, the Northeastern race (*batesii*) has declined dramatically since 1960 and may be gone entirely from our area. The remaining race (*maconensis*) persists at high elevations in the southern Appalachians (Gatrelle, 1998d). Affiliated with the "*thaos* complex" of northland crescents, Tawnys resemble Northern Crescents in many respects.

The Tawny's virtual disappearance from the Northeast prevents investigation of its little-known lifestyle there. Uncertainty even exists as to the original type locality (New Jersey vs. Virginia). *P. b. batesii* should still be pursued in former haunts, where remnant colonies could linger among masses of commoner Pearls and Northerns.

The handsome "Appalachian" Tawny Crescent is a large race, nearly checkerspot-sized. It patrols the highlands of western North Carolina, nectaring regularly and often puddling.

Ventral (*maconensis*): 5/23/99, Clay Co., NC

Identification The southern race is large and distinctive, but the northern *batesii* race is about the size of a Pearl Crescent and harder to identify—especially females. *P. b. batesii* is intermediate in appearance between *maconensis* and the specifically separate Northern/Pearl Crescent. Populations to our west pose even more of a challenge.

Dorsal Male: Orange on FW is reduced to bands, especially in *maconensis* (*compare* other crescents). A rough-edged black patch appears at the outer/trailing FW edge. Dark HW borders are very wide. Female: Yellowish mid-FW band is well-defined; submarginal orange band has some distinctly yellow-centered spots.

Ventral Dark spot near FW apex is usually small, but *black mid-FW mark is large*. HW has *reddish lines* and small postmedian dot row. In *batesii*, HW is pale yellow with modest outer darkening and usually a silvered crescent. In *maconensis*, the crescent is *rarely silvered*, and the "cloud" is *vague or absent*.

Habitat Canadian *batesii* favor alvars, dry boreal scrub; but also damp, low-growth rocky sites. *Maconensis* is found above 4,000 feet in the s. Appalachians, near serpentine barrens, balds, clearings. May wander through open woods.

Hostplants Wavy-leaved Aster (*Aster undulatus*) in NC. Related asters (*A. ciliolatus* and *cordifolius*) in Ottawa (Catling, 1997). Eggs laid in clusters; young caterpillars feed on the underside in silk nests (*compare* Pearl/Northern).

Occurrence Local and colonial in w. NC and n. GA. Also west across Canada, sparsely in US West. One brood, May-early Jun in NC, variably Jun-Jul northward. 3rd instar caterpillars overwinter.

Ecology Medium specialist. Limited in local host plant and habitats.

Wavy-leaf Aster (*Aster undulatus*), not blooming during flight period

Hillside habitat with bare patches, western NC

av wingspan: 1.35" (*batesii*); 1.45" (*maconensis*)

Phaon Crescent   *Phyciodes phaon*

Dorsal: 9/24/95, Homestead, Miami-Dade Co., FL

Ventral: 9/24/95, Homestead, Miami-Dade Co., FL

Pearl and Phaon Crescents seem to be close relatives where their East Coast ranges overlap (in Florida and parts of the Deep South). Similar in appearance and behavior, they often fly directly alongside each other in disturbed, low-growth habitats, such as roadsides, weedy lawns, and grassy fields.

Yet the more closely we study these two species, the less similar they appear. Pearls feed on a wide variety of asters, e.g., while Phaons are limited to hostplants in a single genus of the Vervain Family (*Phyla*: in Verbenaceae). *Phylas* contain toxic iridoid glycosides (Wahlberg, 2000) which Phaons may employ defensively as adults or caterpillars—although direct evidence of such use is lacking.

The Phaon is a southern crescent. Its foodplant belongs to a neotropical genus that cannot tolerate severe cold. (Summer immigrants to the north-central plains are not able to form permanent colonies.) In a genus thought to have originated in the North, the Phaon is noteworthy for having relinquished its cold-hardy lifestyle in favor of a neotropical orientation (*see also* next two species). Indeed, Phaons still continue to extend their range southward occasionally: they reached Cuba in 1930 (Smith et al., 1994), possibly with human assistance.

Phaons are exceptionally low-flying. Males patrol actively during the day, often mating around noon. They nectar on low-growing flowers, including their hostplants. Small, coastal colonies north of Florida are at risk from hurricanes and rampant seaside development (LeGrand & Howard, 2004).

Identification	Our smallest crescent, especially males, though bright colors and active movements make them quite conspicuous. Wings are rounded, with a strongly checkered fringe. Winter phenotype (alternately called "marcia" or "hiernalis") is easily confused with the equivalent phenotype in Pearl Crescent when viewed from below.
Dorsal	Dark orange. The orange outer FW band is preceded by a *pale median FW band* (variably white, yellowish, or creamy).
Ventral	FW *darker orange* than the Pearl's. White mid-FW bar is visible below. The HW is variably chalk-white to creamy, with a light, marginal crescent surrounded by a dark outer-HW field. Note the dusky central HW spot and distinct HW reticulations. In the winter form ("marcia"), HW is brownish with a pale central band.

Habitat	Moist, open, low-growth areas with hostplant. Fields, roadsides, dunes, vacant lots, parks, etc. Mainly a coastal plain species.
Hostplants	Fogfruit, both Common (*Phyla* [= *Lippia*] *nodiflora*) and Lance-leaved (*P. lanceolata*). Also called Mat Plant, Creeping Charlie, etc. *P. lanceolata* grows in wetlands, but *P. nodiflora* is widespread on lawns and roadsides. Eggs laid in clusters of 50–100 beneath host leaves.
Occurrence	Common to abundant in peninsular FL and north along immediate coast to Dare Co., NC. Also across US "south rim," south to Guatemala. Strays north in Midwest during summer. Multiple broods, May-Oct in NC, all year in s. FL. Adults enter reproductive diapause in the Deep South during winter. Overwintering stage in Carolinas not reported.
Ecology	**Medium Generalist.** Generalist-type lifestyle, but limited geographically.

Fogfruit, or Carpetweed (*Phyla* [*Lippia*] *nodiflora*)

av wingspan: 1.1″

Cuban Crescent   *Phyciodes* [= *Anthanassa*] *frisia*

Dorsal male: 6/22/98, Everglades NP, Monroe Co., FL

Dorsal female (rejecting courtship): 8/31/99, Everglades NP, Monroe Co., FL

The range of this attractive crescent has fluctuated dramatically in recent decades. A regular resident of south Florida in the 1950s, Cuban Crescents were "common along roadsides in the Florida keys in winter" (Klots, 1951, p. 101). Strays wandered as far north as Gainesville. But populations crashed in the 1990s, leaving just a few known colonies. Promising signs of a rebound have appeared in recent years, coincidental with increased rainfall in the region, but it may be difficult for Cuban Crescents to fully reclaim their former range, owing to widespread ongoing habitat destruction and the use of increasingly toxic mosquito sprays.

Cuban Crescents belong to a southern crescent group thought to have diverged from their northern ancestors eons ago. Some authorities place them in the neotropical genus *Anthanassa*. Until recently, the Cuban Crescent was lumped with the Pale-banded (*A. tulcis*), a related but visually dissimilar neotropical butterfly. (*Note:* Be aware that earlier accounts discuss them as one entity.)

Cuban Crescents are strongly colonial, plentiful in one patch but absent from similar ones nearby. Adults fly close to the ground—although not as low, in our experience, as Phaon Crescents. Courtship involves much swirling flight, dominated by swarms of ardent males. Adults nectar on *Bidens* and other flowers along weedy hammock or canal edges.

Ventral: 8/31/99, Everglades NP, Monroe Co., FL

Identification *Dorsal*	A small, brightly marked crescent. Outer FW edge is slightly indented (concave), with *light margin* in concavity FW has several *large, separate orange spots* on a mostly dark field (deep orange in males, some more yellowish in females). HW shows considerable orange. One to two rows of *dark, zigzag submarginal lines* on the outer HW are distinctive.
Ventral	Rich, versicolored pattern when fresh, but fades rather quickly to a uniform, pale beige. White zigzags on central HW may be indistinct when worn. Base of FW is a deep, brownish red, the apex smudgy.

Habitat   Weedy hammock edges or canal banks with hostplant. Favors the coastal ecotone between mangrove forests and interior habitats. Additional sites reported from Caribbean, including open scrub and dry limestone areas.

Hostplants   Crimson Dicliptera (*Dicliptera sexangularis*) is main host, but another common Acanthus, Green Shrimp Plant (*Blechum pyramidatum*), a known hostplant in Cuba, grows alongside *Dicliptera* at some US colonies. Eggs laid in clusters of 100+; young feed communally at night.

Occurrence   Usually rare and local, but variable. South FL and W.I. (common on Cuba, probably the origin of US populations). Flies throughout the year, but broods are discontinuous, especially during the winter dry season.

Ecology   Specialist. Can occur in second-growth scrub, etc., but typically a subtropical hammock-edge specialist with narrow feeding preferences.

Crimson Dicliptera (*Dicliptera sexangularis*)

av wingspan: 1.25"

'Seminole' Texan Crescent   *Phyciodes* [= *Anthanassa*] *texana seminole*

Dorsal: 9/30/95, Newnans Lake, Alachua Co., FL

Ventral: 9/30/95, Newnans Lake, Alachua Co., FL

The 'Seminole' Texan Crescent is unusual among East Coast butterflies in having its main population center in northern Florida. But even 'Seminoles' are not confined to this region: a population spike occurred northward along the Atlantic coastal plain during the warm summer of 2002, in areas where the species was seldom previously recorded. Some individuals even reached southeastern North Carolina (LeGrand & Howard, 2004).

The 'Seminole' Texan Crescent is a distinctive butterfly, possibly a different species from its geographically isolated western relative, the 'Texan' Texan Crescent (*P. t. texana*). The case for separating the two is supported to some degree by pattern differences, but equally by lifestyle variations. *P. t. texana* favors dry habitats such as arroyos and thorn-scrub woodlands, whereas *seminole* usually is found in damp surroundings. And *P. t. texana* has a tendency to move northward each season onto the Great Plains, while *P. t. seminole* is comparatively sedentary (notwithstanding its recent resurgence in the Carolinas).

Some experts place Texan Crescent in the genus *Anthanassa*, along with the Cuban Crescent. The "mainline" *Phyciodes* crescents are primarily northern butterflies that sometimes push southward (*see* Phaon account). But *Anthanassa*-type crescents appear to be southern-based species that make occasional incursions northward.

Males perch territorially for mates, but apparently also sometimes patrol. Flight is slower and less jerky than in some smaller crescents. Courting males engage in energetic, circular mating flights around the target female.

Identification	The largest crescent in its range (about equal in size to the Northern). Size is variable, however, with some very small males and some very large females. Note a characteristic *indentation on the outer FW margin*, which gives the wing a slightly falcate appearance (similar to Cuban).
Dorsal	Dark with *orange wing bases* (especially in FW cell; *texana* darker overall). FW has a series of *conspicuous white spangles*. Most important is a series of *squarish white marks* that form a *continuous median line* on the HW. Also note a row of white inward-facing scallop marks near the trailing HW margin.
Ventral	Complexly patterned. FW deep *red-orange at base*, dark apically with white spots mirroring the upper surface. HW contrasting dark and white, with a *white median stripe* and a *white "crescent"* near the margin.

Habitat	Lakesides, bottomlands, and riparian corridors, especially with open, low sites for hostplants.
Hostplants	Mainly Water Willow (*Justicia ovata*) in our area, though other small Acanthaceae suspected. Additional hosts in Texas. Females lay eggs in large clusters beneath hostplant leaves.
Occurrence	Rare to uncommon except in colonies, e. LA to n. FL and sw GA; occasionally e. SC/NC. Also sw US to Guatemala (*texana*). Metapopulations may shift rapidly, especially on lakeside roads with heavy mowing. Two to three broods, late Mar-Nov. Diapause stage, if any, not reported.
Ecology	Medium specialist. 'Seminole' race colonial, thinly distributed, uses regionally specialized hosts.

Water Willow (*Justicia ovata*)

av wingspan: 1.4″

Gorgone Checkerspot   *Chlosyne gorgone*

Dorsal (*carlota*): 5/10/01, Clay Co., NC

Ventral (*carlota*): 5/10/01, Clay Co., NC

Few field observers have been lucky enough to encounter a Gorgone Checkerspot in our region. This Great Plains species is mostly a rare opportunist in the East, with only a few permanent colonies known. And these tend to be thin and ephemeral—or at the very least inconspicuous. The interval between sightings at one historic location in Georgia extended for almost 200 years (Gatrelle, 1998c).

Gorgones may have been more common in the East prior to European settlement, when open sandhills dominated the southeastern landscape. A single-brooded form on the upper coastal plain of Georgia and South Carolina seems well-adapted to this habitat (Gatrelle, 1998c). John Abbot first painted this population, from Burke County, Georgia, in 1810. It was later described as the nominate race (*gorgone*), but the original "type" population was then lost until the mid-1990s. Better-documented populations (still sparse) occur in the southern Appalachians and upper Piedmont (Harris, 1972) but these belong to the multibrooded Midwestern race (*carlota*), as do rare strays and colonists farther north (12 *carlota* colonies were discovered in eastern Ontario in 1996).

Most behavioral notes on Gorgones are based on western observations. Prairie/foothill males patrol most of the day for females, but also perch on ridgetops, especially in late afternoon (Scott, 1975). Courting males fly in an irregular pattern above females, who may in turn flutter their wings or dash away if mating is not desired.

Identification	Small for its genus (about *Phyciodes*-sized). Note *strongly checkered black-and-white fringe*. Nominate race sometimes mistakenly described as large and pale, based on poor reproductions of an Abbot painting (Calhoun, 2003). It is actually darker, with less light patterning (Gatrelle, 1998c).
Dorsal	Similar to Silvery, with concentric black markings on orange field. But Silvery has darker FW apex and prominent white-centered submarginal HW dots (white center reduced or absent in Gorgone, especially *carlota*). Pale HW marginal crescents are arched, again more so in *carlota*.
Ventral	Distinctive *brown-and-white zigzag pattern* (zigzag less pronounced in *gorgone*). Note white mid-HW band, white chevron at mid-HW margin, white bar at HW base, generally "busy" pattern.

Habitat	*C. g. carlota* favors grassy burns, old fields, second-growth scrub, serpentine barrens (NC), open woods. *C. g. gorgone* known from dry sandhills and open hardwood forest. As preferred low-growth habitats regrow, dispossessed individuals may adopt unusual, alternate sites.
Hostplants	Various local composites, e.g., Woodland Sunflower (*Helianthus divaricatus*) (especially *gorgone*), but in CAN *carlota* uses Black-eyed Susan (*Rudbeckia hirta*), refuses *divaricatus* (Catling & Layberry, 1998). Female lays eggs in clusters, young feed communally in early stages.
Occurrence	Occasionally abundant in West, but rare and local here; nw GA and w. SC/NC; isolated records from WV, ne PA, w. NY. Also Gr. Plains, n. Gr. Lakes. *Carlota* multibrooded (2–3), Apr/May-Sep; *gorgone* reportedly single-brooded, Apr-May. 3rd stage caterpillar overwinters.
Ecology	Medium Specialist. Locally adaptable in habitat and hostplant use.

Appalachian burn site, Clay Co., NC

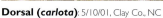

av wingspan: 1.35″

Silvery Checkerspot   *Chlosyne nycteis*

Dorsal: 6/15/95, Van Cortlandt Park, Bronx Co., NY

Ventral: 6/15/95, Van Cortlandt Park, Bronx Co., NY

An interesting pastime on a June afternoon is to study a fresh Silvery Checkerspot as it flits along a dappled forest path. With diligence, you should be able to see the faint, pearly sheen on the lower hindwing for which the butterfly was named. (Some hindwings have dusky scales, though, which can limit reflectance.) An attractive butterfly, the Silvery is a good find in most of our area, as its occurrence here tends to be local and rather sparse.

Despite limited population density, the Silvery Checkerspot is a lifestyle generalist, using a variety of habitats and hosts across its relatively broad home range. At one location in Louisiana, it feeds exclusively on Great Ragweed (*Ambrosia trifolia*) (Ross, 2001). Why the use of this abundant "trash plant" is reported only here is unknown. The Silvery has several races, including the dark western *drusius* and the pale northern *reversa*. The typical eastern race is *nycteis*. A recent proposal to change the scientific species name to *ismeria* seems mistaken (Calhoun, 2003).

The composites on which Silvery Checkerspots feed have no identified toxins, leading taxonomists to wonder why they were originally adopted. One interesting theory is that Silverys' ancestors fed on a class of plants (*hemiparasites*) that derive some of their nutrition from the roots of other plants—such as asters. Perhaps the larvae became acquainted in this manner with the chemical signatures of certain future hostplants (Wahlberg, 2000).

Male Silverys patrol for females, and sometimes perch. Flight is low and slow, with alternate flaps and glides. Adults take nectar and sip moisture from the ground.

Identification	The largest of our *Chlosyne* checkerspots, and females can be much larger than males. Resembles Harris' Checkerspot above, a crescent below. Markings relatively clean-cut and tidy. Margin moderately checkered.
Dorsal	Orange above (tawny to yellowish), with distinct dark borders. Note *dark FW tips*, usually with little orange intrusion. One or two dark spots in the HW submarginal band have *white centers*.
Ventral	Crescent-like. At least one of the dark spots in the submarginal HW band has a *white center*. HW color variable; may be pearly white, or have variable dark, basal scaling.

Habitat	Mainly open areas: streamsides, bottomlands, shores; but also meadows, shaded wood edges, even dry ridges and serpentine barrens. And occurs on the summit of Mt. Jefferson in s. Appalachians. A well-known colony at Van Cortlandt Park in NYC is in mature hardwood forest.
Hostplants	Varied composites. Sunflowers, Wingstem (*Verbesina* [*Actinomeris*] *alternifolia*), rosinweeds (*Silphium*), asters. Coneflowers in West. Females lay 100+ eggs beneath host leaf. Early-stage caterpillars skeletonize leaves in communal forays, then move to other leaves.
Occurrence	Widespread but sparse. Inland from ME to central GA. Isolated populations in FL panhandle. Also s. CAN, Gr. Plains, and Mississippi/Ohio River Valleys; separately in Rockies and central AZ. One brood in North (Jun-early Jul), 2–4 in South (Apr/May-Sep/Oct), some later broods partial. 3rd stage larvae overwinter; diapause generation has extra instar.
Ecology	**Medium generalist.** Seemingly adaptable, but nonetheless declining.

Woodland Sunflower (*Helianthus divaricatus*)

av wingspan: 1.6″

Harris' Checkerspot   *Chlosyne harrisii*

Dorsal: 6/22/96, Pound Ridge Reservation, Westchester Co., NY

Ventral: 6/21/97, Pound Ridge Reservation, Westchester Co., NY

This handsome butterfly is named for Thaddeus Harris, a 19th-century entomologist from Massachusetts. It is wide-spread in northern New England, at times abundant during its early summer flight. But farther south it becomes local and irregular. An outpost in eastern West Virginia is entirely disjunct.

Harris' is unusual among our melitaeines in having just a single documented hostplant, Flat-topped Aster (*A. umbellatus*). This plant manufactures a distinctive, complex hydrocarbon (atractylon) that lures Harris' but deters most other browsers—including larvae of the closely related Pearl Crescent (Williams, 2002). Yet the substance seems not to be used defensively by this edible butterfly (Bowers, 1983), so its exact role remains unclear, except as an attractant. It has long been suggested that Harris' mimics the ventral surface of the Baltimore, a species of known toxicity that often lives nearby. But equally important, some now propose that the *caterpillars* of Harris' Checkerspots may mimic those of the Baltimore, since the two appear nearly identical and build similar, conspicuous leaf nests in the same general habitats (Williams, 2002).

Harris' flight is comparatively slow, with an alternating flap-and-glide style familiar in larger crescents and checkerspots. Males patrol for females, but pause often to perch. Adults take nectar and also puddle. Adults emigrate from overcrowded habitats (Dethier & MacArthur, 1964), often traveling along roadsides or other low-growth corridors as they relocate.

Identification	Slightly smaller on average than Silvery Checkerspot, but with much overlap. Doral surface often difficult to distinguish from those of other melitaeini, especially Silvery Checkerspot. But a glance at the ventral surface should be enough to dispel any confusion.
Dorsal	Orange field with wide, dark borders—darkest farther south ("liggetti" form), more orange northward ("albimontana"). Black submarginal HW dots are at least slightly white-centered, as in Silvery (same on ventral surface).
Ventral	Bright red-orange, black, and white pattern on ventral HW is distinctive.
Habitat	Moist areas: bogs, marshes, brushy wet pastures, and meadows. Wanders for nectar and during dispersal.
Hostplants	Flat-topped Aster (*Aster umbellatus*). Female lays about 25 eggs on the leaf. Young highly communal, build silk nest; after defoliating one plant, move en masse to another (or starve).
Occurrence	Range mirrors that of hostplant: CAN Maritimes to Manitoba, south from OH to n. NJ; separately in WV. One brood, Jun-Jul. 3rd stage larvae overwinter near base of hostplant.
Ecology	**Medium specialist.** Specialized lifestyle, but host and habitats are widespread.

Flat-topped Aster (*Aster umbellatus*)

Early-stage caterpillars in web on hostplant; no nest built second year (after diapause)

av wingspan: 1.5″

Baltimore Checkerspot   *Euphydryas phaeton*

Dorsal: 6/19/94, Errol, Coos Co., NH

Ventral: 7/11/93, Sussex Co., NJ

This striking butterfly displays the heraldic colors of George Calvert, first Baron of Baltimore in the early 1600s. In honor of this historic association, Maryland installed the Baltimore as the state's "Arthropodic Emblem" in 1973. The designation was changed to "State Insect" in 1989, so as not to overshadow another locally popular arthropod, the Blue Crab. (The crab, meantime, was given the vaunted title of "State Crustacean.")

Euphydryas checkerspots are among the best studied butterflies in the world, thanks largely to the efforts of Paul Ehrlich and his students at Stanford University. Mostly sedentary, they tend to form small, loosely connected colonies (in metapopulations). Baltimores are generally distasteful, having incorporated iridoid glycosides from early-stage larval hosts, such as Turtlehead (Bowers, 1980). A segment of the population continually feeds on hosts that do not provide such chemical protection, however. These individuals may be "automimics" of their unpalatable relatives. Baltimores living in dry, upland forest of the Ozark Mountains were formerly considered a separate race—even a sibling species—but more recent investigations failed to support this distinction (Vawter & Wright, 1986). Indeed, "Ozark-like" upland populations are also now known from New England and New York.

Male Baltimores await females on low perches. Despite rapid wingbeats, flight is slow. Adults feed on nectar, but their proboscises are too short to effectively pollinate Turtlehead.

Identification	Fairly large. Unlike any eastern butterfly (although it may be mimicked ventrally by Harris' Checkerspot). Variable geographically; sometimes nearly all-black dorsally. Antennal clubs are *bright orange*. Upper abdomen is marked with a *series of white dots*, lower abdomen has *orange dot-rows*.
Dorsal	Black with red-orange patches along both wing margins. Two or more concentric rows of white dots appear inside the margin, and the inner wing areas have scattered orange markings.
Ventral	Similar to dorsal, but more vividly marked, with creamy crescents toward margin and heavy orange blotches basally.

Habitat	Usually in moist low-lying or riparian habitats with Turtlehead, including wet meadows, bogs (especially in Midwest). But also inhabits dry upland woods, more so than once thought, perhaps expanding ecologically?
Hostplants	Young caterpillars feed mainly on Turtlehead (*Chelone glabra*) in lowlands, on gerardias upland, e.g., Smooth False Foxglove (*Aureolaria* [= *Gerardia*] *flava*). After diapause, may shift onto plantains (*Plantago*), ash (*Fraxinus*), etc. Eggs laid in clusters of several hundred (100–700); young larvae feed gregariously in silk webs, woven by some of the members. Young thrash violently and regurgitate to deter predators and parasitoids.
Occurrence	Widespread but "intensely local" (Klots, 1951). Occasionally abundant. Ranges south to n. GA, mainly in uplands. Also west to MN and south to e. TX. One brood, May/Jun-Jul/Aug. 4th stage caterpillars leave hostplant en masse, separate and hibernate in leafy debris, silked together, some 3 feet from host.
Ecology	Medium Specialist. Somewhat adaptive, but still a highly targeted lifestyle.

Turtlehead (*Chelone glabra*)

av wingspan: 2.1"

Question Mark *Polygonia interrogationis*

Dorsal (summer brood): 5/26/96, Gr. Dismal Swamp NWR, Nasemond Co., VA

Ventral (less mottled form): 9/11/94, Jamaica Bay NWR, Queens Co., NY

This large nymphalid is common in the East. Once called Violet-Tip, its longstanding common name refers to the silvery punctuation mark on its ventral hindwing. Unlike most commas, which are northern in distribution, even boreal, the Question Mark favors temperate, deciduous woodlands. It is uncommon north of central New England, and ventures well south, at least once as far as the Florida Keys.

Question Marks are migratory butterflies, with a strong coastal flight in the fall and a definite—though usually less obvious—northern movement in spring. Some individuals winter in the North, but numbers decline significantly northward in most years.

Question Marks bask conspicuously in the morning. Males display in the afternoon, aggressively confronting passers-by. Some adults nectar, but most favor decaying fruit, sap, dung, and puddles. Several observers have remarked that Question Marks feeding on fermented substances seem to become "intoxicated," showing signs of disorientation and even allowing themselves to be touched.

Dorsal (fall/spring brood): 9/11/94, Jamaica Bay NWR, Queens Co., NY

Identification	Our largest anglewing. The FW is *strongly hooked* and the wing margins relatively smooth (for an anglewing). Question Marks have distinct seasonal forms: HW tails are *short in summer, longer in fall/spring*. Some authors regard the HW pattern as sexually dimorphic (Iftner et al., 1992; Layberry et al., 1998), but others describe it as non-sex linked (Opler & Krizek, 1984; Tveten & Tveten, 1996). We note seasonal associations.
Dorsal	Orange or red-orange with dark spots. A *"dash mark" at the outer end of the FW dot row* is distinctive—although a few Eastern Commas also show a faint mark here. *Summer form* ("umbrosa"): *HW almost entirely black*, with short tails and thin violet edging. *Fall/spring form* ("fabricii"): Bright orange, including upper HW, with longer tails and conspicuous *violet-gray wing edges*.
Ventral	Light gray-brown, either striated/mottled or else relatively unmarked. Note the silvery HW comma with a *separate dot*, forming a "question mark."

Habitat	Deciduous woods with some open space; glades, edges, streamsides, etc.
Hostplants	Elms, including American (*Ulmus americanus*), Slippery (*U. rubra*), Chinese (*U. parviflora*); also hackberries (*Celtis*); sometimes nettles (*Urtica*). Eggs laid singly or in "chains" of up to eight, hanging down from a leaf; or sometimes on a plant near the host. Caterpillars usually feed alone, without nests.
Occurrence	Variably common. Eastern 2/3 of the US. Throughout our area, but rare in the far north or south. Adults emerge in spring to produce a long summer brood (May-Sep); may aestivate. Fall brood begins in late Aug. Adults overwinter in North, or after migrating (many reported at Congaree Swamp, SC).
Ecology	**Generalist.** Broad tolerance in hostplants and habitats, quite mobile.

American Elm (*Ulmus americana*)

av wingspan: 2.5″

Eastern Comma *Polygonia comma*

Dorsal (summer brood): 5/26/96, Gr. Dismal Swamp NWR, Nasemond Co., VA

Ventral (mottled HW pattern): 7/10/94, Tottenville, Staten Island, NY

This common butterfly was formerly called Hop Merchant. Farmers believed they could determine the value of their crops by studying the silver-and-gold markings on the butterfly's pupa, which they found dangling from Hops leaves (Weed, 1917). Eastern Commas closely resemble their near relative, the Question Mark, in both lifestyle and appearance. The Eastern has a more northerly "center of gravity" and its seasonal movements tend to be less extensive—although in certain years a conspicuous fall exodus occurs along the coast.

As with many other nymphalids, Eastern Comma adults feed mainly on nonfloral foods, including sap, mud, rotting fruit, even dung containing fruit (as in ventral photo). Behavior is similar to that of the Question Mark. Tends to be jumpy and difficult to approach.

Dorsal (spring/fall): 8/31/93, Mt. Greylock, Berkshire Co., MA

Identification	Medium-sized, females larger. Smaller on average than Question Marks, although size is variable and identification on this basis can be difficult in the field without a side-by-side comparison. Wing edges are somewhat irregular, but less ragged than those of the "northern commas" (Gray, Green, Hoary, and Satyr).
Dorsal	*Summer brood ("dryas")*: FW is relatively dark orange. The mid-FW dot row consists of three simple, medium-sized spots. The inner spot (nearest the body) is *usually single*, but may be doubled slightly across the vein. HW is *mostly black. Fall/spring brood ("harrisi")*: The upside color is reddish-orange, *HW orange with dark spots.*
Ventral	Yellowish- or ruddy brown, mottled or somewhat plain, as in Question Mark. The silvery "comma" is more smoothly rounded than in Gray Comma, usually with swellings at the ends and a "fish hook" pointed toward body. Lacks green mottling (but has purplish tones).
Habitat	Deciduous woods, openings and edges, often near water. Less frequent away from woods than Question Mark. But like that species it may be found in settled areas. Habitats more varied when migrating.
Hostplants	Nettles (*Urtica*) and elms (*Ulmus*); Wood nettle (*Laportea canadensis*); Hops (*Humuus*) where still cultivated. Egg-laying is similar to the Question Mark's; older caterpillars make silk nests, feed at night.
Occurrence	Usually common, but less frequent in the South. Annual populations variable. Found across e. US, in our area ME to n. FL. Short day lengths in Sep-Oct cause fall form to hibernate or enter reproductive diapause (often after withdrawing some distance south). Reemerges early spring; summer brood follows in May-Sep, may aestivate.
Ecology	**Generalist.** This species and the Question Mark represent minor variations on a common, successful adaptive theme.

Wood Nettle (*Laportea canadensis*)

av wingspan: 1.9″

Satyr Comma   *Polygonia satyrus*

Dorsal (western): specimen photo

Ventral (western): specimen photo

Also called Golden Anglewing. This is among the most rarely seen of East Coast butterflies, barely entering our region from the North. West of the Great Plains, Satyrs occupy a niche roughly analogous to that of the Eastern Comma. There, Satyrs are widespread and more familiar. They even resemble the Eastern Comma in many particulars of their appearance. But at the Atlantic edge of the Satyr's broad range (in eastern Canada and northernmost parts of New England) Satyrs have few predictable population centers and tend to be seen only sporadically (R. Webster, pers. comm.).

In a recent taxonomic study of the Nymphalini, based on a "total evidence" approach (combining a broad set of physical, genetic, and lifestyle traits), researchers found evidence of a close relationship between Eastern and Satyr Commas (Nylin et al., 2001). One can readily imagine how populations of an ancestral *Polygonia* might have become isolated on an east-west basis long ago. If so, today's meager ranks of Satyr Commas in the East could represent a secondary recolonization by members of the newly separated entity. This would help account for the Satyr's more marginal and reclusive existence here.

Male Satyr Commas perch in the afternoon awaiting mates. Aerial "duels" between males result in spiraling dogfights high in the air. Adults nectar frequently in the West, but few behavioral details are available from the East. We once saw a male puddling with other comma species in northern New Hampshire. All flew off when disturbed by a passing vehicle, but only the Satyr failed to return. Such particularly skittish behavior is likely to be random in any given instance, but this is consistent with the experience of other eastern observers.

Identification	Medium-sized, females larger. Appearance is variable, but in this species differences are not consistently seasonal or geographic. There is a definite risk of confusion between male Satyr Commas and fall-brood Easterns. But note the Satyr's *ragged wing edges* (the Eastern's are smoother) and its *lack of pale or violet wing edging*.
Dorsal	Bright *yellow- or golden-orange* ("tawny" vs. "orange"). The inner dot on the FW dot row is *consistently doubled*. The FW has a dark outer border, but the *HW border is frequently vague or nearly absent* (especially in females), giving a bright, golden appearance overall. Yellow marginal HW spots are indistinct or blurry. There is often an isolated dark spot or patch on the central HW.
Ventral	A mix of light brown and golden brown patterns, usually striated and finely dotted. Note *relatively straight central band* across both wings. The FW apex is *single-colored* in Satyr (vs. bi-colored in Eastern) (Glassberg, 1999).
Habitat	Wooded areas near streams, valley bottoms. Canyons, riparian corridors, and prairie ravines in the West.
Hostplants	Nettles (*Urtica*). In eastern CAN, Stinging Nettle (*U. dioica*) growing in shaded areas (R. Webster, pers. comm.). Caterpillars make silk nests.
Occurrence	Rare and elusive; common in the West, but even there declining with disturbance of riparian corridors. Apparently one brood in our area, mid-Jul-Sep. Two to three possible in West. Adults overwinter.
Ecology	Specialist. Eastern populations are scarce and elusive.

Stinging Nettle (*Urtica dioica*)

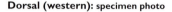
av wingspan: 1.9"

Green Comma   *Polygonia faunus*

Dorsal: 7/17/99, CT Lakes Region, Coos Co., NH

Ventral: 8/19/00, Pondicherry Wildlife Refuge, Coos Co., NH

The Green Comma is a handsome resident of northern forests. A brightly marked Appalachian race (*smythi*) occurs in isolated mountain pockets as far south as northern Georgia, but the species is prevalent mainly in the North. And even there it is subject to marked seasonal population swings. Almost none may be seen in "crash" years.

Cool weather is a regular feature of the Green Comma's existence, even in mid-summer. We have seen adults in northern New Hampshire "shiver" on cloudy days in late August to boost their body temperature. Adults act territorially, darting out to inspect passers-by. It is not unusual for them to land on clothing or exposed skin, where they probe for minerals. Males perch in midday to await females. Early observers such as Scudder reported a faint "clicking" noise emitted in flight (Weed, 1917).

Ventral (chocolate morph): specimen photo

Given significant variations in regional appearance and hostplant use continent-wide, the "Green Comma" may in fact be a complex with more than one sibling species. The northern subspecies in our region is *faunus*.

Identification	Often identifiable at a distance by the *ragged outline of its wing margins.* Yet some individuals have a rather smooth shape (*see* dorsal photo). Virtually all markings are variable, both geographically and within populations.
Dorsal	Intense, rusty-orange base color with inky black spots. Bright yellow dots rim the outer wing margins, set in a dark field. The *innermost black spot on the trailing FW dot row is doubled across the vein.* Outer wing margins variably edged with yellow, especially conspicuous in fresh individuals.
Ventral	Heavily mottled ("dead leaf" pattern) with two rows of *conspicuous green markings* toward the margin. Small, silvery "comma" usually ends in a "fishhook." The Appalachian *smythi* race and rare "chocolate" northern form (silvius) are both much darker below, with little mottling.

Habitat	Northern forests, often a mixture of conifers (spruce, fir, or hemlock) and northern hardwoods (beech, maple, birch, poplar). Frequents forest glades and streamsides, but most often seen along dirt roads and trails.
Hostplant	Varied hostplants, mainly birches (*Betula*), willows (*Salix humilis*, etc.), alders. Elsewhere, rhododendron (CA), blueberries (MI, CAN, etc.). Caterpillars reportedly accept a range of foodplants in captivity. Eggs are laid singly on the top of hostplant leaves. Young caterpillars crawl beneath and feed on the undersurface.
Occurrence	Fairly common in the North (abundant some years); rarer and more local southward. Also across CAN to w. AL; south in the mountains to NM and n. CA. Probably single-brooded throughout our region. Adults hibernate, emerging in spring to mate. Fresh adults appear from late Jun to early Aug, then fly into the fall, storing fat for diapause. May aestivate during summer heat. Adult life span estimated at 9–10 months.
Ecology	**Medium specialist.** Though adaptable in terms of hostplant use, Green Commas are strictly limited to northern/boreal habitats.

Northern forest road, Aroostock Co., ME

av wingspan: 1.9″

Hoary Comma   *Polygonia gracilis*

Dorsal: 8/4/97, Scott Bog, Connecticut Lakes, Coos Co., NH

Ventral: 8/1/94, Scott Bog, Connecticut Lakes, Coos Co., NH

East Coast observers are unlikely to see this species without a targeted trip to one of its few regular haunts, such as the Connecticut Lakes region of northern New Hampshire. Even there, it is local and often difficult to find. (Not nearly as much so as the Satyr Comma, however, which is seldom seen in the region at all.) In some years, Hoary Commas emigrate (or disperse) beyond the boundaries of their normal home range, but they rarely become established and are likely to be absent from these "overflow" sites in subsequent years (Pyle, 1981).

The taxonomy of the Hoary Comma is rather intricate. The Satyr Comma has a generally similar distribution, but no defined subspecies. Hoaries, by comparison, have a distinctive eastern race (*gracilis*) and another well-defined subspecies (*zephyrus*), common in the West. In fact, this was long considered a full, separate species—the Zephyr (*P. zephyrus*). But the two were lumped after a broad Canadian "hybrid zone" was identified (*see* Layberry et al., 1998). The Hoary's closest interspecific relative appears to be the Gray Comma (as discussed in that account).

Male Hoary Commas perch in the afternoon to await females. The flight style is comparatively low and slow. Hoaries are more inclined to take nectar on a regular basis than other northern commas, often from Pearly Everlasting or Joe-Pye Weed. (Nectaring is not diagnostic, however. In dry summers, especially, any of the northern commas can be seen at flowers.)

Identification	Our smallest anglewing, though identification based on size alone generally is not reliable. The HW tails are long, the FW *distinctly falcate*, and the wing margins *very ragged*.
Dorsal	*Deep rust-orange* above with a *heavy black border* around both wings. Small yellow dash-marks or chevrons are set in the dark, submarginal area of the HW. The trailing portion of the mid-FW band, projecting back from the costa, is *curved on both sides* (usually straighter in Gray).
Ventral	*Strongly two-toned*, dark gray-brown toward the base, hoary or white-frosted toward the margin on both wings. If seen well, this pattern is diagnostic. The HW comma is usually tapered at both ends.

Habitat	Moist northern woods and river edges. Also boreal scrub (e.g., when seeking nectar). Hostplants grow at considerable elevation, however, and Hoaries have been reported from granite balds and mountain summits in NY State (Shapiro, 1974).
Hostplants	Currants and gooseberries (*Ribes*), others possible. In the West, azaleas reported, underscoring the distinctiveness of the *zephyrus* race. Life history is poorly known.
Occurrence	Usually rare and very local at the northern edge of our region; Most of boreal No. Amer. below the tundra and south into w. US. One annual brood in our area (*zephyrus* race has more). Overwintering adults emerge in spring and mate; new adults usually not before early Aug in n. NH, but earlier reported Jun-Sep in NY.
Ecology	**Specialist.** Local and restricted. Lifestyle details unclear.

Skunk Currant (*Ribes glandulosum*)

av wingspan: 1.7"

Gray Comma   *Polygonia progne*

Dorsal: 7/23/94, Moose River Plains, Hamilton Co., NY   **Ventral**: 7/23/93, Moose River Plains, Hamilton Co., NY

This bright *Polygonia* is the most southerly of our four "northern commas," and usually the most numerous where several occur together. Yet its distribution remains a bit of a puzzle. On the one hand, it is primarily a northern-boreal species (this is certainly true in our area). Yet it sometimes moves south in the fall, and at least some southern resident populations are known (e.g., in Ohio) that do not fit the boreal-dwelling model (Iftner et al., 1992). Finally, it seems to have been considerably more common and widespread in the late 1800s, suggesting either a population surge at that time or a significant northward withdrawal since.

As with other northern commas, the Gray has a number of distinct geographic forms. The race found in the Pacific Northwest (*oreas*) was once considered a separate species (and still is by some). A recent taxonomic study of the Nymphalini found suggestive evidence of a close phylogenetic link between the Gray and Hoary Commas (Nylin et al., 2001). And while this association is not conclusive, it certainly seems plausible.

Male Gray Commas perch in mid-afternoon to await females—sometimes fairly high off the ground—but more often on low, broad trailside leaves. Several authors describe their flight as relatively slow, but observers should not expect a languorous response from a startled individual. Gray Commas are most likely to be seen at puddles or moist spots on northern dirt roads. They are also drawn to sap flows and decaying material, and sometimes nectar. Some individuals perch repeatedly on clothes and forearms looking for minerals.

Identification	Medium-small. *Wing margins are ragged*, though somewhat less so than in other northern anglewings. Two seasonal forms occur, but most observers encounter summer-brood individuals (sometimes referred to as "l-argentum"). This brood typically emerges a bit earlier than the summer flight of Green or Hoary Comma.
Dorsal	*Bright orange* above, less rusty than some others. The FW dot row is made up of comparatively small, neat spots. The innermost FW spot is *single*, and the outermost spot is *particularly small*, at times almost absent. The HW is dark marginally (more so in summer, less in the fall/spring brood), with small, yellow markings.
Ventral	*Dark, charcoal-gray* (sometimes brownish-gray) with "wood grain" effect. *Outer half of FW (but not HW) is frosted* (the Hoary, by comparison, has a strong, two-tone effect overall). The HW "comma" is thin and tapered at the ends, with one "leg" shorter than the other in the summer brood, creating an *L*-shaped appearance.

Habitat	Openings in rich northern deciduous, or mixed/conifer forest. Cove forests in NC. Medium or wet temperate deciduous bottomland forest in OH.	
Hostplants	Mainly currants/goodeberries (*Ribes*); but Pinxter Azalea (*Rhododendron nudiflorum*) and elms also reported. Caterpillars feed on leaves without nests, using physical and apparently chemical defenses.	
Occurrence	Often common or very common, in the North and Appalachian highlands. Also across CAN and in Pac. NW, Gr. Plains, and Ohio Valley. Two broods, first in Jul–early Aug; fall/spring brood appears late (to Sep–Oct), overwinters.	
Ecology	**Medium Specialist.** More widely adaptable than its relatives, but still mainly a northlands specialist.	**Currants** (*Ribes* sp.): leaves (left), berry (top), and flowers (bottom)

av wingspan: 1.75"

Compton Tortoiseshell   *Nymphalis* [= *Roddia*] *vau-album*

Dorsal: 7/3/95, Amity Marsh, Orange Co., NY

Ventral: 7/4/95, Amity Marsh, Orange Co., NY

This elegant butterfly is among the earliest on the wing each spring in northern habitats. Its adult life span is also among the longest in our region (10 months or more). Yet it is seldom singled out for recognition in either of these respects. Such superlatives are generally reserved for more familiar species, such as the Mourning Cloak. Still, Comptons are not easily upstaged in the field, and those happening upon them may simply be more inclined to marvel at their appearance than to speculate about brood characteristics.

The North American race of this holarctic butterfly is *j-album*, named for a small, silver mark on the ventral hindwing. It was once considered a distinct species, the White J Butterfly. Its common name refers to the town of Compton in Quebec, where it was studied in the 1800s by English naturalist Philip Henry Gosse (Layberry et al., 1998). Note that the butterfly's correct common name is therefore "Compton Tortoiseshell," not "Compton's." Recent taxonomic work in Europe suggests that this species may be more closely allied to the anglewings than to *Nymphalis* (Nylin et al., 2001), and it has been placed in its own monotypic genus (*Roddia*) by some investigators.

There are some reports of Compton Tortoiseshells nectaring at willows or other flowers. But they are far more commonly seen on the ground taking minerals, or on tree trunks drinking sap. This species is exceptionally skittish and hard to approach—except, paradoxically, when it is boldly landing on one's skin or clothing to feed. Males perch during the warm part of the day for females, hilltopping in some cases (especially in spring?).

Identification	Medium-large, our biggest anglewing (if we accept the revised classification noted above; but if the Compton is truly a *Nymphalis*, then it is a distant second to the Mourning Cloak). Wing edges are wavy and irregular.
Dorsal	Handsomely mottled brown, yellow-gold, white, and black—all arranged in a complex, concentric mosaic. Rich, orange-brown basally (both wings), deep, golden-orange toward the margins. Note *a single white dash-mark* at the leading edge of each wing (HW marks often concealed).
Ventral	Well-camouflaged gray-brown, darker toward the base. A small, *"v" or "j"-shaped silver mark* on the central HW is often difficult to see. Surprisingly, with the FW retracted, three vague white marks at the leading FW edge can resemble those of a resting Red Admiral (but the Compton is larger and lacks checkered fringe).

Habitat	Openings and edges of moist deciduous or mixed conifer forests. Usually seen along trails or dirt roads, or at sap runs.
Hostplants	A variety of trees, including aspen and cottonwood (*Populus*), willows (*Salix*), birches (*Betula*); elms also reported. Eggs laid in clusters on hostplant leaves; caterpillars feed communally in a silk-web nest.
Occurrence	Variably common (perhaps nomadic?), from N. Eng. south to w. MD. Occasional coastal movements in fall (e.g., in 1995) may carry individuals to s. NJ, the NC coast, and even FL (once). Also w. US, Eurasia. Overwintering adults emerge in spring; summer brood flies beginning in Jul, aestivates until fall, overwinters in Oct-Nov.
Ecology	Medium Generalist. Northern climate preference, but fairly adaptable.

Quaking Aspen (*Populus tremuloides*)

av wingspan: 2.6″

Mourning Cloak *Nymphalis antiopa*

Dorsal: 5/29/95, Gr. Dismal Swamp NWR, Nasemond Co., VA

Ventral: 7/4/95, Pound Ridge Reservation, Westchester Co., NY

Here is a species for which the familiar epithet "harbinger of spring" is not misapplied, although it is not always the earliest butterfly on the wing. Other adult-overwintering nymphalids can appear as soon, followed closely by some early emerging pierids and lycaenids. But its large wings and a graceful, gliding flight cause it to be noticed even by casual visitors to the outdoors, in still wintry woodlands, parks, and gardens.

Hibernating adults emerge from sheltered nooks on warm days as early as February to feed briefly on running sap. These dramatic flights, through still snowy woodlands, require a good deal of specialized biology. Hibernating adults tolerate winter cold by means of "antifreeze" chemicals (glycerols) in their blood. Before emerging in mid-winter, they use isometric shivering to raise their body temperature (to 15 degrees or more above ambient) (Douglas, 1989, pp. 74–75). At dusk, they return to the entrance of a sheltered crevice and quietly walk in.

This wide-ranging holarctic species shows little geographic variation. In England, where it appears rarely as an immigrant, it is called Camberwell Beauty. Also formerly called Yellow-edged Butterfly.

Males perch from late morning, on hilltops if available. They reportedly move from perch to perch over a territory as large as 300 square meters (Blitzer & Shaw, 1983). When taking flight, males produce a soft "click" with their wings. Adults feed mostly on sap and decaying matter, though summer adults may seek occasionally nectar. If attacked, this and other nymphalines may drop lifelessly to the ground, "playing dead" to fool predators.

Identification	Large, especially southward. Usually simple to identify, especially in early spring when few other butterflies are yet active. In flight, it may briefly be confused with a swallowtail or Compton Tortoiseshell. Wing edges are irregular, but not ragged. Short tail projections may erode with wear.
Dorsal	*Rich, velvety maroon* above with a *bright, yellow marginal band* and a submarginal row of oblong *blue spots*. Marginal band is pale when larvae cold-stressed or in spring (generally so in n. Europe).
Ventral	Uniformly dark, with a crepe-paper like appearance, pale borders.

Habitat Wandering and adaptable, in woodlands, parks, suburbs, riparian corridors, swamp edges, etc.

Hostplants Widely variable. Willows (*Salix*), elms (*Ulmus*), especially near towns, Cottonwoods and aspens (*Populus*), birches (*Betula*), hackberry (*Celtis*), and other broadleaf trees. Eggs laid in clusters of up to 30–50. Early stages live communally, thrashing in synchrony if disturbed to deter predators/parasitoids (even pupae). Older caterpillars wander some distance from the host before pupating.

Occurrence Widespread and familiar but not profuse, especially in SE. Most of No. Amer., including all our area except peninsular FL. Much random-seeming seasonal movement, and some migration. Summer brood members emerge in Jun-Jul, flying as far as 70 km from hatch site before aestivating. Some fresh adults noted in fall. Prediapause adults accumulate body fat to sustain them during hibernation.

Ecology **Generalist.** Varied hostplant and habitat use, with much seasonal vagrancy; adapts well to settlements (if unsprayed).

Cottonwood (*Populus deltoides*)

av wingspan: 3.0"

Milbert's Tortoiseshell   *Nymphalis* [= *Aglais*] *milberti*

Dorsal female: 8/2/94, Mt. Washington, Coos Co., NH

Dorsal female: 8/2/94, Mt. Washington, Coos Co., NH

A familiar resident of the boreal North, this colorful butterfly announces its presence dramatically when dorsal-basking. It is highly cryptic with its wings closed, on the other hand, dropping quickly from sight. A rare fall emigrant, Milbert's historically strayed south as far as western Long Island (Shapiro, 1974). But it was a confirmed northerner during most of the 20th century, with a well-defined southern range limit. Indeed, a sighting in April 2002 in the lower Hudson Valley (presumably an overwintering adult) was the first New York City area record since 1910 (Fiore & Wallstrom, 2001). Others followed in 2003, suggesting a modest, possibly cyclic recurrence.

The species' common name honors a friend of a 19th-century investigator (Godart) who described the species. Populations in our area belong to the nominate race (*milberti*). A less colorful race (*viola*) has been described from Newfoundland. Superficially similar to the Compton—and also called a "tortoiseshell"—the Milbert's is actually a close relative of the Small Tortoiseshell (*Aglais urticae*) of Europe. Some now place both species in the same genus *Aglais* (Nylin et al., 2001). In recent years, the Small Tortoiseshell itself has made increasingly regular fall appearances in the Northeast. It is not clear whether these individuals are international hitchhikers, local escapes, or true colonists with a population center somewhere in North America.

Male Milbert's perch in the afternoon on the ground or low foliage. May hilltop. Nectars actively, unlike some Nymphalini. But it is not averse to puddling or feeding on scat and carrion. Flight is fast and energetic.

Identification	Medium-small. Unmistakable when seen clearly from above (also extremely handsome). FW tip falcate, with concave outer FW edge. The wing margins are more scalloped than ragged.
Dorsal	Dark base color with dramatic, *flamelike bands on both outer wings*, variably yellow basally, then blending to bright reddish-orange. Single black and white marks appear at the FW tips. A thin, clearly outlined dark band runs just inside the margin of both wings, studded on the HW with a row of *chevron-like blue markings*.
Ventral	Both wings are *dark basally*, with iridescent blue-purple overtones when fresh. A pale median band is followed by a darker region, then a fine line blue submarginal line at the wing edges. Since this species flies in the same area as Hoary Commas, some care is needed not to confuse the two when viewed ventrally, although even here the resemblance between them is not exceptionally close.

Habitat	Wet or damp areas near northern woodlands, at higher altitudes in southern parts of the range. Tolerates a variety of habitats, including settlements, urban parks, fields, in Canada.
Hostplants	Nettles (*Urtica dioica & procera*), usually growing in the open (R. Webster, pers. comm.). Eggs laid in clusters of up to 900; young caterpillars feed gregariously in silk nests; older ones in folded leaf nests.
Occurrence	Often common in boreal No. Amer. south of tundra, reaches WV rarely. Apparently two broods. Overwintering adults emerge in spring and mate. Summer brood flies May-June, followed by a fall brood in Aug.
Ecology	Medium specialist. Broad habitat use, but limited in hostplant selection and climate tolerance.

Peter W. Post

Small Tortoiseshell (*Nymphalis* [*Aglais*] *urticae*) vagrant or escape, Manhattan, New York Co., NY

av wingspan: 1.8″

Red Admiral   *Vanessa atalanta*

Dorsal: 7/1/94, Pound Ridge Reservation, Westchester Co, NY

Ventral: 9/12/92, Jamaica Bay NWR, Queens Co, NY

The distinctive color pattern of the Red Admiral is identifiable from "as far as one can see it" (Klots, 1951, p. 107). This fast, lively butterfly is highly familiar on the East Coast, and indeed throughout its broad, holarctic range.

The Red Admiral's life strategy falls somewhere in between those of the Painted and American Ladies (the other two *Vanessa*s in our area). Painted Ladies are freeze intolerant and have no permanent colonies in the East; American Ladies are hardier and can overwinter relatively far north. Red Admirals withdraw largely (but not entirely) from the North each fall, maintaining thin, permanent resident populations to about the Carolinas. In a recently published long-term study, observers in central Florida noted recurring autumn buildups of overwintering adults, beginning in late October (Swanson & Monge-Najera, 2000). A "rear guard" of Red Admirals does seem to linger farther northward, some apparently diapausing as pupae. In mild years, a portion of these lingerers may survive until spring. The *Vanessa* life strategy is nothing if not diversified, and this is plainly a key to its notable success.

While the Red Admiral's southbound migration is incompletely documented, northern movements in the spring and early summer are well-known, both here and in Europe. Massive spring flights on the East Coast have occurred about once a decade recently (e.g., in 1981, 1990, 2001), often followed by precipitous population crashes. In non-peak years, a steady but less dramatic northbound movement is generally noted.

Male Red Admirals perch on bare ground, tree trunks, and hilltops in the afternoon seeking mates and confronting passersby. Both perching and ovipositing can continue well into the evening. Adults feed on sap and decaying material, but are also fond of nectar (more so, we think, than is indicated in some accounts).

Identification	Medium-sized. FW is somewhat falcate. Wing edges are slightly scalloped with loosely checkered fringe. The *winter/dry season form* is smaller and duller than the *summer/wet season form*, which has brighter colors and a variably *interrupted red FW band* (*see* ventral photo, above).
Dorsal	Blackish-brown base color with *red-orange FW bands* and *white FW apical spots* (larger in European race). HW dark with a red-orange submarginal band, studded with blue-centered dots near the body (at the anal angle).
Ventral	FW mirrors dorsal surface, but with a *pinkish-red band* and blue markings. HW is complexly mottled black and brown. *Compare* Compton Tortoiseshell.

Habitat	In migration, anywhere from subtropical hammocks to the summit of Mt. Washington to the corner of Wall and Broad Streets in New York's financial district. Breeding individuals often select moist wood edges or damp meadows, but also fields, gardens, orchards, or nearly any accessible open space.
Hostplants	Mostly nettles, e.g., Stinging (*Urtica doica*), Wood (*Laportea canadensis*). Also Pellitory (*Parietaria pennsylvanica*) and False Nettles (*Boehmeria cylindrica*), especially in FL, where others less available. Single green egg laid on top of host leaf. Caterpillars build progressively larger leaf nests.
Occurrence	Widespread, usually common throughout, but sporadic on Keys. Also all of subarctic No. Amer., south to C. Amer. highlands, and Eurasia. Usually two to three broods, depending on location. Adults, maybe occasionally pupae, overwinter.
Ecology	Generalist. Extremely versatile. An active colonizer, found across the Northern Hemisphere. Though seemingly aposematic, it is evidently palatable.

False Nettles (*Boehmeria cylindrica*)

av wingspan: 2.1″

American Lady   *Vanessa virginiensis*

Dorsal: 5/25/96, Great Dismal Swamp NWR, Nasemond Co, VA

Ventral: 7/10/94, Lakehurst, Ocean Co, NJ

In most respects, our knowledge of this common and exceptionally pretty butterfly is fairly detailed (early stages, hostplants, etc.). But its seasonal behavior still involves an element of guesswork. Field observations suggest that it winters farther north than other *Vanessa*s. Yet it does migrate with some consistency—albeit on a smaller scale than Painted Ladies. Its adaptive strategy may be one of "belt-and-suspenders," reinforcing stalwart resident populations in the North with regular southern emigrants. Large, early season concentrations of American Ladies have recently been reported on the Edwards Plateau of Texas (Tveten & Tveten, 1996). Overflows from such staging points may seed northern colonies. Some southbound fall migration has been noted.

The American Lady was first described as *V. virginiensis* by Drury in 1773. Two years later, Fabricius separately described the species as *V. huntera* (Hunter's Butterfly), and this name was mistakenly used until well into the 20th century. In the past, it was often commonly called the Painted Beauty.

Among the earlier butterflies seen in the spring, American Ladies exhibit a territorial mate-seeking strategy similar to that of Red Admirals. Males often perch on patches of bare ground, where they can display or absorb minerals. Females fly in a low, zigzag pattern looking for hosts on which to oviposit (Allen, 1997). Although sap and decaying materials are sought, this species is an avid nectarer. Curiously, it is singled out by entomologists for its poor antennal hygiene—American Ladies do not scrape the stalks clean with their legs (Robbins, 1997).

Identification	Medium-sized, slightly smaller on average than Painted Lady. Trailing FW edge is concave, giving a *falcate* appearance. The American and Painted Ladies are superficially similar, so a careful look is often needed. It is odd that these species should appear so much alike while the Red Admiral, another close relative with a highly similar lifestyle, is completely different-looking.
Dorsal	Variable shades of orange-brown. The *darkened FW tip* (less extensive than in Painted Lady) bears several *white spots*. Note a *single, isolated white dot on an orange background* at the midpoint of the outer FW; while not always present, it is distinctive when seen. Mid-FW lacks connected band of dark blotches as in Painted Lady.
Ventral	FW mirrors dorsal surface, but field color is pinkish. HW has a lovely, complex cobweb pattern and *two large, blue-centered eyespots* near outer margin (vs. a series of four smaller ones in Painted Lady).

Habitat	Nearly anywhere. Low-growth open areas, fields, vacant lots, dunes, gardens, etc.
Hostplants	Pearly Everlasting (*Anaphalis margaritacea*), Sweet Everlasting and cudweeds (*Gnaphalium*), pussytoes (*Antennaria*), etc. Caterpillar builds first nest from leaf hairs, then larger, silk-bound ones as it grows.
Occurrence	Common throughout our area (but rare in s. FL/Keys). Also CAN to n. So. Amer. Strays rarely as far as Europe. Two to four broods, Mar/Apr to Nov. Adults overwinter.
Ecology	Generalist. Extremely widespread and adaptable, though seldom abundant.

Cudweed (*Gnaphalium* sp.) (top). **Mature caterpillar** on host (below)

Pearly Everlasting (*Anaphalis margaritacea*)

av wingspan: 2.0″

Painted Lady   *Vanessa cardui*

Dorsal: 6/15/95, Van Cortlandt Park, Bronx Co, NY

Ventral: 6/15/95, Van Cortlandt Park, Bronx Co, NY

It is a tribute to nature's untiring inventiveness that the Painted Lady—a species entirely lacking in freeze tolerance—should be the world's most widespread butterfly. Painted Ladies breed on all continents except Antarctica and on many islands in between. The keys to their success are migratory ability and an extremely broad palate. This combination allows them to reach far-flung destinations each summer and survive on the hostplants found there.

Scientists long thought that Painted Ladies hibernated as adults at northern latitudes. As late as 1917, Weed exhorted young naturalists to tear apart wood piles, etc., in search of overwinterers. But each year's North American crop can actually be traced to the Mexican Plateau, from which hordes of individuals stream north in spring. In rainy years, northbound migrants fly in dense hordes that can literally stop traffic. But at other times none at all reach the East. Isolated sightings in such lean years may be of escapes, since pupae are widely available through the mail for science projects and commercial breeders sell adults for release en masse at "special events."

Migrants fly fast and straight, rising over obstacles in their path (from trees to six-storied buildings), rather than detouring around them. Migrants are encountered with some regularity at sea. A less conspicuous return migration has been regularly documented, but none as large as springtime flights. Southbound migrants may fly at higher altitudes, however, thereby avoiding notice (Scott, 1986).

On breeding grounds, males perch to find mates, much like other *Vanessas*, and also sometimes patrol. Both sexes actively seek nectar. Often found in association with American Ladies when not migrating.

Identification	Medium-sized. FW apex rounder in female than in male. The *winter/migratory form* is comparatively small and dull, the *summer/breeding form* larger and brighter with blue-centered dots on the dorsal HW. A white-checkered fringe, conspicuous when fresh, runs along the wing edges.
Dorsal	Bright orange-brown. The *broad, dark FW tip* bears white spots. Note the *nearly continuous line of black blotches* crossing the central FW. Eyespots on the submarginal HW are either blue or black.
Ventral	Complex, multicolored mottling on HW. *Four small, blue-centered eyespots run along the HW margin* (vs. two larger spots in American Lady). FW base color is a mixture of pink, rose, and orange.

Habitat	Nearly any open or disturbed habitat—sea level to timberline, urban lots to open prairies, coastal wetlands to arid thorn scrub (in the West), etc.
Hostplants	More than 100 hosts recorded, from about 10 families. Mainly thistles, mallows, goosefoots, also sunflowers, legumes, etc. Caterpillars construct progressively larger leaf nests as they grow.
Occurrence	Widely variable in abundance, tends to arrive in a spring wave along the central East Coast (Apr-May); but later in FL, which is not along a linear travel route. Usually rare in FL, but had a peak year in 2000. Unclear why warm areas in s. FL and Caribbean are not permanently colonized. Two to four broods locally per year, depending on travel details. Sep-Oct brood triggered into reproductive diapause by declining day length (Scott, 1986).
Ecology	Generalist. Formerly called "the Cosmopolite." A more fitting name is hard to imagine for this embodiment of the generalist lifestyle.

Pasture Thistle (*Cirsium pumilum*)

av wingspan: 2.2"

Common Buckeye   *Junonia coenia*

Dorsal: 9/25/94, Higbee Beach, Cape May Co, NJ

Ventral: 7/10/94, Lakehurst, Ocean Co, NJ

In this genus of tropical butterflies, the Common Buckeye is the lone northern sojourner. A related, but visually dissimilar species, the Mimic (*Hypolimnas misippus*), also strays north, but only very rarely. Lacking strong cold tolerance, Common Buckeyes are true migrants, moving directionally north in the spring and south in the fall, sometimes in large numbers. Their seasonal movements often track coastlines and river corridors.

Now usually split from the Old World genus *Precis*, the buckeyes of the New World (genus *Junonia*) have long challenged taxonomists. Until fairly recently, this and the following two species were considered races of a single entity, simply called The Buckeye (*Junonia* [*Precis*] *lavinia*). Even now, it is not clear whether the dark, southwestern *nigrofussusa* form is (1) a race of Tropical Buckeye, (2) a race of the Common Buckeye, (3) a temperature-induced phenotype of the Common Buckeye, or (4) still another species.

Male Buckeyes await females on low foliage or patches of bare ground in bright sun (McDonald & Nijhout, 2000), where they defend small display territories. Buckeyes are wary and fast-flying, usually keeping within a foot or two off the ground. Caterpillars obtain toxins (iridoid glycosides) from larval hostplants, which deter predatory ants (Camara, 1997; Dyer & Bowers, 1996). But this protection is not passed along to adults (Schappert, 2000). Males sometimes puddle, and both sexes actively nectar.

Identification	Medium-sized, females larger. FW apex is rounded in female, more falcate in male. Seasonal forms have distinct ventral HW colors (*see* below). HW margin slightly scalloped. Nijhout (1991) details the developmental factors governing the formation of the Buckeye's eyespots. Wing colors are brightest in late-season individuals.
Dorsal	Brown (or fuscous) with blackish suffusion. Two red-orange bars across the FW costa (in the cell) are common within the genus. Note *white FW band* nearly (or sometimes fully) encircling the FW eyespot. HW has a bright red-orange submarginal band and two eyespots with multi-toned centers. The outer HW eyespot (often hidden beneath FW) is *distinctly larger* than the one nearer the body, and has a *purplish-red internal crescent*.
Ventral	FW pattern mirrors dorsal surface, but with lighter base color. HW is *yellowish-tan with a reddish postmedian smudge* in summer, or all *rose-colored* in fall (Smith, 1991). Two small, blue-centered eyespots mark the HW.

Habitat	A wide variety of open habitats, mainly open, sunny spots with some bare ground. Also tropical pinelands. Does well in patches of low-growth habitat with connecting "corridors" (Haddad & Baum, 1999).
Hostplants	A variety of plantains (*Plantaginaceae*), figworts, or "scrophs" (*Schrophulariaceae*), vervains (*Verbenaceae*), etc. A single, dark green egg is laid on an upper leaf surface. Caterpillars are solitary.
Occurrence	Resident across s. US; pushes north each spring, eventually reaching subtaiga CAN. Emigrant populations highly variable. Also W.I., south into MEX. Two to four broods locally, depending on arrival dates. Adults overwinter in Deep South.
Ecology	Generalist. Another consummate generalist.

Mature caterpillar   **Purple Gerardia** (*Agalinis* sp.)

av wingspan: 2.0"

Tropical Buckeye   *Junonia genoveva*

Dorsal female: 3/27/02, Miami-Dade Co, FL

Ventral: 3/26/02, Miami-Dade Co, FL

Early strays of Tropical Buckeye in southern Florida went largely unnoticed, despite their colorful appearance. The species bears a close resemblance to the Common Buckeye and confusion in the early literature made proper identification difficult. Even more important was the fact that until recently all three buckeye species in Florida were considered races (or seasonal forms) of a single species, *Precis lavinia* (Turner & Parnell, 1985).

A number of what we now call Tropical Buckeye were found on Key Largo in 1978, and an extant population had taken hold by the early 1980s. Some of these individuals reached the southern fringe of the mainland, forming small, widely scattered colonies (Minno & Emmel, 1993). But while the Tropical Buckeye may be "one of the most widespread and generally common butterflies throughout the West Indies" (Smith et al., 1994, p. 85), its foothold in Florida has proven precarious. It is apparently gone again from the Keys, and is very rare in Miami-Dade County. This may be part of a recurrent colonization cycle that we are just now starting to document.

The Florida/West Indies race of Tropical Buckeye is *zonalis*. A dark, southern/western form (*nigrosuffusa*) is presently listed as a race of the Tropical, but its correct taxonomic treatment is still under debate.

Male Tropical Buckeyes perch from late morning on, using low twigs and foliage, or open, bare ground. They are among the most skittish and unapproachable species we encountered in preparing this book. Females bask late in the day. Their flight is low and direct (more like the Common than the Mangrove). Adults regularly nectar.

Identification	Medium-sized. Closely resembles the Common Buckeye, less so the Mangrove. Male FW tip somewhat falcate.
Dorsal	Dark brown with two red bars ("racing stripes") on the leading FW edge, typical of the genus. A relatively *broad, white band* on the outer FW *half-encircles* the large FW eyespot; the side of the eyespot nearest the body is bordered with brown or (especially in females) a blackened line, but *not by white*, as in the Common. The band itself may have slight brown or orange suffusion, especially in males (in the females we have seen, the FW band is snowy white). Two *HW eyespots are usually close to the same size*, especially in males (*see* next account).
Ventral	Pale beige to rich, chestnut-brown, *darkest beyond the well-defined mid-HW line*; two to three blue-centered eyespots.

Habitat	Open, subtropical fields and grasslands, often disturbed sites. Elsewhere shoreline scrub, roads, Caribbean highlands.
Hostplants	Porterweed (*Stachytarpheta jamaicensis*); probably ruellias, etc. But refuses *Lippia* (eaten in TX by *nigrosuffia*). Solitary caterpillars feed on leaves.
Occurrence	Rare and currently declining, southernmost FL and Keys. Also throughout W.I. and neotropics (including *nigrosuffusa*). Flies all year, three to four broods, including longer winter adult generation.
Ecology	**Medium generalist.** Though rare, has a generalist lifestyle, successful elsewhere. Thin colonies of this species appear vulnerable to extirpation in our area.

Disturbed subtropical habitat. Right: **Blue Porterweed** (*Stachytarpheta jamaicensis*)

av wingspan: 2.0"

Mangrove Buckeye   *Junonia evarete*

Dorsal: 12/29/96, Eco Pond, Everglades NP, Monroe Co, FL   **Ventral: 8/31/99, Everglades NP, Monroe Co, FL**

It is peculiar, in retrospect, that a species as distinctive in appearance and lifestyle as the Mangrove Buckeye took so long to achieve proper taxonomic recognition. It was the last of the Florida *Junonia*s to be singled out as a separate form, despite its specialized association with Black Mangrove habitats. (This hostplant selection is not as much of a departure as it might seem, incidentally: until fairly recently, the Black Mangrove was classified with vervains [*Verbenaceae*], a popular host family among the buckeyes.)

The literature on Mangrove vs. Tropical Buckeyes can be very confusing. Even after their split from the Common Buckeye was accepted (around 1940), experts still disagreed on whether the remaining forms were also distinct. And when species status was finally accorded in the 1980s (Turner & Parnell, 1985), the authors observed that the scientific species names (*evarete* and *genoveva*) had become reversed over time, relative to labels on the original drawings published by Cramer in 1779. Thus, beware that in much of the older literature still in use, written before this correction, we see Mangrove Buckeye referred to as *genoveva* and Tropical as *evarete*.

The flight style of the Mangrove Buckeye is adapted to its high-growth surroundings. In contrast with the low, "scuddering" flight of the two open-habitat species, the Mangrove usually lofts up (to clear, tall foliage), and then glides or flutters down. Males perch during the day but females spend much of their time until late afternoon in the seclusion of mangrove thickets, at which point they emerge to bask. (Some writers consider this distinctive, but we have observed the same late-day appearance and basking behavior in Tropical Buckeyes.)

Identification Slightly larger than the Common and Tropical, with long, fairly full-cut FWs. Antennae are tawny or brown, vs. cream or white in Tropical (Turner & Parnell, 1985), but this is a fine point not required for identification.

Dorsal Dark brown (darker in winter). A relatively *narrow, orange-toned FW band* appears to encircle fully the large, FW eyespot, which has *no white* on the side facing the body. The red-orange HW submarginal band is prominent, and the HW eyespots are very nearly the *same size* (the anterior spot is 1/5 to 1/3 times bigger than the one near the body, vs. 1/3 to 2 *times* bigger in the Tropical [Turner & Parnell, 1985]).

Ventral Variably pale or brown (more rosy after cold weather); markings usually *less well-defined* than on Tropical; *eyespots often very indistinct.*

Habitat Black Mangrove swamps and associated flats and coastal scrub. Strays to gardens and other places for nectar, also sometimes disperses (though range limited by need to maintain breeding contact with hostplant).

Hostplants Black Mangrove (*Avicennia germinans*). Grows in the mid-zone of FL mangrove swamps, between Red and White mangroves; sends up vertical "breathing roots" (pneumatophores). Cold intolerance limits northern range.

Occurrence Can be locally very common along FL coasts. Also W.I., Atlantic coast from s. TX to BRA. Multiple broods, all year. Adults overwinter.

Ecology Specialist. This butterfly has taken the general *Junonia* hostplant strategy into a very specialized niche that allows it to live in a space without butterfly competitors.

Black Mangrove (*Avicennia germinans*). **Inset,** flowers

av wingspan: 2.1″

White Peacock   *Anartia jatrophae*

Dorsal (summer form): 6/24/93, Loxahatchee NWR, Palm Beach Co, FL

Ventral: 3/23/94, Key West, Monroe Co, FL

"White Buckeye" might be a better name for this lively—and lovely—southern butterfly, a familiar inhabitant of wet subtropical lowlands and low-growth open spaces. Its low, gliding flight and restless disposition are indeed reminiscent of a *Junonia*, although it does not lack a distinctive identity of its own.

Taxonomically, two uniform subspecies dominate in the mainland tropics: *luteipicta* (south Texas/Central America) and *jatrophae* (South America). But in the West Indies there is much complexity: "The infraspecific taxonomy of this species is perhaps more complicated and open to argument than that of any other butterfly in [the Caribbean]" (Smith et al., 1994, p. 86). Luckily, our form (*guantanamo*) shows little variation. A very rare Keys stray, the Cuban Peacock (*Anartia chrysopelea*), is small and dark with white markings—entirely different-looking.

Males engage in an interesting (and not fully analyzed) mix of patrolling and perching behaviors. Flight is low and rapid. White Peacocks remain active into the early evening, after most other species have roosted. They stray north with some regularity, but do not colonize beyond the Deep South—for lack of both suitable local hostplants and any appreciable degree of cold tolerance. Their bold, noncryptic coloration, plus certain larval behaviors, suggest White Peacocks could be toxic (aposematic). But their jumpy behavior is not consistent with this view and birds readily accepted them in an early series of palatability studies (Brower, 1985).

Identification	Medium-sized; females larger. The *only completely white nymphalid* in our area, White Peacocks pose no significant identification challenge. The *winter/dry season form* is generally larger and whiter than the *summer form*, which is more color-saturated with a good deal of brownish suffusion on the upper wings. FW tips fairly elongate and rounded. Antennal clubs are yellow-orange. The HW is slightly scalloped, with a hint of two stubby tails.
Dorsal	Mainly white (or dirty white) with *light orange wing margins and FW costal bars*. Variable dark lines mark both wings. Note *three dark eyespots*, one on the FW, two on the HW.
Ventral	Similar to dorsal surface, but with reddish-orange highlights on markings. Eyespots mirror those on dorsal surface; HW eyespots are blue-centered in proper light.

Habitat	Usually open, wet areas, including swales, ditches, swamp, and marsh edges, etc. But also open, low-growth disturbed areas, weedy fields, parks and lawns, shorelines, sometimes rimrock pines.
Hostplants	Fogfruit, or Creeping Charlie (*Phyla [Lippa] nodiflora*), Water Hyssop (*Bacopa monnieri*), ruellias (*Ruellia*), perhaps others. It does *not* use the Brazilian plant, *Jatropha manihot*, after which it was wrongly named. Caterpillars feed in the open, sometimes in concentrations.
Occurrence	Common to abundant in s.-central FL, extends north in summer, very rarely to NJ and MA (once). Also throughout neotropics. All year in South, n. emigrant usually Jul-Nov. Adults undergo reproductive diapause.
Ecology	**Medium generalist**. Lacks cold tolerance. Otherwise a definite generalist.

Water Hyssop (*Bacopa monnieri*)

av wingspan: 2.1″

White Admiral *Limenitis arthemis arthemis*

Dorsal: 6/18/94, Pondicherry Wildlife Refuge, Coos Co, NH

Ventral: 8/3/94, Pondicherry Wildlife Refuge, Coos Co, NH

Among the most stately of our northern butterflies, the White Admiral is now recognized as merely a race of *Limenitis arthemis*—an unlikely, polymorphic species that the North American Butterfly Association (NABA) recently dubbed "Red-spotted Admiral" (2nd ed. checklist, 2001). But it is actually the *two races* of this butterfly—White Admiral (*L. a. arthemis*) and Red-spotted Purple (*L. a. astynax*)—that are routinely familiar to field observers, so we will continue to present them separately here, even though they undeniably belong to a single biological species.

There are 29 admirals in North America (subfamily Limenitidinae), four in the genus *Limenitis*. Almost none of our *Limenitis* bear much resemblance to each other as adults—a testimony to their effectiveness as mimics (*see* following accounts). The appearance of the two eastern forms of Red-spotted Admiral is controlled by several genes, located on different chromosomes. This allows for a number of intermediate genetic combinations, often loosely referred to as "hybrids." The White Admiral's white band is a recessive trait that might appear in any population—even those completely isolated from nearby White Admirals. The forms freely interbreed when they occur together (Platt & Allen, 2001), and the same is true of the more red-toned western race (*rubrofasciata*).

White Admirals have a strong, graceful flight. Males perch for mates, but they reportedly do not use the same perch on successive days (Lederhouse, 1993), and they sometimes patrol. Adults feed on old fruit, sap, and decaying matter, but females in particular may visit flowers with some frequency (more so than Red-spotted Purples).

Identification	Large, especially females, though slightly smaller than Red-spotted Purples. FW long, rounded, HW *slightly scalloped*. Note white dash marks along all wing margins. This species is unlikely to be confused with any other in our area.
Dorsal	*Satiny black* above with a dazzling *white bands* across both wings, and on the HW a row of rectangular, light blue spots. Submarginal blue dash marks occur on both wings.
Ventral	*White band* is mirrored below. Rows of *submarginal red dots* on both wings, also *basal red-orange spots on HW*, two red costal bars on inner FW.
"Hybrids"	*Proserpina* has faint white bands, *albofasciata* better defined ones.

Habitat Transition and boreal forests, especially at edges and in scrubby habitats nearby with nectar, mud, etc. Visits open spaces somewhat more regularly than Red-spotted Purples.

Hostplants Extremely varied; *see* Red-spotted Purple account. Female lays single egg at extreme tip of a fresh hostplant leaf. Caterpillars mimic bird droppings, feed mostly at night.

Occurrence Often common, from ME south to central N. Eng., and in a broad mix zone to w. VA and n. WV. Also AK, southeast across CAN. One to two broods, mid Jul-Aug. Second brood often partial. 3rd instar caterpillars overwinter in a hibernaculum made from a folded leaf. Some 1st brood caterpillars overwinter (Opler & Krizek, 1984).

Ecology *Medium generalist*. Wide use of hostplants; but obligate northerner.

Boreal scrub, Coos Co, NH

av wingspan: 3.0"

Red-spotted Purple   *Limenitis arthemis astyanax*

Dorsal: 7/25/94, Turkey Swamp WMA, Monmouth Co, NJ

Ventral: 6/20/98, Fork Creek WMA, Boone Co, WV

If the adaptive value of the White Admiral's bands is disruptive (to obscure its true wing shape), then the colors of the Red-spotted Purple are almost certainly mimetic, causing predators to mistake it for the toxic Pipevine Swallowtail. As often noted, the Pipevine's range corresponds well to that of the *astynax* race of Red-spotted Admiral. It is a marvel of biology that a single species can maintain two stable forms with such different life strategies and genetic appearances. And it is no surprise that a broad contact zone should occur, with intermediate forms, given the complexity of the genetic coding and the subtle selective factors involved.

Scientists have focused on the bright iridescence of the upper hindwing and the curved, red-orange submarginal dot row on the ventral surface, as important mimetic traits. Other members of the *Limenitis* genus can still hybridize to various degrees with Red-spotted Purples, yet they have developed widely different appearances. The closely related Viceroy, for example, looks nearly identical to a Queen or a Monarch.

In our experience, Red-spotted Purples often perch on the ground, or to about 10 feet up in forest undergrowth. When disturbed, they may fly to treetop perches—and some do this without evident provocation. Red-spotted Purples often visit puddles and wet trails in search of minerals. It has been observed that they raise and lower their wings when drinking, perhaps to mimic the fluttery behavior of Pipevine Swallowtails when nectaring (Opler & Krizek, 1984). Adult behavior is generally similar to White Admiral's.

Identification	Large, especially females. Similar in shape to White Admiral. Dorsal surface can be confused with the female Diana Fritillary (in the limited Appalachian range where the two occur together). All surfaces resemble those of the Pipevine Swallowtail, which Red-spotted Purples are believed to mimic.
Dorsal	Black with *bright blue iridescence*, especially on upper HW (similar to Pipevine). Light blue marks run along the outer edges of both wings. A series of *red-orange marks* near the FW apex is more prominent in the female (Allen, 1997).
Ventral	Sooty black with reddish overtones; two red bars on basal FW. Ventral pattern similar to White Admiral, except white band is missing.

Mixed deciduous forest, Pickens Co, SC

Habitat	Mainly broad-leaved, mixed deciduous forests, often moist. Also coastal plains, valley bottoms. Remains in or near woodlands more than White Admiral (latter is common in open, northern scrub).
Hostplants	Highly variable. For both races, Wild Cherry (*Prunus serotina*), aspens, poplars and cottonwoods (*Populus*), Birches (*Betula*), willows (*Salix*), hawthorn (*Crataegus*), serviceberry (*Amelanchier*), basswood (*Tilia*), Deerberry (*Vaccinium stamineum*), etc.
Occurrence	Fairly common to common, central N. Eng. south to central FL, including a broad mix zone with White Admiral. Also west to central plains; separate race in AZ/NM. Two to three broods, usually Apr-Jun, Jul-early Aug, late Aug-mid Oct. 3rd stage caterpillars overwinter.
Ecology	**Medium generalist**. Wide hostplant use, and broadly adaptable, but has specialized mimicry link with Pipevine Swallowtail.

av wingspan: 3.2″

Viceroy   *Limenitis archippus*

Dorsal (northern): 8/7/94, Clay Pits Pond, Staten Island, Richmond Co, NY

Mature Caterpillar, on willow leaf

Sly magician that it is, the Viceroy engages in many deceptions. Its eggs mimic hostplant leaf gauls, its caterpillars resemble bird droppings, its overwintering chamber (hibernaculum) looks like a dead leaf, and adults closely mimic toxic Danaids (Monarchs and Queens). Birds trained to avoid Monarchs also avoid Viceroys. But more recent studies have found that Florida Viceroys are nearly as unpalatable as Monarchs, and more so than Queens (Ritland & Brower, 1991). This would indicate they are Müllerian rather than Batesian mimics (*see* Introduction, p. 47). Also, the white "saddle patch" on caterpillars is now suspected of doubling as a solar sensor to detect short day lengths, indicating it is time to prepare for winter diapause, i.e., to produce "antifreeze" glycerols in its body and build an overwintering shelter (Platt & Harrison, 1988).

Still, it is hard to give up the idea that the Viceroy is a mimic rather than a model. Each of our two *Limenitis* species mimics an entirely different butterfly, and Viceroys can lose their Monarch-like appearance by simply hybridizing (backcrossing) with other genus members, such as Red-spotted Purples. Also, Viceroys remain at generally low population densities, a good tactic for a mimic. Yet Monarchs and Queens vary greatly in toxicity at different times and places, and it now seems likely that all three may mimic the others in certain circumstances.

Male Viceroys perch around eye level on shrubs, often using the same perch in successive generations (Opler & Krizek, 1984). Flight is a rapid, yet graceful "flap-and-glide," with wings held flat in the glide phase (vs. in a *V*-shaped dihedral, as in Danaids). Adults feed on nectar as well as sap, aphid honeydew, and putrefying matter.

Identification	Medium-large, but smaller than Monarch or Queen. Similarity to supposed "models" is not perfect, as a number of minor differences can be cited. Yet the ruse seems to work well enough to confound predators, who may be trying to sort out "Monarch-like butterflies" from a vast array of potential prey, causing them to use rather crude recognition categories (Ruxton, 1998).
Dorsal	Orange with thick black lines, as in Monarch. Rich russet in North (Monarch-like *archippus* race), a deep brick red or mahogany far South (Queen-like *floridensis* race). Distinguished from Monarch principally by *dark, concentric postmedian line* running through HW. Also note dark wedge with white dots near FW apex.
Ventral	Basically same as dorsal surface.

Habitat	Moist, open, shrubby areas, such as water edges, willow thickets, sloughs, wet meadows, canals, freshwater marshes, sawgrass, etc.
Hostplants	Mainly small willows (*Salix*), also poplars, aspens, and cottonwoods (*Populus*), other woody plants. Eggs laid singly on leaf tips, caterpillars feed at night.
Occurrence	Common throughout, but only a rare stray on the Keys. Most of No. Amer. east of Rockies; riparian areas in Southwest (threatened). Two to three broods, May-Oct in North. All year in FL. 3rd instar larva overwinters, triggered by short days, but some early brood members diapause.
Ecology	Medium generalist. Some habitat restrictions.

Dorsal (dark southern 'floridensis' race): 9/27/97, Gainesville, Alachua Co, FL

av wingspan: 2.8"

Malachite *Siproeta stelenes*

Dorsal: 12/29/98, Tree Tops Park, Broward Co, FL

Ventral: 12/29/98, Tree Tops Park, Broward Co, FL

Before the late 1960s, the Malachite was at best an occasional Cuban stray to south Florida and the Keys. By 1970, however, it had become an established breeder. It gradually extended its range northward, mainly near the coasts, and became locally common by the early 1990s, feeding on fallen fruit in citrus and avocado groves.

Malachites all but disappeared in the wake of Hurricane Andrew in August 1992. It is not entirely clear why the hurricane had such a pronounced impact on many of south Florida's specialty butterflies, even disturbed habitat/secondary-growth specialists such as this, but the effect was decided and widespread. Malachite populations began recovering toward the end of the 1990s, and colonization is again progressing.

Because it looks nothing like any other butterfly in our region, we do not tend to think of the Malachite as a mimic. But in its principal, neotropical range (where it is very common) the issue is not whether it is a mimic, but rather which of several toxic Heliconian species it emulates. The very similar Bamboo Page (*Philaethria dido*) is often cited, but the exact model is not clear (*see* DeVries, 1987). The Malachite bears absolutely no resemblance, in any case, to the Rusty-tipped Page (*S. epaphus*), the only other member of its small genus that sometimes strays into the United States.

Malachite males perch at varying heights, often with their wings partially cocked. They periodically patrol their range with graceful, languorous glides, and sometimes fly well into the canopy of hardwood hammocks for nectar. They feed regularly on fallen fruit and decaying matter, but nectar as well. In the neotropics, Malachites undergo regular altitudinal migrations, and group emigrations have been recorded in Cuba (Smith et al., 1994).

Identification	Large. Note short tails, extended FWs, and toothlike dentations along trailing wing edges.
Dorsal	Brownish-black with *stunning green translucent markings* (color varies from whitish to blue-green to deep emerald; fades with wear); green marks smaller in wet season/summer form (*oscuro*) than in dry season/winter form (*claro*). Note *small reddish mark* at inner base of HW. Our race (*biplagiata*) has two green spots in FW cell (along leading edge).
Ventral	A complex mosaic of reddish-brown, pearl white, black, and pistachio green. A conspicuous white-and-tawny stripe crosses the HW.

Green Shrimp Plant (*Blechum pyramidatum*); **Inset**: flowers

Habitat	Disturbed habitats, often near hardwood hammocks or groves. Will follow fruit and avocado plantings into settled areas, where the hostplant can be a common weed.
Hostplants	Mainly Green Shrimp Plant (*Blechum pyramidatum*), but other Acanthaceae, e.g., ruellias (*Ruellia*), also noted. Caterpillars rest below hostplant leaves, spitting green fluid when disturbed (Scott, 1986).
Occurrence	Can be locally common in far s. FL, active all year, with three or more broods. Strays sparingly to central FL in late summer. Also W.I., s. TX, and south to ARG. Freeze intolerant, no full diapause.
Ecology	**Medium specialist**. Though common throughout the neotropics, the Malachite is confined to subtropical habitats with appropriate hostplants and adult food in our area. Sensitive to loss of required conditions, as after Hurricane Andrew.

av wingspan: 3.4"

Dingy Purplewing   *Eunica monima*

Dorsal: 4/18/01, Deering Estate, Miami-Dade Co, FL

Ventral: 4/18/01, Deering Estate, Miami-Dade Co, FL

Until the 1970s, the Dingy Purplewing was regarded as a stray in southern Florida, a restless tropical emigrant whose northern range limit occasionally brushed our shores. (This was in contrast to the Florida Purplewing, which was a long-established resident.) Dingys were first noted "in numbers" on the upper Keys in the early 1970s, when they began forming colonies rather than fading back to the Caribbean. And in the early 1980s, Dingys took hold on the southern mainland. The vanguard populations on the Keys faltered, but the species kept a foothold in Miami-Dade County (Smith et al., 1994).

Hurricane Andrew (August 1982) seems to have nearly obliterated the Dingy Purplewings of Miami-Dade County, as it did with so many other tropical specialties. Some apparently persisted, however, and by the late 1990s sightings began to recur. Dingys can be seen fairly reliably today at the Deering Estate in Miami and in Homestead-area hammock preserves (especially Castellow).

Dingy Purplewings are butterflies of tropical hammock canopies. They descend with some regularity to lower levels in mid-morning, perching on trunks and leaves. Males freely dorsal bask or display at such times—contrary to some literature reports. Dingys nectar in the afternoon in scrubby areas outside of hammocks, and occasional, premigratory clusters reportedly amass in such locations in the tropics. Observers recently reported Dingys being stalked by Smooth-billed Anis while taking minerals at river banks in Brazil (Burger & Gochfeld, 2001).

Identification	Resembles a small satyr or Hammock Skipper at first glance. Lacks distinctive, falcate FW shape of Florida Purplewing.
Dorsal	Brownish-black, especially dark on FW apex. Males have a *dull, blue or purplish sheen* and a small number of *indistinct white apical spots*. Females may lack iridescence altogether.
Ventral	Dull, gray-brown with irregular dark HW lines and a few *postmedian eyespots*. The eyespot nearest the HW leading edge has a double pupil, the upper one usually being *pale white*. Fresh individuals show a faint, but definite *purplish-blue iridescence*.

Habitat	Diverse habitat across its range. In our area, tropical hardwood hammocks, especially along interior streams (where present). Less tied to dark hammock interiors than FL Purplewings. Noted seeking nectar in open, scrubby areas.
Hostplants	Gumbo Limbo (*Bursera simaruba*) in our area; prickly ash (*Zanthoxylum* sp.) reported in C. Amer.; possibly others. Caterpillars live communally in silk nests, have defoliated trees in Costa Rica (De-Vries, 1987). Tropical *Eunica* caterpillars use a number of techniques to avoid predation by ants (*see* next account).
Occurrence	Local but currently increasing in Miami-Dade Co, FL. Also in W.I. and from MEX patchily to n. So. Amer. Multiple broods, all year. Reproductive diapause only.
Ecology	Specialist. Not clear what specific requirements limit the spread of this species, which uses a very common hammock tree as its host.

Gumbo Limbo (*Bursera simaruba*)

av wingspan: 1.6″

Florida Purplewing   *Eunica tatila*

Dorsal: 7/5/03, Lignum Vitae Key, Monroe Co, Fl

Ventral: 7/5/03, Lignum Vitae Key, Monroe Co, FL

Unlike the Dingy Purplewing, which only recently established permanent colonies in south Florida, the Florida Purplewing is a traditional resident, once reasonably common. But as often happens with species pairs in this region, the tables turned in the early 1990s. After Hurricane Andrew, neither purplewing was recorded for a number of years. In the late 1990s, it was the Dingy Purplewing that first reappeared in a number of traditional sites, while the Florida Purplewing continued missing and unaccounted for, despite diligent searches of its former preferred haunts. Mosquito spraying and rampant destruction of hardwood hammocks throughout the Keys may be responsible for the Florida Purplewing's decline in our area, more so than hurricane disruption. If this is true, it is a poor omen for the future. Recent "finds" on Lignum Vitae Key, a protected island preserve where no spraying is allowed on upper Key Largo, are encouraging, but do not ensure a successful longterm outcome.

There are about 60 species of *Eunica* in the neotropics, many dazzlingly pretty. Most are sedentary, our species being two exceptions. The Antillean race (*tatilista*) occurs in Florida, undoubtedly established here through prior dispersals.

In our region, Florida Purplewings keep to the interior of hardwood hammocks more than Dingys, making short, bounding flights between perches on trunks and overhanging leaves (which they often perch beneath). As with Dingys, Floridas occasionally dorsal bask, but most often perch with closed wings. Occasional nectaring and puddling have been reported, but little feeding behavior of any type is regularly observed.

Identification	Larger than the Dingy Purplewing. The convex trailing edge of the FW accentuates a *falcate tip*. The HW margin is scalloped. Can easily be mistaken for a Hammock Skipper in dense understory, but flight is less frenetic.
Dorsal	Dark, blackish-brown; especially dark on the outer FW, with *six to seven clear, white spots*. A heavy *iridescent suffusion of purple* (males) *or blue* (females) is visible only in good light with the wings held apart—not a frequent combination.
Ventral	Varies from a *plain, lightly marked grayish-brown* to individuals more definitively marked with lines and eyespots. FW has a *dark, subapical spot surrounded by two white marks*. (In the continental *tatila* race, found in TX, faint eyespots on the ventral HW have tiny white centers.) Often some blue-purple iridescence.

Habitat	In our area, confined mainly to interior sections of tropical hardwood hammocks. Elsewhere, in evergreen tropical forests near rivers.
Hostplants	Long unknown. Minno and Emmel (1993) reported Crabwood (*Gymnanthes lucida* [*Ateramnus lucidus*]), a common understory shrub of FL hardwood hammocks. Some *Eunica* larvae are known to deter ants with frass rings around feeding areas and noxious regurgitations when attacked, etc. (Freitas & Oliveira, 1992).
Occurrence	Extremely rare and local, s. FL and the Keys. Also sw US to ARG. Multiple broods, all year, probably tied to wet-dry season cycle.
Ecology	Specialist. Limited to hardwood hammocks; one of FL's scarcest and most endangered habitats.

Crabwood (*Gymnanthes lucidus*), growing in subtropical hardwood hammock interior

av wingspan: 1.85"

Ruddy Daggerwing   *Marpesia petreus*

Dorsal: 5/3/96, Matheson Hammock, Miami-Dade Co, FL

Ventral: 6/24/98, Hugh Taylor Birch Park, Broward Co, FL

There is hardly a more engaging symbol of tropical beauty in Florida than the Ruddy Daggerwing—graceful, exotic, and brilliantly colored. The only resident member of its neotropical genus in our area, it has nevertheless been joined in the Keys, on very rare occasions, by two relatives, the Cuban Daggerwing (*M. eleuchea*) and Many-banded Daggerwing (*M. chiron*).

The world's dozen or so daggerwings all feed on figs (DeVries, 1987). The species used in Florida are semideciduous, dropping many of their leaves at intervals, especially during dry periods. Daggerwings lay eggs singly at the base of reemerging foliage. This provides young caterpillars with a meal of fresh growth, which is more tender and may have desired chemical concentrations.

Ruddy Daggerwing males perch rather high in forest edge trees, some 15 to 30 feet up, when awaiting females. They fly out from these perches in elegant, gliding flights. Males frequently puddle (often reported from the tropics), and may visit rotting fruit and dung, but this species is also quite fond of nectar. Females tend to remain in woodland interiors or canopies more so than the males. Males are more commonly seen.

The Ruddy Daggerwing, like others in its genus, is prone to fairly regular seasonal movements, related to wet and dry seasons. This may sometimes carry them quite far north in Florida.

Shortleaf Fig (*Ficus citrifolia*)

Identification	Fairly large. May be difficult to distinguish in flight from Julia, Florida Leafwing, or even Goatweed Leafwing (rarely, at the Ruddy's northern range limit). This may constitute a form of mimicry, diverting aerial predators to nonorange prey. Ruddy Daggerwings are distinguished by *long, falcate FWs* and *long, dagger-like tails*. The legs and undersurface of the body are *gleaming white* (vs. dark in Cuban Daggerwing).
Dorsal	*Bright orange*, with *three thin lines* running parallel to wing margins. Females duller with heavier wing markings.
Ventral	Brown and leaflike, but with considerable purplish-blue iridescence when fresh (sometimes greenish-bronze). A *dark line* runs parallel to the body along both wings.

Habitat	Trails and edges of subtropical hardwood hammocks, also parks and wooded areas with figs.
Hostplants	Strangler Fig (*Ficus aurea*) and Shortleaf Fig (*F. citrifolia*).
Occurrence	Fairly common (peaks in Apr: 100+ seen on a single tree at Deering Estate 4/18/01); s.-central FL and Keys, also W.I, sw US to BRA. Three-plus broods, all year. Reproductive diapause in dry periods only.
Ecology	**Specialist**. Limited to one subtropical plant family, mainly in specialized local woodland habitats.

Strangler Fig (*Ficus aurea*), widespread host

av wingspan: 2.8″

The Leafwings and Emperors: Subfamily Apaturinae (includes former Subfamily Charaxinae)

Taxonomists have taken varied approaches over the years to classifying leafwings and emperors—from treating them as separate families to lumping them together with a large cluster of "typical" nymphalines, from commas to admirals. In recent years, authorities have placed them in a single nymphalid subfamily, the Apaturinae. This is a bit of an "odd couple" matching, since the two component groups have a number of distinct differences, apparent to any field observer. Yet the new classification also highlights some significant underlying commonalities, which are worth appreciating.

The Apaturinae have an interesting pedigree. Fossils of modern-looking emperors have been found in association with 30-million-year-old fossilized hackberry leaves (Emmel et al., 1992). This suggests an exceptionally long association with this hostplant family. Leafwings, on the other hand, may have radiated more recently: although worldwide in occurrence, no single genus is found in both the Old and New Worlds, implying that present-day genera may have formed after the continents drifted apart.

The two branches of the Apaturinae subfamily are linked by certain behavioral traits. Both feed predominately on sap, decaying materials, and dung. They rarely visit flowers, and when they do often choose flowers high in nitrogen, a substance coveted by gravid females when not otherwise available (Douglas, 1989, p. 69).

Both members of the subfamily also share common territorial and flight styles. Male leafwings and emperors perch on low foliage and tree trunks, usually 5 to 15 feet off the ground. Both are commonly described as pugnacious, dashing out to challenge intruders, possibly seeking to assert their position, but certainly on the lookout for a mate. They land on human interlopers frequently, pausing to sample sweat or other minerals available on exposed skin. Some would deny such behaviors are truly territorial. But males do consistently occupy a specific display perch, often throughout their entire adult lives. This suggests that at least some proprietary sense of place is involved. (As the dynamics of these behaviors become better understood, the correct descriptive label to employ may become more obvious.)

Male Hackberry Emperor perched on tree trunk, alert for intruders

Typical emperor caterpillar (Tawny Emperor, late-instar)

The flight style of the Apaturinae is at once swift and graceful, a combination of rapid, bursting surges of directional flight, mixed in with long, swooping glides—during which the butterfly floats gracefully from one aerial plateau to another in a stair-step fashion.

The leafwings include about 400 species, mostly tropical, of which six have occurred in North America north of Mexico (two in our region). The exact number of species involved is subject to vigorous debate, since populations are highly variable and many occur in isolation from each other (*allopatric*). Leafwings are medium to large butterflies, usually with small tails and hook-shaped forewing tips. In a rare adaptation, the shape of the forewing hook varies phenotypically with the season (*see* species accounts). The group's name is derived from the uncanny resemblance that the ventral wing surface often bears to a dry leaf, including shape and many specific surface details. The pattern evolution needed to produce this similarity is extremely sophisticated (Nijhout, 1991, p. 60).

The emperors are a small group, with only about 50 species, six in North America, two in our area. But here again the exact species count is definitely a matter of opinion. Both East Coast emperors are in the genus *Asterocampa*. Farther west, two members of the genus *Doxocopa* are found, which are visually much more similar to admirals in the genus *Adelpha* (the "sisters") than to our emperors. Mimicry may be involved, based either on distastefulness or on the "sisters" being hard for predators to catch, and thus not being worthwhile to pursue. Emperors in our area are smaller on average than leafwings. Many have eyespots, including on the triangular-shaped forewing.

The caterpillars of leafwings and emperors are similar. Both have a bulging appearance, tapered at either end. Emperor caterpillars have forked tails, longitudinal stripes, and small "horns" on the head. Leafwing caterpillars may live suspended from a leaf rib, having eaten away the rest of the leaf surface. Emperor caterpillars are communal in feeding behavior and overwintering (which takes place during the larval stage, in rolled hostplant leaves).

Some observers describe leafwings as sedentary. But while local colonies persist for many years at a time, significant seasonal movements of Tropical Leafwings are also noted in the fall through southern Texas (across the Rio Grande). Eastern emperors are highly local in the vicinity of hackberry trees, but they too undoubtedly engage in at least occasional dispersal flights, especially along the coast or in river corridors, although there is little systematic evidence to document this behavior.

Florida Leafwing *Anaea floridalis*

Dorsal female (summer): specimen photo

Ventral (winter): 12/22/94, Long Pine Key, Everglades NP, Miami-Dade Co, FL

The Florida Leafwing is a southern Florida endemic. It is the only butterfly species limited entirely to the tropical zone of the East Coast states—if it is a species at all, that is. Some taxonomists consider the Florida Leafwing to be an isolated (or allopatric) race of the Tropical Leafwing (*A. aidea*). Absent further significant evidence, however, it appears justified to maintain the current separation.

More urgently important than its precise taxonomic status, however, is the fact that the Florida Leafwing is rapidly declining. It was reasonably common in southern Florida as recently as the early 1990s. Since then, much of its prior habitat has been lost to development (including pineland sites cleared by Hurricane Andrew in August 1992). On the lower Keys, it appears highly sensitive to strong, new chemical sprays used to combat mosquitoes (Salvato, 2001). And, members of the growing resident community on Big Pine Key have opposed controlled burns needed to keep the leafwing's pineland dwellings from becoming overgrown, with associated crowding-out of its hostplants. Strong measures are needed if this striking yet vulnerable butterfly is to persist.

Male Florida Leafwings are highly territorial, perching on trunks and in pineland scrub to await passing females. They fly out in dramatic fashion, then all but disappear when returning to their perch and closing their wings. Yet while bold and confrontational on initial contact, they are skittish once disturbed, diving into the underbrush for cover. Most feeding takes place on dung, rotting fruit, and other decaying material. Occasional nectaring has been recorded. Adults may be distasteful to at least some predators, though further investigation of this possibility is needed.

Identification	Fairly large. The only leafwing in its range. (Even in earlier times, potential for overlap with Goatweed appears slight.) Wings are large and full-cut. FW slightly hooked, especially in winter/dry season form (*morrisonii*). Wing edges are slightly scalloped, vs. smooth in Goatweed. Note short, pointed tails. Resembles Ruddy Daggerwing in flight.
Dorsal	Male is *bright, red-orange*, with dark dash marks near the FW midcosta and *variable dark wing borders*. Females are deeper red, with more *extensive dark wing marks* and *darker borders*.
Ventral	Mottled gray with dark, irregular midwing striations. FW is red-orange basally (visible when FW is raised). Ventral surface blends well with the flaking bark of Slash Pines (*Pinus elliottii*).

Habitat	Subtropical pine rocklands.
Hostplants	Exclusive host is Narrow-leaved Croton (*Croton linearis*), of the spurge family (*Euphorbiaceae*). Caterpillars rest on isolated leaf veins, after chewing away surrounding leaf material. Older caterpillars live in rolled leaf shelters.
Occurrence	Threatened and declining. Miami south on mainland, also lower Keys (especially Big Pine). Three or more extended broods, most of the year. Reproductive diapause occurs in winter.
Ecology	Specialist. Almost never found far from its niche habitat.

Pine Rocklands, Everglades NP, Miami-Dade Co, FL.
Inset: Narrow-leaved Croton (*Croton linearis*)

av wingspan: 2.9"

Goatweed Leafwing   *Anaea andria*

Dorsal male (winter form): 3/20/00, San Felasco Hammock, Alachua Co, FL

Ventral male (winter form): 3/20/00, San Felasco Hammock, Alachua Co, FL

The Goatweed Leafwing is the only member of its mainly tropical clan able to endure cold, northern winters. It is locally common in parts of the Midwest, but rare and local in our area. The plant from which it takes its name, Goatweed (*Croton monathogynus*), is not found regularly on the East Coast. Nor are some of its other normal hosts. Here, Goatweeds are limited to a relatively narrow habitat niche, mostly dry, pine-oak scrub where local species of croton are present. Observers should note that most published accounts concerning the Goatweed refer to midwestern populations, which differ in several respects from those in our area (in terms of mobility, number of broods, hostplants, etc.).

As with other *Anaea*, male Goatweed Leafwings perch at medium height on twigs or trunks, darting out aggressively to challenge intruders—including wind-blown leaves as well as potential mates or rivals. Their distinctive flight style mixes strong directional surges with extended, wavering glides. Adults feed on sap, dung, and rotting material (the short proboscis is ill-suited for nectaring). Individuals can be baited-in using a mix of sugary and fermented liquids. Chemical esters in sap are a known attractant (Scott, 1986, p. 68). Seasonal forms, described below, are determined by the photoperiod (day length) during the larval stage (Riley, 1975). May play dead if captured by a predator.

Identification	Medium-sized; smaller than Florida Leafwing. The only leafwing to be found in n. FL or the Appalachians. In the winter form, colors are deeper and the FW tip more strongly hooked. Note the *short tail* (may be lost with wear, especially in overwintering individuals in reproductive diapause). *Wing edges smooth.*
Dorsal	Male is *brilliant red-orange*, with vague dark borders and a *single cell-end spot*; often with purplish iridescent overtones. Female duller red, or brownish-red, with jagged, concentric midwing marks and a yellowish submarginal zone.
Ventral	Pale gray-brown with faint markings (male) or moderate markings (female). Note vague but distinctive *white spot* near leading HW edge.

Habitat	Dry, open woods and scrub, especially Turkey Oak (*Quercus laevis*) scrub with pines. Habitats more varied in the Midwest—usually dry or disturbed open areas, but at times riparian corridors.
Hostplants	Various crotons, a half dozen of which occur in dry, sandy habitats in n. FL. Silver Croton (*Croton argyranthemus*) often used here.
Occurrence	Local, declining with habitat loss; w. Appalachians to n. FL. Also s.-central Midwest to CO and WY, and along southern rim to AZ; south into MEX. Seasonal movement in central populations. Usually two broods, but likely more in FL. Adults hibernate in sheltered crannies (in North), or undergo reproductive diapause (farther south).
Ecology	**Medium specialist**. More of a generalist elsewhere, but a niche player on the East Coast. Sometimes seen in suburban areas where development has encroached on favored habitat.

Turkey Oak scrub habitat. **Inset**: *Croton* sp.

av wingspan: 2.5″

Hackberry Emperor   *Asterocampa celtis*

Dorsal male: 9/13/92, Tottenville, Staten Island, Richmond Co, NY   **Ventral**: 9/3/99, Palm Pt. Park, Gainesville, Alachua Co, FL

Many butterflies that use a single hostplant end up as lifestyle specialists, confined to a specific and usually narrow habitat zone. Not so *Celtis*-feeders in the genus *Asterocampa*. Their principal host, American Hackberry, can tolerate a wide range of habitat extremes, and it takes a hardy and adaptable butterfly to keep pace—not a narrow specialist. The required tolerance zone is extended even farther, moreover, when the additional habitat preferences of two other common hosts, Dwarf Hackberry and Sugarberry, are included. Only a few circumstances, such as permanent standing water and severe seasonal cold, seem to limit the range of this host-and-hostplant combo.

Perhaps because of the varied and disparate ecological niches they occupy, Hackberry Emperors show considerable regional variation. Local forms have often been considered full species. At present, there are two recognized races in our area, the northern *celtis* race and the Florida/Deep South *reinthali* race (formerly included in the more western *alicia*). Some investigators believe that *reinthali* should be regarded as a separate species (Gatrelle, 1999a), and this may be correct.

Exceptionally faithful to a chosen perch site (usually 5 to 15 feet up on a tree trunk or on foliage), male Hackberry Emperors stand guard from late morning until late afternoon—and sometimes into the evening. Their rapid, flap-and-glide flight is less elegant than that of the leafwings, since, being smaller, the glide phase of their flight is less extended and graceful. Adults feed on sap, rotting fruit, mud, and various putrefying materials. Nectaring is rare, except perhaps among females seeking nitrogen (Neck, 1983). Females are seen far less often than boldly displaying males, and are best drawn from treetop roosts by fermented mixtures of bait.

Identification	Medium-sized; *reinthali* race distinctly larger. Males smaller than females, with more pointed/triangular wings.
Dorsal	Complex patterning. *FW apex dark* with a spangling of *pure white dots*. Note *single dark eyespot* at midpoint of outer FW margin (two in the western *antonia* race). HW brown or gray-brown, rimmed with black chevrons and eyespots. Base color paler on *reinthali*, apex spots larger.
Ventral	A variable and complex mix of grays, browns, whites, black; also subtle iridescent purple and blue or green overtones, especially when fresh. Note *single, light eyespot near FW apex* and *single, dark spot at midwing* (same as dorsal). An *irregular row of eyespots on outer HW* is highly variable; often iridescent green in the center.

Habitat	Depends on occurrence of hostplant. Mostly low-lying woods or river corridors-often near water, but rarely within permanent wetlands. Also dry upland sites (hackberry is planted for erosion control where drought tolerance is important); limestone outcrops; sandy maritime forests. Also in suburbs and urban parks with hostplant.
Hostplants	American Hackberry (*Celtis occidentalis*), Dwarf Hackberry (*C. tenuifolia*), Sugarberry (*C. laevigata*). Eggs laid in small clusters (1–20).
Occurrence	Locally common. Central N. Eng. south to central FL. Also Gr. Plains, and south to n. MEX. One to three broods, Jun-Sep in North, Mar-Nov in FL. Mid-stage caterpillars overwinter (3rd instar); sometimes from 1st brood.
Ecology	**Medium generalist**. Adaptable in pursuit of its ubiquitous hostplant.

American Hackberry (*Celtis occidentalis*)

av wingspan: 2.0"

Tawny Emperor *Asterocampa clyton*

Dorsal: 8/25/96, Jamaica Bay NWR, Queens Co, NY

Ventral ('Empress flora' form): 9/22/96, Newnans Lake, Alachua Co, FL

Commentators have pondered why—and how—two similar butterflies such as the Hackberry and Tawny Emperors came to be separate species, only to live in close proximity afterwards. The two commonly fly together in close association in hackberry groves throughout their largely overlapping ranges.

Mimicry does not seem to be at work, however, since hackberry trees, while not especially nutritious as forage, are at the same time not toxic. Numerous birds and animals feed on their fruits and leaves. And the emperors' skittish behavior and cryptic ventral patterning are also consistent with palatability.

It has also been suggested that Tawny Emperors display earlier in the day than Hackberries, offering a degree of ecological separation. But while this may be true statistically, it is by no means an absolute. Our own field notes include many cases of early-displaying Hackberries, late-displaying Tawnys, and overlapping displays by both species within visible sight of each other.

A better explanation, in our view, centers on larval specialization. While Hackberry Emperor caterpillars concentrate on fresh growth when foraging—feeding singly or in small, nongregarious groups (Scott, 1986)—communal-feeding Tawny Emperors team up in early instars to skeletonize mature leaves, moving on as a group after finishing along a communally secreted trail of silk. While such concentrations are attractive to parasitoids, dense egg clusters provide some degree of shelter, at least for those eggs located in the interior. Postdiapause Tawnys feed individually on new growth, but they may avoid competition with Hackberry caterpillars by emerging earlier in the spring (Allen, 1997). Adult lifestyle is generally similar to the Hackberry's.

Identification	Slightly larger than Hackberry. The northern race (*clyton*) is smaller and darker than the southern race (*flora*), formerly considered a separate species ("Empress Flora"). Complex and variable, our two emperors can be difficult to identify. It helps that the species took diametrically opposite paths at a number of pattern-related design junctures (*see* below); and this may in fact be an adaptive feature, allowing the two to coexist in close quarters.
Dorsal	*Lacks FW eyespots*. Generally more reddish-brown than the Hackberry. FW apex is *tawny or darkish*, not black as in Hackberry, and FW spots are (a) mainly *yellow* rather than pure white and (b) located farther from the apex. Note *two complete black bars* on inner FW costa (first bar is separated as two dots in the Hackberry.) HW may be a light, tawny color, but in East it is usually dark or smoky (form *geneumbrosa*).
Ventral	Extremely variable, sometimes with bold, contrasting patterns, but often little-marked, *especially the basal HW*. Small *blue-centered eyespots* usually inconspicuous.

Habitat	Similar to Hackberry Emperor, but with greater affinity for maritime forests (at least in the Northeast).
Hostplants	Same as Hackberry Emperor. Female lays several hundred eggs in clusters beneath leaves or on bark.
Occurrence	Abundance fluctuates. May be locally common, but usually less so than Hackberry. Mid-Northeast south to s. FL. Has lately followed wild/planted hackberries up rivers into VT/NH. One to three broods, Jun-Aug in North, Mar-Nov in FL. 3rd instar caterpillars overwinter in curled leaf.
Ecology	**Medium generalist**. Similar to Hackberry Emperor.

Sugarberry (*Celtis laevigata*)

av wingspan: 2.15″

The Satyrs: Subfamily Satyrinae

Typical caterpillar (Georgia Satyr)

The satyrs are a familiar subfamily to East Coast field observers. Each region has at least one common representative—except for the Keys, where only strays are found.

Yet familiarity offers no guarantee against disorder. This nonconformist subfamily proves the exception to many nymphalid family "rules," casting considerable disarray into the overall family profile. For instance, virtually all our nymphalids eat dicotyledons (flowering plants with two seed leaves)—except for the browns and satyrs, which feed exclusively on monocotyledons (grasses, sedges, cane, etc.). Likewise, nearly all of the other subfamilies have at least some members whose adaptive strategies depend on toxic body chemicals, or on mimicking species that carry such toxins. Distastefulness is in fact a defining trait of several important species groups, notably the heliconians and milkweed butterflies. But the Satyrinae appear to be edible at all life stages (Brower, 1985, p. 128), and have in fact made an art out of surviving despite this vulnerability. None is known to be a mimic. Another disparity is that female browns and satyrs tend to lay few eggs together, while species in a number of other subfamilies lay eggs in large clusters and feed communally as caterpillars. Quite a few other nymphalids engage in dramatic seasonal movements, moreover, but browns and satyrs are consistently sedentary. Finally, most nymphalids are brightly colored, but temperate-zone satyrs are generally drab—or subtly muted at best.

There are around 2,000 satyr species worldwide, including nearly 50 on the North American continent north of Mexico, and 15 in our area. Taxonomically, some commentators feel the Satyrinae should continue to be placed in their own, separate family (Tveten & Tveten, 1996). They appear to have diverged from the nymphalid main line at an early point in evolutionary history, and have developed well-differentiated lifestyles, with many distinctive traits. The group merits careful study for its adaptive innovations, although at least one commentator appears unimpressed: "Satyrs specialized as grass-eating, dung-sucking cryptics" (Scott, 1986, p. 100).

The lifestyles of East Coast Satyrinae fall into several broad categories. *Woodland satyrs* (pearly-eyes, wood-satyrs, etc.) are among the best adapted of any East Coast butterflies to dark, woodland interiors. Unlike the many tree-feeding butterflies that spend most of their time in or near the canopy, woodland satyrs tend to stay close to the forest floor, where they feed on sap, dung, and putrefying materials. One group, the wood-satyrs, reach their greatest diversity in the neotropics, where interior woodland temperatures are comparatively accommodating. But there is also the Northern Pearly-Eye, a hardy species that flies at twilight in dim, northern forests. *Wetland satyrs*, as the name implies, favor more open, sedgy habitats, where they patrol persistently (e.g., Georgia and Mitchell's Satyrs, Eyed Brown, and to a lesser extent Appalachian Brown). *Dry field/prairie* species have taken advantage of human disturbance to establish large regional populations. Recently, the Common Ringlet has extended its East Coast range rapidly southward. The *arctic/alpine satyrs* are epitomized in our area by two relict species of arctics, each stranded on high ridge lines in the New England mountains, and by a single boreal bog-dweller. These species have a particularly specialized and intriguing biology. While all of our diapausing satyrs overwinter as caterpillars, the arctics do so twice, requiring two seasons of slow growth to reach adulthood.

Satyrs as a group are medium-sized butterflies with broad, full-cut wings and slender bodies (although the bodies of northern species have a thick covering of hairlike scales). Among East Coast satyrs, the base colors are mainly cryptic browns and grays, but more colorful tribes exist elsewhere, especially in the neotropics and Asia. And even our comparatively plain satyrs tend to have subtle, often handsome markings, including lines, ripples, "gem markings," and mottling. And their eyespots (ocelli) are finely etched and colorful. Each eyespot forms in a single wing cell (i.e., the flat area between wing veins).

Anatomically, satyrs have short proboscises and a characteristic enlargement (swelling) near the base of forewing veins. The proboscis length is consistent with feeding on mud, dung, and decaying matter rather than nectar. The forewing vein swelling is thought to function as an auditory sensor, allowing browns to detect the approach of predators by sound (Scott, 1986, p. 233).

This is useful for butterflies that lack aposematic defenses (i.e., that are palatable). The adult's flight style, often described as weak or erratic, is actually well-crafted to be unpredictable, with sudden changes of direction and frequent sudden drops, hopefully not worth a predator's while to pursue. Most browns rest and feed with their wings closed, but bask with wings open. Males of many species have pheromone-laden "scent patches" on their wings, used during courtship.

As noted, satyr larvae eat monocotyledons, usually feeding at night. Grass and sedges are eaten in our region, but elsewhere club mosses and palms are used as well (DeVries, 1987, p. 257). Most pupae hang upside down, though arctics pupate in protected silk nests amid debris, often beneath rocks. Broods are often long or overlapping, or occasionally short and discrete (as in Mitchell's Satyr), with many confusing sequences.

Northern Pearly-eye   *Enodia anthedon*

Dorsal: 7/22/95, Pondicherry Wildlife Refuge, Coos Co, NH   **Ventral**: 7/23/95, Pondicherry Wildlife Refuge, Coos Co, NH

When twilight approaches in cool, northern forests, it is remarkable to see small groups of Northern Pearly-eyes appear from the shadows and become active, like skittish, crepuscular moths. They have even been recorded at mercury vapor lights after dark (Iftner et al., 1992). Contrary to some accounts, however, the species is not entirely crepuscular; activity occurs throughout the day, and not just in cold or cloudy weather. But there is undeniably a dusk flight, which in many cases follows a period of afternoon inactivity.

Our three species of pearly-eye (genus *Enodia*) epitomize the lifestyle of "woodland satyrs." Among them, the Northern Pearly-eye is perhaps the ecological archetype. Feeding on a variety of woodland grasses as a caterpillar, this wide-ranging but often local northerner keeps to woodland interiors throughout its adult life. Individuals may be found puddling on dirt roads or forest trails, and will bask (usually with spread wings) in patches of sun to build warmth. But unlike many "woodland" species that are confined to edge habitats, or to relatively well-lit canopies, the Northern Pearly-eye is at home in the deep shade of forest understories. It seldom frequents adjoining fields, since it rarely if ever feeds on nectar. Rather, adults forage on sap (poplars, willows, and birches are cited, but any broadleaf tree wound is likely to draw them), also dung, carrion, fungi, and mud. Males perch, often high on tree trunks, and appear territorial. They emit a distinct scent (Weed, 1917) and have a strong, erratic flight.

Identification	Medium-sized. Wings large, slightly scalloped, and generally roundish. But male FW pointed (consistent with perching style of mate-seeking). The literature includes many identification marks for this species, but only a few have proven to be universally reliable. Most important is the antennal club, which is *black with an orange tip*, distinguishing it from the Southern (but *see* Creole Pearly-eye.) *Satyrodes* browns superficially similar.
Dorsal	Brown with a series of dark, submarginal spots on both wings (reflecting eyespot pattern below). Note roughly rectangular white patch on leading FW edge (not seen in *Satyrodes* species or in male Creole). Also *single small white dot* near FW apex (absent in Creole). Male lacks Creole's dark sex-scaling between FW veins.
Ventral	Brown with opalescent, often vaguely lilac overtones. White patches on outer wing areas generally *less extensive* than in Southern Pearly-eye, though individual variations occur in both species. *Four or five eyespots on FW (four in male, four or possibly five in female), each with a small, white "pearly eye" in the center*, also six eyespots on HW. The *FW eyespots are usually aligned in a straight row* in this species, vs. slightly outward curving in the Southern. *The dark line just inside the FW eyespots comes to a single cusplike point* (*compare* Creole).
Habitat	Damp or mesic northern-style forests, often in hilly or mountainous terrain, often (but not always) near marshes or watercourses. As noted above, shows a rare affinity for shady, dense forest understory, often perched on trunks. Also seen basking at woodland edges, puddling on trails, or ovipositing in grassy patches.
Hostplants	Unlike other pearly-eyes, uses a variety of woodland grasses—but not cane. The list of verified hosts is long, including White Grass (*Leersia virginica*), Japanese Grass (*Microstegium vimineum*) (NC introduced), Bearded Shorthusk Grass (*Brachyelytrum erectum*), plumegrass (*Erianthus*), River Oats (*Chasmanthium* [*Unifola*] *latifolia*), Bottlebrush Grass (*Hystrix patula*), Alta Fescue Grass (*Festuca arundinacea*), False Melic Grass (*Schizachne purpurascens*), panic grass (*Panicum*).
Occurrence	Locally common; not rare, as earlier thought (Weed, 1917). ME to n. VA and in mountains and upper Piedmont to n. GA; absent on s.-central coastal plain. Also Nova Scotia to central CAN, south to n. LA. One brood in North (Jun-Aug/early Sep), two to three broods in South, May-September. 3rd or 4th stage caterpillars overwinter.
Ecology	**Medium generalist.** Widely occurring, uses common habitats and hostplants, but lifestyle is distinctly focused.

av wingspan: 1.9"

Southern Pearly-eye   *Enodia portlandia*

Dorsal: 5/26/96, Great Dismal Swamp NWR, Nasemond Co, VA

Ventral: 5/27/96, Great Dismal Swamp NWR, Nasemond Co, VA

The dim-lit understory of southern bottomland swamps, lined with cane thickets, is the archetypal home of this specialized woodland satyr. It actually occurs in several types of low-lying, shaded southern woodlands. Although it is among our most common cane-feeding butterflies, its restricted habitat affinities usually necessitate a targeted visit to its native haunts, as it rarely strays outside them.

As late as the early 1970s, the Northern and Southern Pearly-eyes were considered to be well-differentiated, but clinal races of a single species, the Pearly Eye (*Lethe portlandia*) (Klots, 1951). Stable morphological and genitalic differences were identified among populations, however, and it was learned that discrete populations coexisted in certain regions without interbreeding (Gatrelle, 1971; Heitzman & dos Passos, 1974). The overlap zone between Northerns and Southerns is actually quite narrow in our region, however. Far greater overlap exists here between each of these species and the Creole Pearly-eye, as discussed in the next account.

As with the Northern Pearly-eye, Southern males perch to await females, either on tree trunks or amid thick, understory foliage. Crepuscular activity (including courtship) is also reported for Southerns, but they too are commonly active by day. Southerns feed on sap, dung, carrion, rotting fruit, etc., not visiting flowers. Heavy tachinid fly infestations have been noted in Texas (e.g., 38 of 40 pupal casings attacked, Tveten & Tveten, 1996).

Identification	Medium-sized. Wings are large, slightly scalloped. Most similar to Northern Pearly-eye and to female Creole. Confusion also possible with Appalachian Brown, but that species has rounder wings and lighter base color. Best distinguished from Northern and female Creole by *all-orange antennal clubs* (vs. black or orange-tipped). Male Creole also differs in shape and scaling on dorsal FW surface.
Dorsal	Similar to Northern. Warm, cocoa brown with a series of yellow-rimmed submarginal spots on both wings, set in a light brown field.
Ventral	Inner half of wings *gray-brown with a violet-opalescent sheen*; in some populations *golden*. Outer wings usually have *extensive white markings*, often *directly enveloping* the HW eyespots (more so than in Northern). Usually *four FW eyespots* (vs. five in Creole), white or bluish-centered, and *curved slightly outward*. The brown line inside the FW eyespots has a *single peak (cusp)*, as in Northern (*compare* Creole).

Habitat	Moist or wet bottomland woods, including maritime and southern bay forests, hardwood swamps, canebrakes. Habitats need not be swampy, but must support cane. Rarely leaves shaded areas.
Hostplants	Giant Cane (*Arundinaria gigantea*); Maiden (or Switch) Cane (*A. tecta*) also reported in GA and FL. Thorough field study of hostplant use still needed.
Occurrence	Fairly common in appropriate habitat, se VA to n.-central FL, mainly on coastal plain and lower Piedmont. Also west to s. IL, AK, e. TX. Three or more broods, Mar-Nov. Late instar caterpillars overwinter.
Ecology	Medium specialist. Far more specialized in hostplant and habitat use than Northern; favors shady woodland interiors.

Giant Cane (*Arundinaria gigantea*)

av wingspan: 2.0″

Creole Pearly-eye *Enodia creola*

Dorsal male: specimen photo

Ventral male (warming on rock in cool air): 5/27/96, Great Dismal Swamp NWR, Nasemond Co, VA

Creole Pearly-eyes fly alongside Southerns in many coastal plain habitats, as well as in Appalachian sites beyond the Southern's range. It is not clear what turn of evolutionary events caused two such similar species to live in close proximity over most of their territorial ranges. Sometimes, in fact, all three pearly-eye species can be found together, e.g., in parts of Arkansas. A similar conjunction of closely related species occurs in cane-feeding roadside-skippers of the Southeast (genus *Amblyscirtes*); both cases may well have common historic antecedents.

The generic classification of the pearly-eyes has flip-flopped over time. Originally placed in *Enodia* (their current genus), they were for a time lumped into *Lethe* (*see* Klots, 1951), a large, Old World genus with representatives ranging from Southeast Asia to Japan (even apart from pearly-eyes, *Lethe* is now being subdivided by some authorities). The closely related eyed browns (genus *Satyrodes*) were also placed in *Lethe*. Today, biochemical analysis supports the separation of *Enodia* not only from *Lethe* but also from the eyed-browns (*Satyrodes*).

The lifestyle of Creole Pearly-eyes resembles that of its congeners, particularly the Southern. Males perch to await females, and adults eat sap, dung, mud, carrion, and other putrefying matter instead of nectar. The species is fast-flying, often active on cloudy days or at dusk (though again not exclusively so). Loss of bottomland habitats in the Southeast needs to be limited if we are to avoid losing populations of this uncommon butterfly.

Identification	Largest of US pearly-eyes. Wing edges are well-scalloped. *Male's FW is distinctly pointed, with a convex trailing edge* that gives it an elongated appearance. *Antennal clubs are black or black with an orange tip.* This allows Creoles to be differentiated from Southern Pearly-eyes, but not from Northerns, which can be an issue in upper Piedmont and mountain populations. Females are less visually distinctive in shape, and can be quite difficult to separate confidently. This may be a source of real difficulty in the field, since East Coast Creoles have more range overlap with the other two pearly-eyes than either of them has with the other.
Dorsal	Similar to the other pearly-eyes, but the *male has dark sex-scent patches between the FW veins*, visible during dorsal basking. The distinctive shape of the FW median line is clearly visible from above in the female, but the male does not mirror the ventral lineation (just the eyespots).
Ventral	Similar to Southern Pearly-eye, though usually a *paler gray-brown. FW has five eyespots (rather than four), all aligned in a row*—but note that the small, fifth trailing eyespot is often concealed by the HW (as in ventral photo). The dark line just inside the FW eyespots has the shape of *knuckles on a clenched fist*, vs. a single cusp, as in the other two species. This is usually the single most reliable wing marking. The eyespots in this species often appear *individually ringed in white* (especially on the HW); in other species, the eyespots are normally either set in a white field or ringed as a group.

Habitat	Similar to Southern, with which it often flies. Moist or wet bottomland woodlands, hardwood swamps, canebrakes. Frequently in dense growth at the periphery of swamps. But also dense, moist Piedmont and Appalachian cove forests. Males reportedly stray to more open woodlands than females (Opler & Krizek, 1984).
Hostplants	Maiden (or Switch) Cane (*Arundinaria tecta*) mainly reported; exact status of Giant Cane (*A. gigantea*) use in the field needs specific study.
Occurrence	Almost always less common than Southern, se VA to s. GA, but not in FL (absence here peculiar). Also west to KY, AR, e. TX. Two to three broods, mid-Apr to mid-Sep. Caterpillars overwinter.
Ecology	**Medium specialist.** Similar to Southern, though comparatively sparse populations raise unanswered questions.

av wingspan: 2.2″

Eyed Brown   *Satyrodes eurydice*

Dorsal: 7/2/95, Amity Marsh, Orange Co, NY

Ventral: 7/5/96, Amity Marsh, Orange Co, NY

Also called Northern Eyed Brown, this medium-sized butterfly is a sedentary inhabitant of open sedge marshes in the northern states. It differs greatly in lifestyle from the pearly-eyes—which it superficially resembles—and also noticeably from its close relative, the Appalachian Brown (*see* following account).

Until 1970, our two *Satyrodes* browns were lumped together as a single species, under the common name Eyed Brown (at various times *Satyrodes canthus* and *Lethe eurydice*). The ultimate separation was based on differences in wing markings, male genitalic structure, and larval characteristics (Carde et al., 1970). As now defined, the Eyed Browns in our area belong to the nominate *eurydice* race. There is also a dark, prairie form, the "Smoky" Eyed Brown (*fumosa*), itself formerly considered a separate species. With the decline of undisturbed prairie habitat, the "Smoky" Eyed Brown has become a form of special concern in several states.

Surprisingly well-camouflaged, even amid wetland greenery, the Eyed Brown engages in floppy, seemingly feeble flight patterns in or just above the level of marsh foliage. Males patrol throughout the day for females, but also perch at regular intervals. Adults feed on sap and decaying material, like other satyrs (we have seen them at sap flows in New Hampshire, jousting for positions with Northern Pearly-eyes), but they are also more inclined than many Satyrines to visit flowers occasionally.

Identification	Medium-sized. Wings rounded, consistent with a patrolling lifestyle. Often separable from potential look-alikes by range and habitat. Few serious identification problems are likely to arise anyway, except relative to Appalachian Browns.
Dorsal	*Pale brown, with nearly a dozen black, submarginal eyespots.* Outer 1/3 of wings paler than basal section (demarcation is especially clear on FW).
Ventral	Pale, whitish brown base color (especially females). *Often looks worn or faded, even when fresh.* Numerous eyespots with white pupils (four on FW, six to seven on HW). *FW eyespots are roughly the same size, touching as if linked in a chain;* a white line encircles all at once (vs. individually). Postmedian HW line is usually jagged. Dark inner HW line (postbasal) has a *characteristic, inward-pointing "tooth"* near the leading wing margin (*compare* Appalachian).

Habitat	Open sedge meadows, edges of freshwater cattail marshes, slow streams, also sometimes open bogs, rarely drier meadows. In OH, often flies with Dukes' Skipper (Iftner et al., 1992). Seldom leaves primary habitat, even when populous.
Hostplants	Many sedges (*Carex lacustris, stricta, lupulina, rostrata, bromoides, trichocarpa; Scirpus rubicosus*). Some caterpillars accept grass in captivity. Eggs scattered near host, larvae crawl to plant.
Occurrence	Colonial, uncommon to locally common. N. Eng. south to mid-PA and NJ. Also west into prairie states (*fumosa*). One long brood, Apr-Sep (mainly Jun-Aug). 3rd/4th instar caterpillars overwinter.
Ecology	Medium specialist. Closely tied to open, northern sedge habitats (prairie swales in West).

Northern sedge marsh, Coos Co, NH

av wingspan: 1.8″

Appalachian Brown *Satyrodes appalachia*

Dorsal: 6/24/95, Lakehurst, Ocean Co, NJ

Ventral: 6/22/96, Pound Ridge Reservation, Westchester Co, NY

Also called Appalachian Eyed Brown, this interesting species straddles the worlds of forest and wetland satyrs, flying in the blend zone (or ecotone) between woodlands and marsh. It is less northerly than its close relative, the Eyed Brown, yet in many areas the two occur in close proximity. Some colonies exist within sight of each other without sustained interbreeding, though occasional hybrids may present vexing identification challenges.

As already noted, the Eyed and Appalachian Browns were split taxonomically only in the early 1970s. For this reason it is necessary to reconsider much of the original life history and distribution data, which can be equivocal as to species. Two races of Appalachian Brown occur in our area, the northerly *leeuwi* (Gatrelle & Arbogast, 1974) and the southerly *appalachia* (Chermock, 1947).

Appalachian Browns look and behave much like Eyed Browns, but with some important differences. First, they do not share the Eyed Brown's "site tenacity" with respect to open sedge habitats. Rather, they are quite prepared to fly through woodlands, and in fact seem to disperse by this means (several "mark-and-recapture" studies have had no recaptures at all; *see* Layberry et al., 1998).

Males intermittently patrol and perch throughout the day, resting in dappled, sunny edges or on trunks. Flight is bounding and erratic. Adults feed mostly on sap and decaying matter, but sometimes also on nectar.

Identification	Medium-sized. Wings rounded. The Eyed Brown is generally paler, the Northern Pearly-eye darker and more richly marked (also, the pearly-eyes' FWs are more pointed).
Dorsal	Brown or olive-brown above, with light submarginal patches. Fresh individuals have a bluish or purplish cast. Dark FW spots usually smaller than in Eyed Brown.
Ventral	Medium brown. Four eyespots on FW, six on HW. *FW eyespots vary in size (one or both of the outer two spots are bigger); they are usually separated slightly, and thus appear to be surrounded by individual white circles.* Eyespots are generally more brightly colored than in Eyed Brown, and the postmedial line less zigzagged. *Dark postbasal line is straight (with no inward-pointed "tooth").* Some populations confusingly marked.

Habitat	Wooded swamps abutting (or interspersed amid) sedge marshes. Adults mostly fly at the edges, although they stray with some regularity both into the woods (where pearly-eyes might be expected) and into open marshes (with Eyed Browns).
Hostplants	Various sedges (*Carex lacustris, squarrosa,* and *stricta* cited), also Giant Sedge (*Rhynchospora inundata*) in FL. Females oviposit near host. Caterpillars feed at night, hiding near hostplant base by day.
Occurrence	Fairly common in appropriate habitat. Central N. Eng. south to SC and n. GA. Disjunct population in n. FL. Also west to Gr. Lakes region and in border states area (se MO, s. KY, n. MS, and AL). One long brood in North, Jun-Sep; two broods in South (Jun-Jul, Aug-Sep/ Oct). 3rd/4th instar caterpillars turn yellow and overwinter.
Ecology	Medium generalist. Less habitat-bound than Eyed Brown, apparently more inclined to regular dispersal.

Sedges (*Carex* sp.)

av wingspan: 1.9"

Georgia Satyr   *Neonympha areolata*

Ventral ('*septentrionalis*'): 6/24/95, Klot's Bog, Lakehurst, Ocean Co, NJ

Ventral ('*areolata*'): 3/21/94, Corkscrew Swamp Sanctuary, Collier Co, FL, with light scaling on inner (anal) margin of HW, at left, typical of sw Florida population

The tendency of Satyrines to evolve complex clusters of closely related genotypes is well-illustrated by grass satyrs in the genus *Neonympha*. The Georgia Satyr, our principal *Neonympha*, was drawn by Abbot in the late 18th century and formally described in 1797. Two races were eventually defined, the southern *areolata* and the northern *septentrionalis* (described from New Jersey). It now appears increasingly likely, however, that two or possibly more sibling species exist (Gochfeld & Burger, 1997; Gatrelle, 1999c). Gatrelle has proposed combining the northern *septentrionalis* with Abbot's original form (type specimens lost) as *Neonympha helicta*, with defined subspecies, and recognizing *N. areolatus* as a separate species. This proposed arrangement may be correct, but we will refer simply to two non-taxonomic "types," one corresponding to *areolata* (SE Type), the other to the proposed *septentrionalis/ helicta* (NJ/upland Type).

The SE Type is said to be confined to open, sedgy marshes, grasslands, and savannas. The NJ/upland Type occurs (1) in open, sandy pinewoods in the Piedmont (proposed race *N. helicta helicta*), and (2) in bogs and wet pine savannas in New Jersey (proposed race *N. h. septentrionalis*—also a species?). Males patrol all day for females; the flight of the NJ/upland Type is reportedly higher and straighter than that of the SE Type, but we have not observed this specifically in New Jersey. Adults rest on grass rather than trees, and consume the "usual" satyr foods (although little feeding of any type is normally observed). Occasionally visits flowers. Largely sedentary.

Identification	At the small end of medium-sized, smaller on average than eyed browns and pearly-eyes. In SE Type, the leading FW edge is usually straighter (less rounded) than in NJ/SW Type.
Dorsal	Uniformly plain brown. But this surface is almost never held open to view.
Ventral	Various shades of brown. NJ/upland Type: FW usually has *two small, faint eyespots*. HW eyespots are *comparatively round*, with *no light flecks* in the center. SE Type: *FW eyespots very faint or absent*. HW eyespots *elongate, with yellow flecks* in one to two of the larger spots. The reddish-orange oval surrounding the HW eyespots is usually continuous. Note light scales along inner (anal) margin of HW in SW Florida segregate (illustrated).

Habitat	Strongly disjunct habitat preferences, but for any given population habitat selection appears quite focused. SE Type is reportedly a wetland or wet grassland/savanna butterfly (Gatrelle, 1999c). It is common in freshwater sawgrass marshes of Everglades NP. Needs high-quality savanna in NC, making it vulnerable to disturbance and development (LeGrand & Howard, 2002). NJ/upland Type inhabits grassy openings in sandy pinelands (sandhills, etc.), but not hardwood forests (*compare* Carolina and Gemmed Satyrs); also bogs and wet pine savannas in NJ. *Note*: earlier accounts placed the *areolata* race (SE Type) in pine flats (e.g., Scott, 1986).
Hostplants	Poorly documented, despite our long acquaintance with the species. Most investigators suggest that sedges are used, probably exclusively, e.g., *Carex stricta, cyperus esculentus*. Caterpillars have accepted several grass species in captivity, but we are aware of no reports of grass being used in the wild. Eggs laid singly on hostplant.
Occurrence	SE Type: local and colonial, but at times common; se VA south to FL (stray only on Keys). NJ/upland Type: disjunct populations in s.-central NJ, NC sandhills south and west to GA/nw FL. A declining population similar to NJ/upland Type occurs south of Miami. The species complex extends west along Gulf Coast to LA. One brood in NJ (Jun-Jul); two broods in VA (Apr-Sep); three-plus in FL, most or all of the year. 4th instar caterpillars hibernate.
Ecology	**Medium specialist.** Uses several regionally common habitats and hosts, but individual populations are quite focused, and intact habitat is usually required.

av wingspan: 1.6"

Mitchell's Satyr   *Neonympha mitchellii*

Ventral ('*francisci*'): 8/8/98, Ft. Bragg, Cumberland Co, NC

Ventral (mated pair, '*francisci*'): 8/8/98, Ft. Bragg, Cumberland Co, NC

The Mitchell's Satyr is among our rarest and most restricted butterflies. Its two races are distinct in lifestyle and geography, yet both face similar survival challenges, and both are federally endangered.

The northern, nominate race of Mitchell's (*N. m. mitchellii*) was discovered in the 1880s and has been well-described in the literature (except for hostplant use, which remains vague). The main *mitchellii* population in our area was extirpated in the mid-1980s by a combination of habitat destruction and intense collecting (Gochfeld & Burger, 1997; Wilsmann & Schweitzer, 1991). *Mitchellii* is now known from a dozen or so sites in southern Michigan and northern Indiana.

The southern race, called St. Francis' Satyr (*N. m. francisci*), was discovered only in the early 1980s. Emergency Endangered Species status was conferred in 1992 after its numbers were greatly reduced by collecting (Federal Register, 50 CFR Part 17 [1/26/95]). At present, one metapopulation is known at Ft. Bragg, NC.

The Mitchell's flight style is unhurried and bobbing, yet it can be difficult to track in its mucky habitat. Males patrol throughout the day for females, pausing frequently to rest on foliage. The mating activity illustrated above took place in mid-afternoon (about 3:00). Mitchell's disperse individually along watercourses and pocosin edges.

Identification	Fairly small to medium-sized, with rounded wings. The races are similar in size. The *mitchellii* race is generally darker than *francisci*, with heavier ventral lines.
Dorsal	Mahogany brown, mostly plain, but eyespots show through vaguely from below on dorsal HW. As with Georgia Satyr, dorsal surface is rarely exposed.
Ventral	Brown with reddish-orange lines. *Four FW eyespots*, usually distinct, the inner two largest; also five to six *round* HW eyespots (*less oblong* than in Georgia Satyr, even NJ/upland Type). The area between two marginal orange lines is often *gray* when fresh.

Habitat	In North, calcareous tamarack-poison sumac fens (alkaline), having an open, central area dominated by long grass and sedges. In NC, suitable habitat was formerly associated with beaver dams. But with beavers extirpated, the species now relies on concussion impacts at the Ft. Bragg bombing range to create open sedgelands. NC habitats acidic, subject to succession.
Hostplants	Sedges; exact species not clear.
Occurrence	Endangered. *Mitchellii* found only in s. MI/n. IN, formerly OH; populations in nw NJ extirpated. *Francisci* only in s.-central NC. New reports from AL and sw VA are of uncertain type. One brood, mid-Jul in NJ (*mitchellii*); two closely timed flights, late May-Jun and late Jul-Aug (*francisci*). Caterpillars overwinter, attached to a stem or leaf near the ground.
Ecology	Specialist. Highly specialized, a declining and embattled species.

Sandhill bog (formed by human disturbance, not a native bog habitat), Cumberland Co, NC

av wingspan: 1.55″

Gemmed Satyr *Cyllopsis gemma*

Ventral (spring variant 'inductura'): 4/23/94, Fork Creek Public Recreation Area, Boone Co, WV

Ventral (summer variant): 6/7/98, Weymouth Woods Sandhill Nature Preserve, Moore Co, NC

This smallish, slow-flying satyr is anomalous in several respects. First, all the other members of its genus inhabit mountain woodlands—either in the southwestern United States (two species) or in Mexico or Central America. This is the only low-elevation member of the clan, also the only one found in the East. Second, it is rare in having seasonal color forms in all of its three advanced life stages (caterpillar, pupa, and adult). And finally, it is rarely common in our area, despite much seemingly suitable habitat. Gemmed Satyrs are encountered most often as pale apparitions—glittery only after landing—that drift quietly through woodland glades en route to indeterminate destinations.

The nominate race of Gemmed Satyr (*gemma*) is found in our area. There is also a redder *freemani* race, in south Texas and Mexico.

Male Gemmed Satyrs patrol throughout the day in search of receptive females. They rest frequently, often on bare ground, dead leaves, or other debris, all of which provide excellent concealment. The flight style is low to the ground, and like that of other woodland satyrs, bouncy and erratic. They move into heavy foliage or thickets when disturbed. Adults seem to feed infrequently, on tree sap, dung, rotting fruit, and other decaying matter. We are not aware of any reports of nectaring, and have not observed it ourselves.

Identification	Medium-small. Wings rounded. Not difficult to identify when perched. Often identifiable in flight based on size, flight style, and pale coloration. Note *white "hairs"* on the body.
Dorsal	Seldom shows dorsal surface. Plain brown, with HW "gems" showing through as dark, smudgy spots.
Ventral	Seasonally variable. Spring form (*inductura*) is darker brown with less distinct cross-wing lines (Scott, 1986). Lacks typical eyespots, instead has a series of *glassy, black iridescent spots whose reflectance gives the impression of silver dots* (the "gems"), set in a frosted marginal HW field.

Habitat
A confusing and variable mixture of habitat types. We have seen the Gemmed most often in openings or edges of rich, moist (mesic) deciduous woods, often in somewhat hilly terrain or ravines. But also occurs in open, grassy sandhill pinewoods, xeric ridgetops with oak (OH), wet bottomland woods, etc.

Hostplants
Bermuda Grass (*Cynodon dactylon*) often listed, but this is absent at many sites. Also uses River Oats (*Chasmanthium* [*Uniola*] *latifolia*) and likely others. *Early-season caterpillars (and pupae) green, later ones brown*, probably for camouflage; feeds at night.

Occurrence
Widespread but rare and local in our area. VA and WV (rarely MD) south to central FL, especially in Piedmont. Also west to se KS, e. TX, south into MEX. Probably three broods, Apr-May, Jun-Jul, Aug-Sep in VA; most of year in FL. 4th stage caterpillars overwinter.

Ecology
Medium generalist. Although usually scarce, this species exploits a diverse variety of habitats and has an extensive range.

Grassy sandhill woods, Moore Co, NC

av wingspan: 1.5"

Carolina Satyr   *Hermeuptychia sosybius*

Dorsal: 9/30/95, Newnans Lake, Alachua Co, FL

Ventral: 6/16/96, Hilton Head Island, Beaufort Co, SC

No satyr is more familiar in the Southeast than the Carolina. In nearly any moist, partly shaded deciduous woodlands in its range, this small, jumpy butterfly will leap into flight from a trail edge as one walks by. It often flits ahead just a short distance before alighting, only to be disturbed again as the hike progresses. If disturbed once too often, however—or too threateningly—it will dart into a nearby thicket for cover.

Few East Coast satyrs are free from taxonomic complexities, and the Carolina is no exception. Once named *Euptychia hermes*, it has now been assigned to a new genus and split from its more southerly (but almost identical) counterpart, the Hermes Satyr (*Hermeuptychia hermes*). Throughout the neotropics, at least one member of the *Hermeuptychia* species complex is likely to be present. The taxonomy of this difficult group is likely to continue shifting, but at least there is only one species here to contend with.

The Carolina Satyr's flight is "low and slow" (unless disturbed), seldom more than a foot off the ground. Activity peaks occur at the beginning and end of the day (Opler & Krizek, 1984). Courtship had been noted in the morning—though patrolling species tend to be fairly nonspecific in this regard. Ovipositing is reported mostly in the afternoon. Adult life expectancy in the field is only 6.25 days in mid-summer (Kilduff, 1972). Adults rarely visit flowers, but feed instead on "the usual" satyr diet of sap and decaying materials.

Identification   Our smallest satyr (the "Little" Wood-Satyr averages nearly 25% bigger). Among the reliable characteristics differentiating the Carolina and Little Wood-Satyrs is eye color: the Carolina Satyr's eye is dark with *vertical, blue-gray stripes*, visible in proper light when alive, whereas the Little Wood-Satyr's is pure black. Fortunately, it is seldom necessary to rely on this subtle field mark, as several others are available to resolve uncertainties.

Dorsal   Flat brown with *no apparent markings*, except for a faint shadow of ventral eyespots on the upper HW (sometimes). Little Wood-Satyr has pronounced dorsal eyespots on both wings. Often dorsal basks in cool weather.

Ventral   *Fairly dark brown*, subtly frosted (Little Wood-Satyr usually more tannish). Numerous submarginal eyespots occur on both wings. *FW eyespots are of approximately the same size*, although the second is often better defined. (By comparison, the Little Wood-Satyr has two dominant FW eyespots, with very small ones in between.) *Six HW eyespots are of variable size (second and fifth are largest)*. Also note brown *cell-end marks*, in between the parallel lines running across each wing.

Habitat   A variety of woodland habitats, usually moist pine woods or oak forests (occasionally dry), also in low-lying wooded river bottoms. Formerly shaded bogs in s. NJ (Glassberg, 1993a). Mainly concentrated in grassy areas with dappled light, or near edges, but at home in quite dark woodland interiors. Often found with Gemmed Satyr, but usually greatly outnumbers that species. Sometimes in suburbs, if suitable conditions are present.

Hostplants   Various grasses, including Centipede Grass (*Eremochloa ophiuroides*) in NC, Carpet Grass (*Axonopus compressus*), and, in captivity at least, St. Augustine Grass (*Stenotaphrum secundatum*) and Kentucky Bluegrass (*Poa pratensis*).

Occurrence   Common and widespread, from MD and s. NJ (formerly, in 19th century), south to FL; stray only on Keys. Mainly on coastal plain, Piedmont, and lower mountains in NC. Also west to se KS, central TX; overlaps with *H. hermes* in s. TX. Two to three broods in North (May-Sep), multiple in South. Reproductive diapause in n. FL (Nov-Jan), probably overwinters as a caterpillar farther north.

Ecology   **Medium generalist.** Wide habitat and hostplant use, but seldom lingers in open spaces; largely confined to woods.

av wingspan: 1.3″

Little Wood-Satyr   *Megisto cymela cymela*

Dorsal: 5/31/98, Voluntown, New London Co, CT

Ventral: 5/25/96, Great Dismal Swamp, Nasemond Co, VA

This energetic species is aptly named, insofar as it is a common satyr that dwells in woodlands. But it is "little" only in relation to other satyrs in its northern range, where it is most common. It is practically middle-sized when compared to East Coast Satyrines as a whole (especially if the large *viola* race is included).

Not to be outdone by others in its subfamily, the Little Wood-Satyr is beset by taxonomic issues. Observers note that in much of its range the Little Wood-Satyr emerges in two distinct, yet closely timed flights early each summer (Flight I and Flight II). In the New York City area, Flight I flies mainly from mid-May until mid-June, whereas Flight II flies from late June to July—sometimes in separate locations (Cech, 1993). The short interval between flights does not leave time for eggs from the first brood to complete a full cycle and emerge as the adults of Flight II. This odd phenology has led several commentators to suggest that two sibling species may be involved. But other explanations are possible, such as the stable variations within a single, phenotypically diverse species. Puzzles such as these keep satyrs from being dull, despite their lack of bright colors.

Little Wood-Satyrs are noted for their bouncy, energized flight. They move fairly slowly unless disturbed (but at such times may fly quite high, an uncommon behavior for this group.) Males patrol throughout the day for females, mainly in shaded glades and forest edges. Ovipositing is reported from late morning to late afternoon. Females land on a grass blade and walk down to its base, depositing a single egg (Allen, 1997). Adult foods include sap and putrefying matter, also aphid secretions and occasionally nectar.

Identification	Small to medium; females larger. Most identification issues arise in connection with the Carolina Satyr, a taxonomically distinct species of similar appearance. Indeed, if either species were toxic, the other might immediately be labeled as a mimic. Face and "eyebrows" are white. Also, when alive the Little Wood-Satyr's eyes (the actual eyes, not the eyespots) are *pure black*, whereas the Carolina Satyr's are dark with *vertical, blue-gray stripes*.
Dorsal	Dull grayish-brown with pale fringe. Note *two clearly defined submarginal eyespots on each FW* (yellow-rimmed with blue-white pupils). Carolina Satyr lacks these eyespots. A variable number of eyespots also appear on the HW (males have one or two HW eyespots, females generally two; also, a very small eyespot often occurs at the anal angle near the body). Dorsal surface is often exposed when basking for warmth, especially in the morning or in cool weather.
Ventral	A variable light brown or tan. Usually lighter below than Carolina Satyr. *Two prominent submarginal eyespots on the FW*, vs. a row of nearly equal-sized ones in Carolina Satyr. HW likewise has *two eyespots, proportionately very large*, and often a number of much smaller ones (the latter are sometimes reduced to disorganized color patches or silvering). Two dark lines cross each wing. The Little Wood-Satyr generally lacks a distinct cell-end bar on the HW.

Habitat	Fairly adaptable in habitat use, especially in comparison with the Carolina Satyr. Mainly edges and clearings in open deciduous woods, but ventures with some frequency into adjoining open habitats (shrubby, overgrown fields, etc.). Also in pinelands, salt bays, and brackish stream sides (occasional). Prairie groves in West.
Hostplants	Grasses, including Orchard Grass (*Dactylis glomerata*), Kentucky Bluegrass (*Poa pratensis*), Centipede Grass (*Eremochloa ophiuroides*), St. Augustine Grass (*Stenotaphrum secundatum*). Caterpillars feed at night.
Occurrence	Common to abundant, all of area except n. N. Eng., peninsular FL (but *see viola*). Also west to Dakotas, e. TX. One double-flighted brood in North (May/Jun-Jul), two-plus reported in TX (Mar-Oct). 4th instar caterpillars overwinter.
Ecology	Generalist. Adaptable to a variety of wooded and open (even disturbed) spaces; a successful species.

av wingspan (*cymela* only): 1.6"

'Viola's' Little Wood-Satyr   *Megisto cymela viola*

Dorsal: 3/20/00, San Felasco Hammock, Alachua Co, FL

Ventral (mated pair): 3/19/00, Bulow Creek SP, Volusa Co, FL

This large, handsome wood-satyr is yet another in a long string of taxonomic oddities that characterize the East Coast satyrs. Long considered a distinctive race of Little Wood-Satyr, *viola* was split off as a separate species in the 1980s, the 'Viola's' Wood-Satyr (*Megisto viola*) (Opler & Krizek, 1984; NABA 1st ed. checklist, 1995). But now, barely 20 years later, it has already been relumped, and is regarded once again as a race of Little Wood-Satyr (NABA 2nd ed. checklist, 2002).

A key factor in the original split was reports that Little Wood-Satyrs (*cymela* race) were multibrooded in the Florida panhandle, whereas the *viola* is an obligate, single-brooded butterfly, appearing in spring (Oliver, 1982). Later evidence contested these facts, stating that *all* wood-satyrs in Florida are single-brooded, and showing that several distinctive visual characteristics of the *viola* vary independently along broad, geographic lines—rather than uniformly, along a single, clearly defined boundary line (Catling & Calhoun, 1997).

Species status aside, there is some curious biology afoot here. In northern populations of Little Wood-Satyr, multiple generations per year will be attempted if conditions are amenable (e.g., in the laboratory), whereas Florida *Megisto*s stop at one brood annually regardless of conditions (Oliver, 1982). It is extremely unusual for a species to have fewer annual flights in the South than in the North.

Identification	Strikingly large for a wood-satyr. As noted above, traits described here may vary geographically. For this account, we describe a "pure type" *viola*, typical in central, peninsular Florida.
Dorsal	Dark brown. *Two prominent FW eyespots, boldly outlined in bright yellow.* HW eyespots variable in number (as in *cymela*). A faint, dark line is often visible across mid-HW.
Ventral	Rich, almost violet gray-brown. *Two prominent eyespots* on both wings, circled in *bright, golden-yellow*, usually with patches of *silver scaling* in between. Outer HW line often *bowed outward* (toward eyespots), outer FW line often *bowed inward*, but this can also occur sometimes in *cymela*.

Habitat	Rich southern deciduous woods with grassy patches.
Hostplants	Needs further study as taxonomy is clarified. Orchard Grass (*Dactylis glomerata*), Centipede Grass (*Eremochloa ophiuroides*), St. Augustine Grass (*Stenotaphrum secundatum*) reported.
Occurrence	Range and status of *viola* uncertain. Common in proper habitat in n. FL. Individual *Megisto* wood-satyrs as far away as NJ can look quite *viola*-like. One flight in FL (mid-Mar to early May). Apparently does not diapause, as do the northern *cymela*.
Ecology	Generalist. Quite common in its range. With all the visual ambiguity in satyr forms on the East Coast, it might be worthwhile to examine differences in mating scents (pheromones) in both sexes. In visually similar species, these can be an important factor in mate selection and reproductive isolation (Scott, 1986).

Southern deciduous forest in spring, Alachua Co, FL

av wingspan: 1.75"

Common Ringlet   *Coenonympha tullia*

Ventral: 6/13/93, Mt. Greylock, Berkshire Co, MA

Short-grass field habitat, Washington Co, ME

The growing familiarity of "Inornate" Common Ringlets (*C. t. inornata*) in our area can obscure their origins as a boreal-holarctic species. A native of northern and montane habitats, the Ringlet has extended its range dramatically southward in recent decades along a broad front extending from New England to central Canada. They were first sighted in our region in Maine, in 1968. Progressing steadily south, the species reached the lower Hudson River Valley in 1990, New Jersey in 1994. The pace of the Ringlets' incursion appears to have slowed somewhat since the mid-1990s, but at least one observer expects them eventually to reach Georgia (Glassberg, 1999).

In the United Kingdom, the nominate race of Common Ringlet—Large Heaths (*C. t. tullia*)—lives in wet moors and raised bog habitats, making it one of Britain's few "truly wetland species" (Thomas & Lewington, 1991). Large Heaths face severe Meadow Pipit predation, and their variable number of eyespots, from north to south in British populations, has been linked to local defensive strategies (in the warmer south, where the butterflies are active, multiple hindwing eyespots draw the pipits' attention, deflecting attacks from the body and head; in the colder north, where ringlets are much less active, eyespots disappear in favor of simple, cryptic coloration; i.e., camouflage is preferred over misdirection). Some populations in northern Europe have a two-year development cycle.

Isolated populations of Common Ringlet in New Brunswick and the St. Lawrence River Valley were once considered a separate entity, the Nipisquit Ringlet (*C. nipisquit*). While the marsh-dwelling ringlets of New Brunswick are likely a good species, the St. Lawrence River Valley populations have been progressively demoted, first to subspecies status under Common Ringlet (*C. tullia heinemani*), then to no defined status at all (Scott, 1986).

In warm weather, male Common Ringlets patrol energetically just above grass-top level. But on cold days, common in the North, they remain mostly perched (Heinrich, 1986). Lacking toxic chemistry, and being highly exposed in open, low-growth habitats, Ringlets rely on local variations in appearance to enhance their camouflage, as discussed in part above. In a departure from the Satyrine norm, adult Common Ringlets feed mainly on nectar.

Identification	Among our smallest satyrs. Variable in appearance both within and across populations. Another taxonomically "baffling" satyr species complex (Opler & Krizek, 1984). Discussion here is limited to eastern populations of "Inornate" Common Ringlet. Note *thick, hairlike scales on the body*, consistent with boreal lineage. Dark St. Lawrence River Valley population, discussed above, may have more FW and HW eyespots than described. Unlikely to be confused with any other butterfly in our area.
Dorsal	Muted orange brown (males) or brighter ochre (females), usually unmarked. Seldom displayed at rest, the orange tones of the dorsal wings are visible in flight, making them useful as a field mark.
Ventral	Mousy gray-brown to gray-green. *Orange inner disk on FW* visible when wing raised. Most individuals (50%-60%) have a *small, dark eyespot near the FW apex*. HW darker basally than toward margin (consistent with thermoregulatory function). *Irregular, white median line on HW*; veins on HW slightly whitish. A partial row of very small marginal eyespots sometimes occurs on HW.

Habitat	Varied open, grassy habitats. Fields, northern lawns, highway verges, tundra in CAN. Sometimes wetland edges.
Hostplants	Kentucky Bluegrass (*Poa pratensis*) and needlegrass (*Stipa* sp.) reported, likely others. Sedges in UK.
Occurrence	Increasingly common (*see* above). N. Eng. south to n. PA, LI. Also in the West, AK south in mountains to Baja, central NM. Two broods, mainly May/Jun and Aug. Some 3rd/4th instar caterpillars from 1st brood overwinter, along with 1st/2nd instar caterpillars from the 2nd brood, sheltered in dense grass mats.
Ecology	Generalist. Apart from its northern orientation, the Common Ringlet is an adaptive generalist.

av wingspan: 1.4"

Common Wood-Nymph   *Cercyonis pegala*

Dorsal (southern *'pegala'* type): 7/8/00, Pound Ridge Reservation, Westchester Co, NY

Ventral (southern *'pegala'* type, mated pair): 7/8/00, Pound Ridge Reservation, Westchester Co, NY

Also called Blue-eyed Grayling or Goggle Eye. The name "wood-nymph" is inaccurate, as this is primarily a butterfly of open country. Many local forms were formerly considered separate species or races; it is simplest now to think of two "basic color types" in our area, *pegala* with a yellow forewing patch and *nephle* without it. Both "types" are quite variable.

The lifestyle of the Common Wood-Nymph closely resembles that of *Speyeria* fritillaries, in that males emerge up to a week before females, then quickly die after mating. Adult females do not oviposit until late summer (long after males are gone). Hatching caterpillars proceed directly into diapause without feeding, then complete their development the following spring.

Males patrol for females, which often stay perched. Flight is bouncy and erratic, though not very fast. Human pursuers respect the Wood-Nymph's evasive abilities, including its dives into thick cover and falling "lifeless" to the ground to avoid capture. Feeds regularly on both nectar and decaying matter.

Ventral (northern dark *'nephle'* type): 8/3/97, Pondicherry Wildlife Refuge, Coos Co, NH

Identification	Our biggest satyr (especially females). Also the only eastern wood-nymph. Wings full-cut, with *scalloped HW edges.*
Dorsal	Brown (light to deep chocolate). *Two large FW eyespots set in oblong yellow field* (or not, in North-dwelling *nephle*), with white or bluish pupils. A variable number of small eyespots on dorsal HW.
Ventral	Variable, ranges from dark brown or reddish-brown (males) to paler gray-brown (females). *Nephle* generally darker-toned than *pegala.* HW has *intricate rippled pattern.* FW as above. HW has a variable number of small, poorly aligned eyespots (usually six).

Habitat	A variety of open, sunny habitats, usually with thickets or tangles for escape and concealment. Scrubby fields, grassy wood edges and clearings, salt and brackish marsh edges, savannas, slow-moving stream edges. Elsewhere prairie groves.
Hostplants	Various grasses, including Purpletop (*Tridens flavus*), Poverty Oat Grass (*Danthonia spicata*), bluestems (*Andropogon/ Schizachyrium*), Kentucky Bluegrass (*Poa pratensis*), maybe wild oats (*Avena*).
Occurrence	Uncommon to abundant. All of s. CAN and continental US except southern rim (including s. FL). One brood throughout range, beginning from mid-May to late Jun/early Jul, extending to Sep/early Oct. Hatchling caterpillars overwinter. Females remain active, but may enter reproductive diapause in mid-summer.
Ecology	Generalist. Widespread and successful, highly adaptable.

Shrubby field habitat, Westchester Co, NY

av wingspan: 2.3"

Melissa Arctic   *Oeneis melissa*

Dorsal ('*semidea*' race): specimen photo

Ventral: 7/22/95, Mt. Washington, Coos Co, NH

The arctics are a highly specialized group of butterflies found only in the most climatically extreme outposts of the East Coast region—mainly on the high peaks and mountain ridgelines of New England, where isolated patches of true alpine tundra occur. As often noted, these patches are leftovers from the Pleistocene Ice Age, stranded by retreating glaciers. They persist in our region today only because of elevation and harsh weather exposure.

Arctics are well-adapted to their icy niche. Their bodies are covered with dense "hair" that provides significant insulation. Their wings are medium-sized (optimal for solar warming, *see* Introduction, p. 22), and their scales are dark and heat-absorbing; they bask with their undersurface perpendicular to the sun. Their caterpillars grow slowly, feeding at night on grass or sedge leaves. They usually remain in the larval stage for two years. Unlike other Nymphalids, arctics place their pupae in a silk matrix under rocks or debris, out of the elements as much as possible. Finally, their caterpillars are freeze tolerant, meaning that they can withstand not just below-freezing air temperatures but can actually survive having their bodies frozen solid and later rethawed (Douglas, 1989, p. 91).

Our race of Melissa Arctic, the White Mountain Butterfly (*O. m. semidea*), lives on Mt. Washington (6,300 feet) and the surrounding Presidential ridgeline in north-central New Hampshire. Melissas are capable of long, fast flights, although they are completely inactive when air temperatures fall below the mid-40s (degrees F) or winds exceed 40 mph. Adults seldom feed, but occasionally take minerals or nectar from small alpine flowers with their short proboscises.

Identification   Medium-sized. Identification is not a practical issue in our area, since this is the only *Oeneis* found above timber line in NH. The N. Eng. race lacks the sexual dimorphism seen in some other populations across the Melissa's wide, circumpolar range. The fringes and leading edge of FW are *checked black-and-white*. Arctics' wings are to varying degrees translucent, although this is usually not apparent in the field.

Dorsal   Dull, dark brown above, with *no eyespots.*

Ventral   HW and FW apex have a *smoky, cryptic mottling of blacks, browns, and white. The central HW band is normally indistinct.* Ventral pattern provides excellent camouflage against lichen-covered rocks. In CAN, different color forms stay on appropriately colored rock surfaces for concealment (Layberry et al., 1998).

Habitat   Alpine tundra; rocky talus slopes above tree line. Accessible on Mt. Washington summit by road or the "cog railway." Elsewhere, on dry arctic tundra.

Hostplants   In our area, Bigelow's Sedge (*Carex bigelowii*). Other sedges elsewhere. Will accept some grasses in captivity.

Occurrence   Usually uncommon. Across North from AK to Labrador; isolated populations in Rocky Mountains. Also Siberia, n. Mongolia, Japan (threatened). One flight, late Jun-Jul. Early-stage larvae overwinter first year, mature larvae second.

Ecology   Specialist. Extremely narrow range, conservation monitoring appropriate.

Melissa Arctic habitat (rocky talus) on Mt. Washington, Coos Co, NH

av wingspan: 1.7"

Polixenes Arctic *Oeneis polixenes*

Dorsal female ('*katahdin*' race): specimen photo

Ventral: 7/2/00, Mt. Katadhin 'tablelands,' Piscataquis Co, ME

The windswept heights of Mt. Katahdin in central Maine (5,269 feet)—though less towering than Mt. Washington—are nonetheless stark and treacherous. And it is precisely these rigors that make them a cozy home for our most range-restricted alpine satyr, the Mt. Katahdin Butterfly (*O. polixenes katahdin*), a race of Polixenes Arctic.

The Polixenes has numerous poorly defined subspecies across its holarctic range. In New England, we avoid nearly all of the confusion associated with *Oeneis* satyrs elsewhere—caused by strong similarities among sympatric species and wide variations within species. Our Polixenes race exists in isolation on just one tiny, isolated patch of alpine habitat. Virtually its entire global range can be surveyed in a single sweep of the eye when standing on the Katahdin summit—assuming clouds are not obscuring the view, which is not a given.

Polixenes Arctics are active, conditions permitting, during their flight period in mid-summer. Males are alternately described as perching or patrolling, but a better description might be coping, since sudden, frequent shifts in weather and available sunlight constantly create new circumstances for them to address. Some writers suggest that arctics loft into the wind when disturbed to avoid capture. Our experience, while very brief, suggests essentially the opposite: as the air warmed at midday, we saw a single Polixenes venture forth, flying low and rapidly over alpine tussocks. When a burst of wind arose, the butterfly was pinned, allowing a brief photo session (about 45 seconds). As soon as the wind quieted, it picked up again and darted over a precipice, out of sight forever.

Identification	Medium-sized. Mt. Katahdin Butterfly is darker than some other Polixenes races. In some locations, it can be difficult to distinguish between resident Melissas and Polixenes, but not here, where the Polixenes flies alone.
Dorsal	*Warm brown or grayish-brown*, may be lighter near wing tips (translucence). Our race may have *one to three small, black eyespots on FW*, especially females.
Ventral	Mottled gray, black, and brown. A comparatively *distinct dark central band* on HW, surrounded by white (especially along the outer edge); but banding is less pronounced in *katahdin* than in some other races. Note small white dots along HW margins when fresh.

Habitat	Alpine tundra, on the wide, flat tablelands of Mt. Katahdin below the summit, also on adjoining ridges. Warning: the hike up Mt. Katahdin is long and in places dangerous. Plan carefully and use caution! This butterfly is fully protected under ME law.
Hostplants	Undescribed, alpine grasses or sedges.
Occurrence	Annual populations are variable. Occurs across the boreal North, from Siberia to Labrador; also isolated colonies in the Rockies. One brood, late Jun-early Aug. Two-year life cycle. Caterpillars overwinter.
Ecology	Specialist. Its small range subjects this hardy race to many threats, from human interference to global warming.

Alpine meadow on Mt. Katadhin 'tablelands,' Piscataquis Co, ME. **Inset**: Alpine sedge (*Carex* sp.), possible host

av wingspan: 1.7"

Jutta Arctic   *Oeneis jutta*

Dorsal male (left half), female (right half): specimen photos

Ventral: 6/19/94, Wilsons Mills, Oxford Co, ME

This large arctic seems almost cosmopolitan in comparison with its isolated alpine cousins. To find it, we need not scale any remote peaks, but rather only slosh through a few particular sphagnum bogs. Originally known only from Stillwell Bog, near Bangor, Maine, the Jutta Arctic has since been found more widely. The recent discovery of an established colony at Bloomingdale Bog in the Adirondack Mountains of upstate New York suggests that additional populations still remain to be discovered elsewhere.

A wide-ranging and variable circumpolar species, the Jutta Arctic has numerous, weakly differentiated regional subspecies. The race in our area is *ascerta*. Juttas range farther south (or, in the case of northeastern populations, farther downhill) than most other arctics. Yet this is a matter of degree, as they are an obligate resident of rigorous, boreal wetlands.

In the savanna-like bogs that they favor, male Juttas set up territorial perches on logs and other stable sites, at strategic junctures in the "open lanes" running naturally amid Black Spruce groves. They will return to favored perches repeatedly when disturbed (Masters, 1972). Juttas fly rapidly and are difficult to pursue, especially in their soggy habitat, strewn with fallen, slowly decomposing trees. Females wander randomly, and are more likely to be seen near bog edges. It is thought that they may emit pheromones to attract displaying males. Juttas have been observed nectaring on bog flowers, such as Labrador Tea (*Ledum groenlandicum*).

Identification	Noticeably bigger than Melissa or Polixenes Arctics—though none of the three fly together anywhere in our region. Best initial indications of identity are size, flight season, and habitat.
Dorsal	Dorsal surface is seldom shown. Gray-brown above. Male plain with *dark patch around basal FW vein*. Female has a variable *yellow submarginal band* surrounding *small, black eyespots* on both wings.
Ventral	Mottled brown and gray with an *obscure median band*, outlined weakly with white scaling (more pronounced toward outer wing margin). *Basal FW is a yellowish orange with eyespots* (visible only when FW is raised). May appear very dark (blackish) with wear.

Habitat	Boreal Black Spruce (*Picea mariana*) bogs, usually younger bogs with moderate tree growth. Elsewhere, wet tundra, tamarack bogs, and Lodgepole Pine forests (Rocky Mountains). Proximity to other suitable habitats appears important in sustaining regional metapopulations, which requires regular dispersal among local colonies.
Hostplants	Cottongrass (*Eriophorum spissum*; also possibly *E. vaginatum*) is the main host, although other sedges are probably used as well, especially in other regions. Eggs are reportedly scattered near the hostplant.
Occurrence	Fairly common in suitable habitat, n. ME south to n. NH/VT and s.-central ME; separately in ne NY/Adirondacks. Also AK east across sub-arctic CAN and the n. Gr. Lakes states; Scandinavia east to Siberia and Korea. One brood, late May-early Jul; every year in our region (most common in odd or even years elsewhere). Generally (always?) has a two-year larval development cycle.
Ecology	Specialist. A strict, highly adapted northern bog specialist.

Cottongrass (*Eriophorum* sp.)

av wingspan: 1.95"

The Milkweed Butterflies: Subfamily Danainae

We find it curious that danaids come directly after arctics in the phylogenetic listing of East Coast butterflies, since the two groups could hardly be less similar. Arctic caterpillars dwell in harsh environments where they are frozen solid during winter diapause. But danaids lack any degree of frost tolerance throughout their life cycle and must retreat from northern outposts in the fall or else perish. The arctics are cryptic and presumed edible, while danaids are notoriously toxic, sporting bright colors that advertise their unpalatability. Satyrs are sedentary, danaids highly mobile. Satyrs have a jerky and erratic flight style, but danaid flight is elegant and gliding. And so on.

The Danaiinae are a small subfamily, with just over 150 species worldwide (most residing in tropical Asia), but their importance in the history of butterfly observation and research is unrivaled. The Plain Tiger (*Danaus chrysippus*) was "described" as early as 3500 B.C., in a painting on an Egyptian tomb wall. More recently, many of our basic models of butterfly biology were derived from studies of this subfamily—Monarchs in particular, but other species as well. Key topics of danaid research include use of secondary chemicals for protection and courtship, mimicry, and long-distance migration.

Linnaeus knew only six species of danaids. An explosion of taxonomic activity took place between 1850 and 1930 (Ackery & Vane-Wright, 1984). During this period, upward of 300 species were thought to exist, although in retrospect many were just polytypic forms. The current classification scheme emerged slowly.

Typical caterpillar (Monarch)

Danaids are large butterflies with a number of distinctive anatomical traits (e.g., males have an abdominal "hair pencil" and hindwing scent patches used in courtship; also distinctive are the unscaled antennae and "clubbed" ends on females' atrophied forelegs). Danaid bodies are tough and flexible, capable of withstanding the hard bite of an uninitiated predator while their noxious taste sinks in. All of our species are tawny reddish- or brownish-or-ange, with black and white lines and spotting.

A key feature of the danaids is their toxicity. In most cases, the caterpillars sequester noxious chemicals (cardiac glycosides) from their hostplants, and pass them on to the adults. These chemicals induce retching among vertebrates in small doses—and disablement or death in larger helpings. Most birds learn quickly from experience to avoid danaids and their look-alikes. Cardiac glycosides are found in about a dozen plant families. Those favored by danaids worldwide have rubbery, latex saps: e.g., milkweeds and dogbanes (*Apocynaceae*) and figs (*Moraceae*). All East Coast species use milkweeds. But not all danaid hosts are equally toxic, even among milkweeds, and researchers have explored other possible sources of deterrence, most important, pyrrolizidine alkaloids (PAs). These slow-acting liver poisons/carcinogens are obtained by adults from dried flowers or the nectar of certain plants, such as heliotropes. PAs have also been found to deter attacks by some invertebrate predators, such as spiders. PAs are passed along by danaid females to their eggs. They are also employed to synthesize pheromones used in courtship.

Danaid caterpillars are boldly marked (aposematic) as a warning to would-be predators. They lack hairs or spines, except for fleshy protuberances on the head or other areas. Pupae are suspended upside down, without "girdles," often away from the hostplant. Pupal color is variable in some danaids, enhancing camouflage.

Many danaids engage in seasonal movements. In the tropics, many are short-range emigrants (often altitudinal migrants, relocating to avoid local dry season conditions). The champion migrant, of course, is the Monarch, whose mobile lifestyle has been accurately described only in the past several decades. Earlier lepidopterists, including the eminent Samuel Scudder, thought that adult Monarchs hibernated (Ackery & Vane-Wright, 1984). The Monarch's graceful, gliding flight is optimized to its long-distance lifestyle (see Introduction, p. 41).

Common Milkweed (*Asclepias syriaca*)

Male danaids patrol for mates. Their courtship generally involves the transfer of pheromones from the male's "hair pencil" or wing patches to the female. But in February and March, when Monarchs overwintering in Mexico engage in mass courtship before heading north, the male's approach is more forceful than cajoling, as he physically subdues his intended mate.

Adult danaids feed on nectar, to the near-exclusion of other foods, even when building fat stores for migration. Still, some taking of minerals can be observed, and in Mexico it is critical for overwintering adults to descend periodically from their roosts to drink at mountain streams (if only because fat metabolism requires an infusion of liquids).

Monarch   *Danaus plexippus*

Dorsal male: 9/10/95, Jamaica Bay NWR, Queens Co, NY

Ventral: 7/6/96, Pound Ridge Reservation, Westchester Co, NY

On 1/2/75, an individual named Kenneth Bruegger discovered a large roost of overwintering Monarchs in the state of Michoacan, Mexico (Urquhart & Urquhart, 1976; Brower, 1977). In a stroke, our concept of the outer limits of butterfly mobility was transformed. Celebrated and popular, the Monarch has attracted more research interest and public attention than any other East Coast butterfly (Pyle, 1999).

Each year, between August and November, nearly the entire Monarch population of North America withdraws south, mostly to a few concentrated overwintering sites. Tens of millions of eastern Monarchs congregate at about a dozen roosts in the Neovolcanic Mts. of Mexico, in dense fir forests at 10,000 to 11,000 feet. (There is a also small, nonmigratory population in Florida.) Torpid adults live off body fat during the winter, then awaken and mate in February. Most individuals move northward a limited distance before ovipositing and dying. Nonetheless, this constitutes a partial return migration. Mortality is high at winter roosts, from periodic freezes (amplified by habitat damage), predation by birds and mice, and starvation. In transit, multiple hazards take a toll. And during summer, the spores of a debilitating protozoan parasite (*Ophryocystis elektroschirra*) are transferred to caterpillars with their eggs (they are also spread by contact at winter roosts).

Monarchs have colonized far-flung regions, including Australia, Hawaii, the Canaries, etc., but strays to the United Kingdom lack suitable local hostplants to colonize. Whether Monarchs reach isolated sites such as the Galapagos Islands naturally or with human assistance is subject to debate. Most western Monarchs in the United States winter in California.

Male Monarchs patrol for mates and carry their partners during copulation in graceful "nuptial flights."

Identification	Very large. Unmistakable when seen clearly. Confusion with Queens, Soldiers, and Viceroys is mostly an issue during flight. *See* Viceroy account for comparisons. *Abdomen dark.*
Dorsal	*Bright orange* (males) or *brownish-orange* (females) with black veins and wide borders flecked with white dots. Males have small *oval scent patches* on HW veins. Female vein markings are darker, blurrier. An aberrant white form (*nivosus*) is occasionally seen (common in Hawaii).
Ventral	Similar to dorsal surface, but HW base color is paler.

Habitat	A huge variety of sunny, open spaces, both natural and disturbed. Fields, meadows, urban and suburban parks and gardens, dunes (fall migrant along coast). Conifer groves when overwintering.
Hostplants	Milkweeds (*Asclepias, Calotropis*), e.g., Common (*A. syriaca*), Swamp (*A. incarnata*), etc. Females lay eggs singly beneath fresh leaves, favoring large plants of medium toxicity. Extracted cardiac glycosides are passed to adults. Some birds and mice can tolerate large doses of toxin, others eat around concentrated areas. In MEX, birds and mice eat some 14% of overwintering adults. Toxicity diminishes during the long winter brood.
Occurrence	Annual populations highly variable. Spring migrants appear Mar-May in our area. One to four asynchronous broods, depending on latitude, continual in s. FL. Adults overwinter as conditions warrant (no diapause in tropics).
Ecology	**Generalist**. Critical concentration of winter roosting sites creates potential vulnerability in an otherwise highly successful, generalist species.

Migrants resting at dusk, Suffolk Co, NY

av wingspan: 3.8"

Queen *Danaus gilippus*

Dorsal male ('strigosa' race): 8/9/96, Baviacora, Sonora, Mexico

Ventral ('berenice' race): 3/25/94, Devil's Garden, Hendry Co, FL

This richly colored butterfly is a southern counterpart of the Monarch, active all year in warmer parts of Florida. Though not as thoroughly studied as the Monarch, it is a good deal more typical of the Danaid subfamily than its larger cousin. Indeed, the Monarch's early ancestors probably bore a closer resemblance to the Queen, with its localized, tropical lifestyle, than to present-day Monarchs, with their regimented, long-distance migrations.

It is often said that Queens, because of toxicity, serve as a "mimicry model" for dark, southern-race Viceroys, while the lighter, northern Viceroys mimic Monarchs. But the "mimicry triangle" among these three species now seems far more complex; each has variable levels of toxicity and each appears to be a mimic of the others at various times (*see* Viceroy account). It is also frequently stated that Queens are nonmigratory. This may be true in comparison with Monarchs, but eastern race Queens (*D. g. berenice*) have strayed north with increasing frequency in recent years (where they are sometimes difficult to separate from butterfly house escapees or "wedding releases"). And the midwestern form (*strigosa*), engages in regular seasonal movements (also found in Cuba).

Adult males collect PAs from nectar or withered leaves of certain plants, e.g., heliotropes (Boraginaceae), using them to synthesize pheromones, which are stored in small hindwing patches. Before courtship, the male collects pheromones on the sticky "hair pencil" extruded from his abdomen and pastes the resulting mixture onto the female's antenna. Females can mate up to 15 times each, a record for butterflies. Successive copulations have nutritional value, as the male transfers various useful chemicals in his spermatophore.

Male Queens patrol for mates, with a gliding flight only slightly less graceful than the Monarch's. Adults often roost communally.

Identification	Smaller than Monarchs, about the same size as Soldiers. The top of the abdomen is typically orange. For comparison with Soldier, *see* that species.
Dorsal	A deep chestnut or mahogany brown or orange-brown. Darker than Monarch. Orange extends farther onto FW apex than in Monarch; *subapical spots white* (vs. yellow-orange in Monarch). Males have dark androconial patches along inner HW vein. *Strigosa* form has grayish-white outlines on HW veins; can stray to Keys and should be looked for in our region. *Lacks dark FW veins.*
Ventral	Two rows of small, white spots in dark HW margin. *Dark HW veins* (not on FW). HW base color *bright orange* (paler in Monarch).

Habitat	Various open spaces, fields, roadsides, marshes, forest edges. Mainly coastal in North.
Hostplants	Various milkweeds (*Asclepias, Caloptropis*) and milkweed vines, e.g., White Vine (*Sarcostemma clausum*), Sand Vine (*Cynachum anustifolium*) in NC.
Occurrence	Common to abundant. Resident in FL, regular stray in summer to NC, rarely to MA. Also sw US, W.I., south to ARG. Three-plus broods, Feb-Dec in n. FL, all year in South. Adults overwinter.
Ecology	Medium generalist. Locally range-restricted, but adaptive.

Lanceolate ('Prairie') Milkweed (*Asclepias lanceolata*)

av wingspan: 3.2″

Soldier   *Danaus eresimus*

Dorsal (mated pair): 12/28/96, Virginia Key, Miami-Dade Co, FL

Ventral (acquiring chemicals from dead flower): 11/26/00, Curry Hammock SP, Monroe Co, FL

Also known as Tropical Queen. Until recently, this species appears to have been a casual emigrant from the tropics in fall, making appearances in the Rio Grande Valley of Texas and, to a much lesser extent, in southern Florida. Then, for reasons not determined, it began to colonize southern Florida actively in the 1970s (Calhoun, 1996). Today, it has established a firm, if somewhat peripatetic, residence in much of the state.

Our understanding of this "third Monarch" in our region is far from complete, despite recent advances. Some remaining gaps in our knowledge undoubtedly stem from the Soldier's close similarity to Queens: identification is not difficult with practice, but it is possible for even careful observers to overlook a Soldier flying in close proximity with other danaids. Florida Soldiers are of Antillean origin, but the designation of this group as a distinct race (*tethys*) is now considered doubtful. A separate, valid race (*montezuma*) is found in Texas, and others occur in the tropics.

Soldiers are opportunistic, tropical colonists. Their numbers vary significantly from year to year, perhaps because they are prone to wander in search of suitable, transient habitats, or perhaps because they are vulnerable to winter cold. Or, as recent colonists in our area, they may simply be gaining biological equilibrium before edging north from their subtropical home base (following the Queen's recent lead).

Soldiers nectar frequently. They are comparatively slow-flying and reasonably easy to approach, but become quite flighty if their "personal space" is encroached too closely, and will continue off for some distance when startled (Calhoun, 1996). Males patrol for mates and (typical of danaids) carry their partners in nuptial flights.

Identification	Fairly large; about the same as Queen. Key identification points can usually be discerned quite readily in perched individuals, but identification in flight is more challenging. Scott notes that any confusion remaining after field observation can be resolved by counting chromosomes (Queens have 29, Soldiers 30).
Dorsal	Variably brownish, rusty, or yellowish-orange. Paler toward the wing margins than at the base (which often creates a somewhat worn or faded appearance). *FW veins lightly black-lined* (vs. heavily lined in Monarch, nearly unlined in Queen). Conspicuous HW scent patches on the male.
Ventral	Dark veins on *both wings* (Queen's FW veins mainly unlined). Key mark is a *concentric postmedial band of pale, squarish spots* on the HW, absent in the other species.

Habitat	Found in a wide variety of open, subtropical habitats, including citrus groves. Usually at weedy water edges where hostplant grows, but also seeks nectar and pyrrolizidine alkaloids (PAs) in varied locations, especially dry, weedy fields.
Hostplants	Milkweeds and relatives, including White Vine (*Sarcostema clausa*) in s. FL, Strangler Vine (*Morreni odorata*) northward (Calhoun, 1996).
Occurrence	Variably common, mostly s. FL, north in summer to Orlando area. Strays in GA, NC. Also W.I., s. TX to BRA. Three-plus broods, all year, especially Oct-Dec.
Ecology	**Medium generalist.** Strongly climate-restricted, but generalist lifestyle reminiscent of other danaids.

White Vine (*Sarcostema clausa*)

av wingspan: 3.2"

The Skippers: Family Hesperiidae

Skippers have long had image problems. Called "hogs" in early-18th-century England (Ford, 1977), many causual observers mistake them for moths. Even serious commentators have considered them as "less developed," with "by no means the general importance" of true butterflies (Weed, 1917, p. 268). And traditional collectors, who found skippers difficult to capture, "refused to collect or pay much attention" to them (Allen, 1997, p. 182). Many species remain poorly studied, with little available life history data.

But for a growing number of field observers, skippers are a favorite group, biologically interesting and challenging to study. Most skippers in our area can be confidently identified by sight, and beginners should not make the mistake of "leaving them to the experts."

Skippers split off from the main line of butterfly evolution early on, in the Cretaceous (Scott, 1986, p. 99). They lack eyespots (ocelli) and produce few wing pigments. But their iridescence can be breathtaking, and their patterns are frequently intricate and subtle. Even the lowly Common Sootywing, on a close look, is quite handsome.

There are around 3,600 species of skippers in the world. Typically small or medium-sized, most (2,300) live in the Americas, with about 85 residing in the East Coast states—a little over 1/3 of our total species. Skippers attain their greatest diversity in the tropics, although a handful do reach northern Canada. Their range may be limited by their inability to assimilate indigestible host plants quickly during short northern summers, as much as by cold intolerance.

Anatomically, skippers are distinguished by their thick, stout bodies, relatively short wings (blurred in flight), the curved or hooked apiculus at the end of their antennae, and certain technical aspects of wing venation and tibial hairs. The eyes and antennae are set well apart, and the large head is as wide or wider than the thorax (except in giant-skippers). They have six functional legs and in most species a long proboscis. Adults feed on nectar, and sometimes also mud, bird droppings, etc.

Most female skippers lay eggs singly on hostplants. Larvae tend to be rather shapeless and saclike, with few notable features other than a distinctive head. Most build leaf nests, secured with silk, from which they emerge at night to feed. They produce cylindrical pupae, often secreting a powder that keeps them dry.

The name "skipper" is based on the family's rapid, erratic flight style—an adaptive behavior, it ends up. The two major skipper subfamilies in our region (spread-winged and grass skippers) eat a variety of hostplants—dicotyledons and monocotyledons, respectively—but few that confer toxicity. Skippers' strong body and short wings are optimized for fast take-offs and maneuvering, helping them to avoid predators. In controlling this rapid flight, skippers have refined the basic visual mechanics found in moths, achieving enhanced tactical vision in flight, though they may resolve image details less sharply than true butterflies (Scott, 1986). But skippers are generally not well-designed for gliding or endurance flight, and few migrate. Most males perch to find mates. When a mated skipper pair is disturbed, the female carries the couple to safety.

Mature caterpillar of Brazilian Skipper on canna leaf

Leaf fold 'nest' of Brazilian Skipper caterpillar, on canna, used for protection

Typical pupa (Brazilian Skipper) on canna leaf where the caterpillar fed

Spread-winged Skippers: Subfamily Pyrginae

About one-third of all skippers belong to the subfamily Pyrginae (spread-winged skippers); 26 of these reside on the East Coast. Pyrgines are well-represented on the East Coast, but far more occur the Southwest. Several technical features distinguish this mostly tropical group, but observers can readily assign them to subfamily by superficial appearance. Ours are usually brown or blackish-brown, often subtly mottled (though checkered-skippers are boldly patterned). They usually perch with wings fully spread. The abdomen is short, not reaching the outer edge of the hindwing, and the antennal club is large, with a sharply angled apiculus. Males usually have a costal fold on the forewing, with the wing edge rolled back over the upper surface, protecting scent cells that dispense pheromones during courtship. This distinctive trait is visible in the field. At rest, some Pyrginae fold their wings, others keep them open, but duskywings curiously fold them around twigs (see p. 22). In most genera, males perch when displaying. But some patrol. All species nectar, and many also imbibe minerals. Caterpillars feed on dicotyledons, mostly peas, mallows, and oaks locally. Several groups (notably the duskywings) are notoriously difficult to identify, especially when faded; habitat or hostplant cues are often important.

Mangrove Skipper   *Phocides pigmalion*

Dorsal: 3/23/94, Cactus Hammock, Big Pine Key, Monroe Co, FL

Ventral: 12/24/96, Everglades NP, Miami-Dade Co, FL

The fast, deliberate flight of the Mangrove Skipper can give the momentary impression of a small bird—or even a day-flying bat—as it courses along a hammock trail or through a mangrove thicket. Its deep, iridescent colors, flashing brightly when the sun's rays reflect at the proper angle, add drama to the moment. At rest, the Mangrove Skipper tends to perch higher than most other East Coast skippers, frequently 8 to 10 feet or more. It often lands upside down beneath a leaf, like many of its tropical relatives (but less routinely than the Hammock Skipper).

Despite a striking appearance, the Florida form of Mangrove Skipper (*P. p. okeechobee*) is actually somewhat plain compared with several of the species' other geographic races, e.g., the Hispanola race (*P. p. bicolora*) and South American forms, which are boldly striped blue and white. Overall, the genus *Phocides* consists of around 20 species, some stunningly marked, that live in Central and South America and the Caribbean. The Mangrove Skipper is the only *Phocides* to breed regularly in the United States, but the dazzling Guava Skipper (*P. polybius*) is an annual visitor to the lower Rio Grande Valley in Texas, and the Rainbow Skipper (*P. urania*) a rare stray from Mexico.

Adults nectar actively, on mangrove flowers when present. They wander at other times to seek *Bidens* and other suitable nectar sources.

Identification   Large, stout-bodied, with long, pointed, triangular wings. Long antennae, white "face." Usually easy to identify.
Dorsal   Mainly all dark-brown to black, no hyaline spots; note variable, cobalt iridescence on body and FWs, creating a general sheen when fresh. Note irregular row of iridescent turquoise chevrons on HW submargin.
Ventral   Dark with some basal iridescent stripes, mirrors dorsal pattern.

Habitat   Normally found in or around Red Mangrove forests of coastal FL, but also visits adjoining hardwood hammocks. Occasionally strays well away from mangroves, e.g., to pine woods at the entrance road to Pay-hay-okee in Everglades NP.

Hostplants   Red Mangrove (*Rhizophora mangle*), a tropical and subtropical shrub adapted to flat, muddy seashores; cannot tolerate freezes. Dark caterpillars form silk leaf nests, feed voraciously, but do not impair host populations (Ellison & Farnsworth, 1996).

Occurrence   Fairly common in proper habitat along s.-central FL coasts, but has discontinuous broods, wide population swings. Also found in Caribbean and So. Amer. Strays north to coastal SC. Multiple broods, flies all year.

Ecology   **Specialist.** Closely tied to Red Mangrove, a plant that is strictly limited to suitable niches along the immediate coastline in warm climates.

Red Mangrove (*Rhizophora mangle*)

av wingspan: 2.2″

Hammock Skipper *Polygonus leo*

Dorsal: 12/23/94, Cactus Hammock, Big Pine Key, Monroe Co, FL

Ventral: 11/24/00, West Summerland Key, Monroe Co, FL

The Hammock Skipper is a "marker species" in subtropical woodlands of south Florida—especially Caribbean-style hardwood hammocks, where its main hostplant concentrates. Smaller than the Mangrove Skipper, the Hammock is nonetheless a powerful flier, zipping frenetically through the dim-lit understory. But for all its exertion, it seldom flies a great distance. Most of its movements follow an irregular, but ultimately circular course. It characteristically lands upside down, resting concealed beneath a low, broad leaf. Males perch upright, however, when seeking females. Hammock Skippers sometimes reside in open, subtropical scrub, but this is the exception. Most are seen in shady openings of hardwood habitats, or along interior trails or edges.

The Hammock Skippers found in Florida are of the Cuban/Bahaman race, *Polygonus leo savigny*, with relatively large forewing hyaline spots (Smith et al., 1994). A number of early accounts contained detailed information on how to differentiate the Hammock Skipper from Manuel's Skipper (*Polygonus manueli*), a closely related, but bluer-toned coastal species found in South America and the southern Lesser Antilles. Early investigators (e.g., Kimball, 1965) thought the range of Manual's Skipper included south Florida. But this seems to have been a mistake, based on misidentified or mislabeled specimens. Currently verified records for Manual's Skipper do not reach even as far north as the Virgin Islands. Thus recent commentators have excluded it from the Florida list (or have demoted it to hypothetical).

Identification	A smaller, less "pointy" version of the Mangrove Skipper, wings shorter and rounder. Outer HW is *slightly scalloped, with two short, stubby "tails."* The Hammock Skipper's patterns provide excellent forest camouflage. It is most likely to be confused in the field with other species having similar underside mottling, especially purplewings.
Dorsal	Dark, blackish overall, but lighter near wing base. FW shows *bluish iridescence* when fresh (especially males). Several *large, prominent, squarish FW hyaline spots.*
Ventral	Ventral HW is lighter than other surfaces; ventral FW is dark, with light apex. HW often shows a pale, purplish iridescence; has two incomplete cross-bands, best defined near the wing base. Leading wing margins are unpatterned, highly reflective in strobe-lit photos. A *single, black dot near the HW base* is distinctive.

Habitat	Subtropical hardwood hammocks. Also rim-rock pinelands, gardens, rarely open scrub. Hurricanes in the Keys can destroy or severely degrade stands of coastal habitat.
Hostplants	Various woody legumes. In FL, Jamaican Dogwood (*Piscidia piscipula*); also Karum-Tree or Pongam (*Pongamia pinnata*), an exotic used in settled areas. In Caribbean, uses a vine that synthesizes Rotenone, a strong insecticide (Scott, 1986).
Occurrence	Regular in s. FL, especially near coasts, but scarcer and more local, in our view, than some accounts indicate. Also in TX and US Southwest, Caribbean, So. Amer. to Peru/ARG. Continuous broods, all year.
Ecology	Specialist. Highly specialized in all respects.

Jamaican Dogwood (*Piscidia piscipula*), flower (upper left); skipper perched beneath leaf (upper right); seed pods (lower left); leaves (lower right)

av wingspan: 1.8″

Long-tailed Skipper   *Urbanus proteus*

Dorsal: 12/30/98 Castellow Hammock, Miami-Dade Co, FL

Ventral (on *Bidens*): 9/28/97, Tiger Bay WMA, Volusia Co, FL

This flashy, neotropical skipper is likely to attract attention—even among nonbutterfliers—when it darts into a suburban garden. It is usually abundant in southeastern strongholds, especially in late summer. A capable but irregular emigrant, its late-season appearances in the North are local and unpredictable.

Long-taileds are conspicuous in their bold, dashing flight and active pursuit of nectar. They frequently hang upside down while feeding, wings cocked at 45 degrees, but will also spread fully to catch morning sun in cool weather (*see* dorsal photo). Individuals with broken or missing tails can be confused with other species—though less so in our area than in southern Texas or Arizona, where more tailless look-alikes occur.

Unusually wide-ranging for a skipper. Winter cold limits the year-round range, mostly to Florida, but individuals start north again in spring. Emigrants progress northward less rapidly than Monarchs or Painted Ladies, though still on a "straight line" migratory course, undeterred by local obstacles (Scott, 1986). Migrants reach North Carolina by July, the northern states by late summer or fall. (At the same time, overflow southeastern populations move southward, reinforcing populations in peninsular Florida and the Keys, where they are uncommon during the summer.)

After being recorded with some regularity in the Northeast between the late 1800s and mid-1900s, Long-tailed Skippers virtually disappeared for some 40 years. They returned north in numbers during the 1990s. This long absence is unexplained, but increased cultivation of soybeans in the Southeast may account for the resurgence.

Identification	Usually mistakable only for Dorantes Longtail, and then only where their ranges overlap. Note the long, broad tails and stout body, typical of the *Urbanus* genus. Sexes similar.
Dorsal	No other skipper in our region has *shimmering, iridescent green-blue* on the upper body and inner wings. Rich brown FW base color with 9 to 10 mostly large, squarish hyaline spots.
Ventral	Several continuous dark bands across HW, and a *thin white dash* abutting the frontmost band. Trailing HW fringes clearly white. Dark *outer* postmedian band on FW is *continuous (unbroken) to apex.*

Habitat	Open, disturbed habitats, especially overgrown fields and weedy woodland edges, with nectar and hostplants. In migration, visits parks, gardens—almost any place where nectar can be found.
Hostplants	Many hostplants, especially climbing peas (*Leguminosae*) and beans (*Phaseolus*). Frequently tick-trefoils (*Desmodium*), butterfly-peas, wisteria, etc. Occasionally a crop pest, e.g., on soybeans (*Glycine max*), but a severe infestation is needed to affect crop yield. Eggs laid singly or in clusters of two to six. Caterpillars construct nests by folding over leaves; they feed at night, developing rapidly.
Occurrence	The most common US long-tailed skipper. Moves northward in summer, mainly along coastal plain, as far as s. N. Eng. Permanent in Deep South. Also US Southwest, W.I. (*domingo* race), C. and So. Amer. to ARG. Three-plus broods in FL, long adult winter generation. Annual populations variable, especially northward.
Ecology	**Generalist.** Lack of permanent northern breeding colonies due to cold intolerance is offset by broad hostplant selection, opportunistic habitat use, and strong migratory tendencies.

Climbing Butterfly-pea (*Centrosema virginianum*), one of many hosts

av wingspan: 1.8″

Dorantes Longtail   *Urbanus dorantes*

Dorsal: 12/22/94, Everglades NP, Miami-Dade Co, FL   **Ventral**: 12/22/94, Everglades NP, Miami-Dade Co, FL

The Dorantes Longtail is a relative newcomer in our region. Klots (1951) knew it only as a Texas specialty. But in 1969 it arrived unexpectedly in south Florida. Surprisingly, these colonists did not belong to the dark Caribbean race, but rather to the nominate, continental subspecies *dorantes* (Minno & Emmel, 1993; Smith et al., 1994). Yet whatever their exact origin, the newcomers adapted quickly to their new environment and became firmly entrenched. Today, they may have begun outcompeting their native cousin, the Long-tailed Skipper.

Comparisons between the two Florida longtails are inevitable, and raise interesting questions. They share common habitats, hostplants, general lifestyle, and appearance. And when present in the same general area, they regularly fly alongside each other. This pair exemplifies a recurring question: how do close species pairs (including many *Urbanus* longtails with overlapping ranges in the neotropics) initially develop as separate species?

Yet these two longtails are not ecologically identical. For example, while both exhibit some degree of migratory behavior, Florida Dorantes are currently much more sedentary. During the summer months, when Long-taileds are scarce in southern Florida, Dorantes remain common.

Active and erratic in the air, Dorantes are likely to dart under a leaf and perch upside down after a flight. Some observers indicate they perch with their wings closed more often than Long-tailed Skippers.

Identification	Slightly smaller, on average, than Long-tailed, with shorter and narrower tails, but these are not generally reliable field marks. Individuals with broken tails can be confused with other species, such as Hammock Skipper, unless carefully observed.
Dorsal	All plain-brown above, *lacks green iridescence* of Long-tailed Skipper. Large hyaline spots on FW are *yellow-toned.* Wing margins are checkered-buff (vs. white on Long-tailed, difference most notable on HW).
Ventral	Dark HW bands more segmented than in Long-tailed (where they fuse into a single pattern). Dark outer postmedian band on FW is *broken near the apex by a "flow-through" of lighter brown* (vs. continuous in Long-tailed).

Habitat   Common in many of the same habitats used by Long-tailed, i.e., open, sunny, disturbed habitats and subtropical scrub, especially where overgrown with viny beans and peas. Thought to favor woodlands and hardwood hammocks during cooler months. Uses a wide variety of flowers for nectar. Has been noted feeding on bird droppings (Minno & Emmel, 1993).

Hostplants   Very similar to those used by Long-tailed Skipper, including many beans and peas. Comparison of local hostplant use by Dorantes and Long-taileds would be instructive, especially as Dorantes lay eggs singly on the hostplant, vs. in clusters. Dorantes is not singled out as a pest species on cultivated beans as are Long-taileds, though cultivated plants are among their hosts.

Occurrence   A widespread neotropical species, like the Long-tailed. Both occur from the US southern rim south as far as ARG. Also W.I. Eastern US range continues to expand slowly northward. Scattered records exist for NC between early Jul and Oct, and recently reached VA. Multiple broods throughout its range. Caterpillars eat hostplant leaves.

Ecology   Medium generalist. Well-defined generalist lifestyle, even in limited US range.

Florida Beggarweed (*Desmodium incanum*)

av wingspan: 1.7"

Zestos Skipper   *Epargyreus zestos*

Ventral: 5/15/72, Islamorada, Monroe Co, FL

Milk Pea *(Galactia pinetorum)*

Only a few decades ago, Klots (1951) described the Zestos Skipper as "not uncommon" on the southern tip of the Florida mainland. Strays occurred at the time as far north as the Ocala area, and the species was seen regularly on the Keys. But even in Klots's time, this Caribbean species was not abundant in Florida. Nor was it even regularly dispersed in the West Indies. Its "most unusual distribution" included mainly smaller islands, where it was sometimes frequent, but it was generally absent on larger ones (Smith et al., 1994). In recent decades, an unwelcome overlay has been added to this already complex picture, as the Zestos Skipper's range in Florida has contracted severely. To our knowledge, it is no longer found anywhere on the peninsula, and has not been seen on the upper Keys since the 1970s (Minno & Emmel, 1993). Only on the lower Keys does it maintain a continued presence, but even there its foothold is precarious. In numerous trips to the Keys since the early 1990s, we have seen this butterfly briefly just once, at Stock Island Botanical Garden near Key West (12/31/98).

The Zestos Skipper belongs to the same genus as the prolific and adaptable Silver-spotted Skipper (*Epargyreus*). It is thus particularly difficult to understand either its rather finicky distribution or the reasons for its decline. Its known foodplant in Florida (which is not necessarily the exclusive host) is a rather common legume. The hostplant's status has not changed discernibly in recent years, and it remains common on the upper Keys, where the butterfly no longer occurs (Minno & Emmel, 1993). Loss of native habitat and widespread mosquito spraying are suspected in the Zestos's decline, but no specific cause can yet be verified. Hopefully this will not be a case in which a species' critical ecological requirements remain unknown until it is extirpated.

The Zestos Skipper is described by one authority as "surprisingly slow" in flight (Glassberg, 1999), but from our encounter we must agree with Smith et al. (1994), who describe it as "fast flying." It reportedly perches beneath leaves with closed wings, like many tropical spread-winged skippers. Males are territorial.

Identification	About the same size as the Silver-spotted Skipper, with large, pointed wings and *two very short, threadlike HW tails*. Similar overall to Silver-spotted Skipper, but *lacks prominent white HW patch* that makes the latter species so readily identifiable. Any potential for confusion is further minimized by the absence—or at least great rarity—of the Silver-spotted Skipper in far south Florida and the Keys. Wing fringes are not checkered.
Dorsal	Similar to Silver-spotted Skipper. Base color brown, often with a red or yellow cast. Four *translucent, golden-brassy spots* form a transverse band across the FW, and there is an additional, small golden FW spot, slightly farther out toward the apex (visible from below, where usually obscured by the HW).
Ventral	Rich red-brown with a *violaceous sheen* (especially in fresh females). Light scaling creates the vague appearance of a band on central HW. *No silver HW spot.* (Interestingly, Scott [1986] noted that Silver-spotted Skippers, when starved as caterpillars, develop reduced white HW spots; perhaps Zestos originated in a harsh nutritional climate.)

Habitat	Generally found along the edges of subtropical hardwood hammocks or subtropical scrub, where it comes to take nectar. May visit gardens near suitable habitat. Little specific field data is available on habitat use, however.
Hostplants	Sole reported host are milk-peas (*Galatica*). Many accounts specify "Keys" Milk-Pea (*G. striata* [= *spiciformis*]), but in this taxonomically complex group *G. regularis* or *smallii* are also likely, as well as other legumes.
Occurrence	Rare and declining, present only in lower Keys. Also in the Bahamas, Lesser Antilles, Puerto Rico. All year (or nearly all year) in the Keys, abundance may be dictated by wet season-dry season cycles more than by date.
Ecology	**Specialist.** Selectively distributed and ecologically sensitive, almost never strays outside the subtropical zone.

av wingspan: 1.9″

Silver-spotted Skipper *Epargyreus clarus*

Dorsal: 6/16/95, Lakehurst, Ocean Co, NJ

Ventral: Big Ugly WMA, Boone Co, WV

Beginning observers—who often find skipper identification confusing—regularly single out the Silver-spotted as one species they can identify with confidence. In most of our region, this familiar butterfly bounds energetically among nectar flowers growing in open or scrubby habitats, or launches from a strategic display perch to confront intruders. On most days, one is most likely to see a series of isolated individuals in the field, as "Silver-spots" seldom gather in large congregations. (Exceptions occur: more than 30 were seen puddling together at one site in West Virginia, 6/20/98.) The largest skipper in most East Coast states, the Silver-spotted is conspicuous because of its size as well as its distinctive markings.

Like other large species, the Silver-spotted Skipper has a strong, rapid flight, and it shares the same erratic determination as many of its Pyrgine relatives. Males are likely to remain in the same general area while displaying for a mate, breaking off to engage in high-flying aerial dogfights, or simply to canvass the territory. A wide variety of nectar flowers are visited (though several authors have indicated that they avoid yellow ones). Individuals sometimes perch beneath leaves, but less regularly than some of their tropical counterparts.

The Silver-spotted Skipper is an extremely successful species, with an extensive range and broad ecological tolerance. It has a long flight season, and is among the first northern skippers to emerge in the spring.

Identification	Easily identified, even in flight, by its large size and bold, white HW spot. FWs are notably long and pointed, the HWs proportionately smaller, with two blunt tails. Wing margins are checkered black-and-buff.
Dorsal	Dark brown base color. *Wide, transverse band of translucent, brassy-golden spots across FW.* In some individuals, this band is duller, less obvious. Also has an isolated, squarish mid-FW dot, as in Zestos. *Compare* Hoary Edge.
Ventral	Both wing margins are *frosted lavender. Bold, irregular white patch* in central HW (*not* silver) is diagnostic.

Habitat	Can be found in nearly any open space with nectar, but favors brushy or weedy second growth, marsh edges, gardens, clearings, roadsides. Less inclined to arid habitats. (In desert regions of the West, usually favors riparian corridors and cooler, moister mountain canyons.)
Hostplants	Caterpillars feed on a wide variety of legumes, notably the young growth of Black Locust (*Robinia pseudoacacia*). Also wisteria, tick-trefoils (*Desmodium*), Hog Peanut (*Amphicarpa bracteata*), even Kudzu (*Pueraria lobata*). Females reportedly investigate a hostplant, then oviposit on a nonhost species nearby, seemingly to avoid early predation. Larvae shift from leaf flap shelters to binding whole leaves together as they mature (Lind et al., 2001).
Occurrence	Common from s. ME to s.-central FL (sparingly to the Big Cypress/Devil's Garden area); a rare stray farther south, to Loxahatchee NWR, Miami, and the Everglades. Also throughout the East and most western US states, avoiding only the most arid regions; north into central and e. CAN, south into MEX. One brood in the far North (Jun-Jul); two to three broods at middle latitudes (Apr-Oct); three to four broods, nearly all year, in s. FL. Pupae overwinter.
Ecology	**Generalist.** Few butterflies, especially skippers, are as widespread or adaptable as the Silver-spotted Skipper.

Black Locust (*Robinia pseudoacacia*)

av wingspan: 1.9"

Golden-banded Skipper   *Autochton cellus*

Dorsal male: 6/20/98, Fork Creek Pub Rec Area, Boone Co, WV

Ventral: 6/20/98, Fork Creek Pub Rec Area, Boone Co, WV

The lifestyle of the eastern Golden-banded Skippers is a complicated puzzle, with several pieces still missing. A combination of striking appearance, extreme rarity, and idiosyncratic biology gives the butterfly tremendous allure among observers.

Burns (1984) carefully described the Golden-banded Skipper as a species with two discontinuous populations, one in the eastern United States, the other in the southwestern United States and Mexico. Over time, he concluded, these two populations have come together more than once and then been reseparated. The latest separation seems fairly recent. The eastern differentiate in this complex is our only large, specialized Pyrgine that is based in the central Appalachians (rather than the Deep South). Though widely distributed, Golden-bandeds are rare and local virtually everywhere they occur. They favor steep, moist woods near running water, probably to exploit hostplant concentrations. Their lone eastern hostplant does not require the degree of specialization observed in the Golden-banded, however; several pyrgines (e.g., Silver-spotted Skipper) use Hog Peanut along with many other legumes.

Golden-bandeds are active in mid-morning and late afternoon, often inactive at other times. Their courtship is a confusing mixture of perching and patrolling (authors differ in their characterizations). Flight is "low and slow" as compared with near relatives. Low-perching males often puddle for minerals.

Identification	A large skipper. Rounder, more evenly proportioned than Silver-spotted. Fringes black-and-white checked. *Compare* Hoary Edge. *Male antenna has white ring just below the club.*
Dorsal	Rich, dark brown, with *gleaming golden bands* across both FWs, more prominent and continuous than in other, similar species. Male FW lacks costal fold.
Ventral	Two indistinct, dark HW bands, slight marginal frosting. FW apex area is *light brown*. Golden FW band clearly visible from below.

Habitat	Mainly steep, moist woodland ravines or hollows, with water nearby, in areas with dappled light. Unglaciated areas in OH.
Hostplants	Only known host in the East is Hog Peanut (*Amphicarpa bracteata*), a twisting vine. This is a regionally common plant, so other factors must account for the species' limited occurrence. Ovipositing females frequently lay clusters of two to nine eggs in a row on the undersurface of a hostplant leaf. No other spread-winged skipper in our area engages in cluster-laying. The adaptive value of this behavior is undetermined.
Occurrence	Rare and local. Se PA, west through central states, mainly in the Appalachians. Isolated lowland colonies in GA and n. FL. Declining in eastern part of its range (e.g., around Washington, DC). Has lost key habitat to coal mining in WV. Two to three broods, May-Aug in North, Feb-Sep in FL. Overwinters as pupa.
Ecology	**Medium specialist.** There is no obvious reason for the restricted occurrence of the species, as host and habitat are widespread.

Deciduous bottomland, Boone Co, WV. **Inset**: Hog Peanut (*Amphicarpa bracteata*)

av wingspan: 1.75″

Hoary Edge   *Achalarus lyciades*

Dorsal: 7/3/93, Suffolk Co, NY

Ventral: 6/26/99, Van Cortlandt Park, Bronx Co, NY

A familiar resident of the central East Coast, the Hoary Edge blends in unobtrusively with the butterflies in a typical, overgrown eastern meadow—better than would some of its more flamboyant pyrgine relatives. It is the only member of its distinctive-looking genus found in the East. All others live to the southwest, mainly in Mexico and Central America. (Four of these species, all referred to as "cloudywings," reach the United States along the southern rim, between Texas and Arizona.)

The Hoary Edge's lineage may help explain some of its observed lifestyle traits. For example, individuals quite often rest on patches of bare ground, even when not taking minerals. Some displaying males center their territory on a patch of dirt, flying up to challenge intruders (Allen, 1997). Others choose an elevated perch, three to six feet up. But regular use of the ground during display is a familiar trait in desert or arid-ground groups.

Hoary Edges likewise prefer dry, upland habitats, or sandy pine-oak scrub, largely avoiding wetlands. It is a measure of the butterfly's adaptability, on the other hand, that it is not confined to such surroundings; instead, it has radiated into a number of less characteristic settings, drawn by hostplant and nectar availability. (Nectaring individuals particularly may stray into highly atypical habitats.) Hoary Edges fly fairly close to the ground, moderately quickly. Wary when approached, they move away—unhurriedly but without delay.

Identification	A good-sized skipper, yet visibly smaller than Silver-spotted in the field. Also less angular in shape. Fringe is *checkered black-and-white* (or gray-and-white)—but not buff, as typical in Silver-spotted. Appears dark in flight.
Dorsal	Dark brown, with an incomplete, *brassy spot-band* across the FWs. (The additional, isolated golden spot on the FWs of Zestos and Silver-spotted Skippers is here wedged into the main spot band. In its place is a tiny white dash on the central, postmedian FW.)
Ventral	The basal HW is very dark, contrasting sharply with a prominent, *frost-white "hoary" patch along the outer margin*. Band does not extend onto the FW, which is mostly pale brown with stray markings (though the translucent yellow band is visible from below). The white HW mark is marginal in Hoary Edge, central in Silver-spotted.

Habitat	A range of habitats, mostly overgrown or brushy clearings, edges, and outcrops; nearly always near upland woods. Also sandy, pine-oak habitats, sandhills. Not in high mountains, including in the Southeast. Seems to require some shade nearby. May wander in search of nectar.
Hostplants	A variety of legumes, mainly various broad-leaved tick-trefoils (*Desmodium*). Accounts also mention Wild Indigo (*Baptisia tinctoria*) and bush-clovers (*Lespedeza*). Ovipositing females leave a single egg underneath the hostplant leaf.
Occurrence	Locally fairly common, from s. N. Eng. to n. FL, west across the lower Midwest and Southeast to e. TX. One brood in North (May-Jul), two in South (Apr-May and Jul-Sep). Overwinters as pupa.
Ecology	Medium generalist. A well-adapted Southwest transplant.

Left: Showy Tick-Trefoil (*Desmodium canadense*). **Right**: Egg on underside of hostplant leaf (different *Desmodium* species)

av wingspan: 1.65"

Southern Cloudywing   *Thorybes bathyllus*

Dorsal: 7/14/96, Cumberland Co, NJ

Ventral: 7/14/96, Cumberland Co, NJ

The cloudywings (genus *Thorybes*) are not a group that invites superlatives. These uncelebrated butterflies are plain in appearance, medium in size, and rather conventional in lifestyle. The one connection in which they attract significant interest is in being exceptionally hard to identify.

This difficulty has been severe enough at times to detract from our understanding of the group. The wording of cloudywing life history accounts is often cautious and nonspecific, and numbers of mislabeled specimens have been discovered even in curated reference collections. Positive identification is possible via genitalic examination (Gatrelle, 2001f), but until recently there have been few reliable guideposts for field observers.

The difficulties encountered in identifying Southern Cloudywings in the field have two main sources: individual variation and confusing seasonal forms. In the South, where the species is double-brooded, spring individuals tend to be rather faintly marked, often resembling Confused Cloudywings. Most summer individuals are boldly patterned, by comparison, and thus readily identifiable. But summer Confused Cloudywings are also strongly patterned (often more so than spring Southerns), which is a source of regular difficulty.

Behaviorally, male Southern Cloudywings are highly territorial, known to hold the same low perch (one to five feet off the ground) throughout their adult life span (Opler & Krizek, 1984). They actively seek nectar, often remaining on a single flowerhead for some time. This can allow close study, as they often feed with their wings partly spread. Flight is generally fast and erratic, though lower and less wide-ranging than in some larger Pyrgines.

Identification	Slightly smaller on average than Northern Cloudywing, about the same as Confused; but size varies considerably in all three species. Wings are full-cut and moderately pointed. The HW is more triangular in some individuals (especially males), giving the appearance of stubby tails. The *palps ("face") are uniformly bright white*, with a white ring also extending over and above the eyes (more pronounced in some individuals than others). Male Southern Cloudywings *lack a costal fold* (*compare* Northern Cloudywing).
	Glassberg (1999) noted that the Southern Cloudywing has a *white patch on the antennal club*, just where it bends downward, which is absent in the Confused and Northern. This mark, while valuable, can be difficult to see in the field, and the possibility of spurious light reflections should be ruled out carefully—especially an errant "strobe highlight" when photographs are used for identification.
	In summer broods (and generally in the North), FW fringes are checkered, *HW fringes snowy white*. Fringe color is less distinctively white in southern spring broods, however.
Dorsal	The mid-FW spots are mostly aligned in a row, usually bold and geometric. The prominent mark near the FW costa (at cell-end) is *hourglass-shaped*, except in southern spring broods. Subapical "bracelet" is characteristically *straight* (or banana-shaped), with *all dots connected*, as if in a single mark (*compare* Confused Cloudywing).
Ventral	Mottled dark brown, two variable dark bands across HW; also note variable marginal frosting, heavy in some cases.

Habitat	Open, scrubby sites, including dry or mesic fields and meadows, savannas, burn sites, upland barrens, etc.
Hostplants	Various legumes, similar to those used by Hoary Edge. Little specific information is available on local hostplant preferences. Female lays a single egg on undersurface of hostplant leaf, typically in the afternoon.
Occurrence	Regularly seen but seldom numerous, especially in northern portions of the range. An eastern species, occurs from se ME south to central FL, and from there westward to se MN and e. TX. One brood in North (Jun to mid-Jul); two to three broods in South (Mar-Nov). Mature caterpillars overwinter in leaf nests.
Ecology	**Generalist**. Southern Cloudywings utilize a range of common habitats and hostplants, including disturbed open areas. Not highly tolerant of areas with extensive development.

av wingspan: 1.45"

Confused Cloudywing *Thorybes confusis*

Dorsal: 7/24/97, Sandhills Game Land, Scotland Co, NC

Ventral (same individual): 7/24/97, Sandhills Game Land, Scotland Co, NC

At the risk of stating the obvious, it is not the butterfly that is confused. Klots observed in 1951—almost 30 years after the Confused Cloudywing was first described—that "information is badly needed about its habits and life history" (p. 213). Today, more than 50 years later, some progress has been made, but the species lifestyle remains poorly documented. With recent gains in field identification techniques, however, it should be possible to develop a more broad-based profile of the species' lifestyle and distribution.

It is interesting that the northern range limit of the Confused approaches the latitudinal dividing line (in central New Jersey), where the Southern Cloudywing first becomes single-brooded, or univoltine. We can imagine that the separation of these two species might have occurred in some past glacial epoch, when southern populations of a common ancestor became temporarily isolated. The Confused Cloudywing does not have a northern, single-brooded variant.

The foregoing scenario, if correct, would vary from traditional accounts, which link Confused Cloudywings more closely to the Northern. The reduced wing patterns of Confused do in fact suggest the Northern, especially in the spring brood. But the alignment of the markings, and their general appearance when well-defined (i.e., in the summer brood), seem very close to the Southern, as do other observable features.

Little has been reported concerning the behavior or lifestyle of this species. In our experience, it is low-flying and inclined to take minerals from wet, sandy puddles in the North Carolina sandhills.

Identification	*Compare* Southern Cloudywing. This species *lacks a white patch* at the bend in the antennal club. Its palps (or "face") are gray or whitish, and the eye is clearly ringed with white. The fringes are checkered brown-and-white (or buff-and-white), not snowy white as in the summer Southern Cloudywing. *Males lack a FW costal fold.*
Dorsal	Plain, dark brown above, with extremely variable white markings (faint or absent in spring). The mid-FW marks tend to be aligned in a row (as in Southern Cloudywing), and the midwing bar is frequently *long and thin* (though sometimes rectangular, especially in summer females). Mid-FW spots near costa may have an hourglass shape in summer, as in Southern Cloudywing. But the dots of the subapical "bracelet" are loosely connected, and the trailing dot is always *offset toward the margin, often barely touching the others (if at all)* (Gatrelle, 2001f).
Ventral	Similar to Southern Cloudywing, again highly variable, probably has less marginal frosting, on average.

Habitat	Seems to prefer sandier, drier habitats than other cloudywings, such as sandhills and dry prairie. But all three species can be found flying together. Reported from fields in VA. Rare or absent in the mountains, at least in NC.
Hostplants	Not reported. Bush-Clover (*Lespedeza*) suspected as a principal host (Scott, 1986). But the usual array of herbaceous legumes (tick-trefoils, Wild Indigo, etc.) also possible.
Occurrence	Rarest of our *Thorybes*, but exact abundance unknown. Ranges from se PA and w. NJ (once), along se coastal plain; to Ozarks, e. TX. Range coincides with lower half of the Southern's. Two to three broods (Mar-May and Jul-Sep). Diapause stage unreported, probably mature larva.
Ecology	**Medium specialist.** Difficult to diagnose, but must have some fairly specific habitat-hostplant associations.

"Confusing" Southern Cloudywing, for comparison, 7/24/97, Sandhills Game Land, Scotland Co, NC

av wingspan: 1.45"

Northern Cloudywing   *Thorybes pylades*

Dorsal male: 6/23/96, Van Cortlandt Park, Bronx Co, NY

Ventral: 6/5/94, Pound Ridge Reservation, Westchester Co, NY

The common names assigned to cloudywings in our area, it has been observed, leave something to be desired (Gochfeld & Burger, 1997). If accurately named, the Southern Cloudywing would probably be called Eastern, the Confused would be Southern (or even Southeastern). And the Northern Cloudywing might be called Common Cloudywing, since it is the most widespread member of its genus, and its continental range extends both farther north (in Canada) and farther south (in Mexico) than the range of the other two. It is also far more common and widespread than any of the western *Thorybes*.

The life habits of the Northern Cloudywing are very similar to those of the Southern or Confused. Certain subtle differences have escaped notice until recently, e.g., a preference for the mountain province in North Carolina vs. coastal plain (LeGrand & Howard, 2002). But these distinctions are truly subtle, since the various species coexist in many locations without apparent distinctions in behavior.

As with other cloudywings, males perch on territory when awaiting females, typically on or near the ground. They display in the late morning. Several authorities mention hilltopping, but this is not obvious in eastern populations. Northerns fly rapidly, low to the ground and usually within a limited area. They actively seek nectar, though emerging males also take minerals at puddles.

Identification	Slightly larger on average than Southern or Confused, but sizes overlap considerably. Unlike other local cloudywings, *males have FW costal fold.* The *palps ("face") are brown* (not white or gray), and any white rim around the upper eye is interrupted at least slightly in the center, not continuous. Antennal clubs lack a white mark at the bend (*compare* Southern Cloudywing). Wing fringes checkered buffy brown-and-black (except in w. TX, where a white-fringed form occurs).
Dorsal	Plain brown, with small, but *highly variable* mid-FW markings, tending to be angular (or triangular), not aligned in a row as in other species. No hourglass shape in midwing marks near costal margin. Mid-FW markings sometimes virtually absent.
Ventral	Mottled brown, two dark bands, often with substantial frosting near wing borders.

Habitat	A wide variety of dry or medium-dry (mesic) habitats, including scrubby fields, open clearings, and wood edges; also sandhills, flatwoods, roadsides, meadows, etc. In the Southwest, also dry mountain woodlands.
Hostplants	A wide variety of legumes, especially diverse given the species' wide distribution. Includes tick-trefoils (*Desmodium*), Bush-Clover (*Lespedeza*), Hog Peanut (*Amphicarpa bracteata*), many others.
Occurrence	Regular, but only a handful usually seen daily at any one site, vs. 100/day in AZ mountains, where it is often the most common butterfly. Not in s. FL. One brood in North (May-Jul), two to three in South (Mar-Dec). Overwinters as mature larva.
Ecology	**Generalist.** Highly adaptable in terms of geography, habitat, hostplant use, and lifestyle.

Dorsal (faintly marked individual): 5/30/94, Pound Ridge Reservation, Westchester Co, NY

av wingspan: 1.5"

Florida Duskywing *Ephyriades brunneus*

Dorsal male: 5/20/94, Big Pine Key, Monroe Co, FL

Dorsal female: 5/17/97, Key Largo, Monroe Co, FL

This distinctive and unusual butterfly is found only in a few specialized habitats in southern Florida and the Keys. Its close association with intact, subtropical pinelands has loosened recently, as some populations now sustain themselves in disturbed or extensively altered habitats, using remnant stands of hostplant. This may be analogous to the ordinary process of succession-and-recolonization that occurs in the wake of natural disturbances, such as fires or tropical storms. But it is not clear whether populations living in uncharacteristic surroundings (as on north Key Largo, *see* below) will continue to thrive. Overall, the Florida Duskywing seems generally in decline, though cyclical variance in its annual broods makes it difficult to estimate trends.

Two genera in the United States share the name "duskywing" The Florida Duskywing is the lone U.S. representative of the genus *Ephyriades*. (All our other U.S. duskywings are *Erynnis*.) *Ephyriades* is distinguished by its shape—including long, pointed wings and a comparatively large hindwing—as well as by a satiny sheen on the dorsal surface in most species. Three or four other *Ephyriades* exist in the West Indies, where they fly with additional races of the Florida Duskywing. The endemic race in our area is *floridensis*.

Florida Duskywings are usually seen flying in the shrub layer of subtropical pinelands, where their hostplant is quite common. Males perch in dappled sun to await passing females. Both sexes nectar actively.

Identification	A large duskywing. As noted above, it has long, pointed FWs and large, full-cut HWs. Strong sexual dimorphism. Wing fringes dark.
Dorsal	Male: dark, satiny blackish-brown, darkest at wing bases; outer margins may show a reflective, brassy-orange sheen, especially with wear. Note circular cluster of small, white dots on outer FW. Female: brown above, from rich chestnut to pale beige; FW variably mottled, white FW dots larger than male's; conspicuous violaceous sheen when fresh. Vague concentric postmedian band on HW.
Ventral	Both sexes brown below; FW dots show through on ventral surface.

Habitat	Primarily intact rimrock pinelands with hostplant, but also sometimes subtropical scrub and successional habitats (e.g., Owaissa Bauer Park in Miami-Dade Co after Hurricane Andrew). Persisting in reduced numbers in certain hardwood hammocks of upper Key Largo (formerly pinelands) where the hostplant remains (Minno & Emmel, 1993).
Hostplants	Locustberry (*Byrsonima lucida*), a common understory shrub of rimrock pinelands, confined to subtropical areas of s. FL. The larva can survive on Barbados Cherry (*Malpighia glabra*), a non-native ornamental in the same family, planted in urban areas. It was once thought to be the main host, based on rearing experiments with a related species from the Caribbean (*see* Smith et al., 1994), but this seems not to be the case, at least in FL. Larvae eat young hostplant leaves, and live in silk nests made of leaves rolled or tied together.
Occurrence	Fairly common but declining. Miami-Dade Co, Everglades NP, FL Keys (mainly lower Keys). Also W.I. Many broods, all year.
Ecology	Specialist. Limited to a single host, restricted in its habitat and geography.

Locustberry (*Byrsonima lucida*)

av wingspan: 1.45"

Dreamy Duskywing   *Erynnis icelus*

Dorsal male: 4/25/95, Romney, Hampshire Co, WV   **Dorsal female**: 5/5/02, Clay Co, NC

This small, northern duskywing is a familiar sight in mid-spring. Its name is derived from the fact that it lacks a series of white forewing "wrist" marks, common to its genus. This trait gave early investigators the impression that it had its eyes "closed" (same with next species). Such an image would be likely to occur, we think, only to someone who had spent a good deal of time inspecting *Erynnis* wing patterns. But attention to detail is critical in studying this genus, and there are rewards to be gained from such close analysis. Such as noticing that Dreamys actually *do* have a single, tiny white mark on the leading edge of the forewing. (It is advisable not to reapply the original naming convention to reflect the presence of this single "eye," however, as the result would be a logical, but unfortunate bit of nomenclature— the "Winky Duskywing.")

The Dreamy Duskywing is one of nine resident *Erynnis* duskywings on the East Coast (17 occur in the United States). The members of this genus often run close from an identification standpoint and individual variation within species can be significant. As a result, several *Erynnis* combinations are highly challenging to separate, and certain individuals—worn ones in particular—are probably beyond hope for purposes of field identification. Fortunately, Dreamy Duskywings pose few such difficult challenges, and can be enjoyed for their sprightly nature.

Males perch to await females on the ground, or on fallen leaves or low twigs. They frequently gather on muddy trails or streambeds to take minerals. Males fly low to the ground and quickly, but with their small size they do not seem to cover much distance. Females are often seen weaving their way through trailside foliage.

Identification	Very small, often tentatively identifiable by size alone. Wings long, relatively pointed, more full-cut in females; held straight out to the sides. Antennae are long relative to the FW and have a pointed tip, or apiculus (*compare* Sleepy Duskywing), but this can be difficult to see in the wild. The most reliable field mark is the *long, forward-pointed palps* (or "snout"), especially on males. Note, however, the palps may become lost with wear.
Dorsal	Dreamys have a *black-edged, gray "chainlike" pattern* across the outer FW, usually less well-defined near the leading edge than in Sleepy. Between this band and the dark basal scaling is a *silvery-gray "forearm" patch*, usually brighter and more obvious than in Sleepy. *Male has a costal fold*, protecting a series of scent cells enfolded underneath. Finally, though commentators state that Dreamy Duskywings lack white subapical (or "bracelet") spots on the FW, it is possible—even in the field—to detect a very small white dot on each wing, near the leading edge. This mark is absent in Sleepy.
Ventral	Plain brown in both sexes, with rows of pale dots on the HW.

Habitat	An adaptable species, found in a variety of open habitats, usually near woods. Includes clearings, edges, trails, low swales, and streamsides; sometimes fields and meadows (especially when nectaring). Often in moist deciduous and transitional woodlands, yet also at home in dry barrens (pine, shale, and serpentine).
Hostplants	Most sources indicate aspens and poplars (*Populus*), willows (*Salix*), and birches (*Betula*). Yet Iftner et al. (1992) suggest the likelihood of oaks as a hostplant in OH, and LeGrand and Howard (2002) also suspect oaks, noting absence of traditional hosts at w. NC sites where the species is profuse. Tom Allen has similar questions from WV (pers. comm.). The issue warrants further investigation, especially in light of the frequency with which other *Erynnis* use oaks. Probably has multiple hostplants. Caterpillars eat leaves and construct silk nests.
Occurrence	Fairly common to abundant in proper habitat. N. N. Eng. to se VA on the coastal plain (but not s. NJ or Delmarva). South in Appalachians to n. GA. Also well north into CAN and mountain areas of the West. One brood, Apr-Jul, possibly a partial 2nd in South. Overwinters as mature larva in silk nest.
Ecology	Generalist. This successful species is adapted to a variety of habitats, with a number of common hosts reported.

av wingspan: 1.2″

Sleepy Duskywing   *Erynnis brizo*

Dorsal male: 4/25/95, Flintstone, Allegany Co, MD

Dorsal female: 4/24/94, Brattons Run Shale Barrens, Rockingham Co, VA

The Sleepy Duskywing is a harbinger of early spring throughout its range. It flies as soon as plants begin flowering and the air turns warm. Like other species on the wing at that season, it will vanish almost instantaneously if a thick cloud passes overhead, chilling the environment below the critical threshold needed for flight. It is difficult to relate this to specific air temperatures, since spring butterflies exploit sun-warmed microclimates to raise their temperature. The effect of clouding is particularly dramatic at this season, since the tiny, reflective solar ovens on which it relies for body heat cool quickly when direct radiation ceases. If pursued carefully at such moments, Sleepys can be found wrapped around a sheltered twig, moth-style.

Though visually similar to the Dreamy, the two species can be reliably differentiated, with practice, based on size and a combination of field marks. And, while it is not uncommon to find them together, they most often fly in separate habitats. The Dreamy, further, tends to emerge slightly later at any given location, so that it is likely to be fresher than local Sleepys. Once familiar with these factors in a particular locale, an observer can usually form a good working hypothesis as to the identity of a particular butterfly from preliminary cues.

Male Sleepys perch to await females, but are also reported to patrol if population density is exceptionally high. Sleepys fly low, but quite fast. They are often highly skittish, especially on warm days.

Identification	Small relative to Juvenal's (the two often fly together); usually noticeably larger than Dreamy. Antennae shorter than Dreamy's with blunt clubs. Palps project forward, but less so than in Dreamy. When perched, FWs are rarely held as far forward as in Dreamy. Endemic peninsular Florida race (*E. b. somnus*) somewhat darker (Howe, 1975).
Dorsal	*Lacks FW "wrist bracelet."* Well-defined, *steel-gray "chain" pattern* across entire outer FW. A second band, often less well-defined, located nearer the body. Males blackish above, with *gray frosting around FW apex and costa* (less silver than in Dusky). Female more boldly patterned, light "forearm" patch is variably frosted.
Ventral	Brown with fairly well-defined spots (mirroring upper surface).

Habitat	Rather specialized, in dry, pine-oak barrens (sand or shale), oak woods, sandhills, alvars. Sometimes in moister settings, though when found with Dreamys, it is usually due to Dreamys occurring in Sleepy habitats, rather than the reverse.
Hostplants	Scrub Oak (*Quercus ilicifolia*) in North, other scrubby oak species to the south, where *Q. ilicifolia* does not grow.
Occurrence	Locally common, at times abundant, in appropriate habitat. S. N. Eng., south to central FL. One brood, Mar-Apr in FL, late Apr to mid-Jun northward. Overwinters as mature larva.
Ecology	Medium generalist. Widespread. Common hostplant, but fairly habitat-specific.

Sand barrens in spring, Ocean Co, NJ

av wingspan: 1.35″

Juvenal's Duskywing   *Erynnis juvenalis*

Dorsal male: 4/30/94, Allaire State Park, Monmouth Co, NJ

Dorsal female (taking minerals): 4/26/98, Greenbrier Co, WV

Our two largest *Erynnis* duskywings (Juvenal's and Horace's) are similar in appearance and habitats, and are both variable enough to generate confusion in the field. Yet distinguishing features have been identified that will allow us to sort out most individuals. The Juvenal's, moreover, is single-brooded, absent in mid-summer.

Most northern butterflies that emerge in early spring overwinter as adults or pupae. This allows them to emerge quickly when the weather turns favorable. A number of spring-flying spread-winged skippers do overwinter as pupae—including some that do not emerge especially early (e.g., Golden-banded Skippers and Hoary Edge). Yet the earliest emerging of all northern skippers, the Juvenal's Duskywing, enters diapause as a final-stage caterpillar, undergoing final metamorphosis in the spring. This seemingly suboptimal strategy is probably the result of chance: the ancestors of duskywings, it would be argued, simply did not evolve pupal overwintering capability, and our modern early-spring *Erynnis* are lucky enough to be able to get by without it.

Like other butterflies that emerge in early spring, Juvenal's spend considerable time taking minerals from puddles and wet trails (while not neglecting nectar sources when available). We have even seen a fair number of females sipping at puddles, joining in a usually male-dominated activity. Males perch when displaying, as high as 10 to 15 feet off the ground. They dart rapidly around the area of the perch, challenging intruders.

Identification	A large duskywing with full-cut wings. Sexually dimorphic, but less so than Horace's. Perches (like other duskywings) with wings held flat, often slightly recurved at the tips. Rests (and spends the night) with its wings wrapped around a twig, antennae folded back, much like a moth. Has uniformly dark-buffy wing fringes (except for *clitus* race of se AZ and MEX, whose fringes are white). Male's costal fold conceals yellow scent cells.
Dorsal	FW mottled dark brown-and-black, with scattered whitish overscaling over most of the wing. This and Horace's have a *white cell-end dot* just beyond midpoint on the FW (sometimes obscured by scales, especially in fresh males), also a *conspicuous set of large, glassy FW dots* aligned with the subapical wrist band ("bracelet"), as if in a concentric pattern with one to two dots missing. All dots more prominent on female. Female's color pattern generally better defined than male's (but less dimorphism than in Horace's), with a light brown "forearm" patch. Male HW dark and plain, female's more patterned, with variable outer rows of light dots and inner dark mottling.
Ventral	*Pair of light dots on outer HW below diagnostic* (may be reduced in male; absent in white-fringed AZ race).

Habitat	Varied open habitats, oak clearings, edges, trails, roadsides, woodland interiors. Also scrub oak barrens, seldom in wetlands or disturbed habitats; usually scarce/absent along immediate coast.	
Hostplants	A variety of oaks and scrub oaks (*Quercus*). Uses White Oak (*Q. alba*) relatively often (Allen, 1997). Caterpillars build nests by folding a flap *below* the leaf surface, or sometimes above.	
Occurrence	Common and widespread, but annual populations vary. E. US from se ME to s.-central FL, west to e. prairies. AZ race described above. One brood in East, Feb-May in South, Apr-Jun/early, Jul in North. Larva overwinters.	
Ecology	Generalist. Common and adaptable.	**Ventral (showing dots on leading HW edge)**: 4/9/95, Cumberland Co, NJ

av wingspan: 1.4"

Horace's Duskywing *Erynnis horatius*

Dorsal male: 9/5/98, Highlands Hammock SP, Highlands Co, FL

Dorsal female: 7/3/93, Suffolk Co, NY

Any really large duskywing seen in the Northeast in mid-summer is likely to be this species. It is also by far the most common duskywing in most localities in the South. In either case, however, it may be necessary to rule out as many as three other similar species before reaching a firm conclusion (*see* below).

The center of gravity of the Horace's population on the East Coast is more southerly than Juvenal's, even though its range extends nearly as far north in absolute terms. By the time Horace's reaches Massachusetts it is scarce, whereas Juvenal's is still quite common there. Similarly, Juvenal's is a butterfly one sees sparingly in Florida—locally fairly common at best—while Horace's is "one of the most common skippers" in the state (Glassberg et al., 2000, p. 118). In the western mountains, on the other hand, the situation is reversed. There, Horace's occurs northward, in the Rockies, while Juvenal's range is centered in Mexico, reaching the United States only in Arizona and southern-most New Mexico. We have no ready explanation for these anomalous distribution patterns, except to point out that individuals who prefer their natural history in neat packages should not study duskywings.

Behaviorally, Horace's follows the same pattern as most other *Erynnis*. Males perch (usually low), seeking females around midday. Hilltopping has been reported in the Rocky Mountains, but occupying a central position in an open clearing seems to be more the rule in our area. Males nectar actively and also puddle.

Identification	Virtually identical to Juvenal's in general appearance and behavioral style, except that Horace's is somewhat less closely attached to woodlands. Palps ("face") are often snowy white in Horace's, vs. darker (or entirely dark) in Juvenal's, but this trait can vary. Fringes are dark. Many accounts stress the distinctions between Horace's and Juvenal's almost exclusively, but other sources of confusion exist as well, i.e., with Zaruccos and Wild Indigos.
Dorsal	Sexes highly distinct. Male is mostly dark blackish-brown, with reduced subapical marks and little gray overscaling. Easily confused with male Zarucco, but latter lacks white ring over eye, is even blacker and less patterned on basal FW; wing shape also varies (*see* Zarucco). Female has highly patterned FW, more so than Juvenal's, contrast increases with wear; note *extremely prominent glassy hyaline spots*. Can resemble female Wild Indigo, but basal FW less oily-dark, spots much bigger, generally a larger butterfly.
Ventral	Distinctive for what it lacks, namely, two leading HW spots that identify Juvenal's. Zarucco plain blackish below.

Habitat	A wide range of sunny, open spaces, including wood edges, clearings, nearby fields, road-sides, and southern swamps. More likely in gardens than Juvenal's. Occurs along immediate coast in NC (LeGrand & Howard, 2002).
Hostplants	Various oaks (*Quercus*), with some preference for red oak group (Allen, 1997). Possibly wisteria. Early stages similar to Juvenal's.
Occurrence	E. MA to s.-central FL, west to edge of Gr. Plains, e. TX; also Rocky Mountains. Two broods in North (Apr-May and Jun-Aug); three-plus in South, mid-Jan to Oct. Mature larva overwinters.
Ecology	Generalist. An extremely adaptable duskywing.

Ventral (mated female; no dots on leading HW edge): 3/30/02, Pasco Co, FL

av wingspan: 1.45"

Zarucco Duskywing *Erynnis zarucco*

Dorsal male: 9/28/95, Ocala National Forest, Marion Co, FL

Dorsal female: 6/15/96, Edisto Island, Charleston, SC

Male Zarucco Duskywings are difficult to tell from Horace's, even when nectaring side by side on a *Bidens* flower. Yet the Zarucco is less closely related to the oak-feeding Horace's than to a trio of smaller, herb-feeding species (Wild Indigo, Columbine, and Persius). And it is especially close to the western Funereal Duskywing—enough so that some authors consider them a single species (Scott, 1986).

Zarucco Duskywings living in the lower Keys have white-and-brown checkered fringe (vs. brown on the mainland), and some are purely white fringed. Rare but persistent sightings of stray Funereals in the Southeast must therefore be carefully separated from atypical Zaruccos. Also, the Keys race resembles Zaruccos living in the Caribbean, and some

Funereal Duskywing (*Erynnis funeralis*), dorsal, for comparison

believe these could be descendants of early, wandering Funereals (Minno & Emmel, 1993). Thus, the population on the Keys may have completely different origins from mainland populations. (We said duskywings were complicated.)

Identification	Large. Wings relatively long and pointed, more so than Horace's or Wild Indigo. Wing fringes brown or buffy, whiter in lower Keys (*see* above). Little or no white behind eye or on eye rim. Male FW has a costal fold.
Dorsal	Both sexes have a pronounced *pale- to reddish-brown "forearm" patch* (especially prominent relative to the Funereal). *Reduced glassy hyaline spots* vs. Horace's; white cell-end spot diminished. Male: basal FW oily black, virtually no patterning or gray overscaling, HW with pale dots (all dark in Funereal). Female: very similar to female Wild Indigo, but Zarucco often darker between wrist marks and apex. Also similar to Horace's, but Zarucco's hyaline spots are considerably smaller, FW is darker basally.
Ventral	*Distinctively dark*, little dotting on HW.

Habitat	Often found in hot southern coastal environments, roadsides, open fields, woodland edges, open scrubby habitats with hostplant or nectar.
Hostplants	Mostly woody legumes, including Black Locust (*Robinia pseudoacacia*), Dwarf Locust (*R. nana*), Sesbans (*Sesbania* [*macrocarpa*] *exaltata*), various others. Larvae tie leaves together, can be numerous on infested host. Female oviposits on young growth.
Occurrence	Fairly common on se coastal plain, se VA to s.-central FL, across Gulf Coast to e. TX; only *Erynnis* in the Caribbean. Apparently strays (once to Toronto) but earlier confused with Wild Indigo. Three-plus broods, Mar-Oct, all year in FL. Larva overwinters.
Ecology	Medium generalist. Less broadly tolerant than others.

Sesbans (*Sesbania herbacea*)

av wingspan: 1.5″

Wild Indigo Duskywing *Erynnis baptisiae*

Peter W. Post

Dorsal male: 5/20/95, Assunpink WMA, Monmouth Co, NJ

Dorsal female: 7/27/85, Purchase, Westchester Co, NY

The task of identifying eastern duskywings would be much simpler if it were not for this single, confusing butterfly. An intermediate species in both size and appearance—and showing great individual variation—Wild Indigos can be mistaken for almost any of our other *Erynnis* species in one way or another.

And the Wild Indigo's disruptiveness is not limited to current sightings. The species was not formally described until 1936. It is clear that much of the still-lingering uncertainty in the literature over duskywing distribution and identification can be traced to Wild Indigo populations that were incorrectly assigned (in retrospect) to some other species, such as Zarucco or Persius.

Lastly (as if this were not enough), Wild Indigos have rewritten their own range map and life history details in the past 30 years, adding an entirely new major hostplant, Crown Vetch. This non-native legume was planted widely along transportation corridors during this period to prevent erosion. It has inadvertently enabled a formerly restricted species to undergo a complete lifestyle makeover and extension (Shapiro, 1979).

Male Wild Indigos perch most of the day near hostplants, but they are known to patrol in late afternoon. In areas of high population density, i.e., near Crown Vetch, structured displays may break down altogether (Gochfeld & Burger, 1997).

Identification	Smaller on average than Juvenal's, Horace's, or Zarucco, usually larger than Persius or Columbine. But many individuals are not distinguishable by size, especially large females or small males. Wings less pointed and angular than Zarucco. Male has a costal fold. Since genitalic characteristics run close in the group (Klots, 1951; Scott, 1986), habitat and hostplant associations use can play an important role.
Dorsal	FW dark basally, especially in males, though some pattern usually evident. Females more mottled, can resemble Zarucco or Horace's. Glassy wrist band ("bracelet") is small and indistinct (especially in males). *Usually lacks mid-FW cell-end dot.* Light patch on "forearm" less distinct, does not form a continuous transverse band (even in females). Innermost dot of wrist "bracelet," when present, is offset toward apex (*compare* Persius).
Ventral	Dark with light dot-rows (*compare* Zarucco).

Habitat	Originally favored barrens, usually dry and sandy, with hostplant. Now an open-space generalist, feeds widely on Crown Vetch.
Hostplants	Crown Vetch (*Coronilla varia*) is now the dominant host. Also Wild Indigo (*Baptisia tinctoria*) and other *Baptisias* and lupines locally.
Occurrence	Originally local, mostly uncommon. Now locally abundant, mostly on account of expanded hostplant utilization. Occurs from s. N. Eng., south to n. FL; reduced or absent se coastal plain. Also west to edge of Gr. Plains. Three-plus broods, Mar-Oct in South, May-Sep in North; sometimes a partial 4th brood in Oct. Overwinters as mature larva.
Ecology	**Medium generalist.** Originally more restricted; quickly becoming a full-scale generalist, as its introduced hostplant is prolific.

Crown Vetch (*Coronilla varia*)

av wingspan: 1.3"

Persius Duskywing   *Erynnis persius*

Dorsal male; specimen photo

Dorsal female (ovipositing on *Baptisia*): 5/29/99, Voluntown, New London Co, CT

Until the Wild Indigo Duskywing was described in 1936 (*see* previous account), most records for that species were attributed to Persius. Thus, Weed (1917, p. 276) thought that the Persius Duskywing occurred "from ocean to ocean along the northern tier of states . . . [and] in the Eastern states as far south as Florida"—an understandable composite of the geographic ranges of Wild Indigo and Persius Duskywings. Weed also thought that Persius fed on "willows and poplars," a widely held view that is now questioned.

Even after Wild Indigo was split off from Persius, it took decades for investigators to realize just how rare the latter species actually is in the East. Confusion was fueled by the fact that Persius appears to have undergone a significant range contraction during this period (based on apparently creditable records from areas where the species is not currently known). For purposes of this account, we indicate currently documented eastern colonies on the range map, and are less certain as to prior range limits; it is hoped that over time other bona fide populations will come to light. Western populations (*borealis* race) differ significantly in habits and appearance, and will not be addressed here.

The Persius's behavior in the East is distinctive—and frustrating. Males can be found on low perches some of the time, but are most often on the move, and will veer erratically into nearby woods at the smallest disturbance, where they are extremely difficult to relocate. We have seen males patrol near hostplants as females oviposited.

Identification	A small duskywing, intermediate between Wild Indigo and Columbine (the two other members of the Persius complex). Visual similarities among the three are daunting. Even male genitalia run close (Klots, 1951), though Persius is best differentiated in this aspect (Layberry et al., 1998). Many authors recommend separating individuals by hostplant associations, but overlap hostplant use in our region negates this strategy. Male has a FW costal fold.
Dorsal	Male darkly mottled with a *small, neat wrist "bracelet" in which all dots are aligned in a row perpendicular to leading FW edge*. Raised, white hairs cover much of FW, but are difficult to observe (at best) in the field; they may be visible in good-quality macrophotos. Female lighter, more patterned; those we have seen have a continuous, transverse band across the entire FW, not interrupted, as in typical Wild Indigo or Columbine.
Ventral	Mainly dark brown with mottling and dots.

Habitat	In the East, dry, patchy pine-oak woods, pine barrens. Often flies with Frosted Elfin.
Hostplants	Known only to use Wild Indigo (*Baptisia tinctoria*) at existing colonies in our area, though midwestern populations use Wild Lupines (*Lupinus perennis*). Others in West, including Golden Banner (*Thermopsis*). Eggs laid beneath young leaves.
Occurrence	Extremely rare. Presently known from a handful of locations in e. N. Eng. (CT, MA, NH). Apparently extirpated in NY, se ONT, e. MD, w. VA. Largest eastern populations are west of Gr. Lakes. One brood, May to early Jun in our area. Mature larva overwinters.
Ecology	Specialist. Succession of patchy, pine-oak habitats may be a partial clue to Persius's marginal status.

Opening in oak-pine woods, New London Co, CT

av wingspan: 1.25"

Columbine Duskywing   *Erynnis lucilius*

Dorsal male: 5/21/94, Limerick Barrens, Jefferson Co, NY

Dorsal female: 5/21/94, Limerick Barrens, Jefferson Co, NY

This miniature, northern duskywing is another enigmatic member of the *Erynnis* clan. It is barely differentiable from Wild Indigo, in the field or the laboratory (Layberry et al., 1998). Traditionally, it has been identified by hostplant association and by its preference for rocky outcrops and alvars. But this approach has limits. Hostplant association is unreliable, since Wild Indigos have reproduced on garden columbines. And a Wild Indigo population, radiating along a highway corridor where Crown Vetch is planted, might well encounter a palatable stand of Wild Columbine on a rock face exposed by construction blasting. (If, that is, the Columbine Duskywings on the face had not already discovered the Crown Vetch below and "jumped" to it, as their close cousins did some generations earlier.) Real possibilities exist for new population overlaps to develop, in which case hybridization is not out of the question.

The Columbine is primarily a northern butterfly, thinly distributed in a range extending from Minnesota to southern New England. Another population, largely discontinuous, occurs southward in the mid-Appalachians. Wild Indigo Duskywings continue southward beyond this point, reaching northern Florida.

We have observed male Columbines in alvar habitats perched on or near the ground, in areas with intermittent stands of hostplant and meandering females. Fast-flying for a small species, usually very difficult to approach.

Identification	A tiny duskywing, often best distinguished from Wild Indigo by size, habitat, and range. But larger-sized populations exist, especially in the Appalachians (David Wagner, pers. comm.), so caution is advised. Males have FW costal folds, and lack the raised, hairlike scales found in Persius.
Dorsal	Highly similar to its close relatives, especially Wild Indigo Duskywing. Male dark basally, with a small, silvery patch inside the "wrist bracelet." Often rather pale along FW margin. Female more highly patterned; the "forearm" patch is brown rather than silver, and two postmedian FW dots are relatively prominent. Columbine wrist marks are intermediate between the tight pattern of Persius and the looser, more zigzagged pattern of Wild Indigo.
Ventral	Mostly dark with two dot rows near the margin.

Habitat	Some affinity for alkaline habitats where hostplant grows, such as alvars, limestone outcrops, shale slopes below limestone cliffs, rocky banks, etc. But the hostplant occurs in varied settings, and the butterfly is also found in rocky deciduous woods, seemingly more acidic.
Hostplants	Wild Columbine (*Aquilegia canadensis*). Survives on garden columbine (*A. vulgaris*), though not certain that females select it in the wild. In one case, however, a Wild Indigo female reportedly did just that (Gochfeld & Burger, 1997). Our only skipper with its host in the Buttercup family (and our only butterfly of any kind except Appalachian Azure).
Occurrence	Range/abundance uncertain, but locally common on alvars. Very local in NY, CT, PA; formerly in NJ, N. Eng. Separately, in ne WV, w. VA. Also s. QUE, west to Gr. Lakes. Two to three broods reported, Apr–early Jun, late Jul–Sep. Mature larvae overwinter in leaf litter.
Ecology	**Specialist.** Limited lifestyle in terms of geography, habitat, and hostplant.

Wild Columbine (*Aquilegia canadensis*)

av wingspan: 1.15″

Mottled Duskywing   *Erynnis martialis*

Dorsal Male: 7/23/93, Albany Pine Bush, Albany Co, NY

New Jersey Tea (*Ceanothus americanus*)

The Mottled is a popular duskywing among observers, attractive in appearance and reasonably easy to identify. It is unfortunately also embattled and declining, gone from most of its northeastern range, conservation-listed in five East Coast states, and threatened in Indiana. Habitat loss is definitely implicated in this decline, but additional factors may be involved as well, such as earlier widespread Gypsy Moth spraying in its localized habitats.

The Mottled Duskywing is an outlier in its genus, distinctive in lifestyle and appearance. In a group dominated by oak- and legume-feeders, the Mottled is the only *Erynnis* to feed on buckthorns (*Rhamnaceae*)—other than its very close western relative, the Pacuvius Duskywing. Indeed, apart from Henry's Elfin, which feeds on the exotic European Buckthorn (*Rhamnus frangula*) in New England, and sometimes azures, no other butterfly in our region uses any plant in this family. The biological advantages of selecting buckthorns have not been determined.

Although the Mottled Duskywing has a wide geographical distribution, it seems never to have been particularly common anywhere in its range. Western populations, living in Rocky Mountain chaparral and prairie hills, have differing hostplants and biology.

Among eastern duskywings, the Mottled is notable as a hilltopper. In the Albany Pine Bush, several males can regularly be found in season perching aggressively at the highest overlook in the reserve. In active flight. they are visually a blur until landing, and once settled will launch into flight again at any slight disturbance. Adults feed on mud and decaying matter.

Identification	A small species, yet with full-cut wings and a bold pattern that seem to amplify its modest proportions. Buffy brown wing fringe. Male FW has a costal fold.
Dorsal	Sexes largely similar, but females have brighter, more contrasting patterns. 1st brood is generally smaller and darker than summer brood. Brown FW is overlaid with a series of irregularly positioned black patches and white markings. Together, these convey a *mottled, or checkerboard appearance.* The only other *Erynnis* in our area approaching this degree of pattern variation on the basal FW is the female Horace's, which is considerably larger and lacks mottling on the dorsal HW. *Fresh individuals show purplish-blue iridescence* in favorable light.

In the postmedial FW (just beyond the midwing), there is a pale patch, sometimes extending into a vague, transverse band. Note the small, silvery patch at the leading edge of this band, near costal wing margin. The FW apex is dusky, with a series of outward-pointed, white chevrons on the submargin. White outer wing markings are more pronounced in the spring brood (Opler & Krizek, 1984).

Despite these several distinctive FW characteristics, we identify this species primarily by looking at the dorsal HW, which is *strongly mottled (or checkered)*, well into the basal area. The extent of this checkering is distinctive in the Mottled.

Habitat	Dry, open woods and barrens, including shale, serpentine, and pine barrens; also sandhills and alvars, sometimes brushy fields. In the West, prairie hills and chaparral (in Rocky Mountains).
Hostplants	After some early confusion (e.g., Klots, 1951), it has been determined that Rhamnaceae is the sole host family. In our region, New Jersey Tea (*Ceanothus americanus*).
Occurrence	Rare to uncommon throughout its range, nearly gone from Northeast except isolated barrens in PA and Albany Pine Bush. Scattered across Southeast n. of FL, west into prairie states, Rockies in CO. Two broods in East, beginning Mar in GA to May in NY, also mid-Jul. Partial 3rd brood sometimes in Deep South. Mature larvae overwinter.
Ecology	**Specialist.** Uses a single hostplant, thinly distributed in barrens and other niche habitats in our region.

av wingspan: 1.3"

"Appalachian" Grizzled Skipper   *Pyrgus centaureae wyandot*

Dorsal: 4/24/80, Green Ridge State Forest, Allegheny Co, MD

Ventral: 4/24/94, Brattons Run Shale Barrens, Rockbridge Co, VA

Few experiences in our region provide greater satisfaction than successfully tracking down this elusive butterfly (Cech, 1995). Even in active colonies, it can be maddeningly difficult to observe.

The Grizzled Skipper is one of very few holarctic skippers in our region (i.e., found in boreal areas across the Northern Hemisphere). Though still fairly common in some parts of its range, it is broadly declining globally. Our local form—the small, dark "Appalachian" Grizzled Skipper (*Pyrgus centaureae wyandot*), believed by many to be a separate species—is in particularly bad straits. It is a federal species of concern in the United States and is conservation-listed in seven separate states (including some where it is probably already extirpated). The definitive causes for this decline are not fully known, but the "Appalachian" Grizzled Skipper appears especially vulnerable to spraying. Population losses have been linked to mosquito control programs in the 1950s (Gochfeld & Burger, 1997) and to Gypsy Moth spraying in more recent decades (Allen, 1997). Climate change may also be a factor.

Males in the genus *Pyrgus* patrol to find females (rather than perch). And while not fast-flying, the butterfly has legendary evasive abilities. Klots (1951, p. 215) recounts that individuals in plain view can "vanish as if by magic to reappear thirty feet away." We have seen a Grizzled drift casually into apparently thin cover, never to be found again despite diligent searching. Even when patrolling in the open, it is difficult to follow the low, winding flight. Readily observed only in dense colonies (but these are exceptionally rare).

Identification	Small, easily mistaken for a moth on the wing (Layberry et al., 1998). The body is *covered with hairlike scales*, giving the impression of fur. This is effective as insulation, a token of its boreal lineage. *Fringes are boldly checkered black-and-white*. Male has costal fold. In its range, can be confused only with Common Checkered-Skipper.
Dorsal	Dark blackish-gray above, with *large, squarish FW spots*, arranged as if in jumbled rows. Light hairs create FW flecking in fresh individuals. Grizzled has fewer FW markings than Common. Note joined dot-pair partway out on FW edge (these are separated in Common). Female Commons even more similar, but FW markings more neatly aligned. Also, Common has a continuous spot-band across central HW; in Grizzled, there is only one significant patch, near leading HW edge.
Ventral	FW pattern mirrors dorsal; HW crossed by patchy, olive-brown to blackish bands.

Habitat Varies globally. In our area, generally confined to open and semishaded shale barrens slopes, especially in clearings or along roads or easements. Elsewhere, oak savanna, tundra, etc.

Hostplants Dwarf Cinquefoil (*Potentilla canadensis*), a common host; uses Wild Strawberry (*Fragaria*), as in Midwest, and other cinquefoils (Allen, 1997).

Occurrence Rare and declining. Up to 40 a day in one WV colony, otherwise one to two (or zero). Gone from NY, NJ. Remains mostly west of I-81 corridor, PA to VA. One brood, mostly Apr. Larvae roll leaves into nest, develop slowly (two-year cycle in arctic). Overwinters as pupa.

Ecology *Medium specialist. Wyandot* form highly local, despite using an extremely widespread hostplant.

Dwarf Cinquefoil (*Potentilla canadensis*), growing on shale barrens

av wingspan: 1.0"

Common / White Checkered-Skipper   *Pyrgus communis / albescens*

Dorsal male: 9/21/96, Gainesville, Alachua Co, FL

Dorsal female:7/11/99, Sterling Forest, Orange Co, NY

Common

Both

A former race of Common Checkered-Skipper (*P. communis*) was recently split off as the separate White Checkered-Skipper (*P. albescens*) (Burns, 2000). At present, these variable species can only be reliably distinguished by dissection, so they will be treated here together.

Formerly unknown in our area, the White Checkered-Skipper rapidly colonized the Florida peninsula beginning in the early 1990s, as Common Checkered-Skipper nearly vanished (Calhoun, 2002). Such local "replacements" have been seen in other areas where the two species coexist, and may reflect differing responses to Florida's recent severe droughts. But it is too early to know.

Males patrol a measured territory seeking mates, mostly in the afternoon, chasing any moving object they might encounter. Emigrates irregularly northward in summer, reaching Canada and southern New England late in the season.

Ventral: 3/31/99, Santa Ana NWR, Hildago Co, TX

Identification	Small but striking. FW fringe strongly checked brown-and-white; HW markings maybe almost pure white. (*Compare* Tropical.)
Dorsal	Male highly patterned with bluish, hairlike scales on the body and inner wings. Female less boldly patterned, lacks blue hairs. Row of white dots along the immediate outer wing margins are *tiny,* and *the final dot at the FW apex is missing* or very faint. Also lacks one large FW dot found on Tropical (*compare* that account).
Ventral	Eggshell white to yellowish on HW, with two major spot bands, variably olive-brown to grayish. At base of leading HW edge, note two connected black dots, somewhat like a barbell. A broad, smudgy mark at midpoint of leading HW edge, found in Tropical, is absent in Common/White (Glassberg, 1999).

Habitat	Open, sunny places with low vegetation, often some exposed soil, including fields, roadsides, easements, lawns, gardens, clearings, etc.
Hostplants	A variety of wild and cultivated mallows (*Malvaceae*), including sidas (*Sida*), Velvetleaf (*Abutilon theophrasti*), Hollyhock (*Althaea rosea*), and others.
Occurrence	Common. CAN south to ARG. Permanent resident to about 40th parallel (Philadelphia), recolonizes yearly northward. Winter stress thins northern populations, which rebuild each summer. Three to four broods, Mar-Nov in South, mostly Jul-Oct in North. *Albescens* may reach far s. FL (*communis* has not). Mature larvae diapause.
Ecology	Generalist. Extremely widespread and adaptable.

Common Mallow (or 'Cheeses') (*Malva neglecta*). **Inset**: Flower

av wingspan: 1.1"

Tropical Checkered-Skipper *Pyrgus oileus*

Dorsal male: 9/24/95, Loxahatchee NWR, Palm Beach Co, FL

Dorsal female: 4/12/92, Palm Beach Co, FL

On a tour of south Florida, this ubiquitous little skipper is one of few butterflies likely to be seen on every day of a trip. It is so widespread, in fact, that its habitats are best described in terms of the few locations where it is *not* likely to be found (such as hammock interiors or salt marshes).

Yet oddly enough the Tropical's range stops abruptly at the Florida border. Apart from a handful of strays reaching Georgia and the Carolinas, the Tropical shows no hint of the Common's strong, emigratory proclivities.

As with other *Pyrgus* skippers, males patrol for females (though they may also sometimes perch). The flight is fast and blurry, low to the ground. It often flies for long intervals, punctuated by occasional nectar breaks.

Ventral: 9/24/95, Tree Tops Park, Broward Co, FL

Identification	Small. Fringes are checkered, as in Common/White (called "Common" from here on), but note *almost continuously dark fringe* on outer FW near apex. Male has costal folds.
Dorsal	Dark brown base color. Male brightly patterned with extensive blue-gray hairs on body and inner wings. Female is less brightly patterned than male, but less dimorphic than Common; female has far fewer blue hairs (most evident on inner HW). Marginal dots on wing-edge larger than in Common, may approach size of adjacent submarginal dots; note *final dot present at FW apex* (absent in Common). On FW, large rectangular cell-end white spot is followed by a distinct oblong or rectangular mark, which is absent or very faint in Common.
Ventral	HW bands less contrasting than in Common. Note *smudgy mark at midwing on leading HW edge*.

Habitat As noted, almost any open, sunny habitat with mallows. Tolerates (even prefers) locations with some shade. More likely in moist habitats near wetlands than Common.

Hostplants Similar to Common, especially Arrowleaf Sida (*Sida rhombifolia*) and other sidas. May use fewer hosts than Common. Female lays a single egg on top of hostplant leaf, often late afternoon. Larva lives in silk-bound leaf web.

Occurrence Very common. N. FL, across panhandle along Gulf Coast to TX. Also W.I., So. Amer. to ARG (So. Amer. race may be a different, sibling species). Many broods, all year in s. FL.

Ecology Generalist. Not as wide-ranging as Common, but uses many habitats and a variety of hosts.

Arrowleaf Sida (*Sida rhombifolia*)

av wingspan: 1.2"

Hayhurst's Scallopwing *Staphylus hayhurstii*

Dorsal: 9/22/96, Newnans Lake, Alachua Co, FL

Alligatorweed (*Alternanthera flavescens*)

Formerly called Scalloped Sootywing. This tiny, duskywing-like butterfly leads an inconspicuous life, concealed under foliage or basking in dappled sun beside woodland paths. But many features of its clandestine lifestyle warrant closer scrutiny. The Hayhurst's Scallopwings that we see in North America are but the tip of a huge, neotropical "iceberg," one plain-brown Pyrgine skipper among scores of others found in the New World tropics. The genus *Staphylus* alone has over three dozen species, many nearly identical. And several additional genera (*Gorgythion, Carrhenes, Bolla, Pellica, Nisoniades*, etc.) contain hosts of confusingly similar "little brown jobs" as well. Interestingly, the visual similarity of many of these plain skippers is offset by certain less obvious traits, such as distinctive genitalia, and most likely by distinctive chemical signatures, which allow for intraspecific recognition.

Still, among this large, neotropical assemblage, why has just one extended its range to southern New Jersey and to the central Great Plains—where it is biologically quite incongruous? And why has it largely eschewed native hosts in favor of Lamb's Quarters, a rank European weed? And why, further, having made this jump to a widespread host, is it nonetheless quite rare and thinly distributed over most of its large range? These are logical questions for which definitive answers remain to be found.

Hayhurst's perch low, often on trailside foliage or other thick vegetation, and will dart into thicker cover when disturbed. They fly rapidly, though rarely for long distances, and visit flowers readily. Typically colonial.

Identification	Among our smallest spread-winged skippers. Sometimes mistaken for Common Sootywing, but the two are in fact only superficially similar. Most easily identified by *scalloped wing edges,* most notable on the HW. Fringe is checkered black-and-tan, though this feature can be hard to see in the field. Males have an evident costal fold. Perches with wings open. Has *dark palps,* or "face" (*compare* Common Sootywing).
Dorsal	Dark base color on wings, overlaid by even darker, *concentric bands on both wing surfaces.* Two small, white hyaline dots on the FW "wrist," another at the cell-end (midwing). These markings vary in expression with wear and sex (more prominent in females; cell-end dot is faint or absent in some males). Fresh individuals are densely flecked with *small gold and silver scales* on both wings. This pattern is very striking when viewed closely.
Ventral	Similar to dorsal surface but paler. The ventral surface is rarely exposed to view.

Habitat	Hayhurst's Scallopwings are usually found in or near shaded edge habitats, with dappled light. Some of their habitats are moist, others quite dry—though in the latter case they are often within easy reach of a wet prairie, lakeside, or marsh. Non-native habitats include roadsides, trail or field edges, even parks and gardens (they visit city gardens in some cases quite readily for nectar).
Hostplants	Mainly Lamb's Quarters, or Pigweed (*Chenopodium album*), an introduced European weed, widespread in disturbed habitats in most of our area. Native hosts include Amaranths, such as Alligatorweed (*Alternanthera flavescens*) and especially Bloodleaf (*Iresine diffusa*) in s. FL. Both fairly close relatives of Lamb's Quarters.
Occurrence	Locally fairly common, but colonial and thinly distributed. Appears to spread along river corridors, though perhaps also coastally. S. NJ (known from Ocean and especially Cape May Cos), south along coastal plain and Piedmont to s. FL, rare in Appalachians. Formerly abundant in s. FL, but declined greatly during 20th century. Also Gr. Plains to e. NE, central TX; one isolated colony at Pelee Island, ONT. Two broods, May-Aug in North, Feb-Dec in FL. Female lays single egg on underside of host leaf, larva builds rolled leaf nest, overwinters as last instar in dead leaf nest.
Ecology	**Medium generalist.** Though uncommon, uses an extremely widespread weed as host, in a broad range of habitats.

av wingspan: 1.1"

Common Sootywing *Pholisora catullus*

Dorsal: 7/3/93, Belmont, Suffolk Co, NY

Ventral: 7/4/97, Lakehurst, Ocean Co, NJ

Despite its rather disparaging name, the Common Sootywing is actually a handsome butterfly when observed at close range in fresh condition (worn individuals, of this as well as many other species, may admittedly lack visual appeal). Sootywings are also quite successful, having colonized a wide range of natural and altered habitats across the continent. In the East, they are especially successful in disturbed habitats and waste areas.

The Common Sootywing is related to a group of mostly arid-country butterflies. U.S species are centered mainly in the Southwest and comprise two closely related genera, *Pholisora* and *Hesoeopsis*, both commonly referred to as "sootywings." Though taxonomists do not consider sootywings especially close to the Hayhurst's Scallopwing phylogenetically (their placement here is based on visual similarity), significant commonalities exist between them. Both of our species use hostplants in the Amaranth and related families and both may have radiated in North America using river corridors and coastal habitats, where hostplants are prevalent. Beyond these interesting parallels, however, the two species have developed distinct lifestyles, as discussed below.

Common Sootywing males patrol nearly throughout the day looking for females, and mate opportunistically whenever one is encountered. The species tends to be skittish, nearly always on the move. Wings usually held flat at rest, but sometimes raised while basking or taking nectar. Adults visit flowers for nectar.

Identification	Very small, with long, rounded FWs and dark brown fringe. Palps project forward, creating a somewhat pointed look. Underside of the palps ("face") and upper thorax are white (*compare* Hayhurst's).
Dorsal	*Glossy black* (especially basally, though fades with wear), with no dark concentric bands or flecking. Variable, crescent-shaped band of *pure white dots* on outer FW, best expressed in female, and an isolated cell-end mid-FW dot.
Ventral	All dark brown with white markings on outer portion of both wings.

Habitat	Use of introduced hostplant obscures original habitat preference, which was probably sandy river plains, perhaps brackish marshes. Occurs today in a wide array of open, sunny places, including badly disturbed habitats and waste areas with low vegetation.
Hostplants	Mainly Lamb's Quarters (*Chenopodium album*); also Cockscomb (*Celosia* sp.) and miscellaneous amaranths—including Tumbleweed (*Amaranthus graecizans*) in the West. Lamb's Quarters is found widely in disturbed habitats in our region. Female lays a single egg on top of hostplant leaf, often at midday. Larvae live in folded leaf shelters.
Occurrence	Fairly common to abundant, but populations vary significantly by year, with occasional multiyear crashes. Range forms a wide band across the central portion of the continent (from s. ONT to central MEX). Generally absent in boreal areas and extreme desert. In our area, central N. Eng. south to n. FL. Two to three broods, Apr-Sep in North, Feb-Oct in FL. Overwinters as mature larva in strongly bound silk nest.
Ecology	**Generalist.** One of few butterfly species to profit from human development, adaptable over its range in various ubiquitous habitats.

Lamb's Quarters, or Pigweed (*Chenopodium album*)

av wingspan: 1.1″

Grass Skippers: Subfamily Hesperiinae

Estimates vary, but there are probably about 3,000 species of grass skippers scattered throughout the world. Of these, around 140 are found in North America and 60 reside in our region. While that is a modest percentage of the whole, it is more than enough to keep local observers well-occupied.

The hesperid subfamily goes by several alternate common names, each reflecting something of the group's character: *fold-winged skippers* (for their typical resting pose, in contrast with that of spread-winged skippers, in the subfamily the Pyrginae); *branded skippers* (because of the male's stigma, described below); and *smaller skippers* (because they are generally smaller than Pyrginae).

Anatomically, grass skippers are distinguished by a number of technical characters. The tip of the antennal club (apiculus) is generally shorter than in spread-winged skippers. The abdomen is long, normally extending to or beyond the hindwing margin. The outer veins of the forewing show certain characteristic patterns. The base color is usually brown or golden-orange, with brown margins. And, perhaps most important, the males in a majority of species have a distinctive dark stigma across the dorsal forewing which holds androconial cells that disperse pheromones, used to influence females during courtship. The stigma is also sometimes referred to as a "brand."

Grass skippers feed on monocotyledons, mainly grasses and sedges. And while many species will accept a number of the 1,400 or more U.S. grasses in captivity, far greater selectivity is typical shown in the field (Layberry et al., 1998). Hostplant selectivity in the wild is thought to be especially high among sedge-feeders (Allen, 1997).

Most grass skippers are medium-sized. They beat their wings rapidly, often making them appear blurred to an observer, and they fly rather fast and straight. When basking, they adopt a characteristic "jet fighter" pose, with the hindwings held flat to the sides and the forewings raised to a 45 degree angle (or sometimes almost straight up, as in roadside-skippers). While they can cover significant distances in daily feeding forays, grass skippers tend to be comparatively sedentary overall. Few migrate. Their exceptionally long proboscises (often exceeding the length of the forewing) allow active and efficient nectaring. When seeking mates, males tend to be strong perchers, though a few also patrol (e.g., Least and European Skippers). And, like the proverbial banker, most start the day later, and quit earlier, than other butterflies.

Grass skipper caterpillars generally construct a nest by folding over hostplant leaves or bundling them together, fastened with silk threads. Compared with other butterfly groups, the early stages of grass skippers have been poorly studied in the field. Many feed nocturnally, but there is also the "difficulty of identifying grasses and of finding larval nests among literally millions of grass stems" (Layberry et al. 1998, p. 47). This is an area where knowledge gaps could be patched by watchful field observers. Some grass skipper caterpillars build nests at ground level, at the base of the hostplant. Those living on or near the ground tend to be dark-bodied with blackish heads, while those building leaf nests tend to be green with light heads, and often have distractive patterns on the body (Minno & Emmel, 1993).

Organizing grass skippers into meaningful subgroups is a difficult task, for professional taxonomists and casual observers alike. Subgroups based on taxonomic classification are cluttered with many small, heterogeneous genera. But if we group by visual appearance or by ecological style, we will frequently need to cut across taxonomic lines. Here, we specify a few *ad hoc* species clusters, purely for illustrative purposes, to convey a sense of the subfamily:

Intermediate Skippers (*Carterocephalus*): One species, Arctic Skipper, formerly in separate subfamily (*see* account).

Small Brown Skippers (*Nastra, Cymaenes, Lerodea*): A relatively nondescript group of skippers with plain brown wings and variable white markings, mostly found in open habitats. Taxonomically diverse.

Small Orange Skippers (*Thymelicus, Copaeodes, Ancyloxypha*): A motley trio of mainly open-country skippers, two of which are distinctive in that males patrol for mates, vs. perching.

Beardgrass Skippers (*Hesperia, Atrytonopsis*): Usually, but not always, beardgrass (*Andropogon*, etc.) specialists with targeted seasonal broods. *Hesperia* build nests near hostplant base, enhancing survival in fire-prone settings such as savannas and pine barrens. Many fly at specific, often extreme points during the season. Wary and elusive.

Meadow Skippers (*Lerema, Atalopedes, Hylephila, Polites, Pompeius, Wallengrenia*): Mainly grass feeders, these are the familiar, "everyday" skippers of meadows and fields.

Marsh Skippers (*Atrytone, Anatrytone, Problema, Poanes, Euphyes*): Mostly dedicated marsh-dwellers, including many sedge- and rush-feeders. This group reaches a high level of diversity in our area, with a number of endemic or near-endemic species. Taxonomy has been unsettled, with significant recent realignments.

Roadside Skippers (*Amblyscirtes*): A small, distinctive genus of cane- and grass-feeders. Much sought-after.

Nongrass-Feeders (*Asbolis, Calpodes*): Two large species that have left grass for palms or other monocotyledons.

Longwings (*Panoquina*): Distinctive in silhouette (long, slender forewings); long side-stripe on abdomen.

Arctic Skipper *Carterocephalus palaemon*

Dorsal: 6/18/94, Pondicherry Wildlife Refuge, Coos Co, NJ

Ventral: 6/15/97, Wentworth Location, Coos Co, NH

Two pleasures await those who observe this boreal (but not really arctic) species—first, the butterfly itself, and second, the picturesque settings in which it characteristically occurs. Its visual appeal is self-evident, and its habitats include many lush, boreal lowlands and clearings in rich northern transitional forests.

Until recently, the Arctic Skipper was classified in the subfamily Heteropterinae (Intermediate Skippers), a small group of butterflies with traits halfway between those of spread-winged and grass skippers. (They bask with wings flat, like a Pyrgine, yet rest with wings folded, like a hesperid skipper.) Authorities now generally lump this subfamily with the grass skippers (Hesperiinae).

Arctic Skippers are holarctic in distribution, occurring in Eurasia, Alaska, and northern Canada. The North American race is *C. p. mandan*. Though population levels in our area seem stable, the species is declining or threatened in areas as widely separated as Japan and Great Britain. (They vanished from the English midlands in the mid-1970s and are now confined to a limited area of western Scotland.)

Flight is weak, by skipper standards, and low to the ground. Males perch to await females, or sometimes patrol in sheltered openings. Females are more mobile and have been tracked for several kilometers over their life span in the United Kingdom.

Identification	Small. This is not a difficult butterfly to identify, but it is worth studying for its attractiveness alone. Wings are fairly long proportionally, with buffy-brown fringe (solid above tannish below with dark cross-markings, at vein-ends). Both wings are held mainly flat when basking, or cocked slightly, up to a 45 degree angle. At rest, or when taking minerals on the ground, the wings are usually closed, in grass skipper fashion.
Dorsal	Blackish-brown above with *large, mostly geometric yellow patches* on the FW and basal HW. Also a line of vague, reddish marks along FW margin. Yellow flecking on body and inner wings when fresh.
Ventral	FW roughly mirrors dorsal, but with *reverse colors*. HW is a rich, brownish-orange with three rows of rather large, *variable-sized cream-white oval spots*.

Habitat	Variable across its wide range, but mainly openings and edges in transitional forest; also trails and grassy areas near meanders or other wet, boreal lowlands. Subarctic tundra in North.
Hostplants	Grasses. No specific reports in our area. But reedgrass (*Calamagrostis* sp.) and brome (*Bromus* sp.) used elsewhere.
Occurrence	Locally fairly common. N. N. Eng., south to n. CT/RI, nw NJ, e. PA; also n.-central NY; western mountains in US to AK; also Eurasia. One brood, mostly Jun. Mature larva overwinters, spring pupae resemble withered grass.
Ecology	**Medium specialist.** Confined to boreal habitats, but found in a variety of settings, probably on common grasses.

Boggy northern stream, Washington Co, ME

av wingspan: 1.0″

Swarthy Skipper   *Nastra lherminier*

Dorsal: 9/8/96, Jamaica Bay NWR, Queens Co, NY

Ventral: 6/22/94, Lakehurst, Ocean Co, NJ

Klots (1951, p. 266) said it well enough: "The Swarthy Skipper is distinguished chiefly by its lack of distinguishing characteristics." If you have studied the dorsal surface of a small, dark brown grass skipper in our region and find it has essentially no field marks at all, it is likely to be a Swarthy. (This rule quickly breaks down in the tropics, where there are numerous qualified entrants in the "least descript skipper" category.)

All of the 12 or so members of the genus *Nastra* reside in the Americas, and most are tropical. Two others occur in the southern United States (Julia's and Neamathla), but the Swarthy is by far the most northerly. The entire genus was formerly included in *Lerodea*, with the Eufala Skipper. These genera include many plain, smallish open-country species, sometimes referred to by birding types as "little brown jobs" (or "LBJs").

The Swarthy's lifestyle is generally retiring and colonial (with exceptions, as discussed below). Males perch on thick grass blades or low-hanging leaves when displaying. Females and foraging males thread their way inconspicuously through grassy undergrowth. They seem to spend comparatively little time nectaring, but may simply visit lower, less conspicuous flowers than other species. Mating has been reported from late afternoon. We have also observed pairing in mid-morning, however.

Identification A small skipper, and as already noted an exceptionally nondescript one. Several observers have emphasized the contrast between the triangular shape of the FW and the rounder, more full-cut HW. The antennal shaft is light below, with alternating black and white cross-bands. As in all *Nastra*, the male lacks a stigma (brand). Sexes are similar. In southern Florida, needs to be distinguished from Three-spotted Skipper, as well as from Neamathla, depending on location. Eufala is generally paler (*see* that species account).

Dorsal Plain dark brown above, often blackish. A variable, bronzy-metallic sheen develops with wear. A faint pair of ill-defined spots is occasionally visible on the mid-FW.

Ventral In fresh individuals, yellowish-brown scales cover the HW. Resembles Neamathla, or even a small Tawny-edged Skipper, but *light HW veins* are distinctive. Becomes darker and less distinctively marked with age.

Habitat Several quite different habitat types are used. Most often associated with dry barrens and grassy spots, including well-drained fields and pastures, powerline cuts, sandy waste areas, coastal scrub, savannas, dry forest clearings, rimrock pinelands, etc. But also found at times on grassy marsh edges. Even known to occasionally colonize wetlands, e.g., an overgrown cranberry bog in Lakehurst, NJ. In boom years, radiates widely, to nearly any open, grassy habitat, including roadsides, lawns, gardens, vacant lots, etc.

Hostplants Not fully reported. The only known hostplant in the wild is Little Bluestem Grass (*Schizachyrium* [*Andropogon*] *scoparius*), though additional species have been accepted by captive reared larvae. Caterpillars build nests in partially rolled leaf blades (Allen, 1997).

Occurrence Widespread but usually uncommon—or just fairly common—and rare in some regions. Only a limited number of individuals are likely to be seen per day. Not cold hardy, so northern populations are reduced by winter stress. This fuels large population swings (along with other, normal factors), causing occasional abundant broods. Regular in PA, s. NY, and CT (in central valley), south to FL (rare in far South). We have seen the species once in Miami-Dade Co, but it is absent from the Keys. Also west to MO, e. TX. Curiously, none of the *Nastra* have colonized the W.I., despite suitable-seeming habitat. Two broods (maybe three in FL). First begins from Mar (in FL) to late Jun (in NY), 2nd brood Aug-Sep/Oct throughout. Mature larva overwinters.

Ecology Generalist. The Swarthy is proof that a butterfly can be a generalist without being overwhelmingly common.

av wingspan: 0.9"

Neamathla Skipper   *Nastra neamathla*

Dorsal: 9/8/98, Gainesville, Alachua Co, FL

Ventral: 9/22/96, Gainesville, Alachua Co, FL

Neamathla was a Seminole chief who resided near present-day Tampa in the early 1800s. He resisted a U.S. Army program to relocate his tribe to an inland reservation located on sandy, infertile soil. The skipper that bears his name is consigned permanently to such habitats, but unlike its namesake the Neamathla Skipper seems to find these surroundings quite agreeable. (If Chief Neamathla's spirit lives on, this adaptation may amount to a final, symbolic requital.)

The Neamathla has been recognized as a species since 1923, but remarkably little has been learned of its life history over the ensuing decades. It is visually similar to the much more common Swarthy Skipper in our area, and to both that and the Julia's Skipper farther west. Indeed, there is some evidence that Neamathla and Julia's may be a single, clinal species (Tveten & Tveten, 1996). For early investigators, the butterfly's "undistinguished" appearance and uncertain identification apparently made it unappealing, so it is poorly represented in collections (Smith et al., 1994). As a result, even the species' basic range boundaries are still imprecisely known.

With the Neamathla's field marks now better described, and with more local observers in the field, additional sites are becoming known. Already, the Neamathla's overall occurrence is becoming better understood. Like the Swarthy, it tends to fly close to the ground, often among low vegetation, where it can easily be missed. It is difficult to distinguish Neamathla from Swarthy in terms of behavior and habitats, based on current knowledge.

Identification	A small skipper, similar in size and general appearance to the Swarthy. Relatively short antennae; banded as in the Swarthy. Fringes appear light above (especially when fresh), darker below. *Head and mantle olivaceous.*
Dorsal	Upperside glossy brown or olivaceous-brown. *Two small, pale dots* on central FW (variably expressed), also a tiny row of *subapical marks ("bracelet")*. Without close scrutiny, or after substantial wear, FW can appear unmarked.
Ventral	HW below yellow-brown, *lacks light veins* of Swarthy. A *faint, pale spot-band* is usually visible past the center of the HW. Underside FW distinctive: inner costal edge covered with yellow-brown scaling, remainder of wing dark brown. FW spot pattern often more visible from below than above.

Habitat	Similar to Swarthy, and both species may be seen together. Some accounts place them at wetland edges, but we have found them in open, grassy pine woods, dry fields, and well-mown roadsides (mainly away from nearby drainage ditches).
Hostplants	Big Bluestem (*Andropogon gerardi*) and Little Bluestem (*Schizachyrium scoparius*). Other grasses eaten by captive-reared larvae, and in TX.
Occurrence	Not clearly documented. More common than once supposed, especially late season in n. FL, but generally infrequent. Peninsular FL to upper Keys (rarely). Also se TX, CA, south to Costa Rica. Three broods in n. FL, Feb-Nov, all year in s. FL.
Ecology	**Medium generalist.** Apart from undocumented, range-limiting factors, adopts a generalist lifestyle.

Underside forewing (showing dots): 9/22/96, Gainesville, Alachua Co, FL

av wingspan: 0.9"

Three-spotted Skipper *Cymaenes tripunctus*

Dorsal male: 12/30/98, Snake Bight Trail, Everglades NP, Monroe Co, FL

Dorsal female: 12/30/98, Snake Bight Trail Everglades NP, Monroe Co, FL

The most difficult thing to explain about this plain-looking skipper may be the origin of its name. It has three small dots in its subapical "bracelet" band, but so do many other species. And the central forewing spots number *either* two (in males) *or* three (in females)—in neither case a distinctive number.

Whatever the source of its name, the Three-spotted can pose serious identification problems, at times catching even experienced observers off-guard. There are reasonable grounds for confusion with a half dozen other species in our area (plus more in the West Indies), and at least plausible grounds with several others. Glassberg (1999) reports that nearly half of the specimens assigned to this species in one large museum collection were incorrectly labeled. If the species' range extended a bit farther north, moreover, it would intersect with an additional group of somewhat similar meadow skippers at their southern limit. But this most northerly representative of an otherwise tropical genus (which includes about two dozen species) does not seem inclined to press its range limits northward, and probably lacks adaptations to manage even modest bouts of cold weather.

The Three-spotted is often described as "low perching," but this is a bit misleading, since a number of visually similar species in south Florida (especially Eufala, Obscure, and Neamathla Skippers) characteristically perch *very* close to the ground, even more so than Three-spotted. Nominate race (*tripunctus*) occurs in Florida and West Indies.

Identification	*Very long antennae*, commonly exceeding half the length of the FW. Note a *distinct white mark* at the base of the antennal clubs. Male's wings are narrower, less fully cut than the female's. Mantle and upper edge of FW costa are variably *dull, tawny orange* when fresh, but this coloration is lost quickly with wear (already faint in the illustration photos).
Dorsal	Both sexes are dark brown to blackish above. The wrist "bracelet" includes three distinct dots, *curved outward* (toward the apex). *Two white FW spots (three in the female)* cross the wing just past its midpoint. The female has an additional, *very small white spot* along the leading FW edge at midpoint, usually visible in the field.
Ventral	Paler brown, sometimes yellowish, with a *faint, but definite, concentric spot-band* on the postmedial HW. Lacks prominent white HW spots and body stripe of Obscure, also lacks white abdomen of Eufala.

Habitat	Mainly found in shady areas, in or near subtropical woodlands, including hardwood hammocks, s. FL pinelands, mangrove forests, etc. Often in disturbed edge habitats. Elsewhere, subtropical grasslands.
Hostplants	Guinea Grass (*Panicum maximum*), Crabgrass (*Digitaria sanguinalis*). Accepts sugarcane in captivity. Larvae build rolled leaf nest.
Occurrence	Fairly common. S. FL and Keys; range extends slightly farther north along coasts. Also W.I., C. and So. Amer. to ARG. Broods not fully described (may be linked to rains), all year.
Ecology	**Medium specialist.** Specialized ecological habitat preference, limited to subtropical regions.

Ventral: 3/24/94, Matheson Hammock, Miami-Dade Co, FL

av wingspan 1.1″

Eufala Skipper *Lerodea eufala*

Dorsal: 9/29/95, Gainesville Airport, Alachua Co, FL

Ventral: 9/24/95, Homestead, Miami-Dade Co, FL

The Eufalas were one of several native tribes that once lived just north of Florida (along with Creeks, Mikasuki, etc.). Members of these tribes eventually moved south to form the Seminole Nation. It is not clear why the name of this particular tribe was chosen for *Lerodea eufala*—or even that the tribe's identity was specifically involved, since the skipper's range extends to Oklahoma, where a city named Eufala is located.

Small and nondescript, the Eufala Skipper is the northernmost representative of a fairly large tropical genus of grass skippers. Two of its relatives, the Violet-clouded Skipper (*L. arabus*) and the Olive-clouded Skipper (*L. dysaules*), also reach the United States. Unlike some emissaries of tropical genera in our region, the Eufala is not simply a northern break-away. Rather, it is a highly successful, cosmopolitan butterfly that has been able to incorporate large regions of the United States into its extensive range, along with most of Central and South America. Yet despite its wide distribution the Eufala lacks recognized subspecies, perhaps because its mobile disposition leads to regular gene-sharing across populations.

Eufalas are low perchers, and by many accounts they remain within inches of the ground at all times. But we have seen them nectaring several feet high at times in Texas and South Carolina. This is a butterfly of flat, open habitats, nonetheless, whose males perch on grass stems throughout the day awaiting females.

Identification	Small and inconspicuous, intermediate in size between Neamathla and Three-spotted, with *long, pointed FWs, short antennae*. Viewed from below, the *body is white* (contributing to the general, light-toned appearance). Eufalas make two quick wing-claps on landing (Glassberg, 1999). Grayish color is often cited as the most reliable primary field mark, but color is variable, with some individuals being rather brown-toned.
Dorsal	*Plain gray-brown* above, normally lighter than most potential look-alikes (Swarthy, Neamathla, Three-spotted, Cloudy, etc.). Obscure Skipper is also light-toned, but not grayish. Note a *U-shaped series of white dots* on postmedian FW (including subapical dots), which resembles a bite mark. Male lacks a stigma.
Ventral	Base color is light brown ("cool tan"), variably covered by *even, gray overscaling*. In some individuals (*see* photo above), *HW has a faint crescent-like dot pattern*, usually less pronounced than in Three-spotted. HW veins are not lightened.

Habitat	In our region, mainly flat, grassy areas with low vegetation, including vacant lots, road edges, agricultural areas, pastures, gardens, lawns, etc. Also coastal lowlands (SC), pine woods (Keys), salt marsh edges, wood margins, savannas. Elsewhere, oak openings, desert valleys, TX hill country. Only one record from mountains in NC.
Hostplants	Various grasses, including Bermuda Grass (*Cynodon dactylon*), St. Augustine Grass (*Stenotaphrum secundatum*), and Johnson Grass (*Sorghum halepense*), a weedy exotic. Sugarcane (*Saccharum officinarum*) in Cuba. Larva lives in rolled leaf nest. Pupa has pointed "head."
Occurrence	Widespread but uncommon. Resident from se SC coast through FL Keys; not cold hardy, rebuilds populations each summer in mid-Atlantic states; strongly emigratory, strays north as far as s. NJ in late summer (especially Cape May). Also Midwest and southern rim of US to se CAN, south to ARG. Two broods in North (Mar-Apr and Aug-Sep), all year in s. FL. Diapause stage not reported.
Ecology	Generalist. A versatile and widespread species.

Ventral (grayer variant): 2/19/98, Bentsen State Park Hildago Co, TX

av wingspan: 1.0"

Clouded Skipper   *Lerema accius*

Dorsal male: 8/22/92, Great Dismal Swamp, Nasemond Co, VA

Dorsal female: 5/23/99, Macon Co, NC

The Clouded Skipper is a common and successful butterfly in the Southeast, with a strong, innate urge to emigrate northward. Like other essentially tropical species living in our area, it faces climatic challenges, most specifically the need to survive winter cold. It does this with only passable success, and must recolonize northern territories anew each summer. Several scenarios can be involved. At middle latitudes, winter stress may thin resident populations without eliminating them. In these cases, winter survivors create small spring broods, rebuilding local colonies from within over the course of the season. Farther north, where Cloudeds are not seen until mid-summer, fresh emigrants are needed each season. After breeding in the North, some of these colonists (or their progeny) appear to return south (Walker, 1991). The exact origins of these summer emigrants, which stray as far north as New England, are not precisely known. But they may represent an overflow from "skipper factories" in the Southeast or mid-Atlantic states. (On potential effects of recent climate change, *see* Sachem account.)

Behaviorally, Clouded Skippers are notable as early risers. They can often be seen basking on still-damp foliage before other grass skippers have stirred. Even mating has been reported from early morning (e.g., we recorded a joined pair at 9:30 A.M. in southeastern Virginia one August). Males perch low while displaying, often in shaded areas near trails or wood edges.

Identification	Larger than most similar-looking species (especially females); otherwise could be a source of significant confusion. And small males can still prove difficult, when seen only from above. Northern-form Dusted Skippers, which have superficial similarities, fly in early spring, usually before Cloudeds, though in some places overlap is possible. Buffy fringes, darkened at wing vein ends (darkening most visible from below). Antennal club end (apiculus) is *long and pointed*, unusually so for a grass skipper.
Dorsal	Dark, blackish brown, with a wrist "bracelet" and *one to two separate white dots on the central FW* (often inconspicuous in male; consistently prominent in female, with innermost dot larger and irregularly shaped). Male's black stigma usually difficult to see against dark FW base color.
Ventral	Brown with *violet-gray frosting along outer wing edges*, and a separate, parallel band of frosting at mid-HW. When frosting is worn, these areas look pale/mottled. *Compare* Dusted, female Zabulon, Common Roadside (smaller).

Habitat	A wide variety of open grassy areas, both wet and dry, often near woods. Fields, easements, gardens, swamp edges, etc. We agree with LeGrand and Howard (2002) that proximity to swamps or rivers is not critical.
Hostplants	Many grasses. St. Augustine Grass (*Stenotaphrum secundatum*), millets (*Echinochloa*), etc. Larvae eat leaf shelter from within, build another.
Occurrence	Common resident to se VA; irregular emigrant farther north, reaching w. CT. Also west to CA, south to n. So. Amer. Not in W.I. Two to three broods northward, Apr-Oct; all year in s. FL. Pupa apparently overwinters.
Ecology	Generalist. Widespread and adaptable.

Ventral: 6/15/96, Edisto Island, Charleston Co, SC

av wingspan: 1.25"

European Skipper   *Thymelicus lineola*

Dorsal male: 6/23/94, Pound Ridge Reservation, Westchester Co, NY

Ventral: 6/21/97, Pound Ridge Reservation, Westchester Co, NY

The European Skipper is native to Eurasia, where it is known as the Essex Skipper (most species in their home territory are "European," after all). When accidentally introduced in Ontario in the early 20th century (discovered 1910 at London), it brought along some foreign customs, largely unknown to local grass skippers, such as cluster-laying, diapause in the egg stage, and patrolling for mates. These fashions continue to distinguish Europeans from native grass skippers, and have allowed them to mount a dramatic range expansion across North America, averaging 20 miles a year. Human commerce has undoubtedly provided assistance, as Timothy grass shipments laden with skipper eggs have spawned satellite colonies as far off as British Columbia and the western United States. In our region, however, much of the dispersal seems natural. Single individuals can cover substantial distances during their short adult lives and seem to be radiating here along a coherent "front line."

All this may seem unlikely for a small skipper whose flight is variously described as "weak" or "feeble." Male Europeans patrol all day in grassy fields, stopping only for nectar and occasionally to bask. Studies have shown that Europeans prefer nectar with high sugar concentration (Pivnik & McNeil, 1986). Without natural controls to rein them in, European Skippers have become locally hyperabundant, at times outnumbering all other butterflies in an area combined. Tens of thousands may sometimes be seen on a "good" day. Hundreds sometimes roost together, clinging to grass stems. Lean years also occur, however, with greatly reduced numbers.

Identification	Small. Most easily confused with small male Delaware, or momentarily with Least Skipper, Southern Skipperling, or Tawny-edged Skipper (ventrally). Wings are pointed and angular, with pale, gray or orangish fringe. Antennae are short, with much rufous color and *blunt tips*. A rare, pale form (*pallida*) is reported. Body hairy.
Dorsal	Uniformly bright, reddish-orange (termed "burnt" or "brassy"). Narrow black margin surrounds both wings, with black extending inward some distance from the outer margins along wing veins. Male has a narrow, black stigma near FW costa.
Ventral	Unmarked. Paler than dorsal surface, HW variably grayish-orange, greenish-ochre. FW uniformly pale orange.

Habitat	Possible in any open, grassy area. Some propensity for damp fields and meadows, but quite tolerant and opportunistic. Common in agricultural fields (where sometimes identified as a pest), and near settlements. Disfavors arid habitats.
Hostplants	Grasses; mainly Timothy (*Phleum pratense*), but also Orchard Grass (*Dactylis glomerata*), Velvet Grass (*Holcus lanatus*), etc. Female lays 30 to 40 eggs in parallel strings on host. Larvae feed on leaves, then flowers. Pupate at base of plant.
Occurrence	Common to abundant. ME south to PA/NJ; across n. MD/VA and south in Appalachians to SC. Also much of CAN, Eurasia, following James Bay Highway into n. CAN. One primary brood, Jun-Aug; possible small 2nd brood locally, Aug-Sep. Eggs overwinter, with developed 1st instar larva inside, only No. Amer. skipper to diapause in this stage.
Ecology	Generalist. A natural colonist. One female may populate an entire field.

Common Milkweed (*Asclepias syriaca*) with European Skippers

av wingspan: 0.95"

Southern Skipperling   *Copaeodes minimus*

Dorsal male: 9/24/95, Homestead, Miami-Dade Co, FL

Dorsal female: 10/23/99, Hildago Co, TX

The common name "Least Skipper" was already taken when *Copaeodes minimus* was described in 1870. By then, the current "Least" Skipper (*Ancyloxypha numitor*)—a hulking insect by comparison—had already owned the title for more than 75 years. Some have suggested compromise solutions to clarify the nomenclature, such as Very Least Skipper, even Leastest, but the critical moment has clearly passed.

We generally avoid referring to butterflies as "easily overlooked" in these accounts, since the phrase tends to address limitations of the observer more than traits of the butterfly. But this is a species that truly *can* be overlooked, with its diminutive size and habit of flying through low foliage very near the ground. Once the eyes have fixed on it, on the other hand, its appearance is so striking that size is immaterial. Displaying males are glistening gems, dwarfed though they may be by the meager blade of grass on which they perch.

Some authorities describe the flight of the Southern Skipperling as weak or fluttering. But it can actually move quite rapidly for a species its size. The failure to cover much distance seems to be mostly a matter of scale. As a perching species, males typically dart abruptly and erratically, more so than Least Skipper males, which spend their days patrolling. Female Southern Skipperlings often move rather slowly through low grass thickets, their wings whirring rapidly in a blur. Adults nectar and visit puddles. The scientific species name was formerly *minima*, but was changed to comply with biological naming standards.

Identification	Tiny. The smallest US skipper. And wingspan statistics fail to convey fully its smallness, since the wings are proportionately quite long. Similar only to the Least Skipper in our area, but it can usually be separated from that species, even in flight, without great difficulty. *Short, orange antennae with no tip* (apiculus). Usually a significant amount of white around the eye. Females slightly larger than males, less angular.
Dorsal	Male is a *clear, bright, metallic golden-orange* above, with minute black wing borders. Base of upper HW has *smudgy black "shoulders."* The thin, slitlike stigma is extremely inconspicuous. Female similar, but somewhat more red-orange, darker wing borders, and *irregular black markings* on both wings.
Ventral	Plain orange, with a distinctive and *conspicuous white ray* running the full length of the HW. White underparts.

Habitat	A variety of natural and disturbed, low-growth grassy areas. Fields, clearings, grassy wood edges, lawns, parks, roadsides, etc. Often drier habitats than Least, but avoids arid settings.
Hostplants	Bermuda Grass (*Cynodon dactylon*), probably others. Not fully documented. Typical larval development cycle.
Occurrence	Fairly common to common (especially late summer). Se coastal plain/lower Piedmont, to s. FL (but on Keys only during one invasion year, 1989). Also Gulf Coast to TX, south to Panama. Several broods, Mar-Nov in NC, all year in s. FL. Diapause stage not reported.
Ecology	Medium generalist. Widely adaptable, with one common host, but only in warm climates.

Ventral: 8/8/96, Sonora, Mexico

av wingspan: 0.6"

Least Skipper *Ancyloxypha numitor*

Dorsal: 6/21/97, Pound Ridge Reservation, Westchester Co, NY

Ventral: 6/12/94, Pound Ridge Reservation, Westchester Co, NY

Least Skippers attract diverse forms of attention as they weave their way distractedly through dense, streamside thickets. A recent study found that male Eastern Amberwings (*Perithemis tenera*) tended to mistake just two nondragonflies for potential mates, the Least Skipper and a certain species of horse fly (Schultz & Switzer, 2001). The Weed Science Society of North Carolina thought Least Skipper caterpillars might be useful in controlling invasive Cogongrass (*Imperata cylindrica*), but concluded they were too small and might spread to other grasses if so used (indeed!) (Clewis, 1999; Bryson & Sudbrink, 1999). The eminent Samuel Scudder even noted that perched Least Skippers rotate their antennae in opposite directions "in a small circle" (Weed, 1917).

The Least Skipper belongs to a neotropical genus of seven small, light-colored skippers, all of which lack male stigmas. Among them, only the Least Skipper occurs in our region. It has no recognized subspecies.

Unlike the males of most grass skippers, Least Skipper males patrol to find mates. They meander through likely habitats with a busy, satyr-like flight, active but unhurried. They usually fly within a stand of grass, or just above it, rarely more than three feet off the ground. Least Skippers mate when encountering a female, and this may occur at any time of the day. Often seen nectaring on wetland flowers, such as Pickerelweed or arums.

Identification	Small, but noticeably larger than Southern Skipperling. *Slender body shape and rounded wings* are distinctive. Abdomen orange with dark stripe along the top. Short, checkered antenna with no hook. Underparts whitish. No male stigma. Sexes largely similar.
Dorsal	FW dark—sometimes completely, sometimes with a confined orange patch. HW orange, surrounded completely with a *broad, uniformly dark band*.
Ventral	Ventral FW dark basally, orange at apex and along leading edge. HW plain orange, *paler at the center* than along the margins (a quick, generally reliable clue to identity). White along HW veins.

Habitat — Usually wet (or moist) areas with tall grass. Ditches, streamsides, wet meadows. Sawgrass prairie edges in FL. Sometimes wanders to drier sites, especially in late summer (Opler & Krizek, 1984).

Hostplants — Grasses. Rice Cutgrass (*Leersia oryzoides*), Marsh Millet (*Zizaniopsis miliacea*), bluegrass (*Poa* sp.), cultivated rice (*Oryza sativa*), Cogongrass (*Imperata*), etc. Typical grass skipper larval lifestyle.

Occurrence — Often very common in colonies. Central ME south to s. FL (absent on Keys). Also west across s. CAN, south to e. TX. Strays to AZ. Two to four broods, May-Oct in North, Feb-Dec in South, all year in s. FL. 3rd or 4th instar caterpillars overwinter.

Ecology — **Generalist**. A highly widespread and adaptable species, but somewhat specialized in habitat preference.

Weedy streamside, Rockland Co, NY

av wingspan: 0.9″

Cobweb Skipper   *Hesperia metea*

Dorsal male: 5/9/95, Chatsworth, Ocean Co, NJ

Dorsal female: 5/15/94, Pound Ridge Reservation, Westchester Co, NY

In classic *Hesperia* style, the Cobweb Skipper is specialized not just in hostplant and habitat preferences, but also in flight season. It vies with Indian Skipper as the earliest emerging hesperid in northern parts of our area. But after a brief, early spring flight, it quickly vanishes.

The Cobweb's dark colors undoubtedly assist in warming its body on cool spring days while perched on open, reflective surfaces (e.g., sand barrens). Sparsely distributed and highly colonial, it is among the species for which a dedicated field trip is usually required each spring.

Males perch throughout the day on or close to the ground near hostplants. This allows them to take full advantage of available warmth while courting, in a season with reduced day lengths. On finding a female, the male induces her to land, then initiates mating.

Ventral: 5/9/95, Chatsworth, Ocean Co, NJ

Identification	Small, distinctive, with short, compact wings. Well camouflaged, Cobwebs are difficult to track in low flight against mats of dead grass. (Their weak flight in fact helps differentiate them from Dusted Skippers, which emerge only slightly later than Cobwebs at many of the same locations.)
Dorsal	Male is brown above with a wrist "bracelet" and *two isolated spots near the outer FW margin* (typical of *Hesperia*). Mid-FW black stigma is surrounded by a *vague, orange field*. HW dark, with a faint trace of ventral pattern. Female darker brown, with a transverse series of *jagged white FW spots*, also one to two white pin-dots near FW costa. HW shows ventral *V*-shaped pattern more clearly than in male.
Ventral	HW variably marked with *white veins and a white, V-shaped chevron* (creating a "cobweb" effect). White FW markings mirror the dorsal surface pattern.

Habitat	Mainly dry or successional habitats with beardgrass. Sandhills, barrens (pine, serpentine, shale), pine-oak savanna, open scrub, outcrops, burn sites, etc.
Hostplants	Little Bluestem (*Schizachyrium* [*Andropogon*] *scoparius*) and Big Bluestem (*Andropogon gerardi*); both have telltale thatch mats, visible in the spring. Half-grown larvae build nests in central host clump, aestivate during summer (Gochfeld & Burger, 1997).
Occurrence	Highly colonial. Patchy distribution from se ME to central GA. Also west to W.I., south to e. TX. One brood, late Mar-Apr (NC), May-Jun (NY). Winters as mature larva, pupating quickly in spring.
Ecology	Medium specialist. Though hostplant is common, lifestyle and habitat selection are fairly specialized.

Bluestem Grass (*Andropogon/Schizachyrium* sp.) growing through dead thatch from prior season

av wingspan: 1.15″

Indian Skipper *Hesperia sassacus*

Dorsal male: 6/3/00, Pound Ridge Reservation, Westchester Co, NY

Dorsal female: 6/3/00, Pound Ridge Reservation, Westchester Co, NY

It is not inaccurate to think of this little-studied butterfly as a widely distributed, spring-flying counterpart of the Common Branded Skipper. Both species subscribe to the *Hesperia* model of a single, seasonally targeted brood (though at different seasons). Neither is host-plant-specific as much as other *Hesperia*. Nor is either confined to dry barrens or other fire-prone habitats. Even their dorsal patterns are comparable.

Family lineage aside, Indian Skippers behave a good bit like ordinary "meadow skippers," nectaring avidly in wet fields alongside Long Dashes and Peck's Skippers. Males perch low on broad-leaved plants to await females. Wary and rather fast-flying, they are difficult to approach closely, except while feeding. In some locations they are the first local grass skipper to emerge. The flight period is longer than Cobweb's.

Ventral male: 5/19/02, Clay Co, NC

Identification	An average-sized Hesperid with long, triangular wings. Often difficult to distinguish from Long Dash in terms of habits or appearance. The two fly together regularly in close proximity. Male antennal club dabbed with orange.
Dorsal	In both sexes, the trailing edge of the FW orange field is *jagged*; both have *two isolated dots* near FW apex. Male's conspicuous FW stigma sets in a rufous-orange field; small FW patch, surrounded by a wide, brown trailing border. Dark, semicircular crescent near FW apex. Female FW patch larger, crescent more prominent. Yellow-orange upper HW marks more delineated and geometric than in male (where looks more like a single patch).
Ventral	Variably yellow-orange or greenish-orange. Both sexes have *chevron-shaped pattern on HW* (vague and ill-defined in males, well-defined and more geometric in females). Isolated dots near outer FW edge extend *entirely past* the imaginary line formed by the FW "bracelet" and mid-FW markings.

Habitat	In the North, a variety of grassy sites with nectar, including fields and pastures, wet meadows, roadsides, clearings; in Appalachians, mountain meadows and grassy balds.
Hostplants	Various grasses. Little Bluestem (*Schizachyrium* [*Andropogon*] *scoparius*), Red Fescue (*Festuca rubra*), panic grass (*Panicum*), crabgrass (*Digitaria*), Poverty Oatgrass (*Danthonia spicata*), etc. Early stages not fully reported, presumably like other *Hesperia*.
Occurrence	Fairly common in Northeast. ME south to PA/NJ, and in Appalachians to w. NC (above 3,500 feet). Also Gr. Lakes area. One brood, May-late Jun. Larva overwinters.
Ecology	**Medium generalist.** Specialized seasonal cycle, but generalized in use of habitat and hostplants.

Ventral: 6/16/97, Carrol Co, NH

av wingspan: 1.2″

Meske's Skipper   *Hesperia meskei*

Dorsal male: 5/7/98, Sandhills Game Land, Scotland Co, NC

Dorsal female: 9/29/95, Gainesville, Alachua Co, FL

The Meske's or "Dixie" Skipper is our only *Hesperia* whose range is limited to the Deep South (though it does have an isolated occurrence on the southeastern plains). The Keys population may be a separate species (Minno & Emmel, 1993), but it is now greatly reduced, and appears to still exist—if at all—only in pine rockland habitats on the lower mainland. Meske's was little-known for nearly 125 years after its original description, but it is now becoming more familiar with increased field study by local observers, e.g., in North Carolina and central Florida.

Meske's Skipper follows the standard *Hesperia* model in habitat choice, living mainly in dry, sandhill pinelands or similar habitats. It diverges from the norm (as does the ecologically similar Dotted Skipper) in having more than one brood annually—though its flights do appear to be ecologically targeted, coinciding with the beginning and end of the southeastern rainy season. And the early stages are well-adapted to fire-prone habitats. Caterpillars construct nests in the center of thick grass clumps, partly underground, which they use as a base when foraging and during diapause. Absent human intervention, sandhill pinelands are subject to regular, low-intensity burns (as often as once a decade), which Meske's can evidently withstand.

We agree with the characterization of Meske's as a solitary species, rarely found in large groups (LeGrand, Cooper, pers. comm.). Individuals may need to travel considerable distances amid dry surroundings to locate nectar. Yet once a desirable nectar source is found, it may be worked for a considerable period of time.

Identification	A fairly large skipper, with long, triangular FWs. Can be confused with Dotted in certain cases, especially as fresh males of both species are largely unmarked below and have fairly similar dorsal patterns. (And they fly in similar habitats as well.) Ventral surface can resemble Delaware or Arogos Skipper (and the Arogos seldom shows its dorsal surface, so care is needed). Remotely confusable with Byssus, but habitats differ.
Dorsal	Rather dark above, with a *limited orange field*. Note *large, shapeless dark FW patch* in both sexes, just before subapical marks (more geometric in male Dotted). Also two dots approach trailing FW edge as in Indian Skipper.
Ventral	Variably yellow-orange or rusty-orange, often with a sooty appearance. A *pale, chevron-shaped spot band* often visible on HW; can be entirely absent in fresh individuals, or very pronounced in worn females.

Habitat	A denizen of sandhill pinelands on the se coastal plain, also pine rocklands in s. FL and the Keys. Wanders to dry fields, roadsides, clearings, etc., for nectar. Not in moister flatwoods or savannas (LeGrand & Howard, 2002).
Hostplants	Life history poorly documented. Little Bluestem (*Schizachyrium* [*Andropogon*] *scoparius*) recorded, Arrowfeather Three-awn (*Aristida purpurascens*) in s. FL. *See* discussion of larva, above.
Occurrence	Usually rare (most common in 2nd brood, e.g., NC sandhills in Oct). S. NC, south very sporadically to central FL in sandhills, also s. FL. Two broods in NC, June and Sept-Oct; all year in s. FL (3rd brood Dec). Larva overwinters.
Ecology	Specialist. Rare and local, habitat restricted.

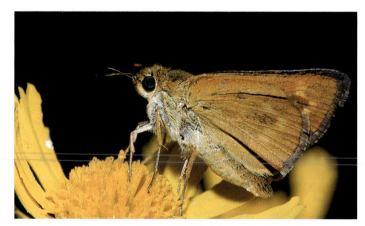

Ventral female (worn individual): 9/29/95, Gainesville, Alachua Co, FL

av wingspan: 1.3"

Dotted Skipper *Hesperia attalus*

Dorsal male: 7/4/96, Double Trouble SP, Ocean Co, NJ

Dorsal female: 7/7/96, Lakehurst, Ocean Co, NJ

It is not clear whether the term "dotted" better describes this skipper's hindwing pattern or its distribution. When sandhills dominated the southeastern landscape (before settlers arrived), this large, often handsome skipper was undoubtedly more prevalent. But today its few populations remind us of remnant moisture patches in an otherwise dry lakebed. Still locally common in places, the Dotted is nonetheless an appropriate candidate for protection wherever it occurs.

MacNeill (1964) argued that, among existing *Hesperia*, southeastern species appear to be the oldest. This would make the eastern race of Dotted Skipper (*slossonae*) an ancient veteran of its narrow, specialized niche. Indeed, it seems unable to rebound from recent, extensive loss of traditional habitats. Males perch on or close to the ground when seeking mates. May wander for nectar.

Ventral male: 7/4/96, Double Trouble SP, Ocean Co, NJ

Identification	Medium (southern females largest, northern males smallest). Much of the variability reported in early life history accounts is attributable to geographic or sexual dichotomies, but even with these factors sorted out significant individual differences remain. FW pointed, especially in male. *FW fringe dark, HW fringe light.*
Dorsal	Two small, isolated spots near the outer FW edge (*Hesperia*-style). Ventral dot pattern is visible on upper HW. Male: prominent FW stigma surrounded by small yellow- or tawny-orange field, otherwise brown. Female: dark brown with 8 to 10 white FW spots; rusty along FW costa when fresh.
Ventral	Male: ochre to brownish-orange, *single inner HW dot* and *V-shaped outer HW dot pattern* (variably expressed, in some cases virtually absent). Female: more greenish- or yellowish-brown, usually with *well-defined HW dots*.

Habitat	Dry, fire-maintained habitats: sandhills and other sunny Longleaf Pine forests; dry, grassy pine-oak scrub in or near sand barrens; grassy cuts, easements, and dry meadows within preferred areas.
Hostplants	Multiple host grasses, even at specific locations, including Switchgrass (*Panicum virgatum*), wiregrass (*Aristida*). Caterpillars build nests partially underground in hostplant clump.
Occurrence	Rare and very local, but can be numerous at colony sites. NJ south to central FL. Formerly s. FL, rarely to coastal N. Eng. (strays). Separate race on s. prairies. One brood in NJ (early July-Aug), 2 broods in NC (May-Jun, mid-Aug-Oct) and FL (Feb-May, Aug-Oct). Caterpillar overwinters.
Ecology	Specialist. Seldom far from a dry, fire-prone habitat.

Ventral female: 9/6/99, Sandhills Game Land, Scotland Co, NC

av wingspan: 1.4"

Common Branded Skipper *Hesperia comma*

Dorsal male: 8/5/01, Scott Bog, Coos Co, NH

Dorsal female: 8/6/01, Scott Bog, Coos Co, NH

All of the 20 or so *Hesperia* skippers in the world are confined to North America—except this one, which is distributed holarctically. A complicated species, its numerous local forms have long bedeviled taxonomists. Several current races were earlier regarded as independent species; e.g., our local race (*laurentina*) was originally known as the Laurentian Skipper (Klots, 1951). In England, confusingly, the species is called "Silver-spotted Skipper."

Important scientific studies have focused on this species in England, where it lives in patchy, short-grass habitats near the northern edge of its range. Over an extremely short time period (one to two decades), populations inhabiting areas with widely separated grassland patches developed large thoraxes, vs. populations with easier access to adjoining plots (Hill et al., 1998). The latter were able to devote more anatomical resources to abdomen development (better for reproduction), whereas the former needed to maintain travel readiness in order to bridge large gaps between suitable grasslands. The rapidity of this evolutionary adjustment was noteworthy.

Scientists have also studied the sex attractants (pheromones) of Common Branded Skippers, and found that a relatively small dose of the complex chemical was adequate for courtship, given sophisticated scent-releasing structures in the wing scales (Tranefors, 1999).

In our area, this is among the wariest and most difficult species to approach. When startled, they may fly rapidly away, not to be seen again. Males perch into the late afternoon, their wings usually closed.

Identification	Relatively short, rounded wings (more pointed in male). Most likely to be confused with Leonard's Skipper in terms of appearance, distribution, and flight time. But even where their ranges converge (in n. N. Eng.), Common Branded usually occurs farther north, and its flight peak is earlier than Leonard's.
Dorsal	Two isolated, orange spots near the outer FW edge (*Hesperia*-style). Male: bright, rusty-orange with a prominent dark stigma. Medium-brown band on trailing FW has uneven, smudgy edge. Female: similar, but less extensive orange field; extensive spotting may create a vaguely checkered effect in some individuals.
Ventral	*Clear white spots* on a background that varies from brown to golden to golden-greenish; note variable black flecking. Leonard's has a single inner spot, vs. *two irregularly shaped marks* in Common Branded. Outer HW fringe darkened at end of wing veins.

Habitat	In our region, natural and disturbed open spaces, including upland northern fields and meadows, easements, boreal scrub. Elsewhere low-growth, alkaline grasslands, tundra, subalpine meadows, foothills, etc.
Hostplants	Various grasses, including fescue (*Festuca*) and oat grass (*Danthonia*). Females scatter eggs on or near hostplant.
Occurrence	Holarctic. Annual populations variable. N.-central ME and far n. NH/VT. West to AK, south to CA/AZ, also Eurasia. Common in much of w. US. One brood in our area, mainly Aug (earlier in the West). Larva overwinters.
Ecology	**Medium specialist.** Adaptable within its range, but strictly limited to boreal areas in our region.

Ventral male: 8/5/01, Scott's Bog, Coos Co, NH

av wingspan: 1.1"

Leonard's Skipper *Hesperia leonardus*

Dorsal male: 9/6/97, Manumuskin River Preserve, Cumberland Co, NJ

Dorsal female: 9/6/97, Manumuskin River Preserve, Cumberland Co, NJ

In the waning days of summer—months after the hardy Cobweb Skipper launched the *Hesperia* flight season—Leonard's Skippers emerge to complete it, sometimes in exactly the same location. This is our only butterfly with a single brood in late summer (August). It emerges first in the North, thereby avoiding early cold snaps.

Leonard's Skipper has undergone a gradual, but persistent decline in its eastern range. In Westchester County, New York, for example, it is gone from key former outposts at Ward Pound Ridge Reservation (probably because of grassland management changes), also from an established upland site in the town of Somers (because of development). It is a Species of Special Concern in Connecticut. These losses are dwarfed by the plight of the Pawnee Montane Skipper (*H. l. montanus*), however. This federally endangered western race of Leonard's is confined to a 38-square-mile range in Colorado—where local officials seek to place a reservoir. Western Leonard's were formerly classified as a distinct species, the Pawnee Skipper (*H. pawnee*), but continuous intergrades have since been found in midwestern contact zones (Scott & Sanford, 1981; Spomer et al., 1993).

A strong, rapid-flying skipper, Leonard's is notoriously difficult to approach. It nectars actively, often at tall, pink or purple flowers. In New Hampshire, we have monitored a stand of garden phlox that was visited daily by Leonard's, always between the hours of 10 and 11 A.M. Males perch throughout the day, or sometimes patrol continuously in areas where females are likely to appear in search of nectar.

Identification	Medium-sized. Not easily confused with any other species in our area (occurs with Common Branded only in a narrow band in n. N. Eng.). Differs visually from Pawnee race, but discussion here is confined to eastern *leonardus* subspecies. Sometimes compared with Indian Skipper, but outside of collections the two are never seen together, given widely varying flight seasons.
Dorsal	As in other *Hesperia*, has two isolated, orange spots near the outer FW edge. Ventral HW pattern mirrored strongly above. Male: dark base color. Orange field on central FW is small and discontinuous. Stigma often difficult to see against surrounding dark patterns. Female: striking blackish-brown with prominent orange FW marks.
Ventral	HW rust-colored (male) to rich chestnut (female) with *conspicuous white postmedian spot-band*. Also *a single, small, isolated white spot near the central HW* (vs. two spots, one elongate, in Common Branded).

Habitat	Variable, often a combination of dry, low-growth upland habitat near a lower-lying area with nectar. Also pine barrens, brushy fields, grassy trails, etc.
Hostplants	Little Bluestem (*Schizachyrium* [*Andropogon*] *scoparius*), bent grass (*Agrostis*), Switchgrass (*Panicum virgatum*), Poverty Oatgrass (*Danthonia spicata*), etc. Eggs scattered on or near hostplant.
Occurrence	Locally common in scattered colonies, but declining. Central ME south to w.-central NC and SC. Also s. CAN to W.I. and south. One brood, late Aug-Sep in North, Sep to mid-Oct in South (flies *later*). 1st instar caterpillars overwinter.
Ecology	**Medium generalist.** Fairly broad hostplant and habitat use, despite targeted lifestyle.

Ventral female: 8/31/96, Somers, Westchester Co, NY

av wingspan: 1.25"

Fall Skipper "Migration": A New Pursuit

Many birders stop by the new Cape May Bird Observatory (CMBO) headquarters in Goshen, New Jersey, to browse for nature books or to check out the latest in spotting scopes. But few come here specifically to watch birds: that activity is reserved for Cape May's numerous "migrant traps" and flyway lookouts, such as the famed Cape May Hawkwatch a few miles to the south.

Still, a growing number of observers now stop by the CMBO gardens for another reason—to see emigrating butterflies. Like southbound land and coastal birds that find themselves concentrated at Cape May in the fall because of its narrow peninsular shape and location (just before a long, over-water flight), butterflies moving up or down the coast also collect here.

We know that many lepidopteran emigrants travel north in the fall, stopping at Cape May to rest (we think) after the long flight over Delaware Bay. Others, like Monarchs and certain nymphalids that spend the winter as adults (ladies, commas, etc.), are in the same position as the birds, moving south each autumn in search of a warmer place to spend the winter. These species pause at "land's end" before pressing ahead.

Beginning in the 1990s, some ardent butterfly enthusiasts associated with CMBO, led by program director Pat Sutton, began promoting nectar gardens specifically to attract emigrant butterflies—which were already known to appear regularly at private gardens near "the Point." These plantings now attract not just emigrants, but also desired local marsh species, such as Rare Skippers (mainly in August) as well as former southern emigrants whose ranges have extended northward and appear to have established nearby colonies—such as Sachems, which have been seen recently at CMBO literally by the hundreds.

The fall buildup of emigrating butterflies at Cape May starts in August, although most of the sought-after southern strays do not appear until September and early October, and sometimes even later. More than 30 species can be seen at the CMBO gardens near Goshen during peak flight days. Gardens are admittedly an artificial setting for butterflies, which do not allow us to study ecological associations as we could in native habitats. Yet they provide an unparalleled opportunity for beginners to learn field marks, for casual observers to become aware of butterflies for the first time, and for serious fieldworkers to monitor trends and patterns in seasonal activity over time.

It is not known why some butterflies make a futile-seeming surge into northern latitudes just as the summer is ending (*see* Introduction, p. 25). But many pierids, notably the Cloudless Sulphur, and a number of southern skippers (Long-tailed, Fiery, Clouded, etc.) definitely engage in such northward movement. In Florida, by comparison, these same species move southward each fall, suggesting that emigration originates in population centers of the Deep South.

Nectar-rich gardens draw emigrating butterflies. Cape May Bird Observatory, Goshen, Cape May Co, NJ

Birds & butterflies. Alongside the traditional legions or birders at Cape May in the fall, butterfly watchers now appear in growing numbers. Cape May Bird Observatory, Goshen, Cape May Co, NJ

Sachem *Atalopedes campestris*

Dorsal male: 9/9/01, Homestead, Miami-Dade Co, FL

Dorsal female: 9/3/95, Cape May Point, Cape May Co, NJ

A prolific butterfly of disturbed open spaces, the Sachem is a regular late summer emigrant in the North, dying back each winter for lack of cold tolerance. In the past decade or so, the year-round range shifted northward. Sachems suffer high mortality in extended cold below 25 degrees F, or after shorter exposures below 15 degrees F (Crozier, 2002). Recent warm winters did not press these limits (lowest New York City temp in 2001–02 winter was 19 degrees F), and Sachems survived well north of their prior range. The trend was deterred by the cold winter of 2002–03.

The Sachem's genus (*Atalopedes*) is related to *Hesperia*. Males perch on or within one to two feet off the ground most of the day to await females. Sometimes a minor agricultural pest (*see* below).

Ventral male: 9/3/95, Cape May Point, Cape May Co, NJ

Identification	Medium-sized, male's wings more pointed than female's. Male upperside can be confused with Whirlabout or Fiery, but ventral surfaces are very different. Female can superficially resemble a number of other species.
Dorsal	Male: wide rufous-orange FW field with *large, stand-alone, stigma.* HW rufous-orange with *dark, branching veins.* Female: tawny costa, two glassy, white mid-FW marks, and two *Hesperia*-style outer FW spots.
Ventral	Male: vaguely marked, yellowish-brown. A pale, chevron-like postmedian pattern and light inner pale patch frame a *vague, squarish dark spot midway out the inner HW.* Leading FW clear orange. Female: muddy olivaceous-brown with a variable, but often *pronounced, white V-shaped postmedian HW chevron.*

Habitat	Almost any imaginable disturbed, sunny, open area with low, grassy vegetation, including fields, pastures, waste areas, roadsides, lawns, parks, gardens, utility easements, etc. Any available nectar source during emigration.
Hostplants	Bermuda Grass (*Cynodon dactylon*), crabgrass (*Digitaria* sp.), St. Augustine Grass (*Stenotaphrum secundatum*), etc. Caterpillars build nests at base of hostplant. TX pest control website includes this amazing description: "Entire fields commonly infested. Once grass is consumed, larvae may crawl armyworm-style across roads and damage adjacent fields or lawns. Infestations are often first detected as masses of caterpillars crawl up buildings and attack lawns around homes." This account, if accurate, is *highly* unusual.
Occurrence	Common to abundant. NJ (maybe LI?) south to Keys. Strays north in summer. Also all of w. US except upper Gr. Basin and n. prairies, south to BRA. Three broods in North, May-Nov, all year in s. FL. Larvae overwinter.
Ecology	**Generalist.** Few skippers exhibit a more generalist lifestyle.

Ventral female: 8/14/94, Eastern Shore of Virginia NWR, Northampton Co, VA

av wingspan: 1.2″

Fiery Skipper   *Hylephila phyleus*

Dorsal male: 9/2/91, Staten Island, Richmond Co, NY

Dorsal female: 9/28/97, Tiger Bay WMA, Volusia Co, FL

The Fiery Skipper's genus name translates as "forest loving," yet this denizen of open spaces is anything but that. Extremely common in the Southeast, Fierys are also regular emigrants northward during the summer. Like the Sachem, their year-round range is gradually extending northward (though perhaps a bit less rapidly).

The lone northern representative of its South American genus, the Fiery is a welcome example of a common species that is also highly attractive. A dedicated generalist, it maintains an open population structure, with frequent vagrancy and genetic intermixing (Ehrlich, 1985).

Fiery Skippers are active throughout the day. Males perch near the ground and actively pursue intruders that happen by. Females are usually more retiring, in grassy spots nearby. Both sexes fly rapidly and actively seek nectar. Mating usually takes place in late afternoon.

Ventral male: 9/24/95, Tree Tops Park, Broward Co, FL

Identification	Medium-sized. Antennae are *very short* (less than 1/3 the length of the FW), but when viewed from differing angles in the field this can be difficult to assess. Sexes are very different, but neither has any glassy, translucent "hyaline" wing spots. Eastern populations are brighter-colored than those in Southwest.
Dorsal	Male: both wings bright orange. *Inner edge of brown FW margin is ragged (toothed or zigzagged).* Prominent stigma (less so than Sachem's), with *additional dark patch* between stigma and FW apex. Female: dark brown with *irregular, tawny FW spots,* some tawny-orange on leading edge of FW. *"Arrow-shaped" pattern on upper HW.*
Ventral	Male: bright, often shiny orange base color with variable, *small black dots.* Female: pale, overlaid with olivaceous scales; *dark spots at end of pale pattern elements,* especially on HW.

Habitat	Wide range of sunny, open habitats, natural and disturbed, including lawns, roadsides, fields, scrub, sandhills, savannas, etc. In settled areas when colonizing on non-native grass.
Hostplants	Bermuda Grass (*Cynodon dactylon*), crabgrass (*Digitaria* sp.), St. Augustine Grass (*Stenotaphrum secundatum*), etc. Larval leaf nests lie horizontally in the sod, where they can survive lawn mowings.
Occurrence	Common to abundant. Resident on se coastal plain and Piedmont, throughout FL, strays north to s. NE. All of s. US except Rockies and Gr. Basin, south to ARG. Appeared in Hawaii 1970. Two to five broods, May-Aug (North), all year in FL. Diapause stage unreported.
Ecology	Generalist. Epitomizes the generalist strategy.

Ventral female: 6/15/96, Edisto Island, Charleston Co, SC

av wingspan: 1.2″

Whirlabout *Polites vibex*

Dorsal male: 9/27/97, Ocala NF, Marion Co, FL

Dorsal female: 9/28/95, Ocala NF, Marion Co, FL

This small, darting skipper, while similar to the Fiery, has a much less expansive lifestyle. Records of northern strays exist from earlier decades, but this species has not joined the recent northward push of other grass skippers. Such reticence is a bit puzzling, since many other *Polites* are already well-established in northern meadows.

The Whirlabout is named for its fast, but ultimately circular flight pattern. The flight style is actually similar to that of its larger relatives, but since it covers a small area observers can see the full trajectory. Males perch for mates, usually in the afternoon. Females are typically found nectaring in woodland edges and clearings.

Ventral male: 3/25/94, Hendry Co, FL

Identification	Fairly small. Males can be confused with Fiery Skippers or Sachems. Females are highly variable and often hard to identify. Indeed, if you have studied a small, nondescriptly mottled grass skipper in the Southeast for some time and still draw a blank, try female Whirlabout. This skipper can at times stump even experienced observers.
Dorsal	Male: bright, yellowish-orange above, vivid when fresh. Broad stigma and dark patch to apex similar to Fiery, but *brown outer FW margin is narrower and straighter* (not zigzagged). Female: dark brown with small spots, similar to female broken-dash or Little Glassywing (even female Dun). Head and "shoulders" a distinctive greenish-yellow when fresh, but this quickly fades.
Ventral	Male: yellow orange with a few *large, smudgy HW marks*. Female: sooty, greenish-gray or grayish-yellow, with vague, pale banding. Appearance varies greatly among individuals and with wear. May resemble worn Sachem.

Habitat	Highly varied. Favors sunny, open spaces, often hot and exposed. Fields, disturbed sites, sandhills, clearings, yards, parks, roadsides, dry prairie.
Hostplants	St. Augustine Grass (*Stenotaphrum secundatum*), Thin Paspalum (*Paspalum setaceum*) Bermuda Grass (*Cynodon dactylon*), etc. Caterpillars live in silked leaf-tubes, feed mostly at night.
Occurrence	At times abundant. Southeastern US, across Gulf Coast to e. TX, south to ARG. Not W.I. Rarely strays north (more so 1960s and 1970s). Rare in NC mountains, Keys. Single records in WV and Toronto (1910). Broods indistinct, two in GA, Apr-Sep, all year s. FL.
Ecology	Generalist. Highly general lifestyle despite limited range.

Ventral female: 9/28/95, Ocala NF, Marion Co, FL

av wingspan: 1.1"

Peck's Skipper   *Polites peckius*

Dorsal male: 8/1/94, Scott's Bog, Coos Co, NH

Dorsal female: 7/16/95, Lakehurst, Ocean Co, NJ

This species belongs to a loosely defined group of butterflies we call "meadow skippers." This designation is not based on taxonomy per se (although close to half belong to the genus *Polites*). Rather, it reflects a set of lifestyle similarities. In particular, meadow skippers are prolific, everyday species found in ordinary grassland habitats (fields and meadows, roadsides, etc.), often in substantial numbers. We can expect to run across them any time we venture into the field during flight season. Their prevalence and lack of restrictive habit preference are in themselves distinguishing. Yet being common they have not always drawn careful attention, and their life histories are not as consistently well-documented as one might expect for such a common group of butterflies.

In the case of Peck's, for example, simple brood sequences are not consistently reported. Klots (1951) wrote that "individuals of a single brood may transform straight through into adults of the next brood, or may slow down their development very erratically and hibernate as 3rd, 4th, or 5th stage larvae or as pupae." Such an unusual generational strategy warrants close inspection, and might account for some differences seen in later accounts: Glassberg (1999) reports two broods south of central New England, one farther north. Allen (1997) cites one full brood in West Virginia, plus sometimes a second. LeGrande and Howard (2002) and Cech (1993) report two full broods plus a partial third in North Carolina and New York City, respectively. Most current sources indicate that Peck's overwinters only in the larval stage, but the possibility of pupal diapause is still raised.

Peck's fly rapidly for a small skipper. Males perch throughout the day on low foliage. They are avid nectarers.

Identification	Small and angular. No other local species closely resembles Peck's, and identification is seldom an issue. Individuals with extensive HW patches (lacking dark central band) can look something like Hobomok Skippers, and the dorsal surface can suggest Tawny-edged or Northern Broken-Dash.
Dorsal	Male: dark with a *reddish-orange stripe* along FW costa, bounded by a dark stigma. Note large brown patch on inner side of the stigma, as if part of central orange field had been scraped away. Female: bright *red-orange FW costa* and miscellaneous tawny marks on both wings; the only eastern *Polites* with *Hesperia*-like outer FW dots.
Ventral	Tawny brown with two large, sharply defined *yellow patches on HW*, separated variably by a brown central band.

Habitat	As noted, common in nonspecialized grassy habitats, dry or wet: fields, meadows, roadsides, parks, prairies, disturbed places (even fairly often in disturbed urban waste areas). Likes Red Clover.
Hostplants	Rice Cutgrass (*Leersia oryzoides*), Kentucky Bluegrass (*Poa pratensis*), probably others. Eggs laid singly on hostplant leaf.
Occurrence	Common to locally abundant, but rare at se edge of range. ME south to w. VA, in mountains/upper Piedmont to n. GA and SC. Also across n. US and CAN (ranges well north). One to two broods in North (Jun-Jul), two to three in South (May- mid July, late Jul-Aug, Sep-Oct). Overwinters as larva, maybe pupa.
Ecology	**Generalist.** Common and prolific.

Ventral: 6/23/94, Pound Ridge Reservation, Westchester Co, NY

av wingspan: 0.95"

Long Dash *Polites mystic*

Dorsal male: 6/5/94, Pound Ridge Reservation, Westchester Co, NY

Dorsal female: 6/5/94, Pound Ridge Reservation, Westchester Co, NY

A familiar meadow skipper with a long stigma (dash). Active in wet meadows, it is strongly attracted to Blue Flag nectar.

Beginning butterfly watchers often puzzle at references to the stigma (or brand) in the descriptive literature. This indented, troughlike feature crosses the upper forewing in many male Hesperines. It emits pheromones during courtship. The apparent length of the stigma is extended in many cases (e.g., in the Long Dash) by dark, non-pheromone producing scales toward the apex. And the female's extensive dark forewing stripe, while highly pronounced, has no stigma at all (they occur only in males).

A pale prairie form (*P. m. dacotah*), found outside our range, is likely a full subspecies (Layberry et al., 1998).

Males perch to await females. Courtship is reported from mid- to late afternoon. Often found in multispecies aggregations with other "meadow skipper" species.

Ventral male: 6/9/96, Sheffield, Berkshire Co, MA

Identification	Large for *Polites*, but barely medium-sized among grass skippers generally. Males can look very much like male Indian Skippers. *Key distinction*: Long Dash lacks two isolated, *Hesperia*-style FW dots, completely separate from the postmedial band (*compare* with Indian Skipper; applies to both sexes, dorsal and ventral).
Dorsal	Male: *long, jagged black line across FW ("long dash")* set in large, rufous-orange field; the inner portion is the stigma. Bright upper HW patch has black veins separating orange segments. Female: blackish base color, with reduced orange field, *large, stigma-like transverse dark patch*. HW clearly reflects ventral pattern.
Ventral	Male: yellowish orange, with a concentric pattern of stacked rectangles and a pale central HW dot. Female: darker base color, reddish-black, with strongly defined postmedial markings.

Habitat	Open, grassy habitats, usually moist or wet, but sometimes upland fields. Often near Blue Flag iris in NY area. Prairie swales in the Midwest.
Hostplants	Bluegrass (*Poa*) cited in the literature; others undoubtedly as well, but not documented. Females lay on or near the hostplant, caterpillars live in tied leaf shelters.
Occurrence	Common in the North. N. Eng. s. to WV, VA, NJ. Also west across n. US and s.-central CAN. One brood, late May-early Aug. Early reports of double-brooded coastal population in NY/NJ seem erroneous. Middle-instar caterpillars overwinter.
Ecology	**Medium generalist.** Common and familiar, yet linked to moist meadow habitats in many localities. Widespread in boreal wetlands.

Ventral female: nectaring on Blue Flag (*Iris versicolor*), 6/5/94, Pound Ridge Reservation, Westchester Co, NY

av wingspan: 1.15"

Crossline Skipper   *Polites origenes*

Dorsal male: 7/3/99, Morris Co, NJ

Dorsal female: 7/3/98, Suffolk Co, NY

Though described more than 200 years ago (in 1793), the Crossline is another common meadow skipper whose lifestyle has not been fully documented. Its common name is derived from the appearance of the male's stigma—or, more specifically, from its diagonal position on the forewing, which gives the impression that the brand is *crossing* the forewing rather than running lengthwise, from "shoulder" to apex. (The stigma's apparent length is increased by a patch of dark, nonandroconial scales extending toward the outer forewing.)

Other matters are less clearly established. For example, several authors describe Crosslines as more "ecologically restricted" than Tawny-edged Skippers, a close relative to which the Crossline is usually compared. The Crossline's favored habitat, they indicate, is various dry grasslands. Yet Crosslines regularly select sedge environments in Canada (Layberry et al., 1998), and in Florida favor "moist grassy areas," including "seepage slopes" and pitcher plant habitats (Glassberg et al., 2000). Neither of these environments fits the standard, dry grassland model. Moreover, even authors who cite a preference for dry habitats (ourselves included) quickly point out that Crosslines often fly with Tawny-edged Skippers, a denizen of mostly moist habitats. These incongruities are not beyond reconciling, but they do suggest ecological subtleties in the Crossline's lifestyle that are not as yet fully resolved.

Males display from a low perch during most of the day. Mating takes place from noon to mid-afternoon.

Identification	A medium-sized *Polites*. Often compared to the smaller Tawny-edged in terms of appearance, but females can also be difficult to separate from female broken-dashes, Little Glassywings, and several others.
Dorsal	Male: dark brown upper surface with a *prominent stigma* that *crosses the FW*. The bright orange FW field largely stops at the stigma, with just two to three marks beyond it (the central poststigmal mark is better defined than in Tawny-edged or Northern Broken-Dash). Female: dark brown, usually with a slight tawny edge on shoulder, though this may vanish quickly with wear. Also a prominent wrist "bracelet" and *two to three conspicuous pale FW marks*, the central one especially large and squarish. Diffuse orange patch on upper HW, variably expressed.
Ventral	*Both wings* orange-brown to olive-buff below, with a variable *postmedian crescent band*; sometimes sharp, elongate spots, sometimes diffuse pale spots, sometimes virtually absent (may be obscured by scales when fresh).

Habitat	Often seen in dry, grassy habitats, sand barrens, sandhills, old fields, power easements, etc. Yet not uncommon in damp meadows with Tawny-edged Skippers, and in other mesic situations.
Hostplants	Purpletop (*Tridens flavus*), Little Bluestem (*Schizachyrium* [*Andropogon*] *scoparius*), Mannagrass (*Glyceria melicaria*) in WV. Has accepted a sedge (*Carex scabrata*) in captivity, possibly also in wild?
Occurrence	Usually fairly common, depending on region. Central N. Eng. south to n.-central FL. West to CO. Occasional coastal emigrant to NY/NJ in early fall. One brood in North (Jun-Aug), two south from Philadelphia (May-Jun, Aug-Sep). Mature larva overwinters.
Ecology	**Generalist.** Widely adaptable, including disturbed sites, with some preference for dry grasslands.

Ventral: 7/11/98, Pound Ridge Reservation, Westchester Co, NY

av wingspan: 1.1″

Tawny-edged Skipper *Polites themistocles*

Dorsal male: 5/27/95, Wakefield, Sussex Co, VA

Dorsal female: 8/15/98, Green Swamp, Brunswick Co, NC

A small, widespread skipper, the Tawny-edged is common over much of its range, at times very common. It is the epitome of a meadow skipper, occurring regularly in a wide variety of generalized, grassy habitats. Tawny-edges associate regularly (but not invariably) with damp meadows and water edges—settings that are neither fully wetlands nor dry upland meadows (*see* habitat description, p. 14).

Early commentators considered the Tawny-edged "one of the commonest and most widely distributed of all our Skippers" (Weed, 1917; *see also* Klots, 1951). Current reviewers have tempered this assessment, especially in the South, where the species is only moderately common—a fact that has been cited as evidence of a northern slant in the literature (LeGrand & Howard, 2002). That may be, but even northern commentators have recently been more circumspect in describing Tawny-edged population levels (e.g., Gochfeld & Burger, 1997). We suspect that (1) the reversion of open habitat to woodlands in the Northeast during recent decades, (2) widespread destruction of meadow habitats throughout the region, and possibly (3) increased use of environmental chemicals may simply have caused the species to be less widely abundant than it once was.

Males perch most of the day on grass stems or other low roosts, mating from noon to mid-afternoon. Females oviposit in late afternoon. An active nectarer, the Tawny-edged Skipper uses a wide variety of flowers.

Identification	Noticeably smaller usually than Crossline or Long Dash, size of Peck's. Wings pointed. Markings variable geographically and within populations.
Dorsal	Male: brown or olive-brown above, with a *bright rufous-orange costa*, ending abruptly at the stigma. The stigma generally runs more parallel to the costal FW margin than in Crossline. A solitary, orange "keyhole" mark appears at the end of the stigma, as in male broken-dashes and Peck's. Female: variable "tawny edge" on FW costa, usually less extensive than male's; note *rectangular, cell-end FW dots* and wrist "bracelet."
Ventral	Plainest of the *Polites* skippers below. HW usually *unmarked brownish or brassy orange, or mustard-colored*. Occasionally has a pale crescent-shaped spot band, as in Crossline (but usually fainter, when present). Submarginal FW below *contrasts with HW color* (uniform in Crossline). Female's wrist "bracelet" shows ventrally.

Habitat	Variable. Favors a range of grassy habitats, usually damp, including moist meadows, swales, steam edges, savannas, sometimes vacant lots. Not especially tolerant of heavy disturbance. Forest glades and alpine bogs in the West.
Hostplants	A variety of grasses, maybe sedges occasionally. Panic grasses, especially smaller species (*Panicum microcarpum* in WV), Slender Crabgrass (*Digitaria filiformis*), bluegrasses (*Poa*), etc. Slow-growing larvae live in silk shelters.
Occurrence	Mostly common, especially northward (but *see* above). ME south to central FL; also NS to BC in CAN, scattered in moister areas of w. US. One brood far north (Jun-early Aug in ME), two elsewhere, May-early Nov, with a gap in mid-Jul. Pupae reportedly overwinter.
Ecology	**Generalist.** May have lost some ground here and there, but still going strong.

Ventral (mated pair): 8/15/98, Green Swamp, Brunswick Co, NC

av wingspan: 1.0″

Baracoa Skipper   *Polites baracoa*

Dorsal male: 9/9/01, Homestead, Miami-Dade Co, FL

Dorsal female: 9/26/95, Redlands, Miami-Dade Co, FL

This miniscule skipper can be found throughout the Florida peninsula, but it is typically common only in the far South, and even there it tends to live in discrete, somewhat local colonies.

The Baracoa has a small global range (just Florida, Cuba, and Hispaniola), yet it has a generalist lifestyle, adapting well to human settlement. With such flexibility, it is not clear why its distribution is so limited. But the species is a poor emigrant, so much so that different races can develop on nearby Caribbean islands.

The Baracoa was formerly thought to inhabit "grassy flats and streambeds" (Pyle, 1981). But we now know that the main native habitat is actually open pinelands, with a recent extension to short-cropped lawns in settled areas.

Fast-flying and nervous, Baracoas are difficult to see because of their smallness and very low flight; difficult to approach.

Ventral: 9/26/95, Redlands, Miami-Dade Co, FL

Identification   Our smallest *Polites*, nearly our smallest grass skipper of any kind (except Southern Skipperling). Somewhat resembles a small, more colorful Tawny-edged Skipper. Fringes buff. Darting, ground-level flight style distinctive.

Dorsal   Sexes distinguishable but not dissimilar. Male: tawny-orange field on dark FW, stigma is short and straight (but dark patch near apex may be confused as an extension). Female: orange field on dark FW with dark patches in much the same location as male's.

Ventral   HW variably olive-brown (often with tawny overcast in female). Note *pale, concentric spot band on central HW* (comparatively well-defined, especially in female). FW mirrors dorsal pattern.

Habitat   As noted, native habitat appears to be clearings in or near open pine woods (pine rocklands, etc.). Said to favor habitats near wetlands, but we have not noticed such an association. Radiating in settled areas of s. FL, where it inhabits open, close-cut lawns.

Hostplants   Apparently adapts to a variety of low-growing grass species (Smith et al., 1994).

Occurrence   Locally common on FL peninsula, especially s. mainland, but not Keys. Ephemeral colonies to s. GA. Also W.I. Several broods in n. FL, Mar-Oct, all year in South. Mature larvae overwinter.

Ecology   Medium generalist. Undoubtedly more specialized before lawns invaded s. FL.

Close-cropped lawn, Deering Estate, Miami-Dade Co, FL

av wingspan: 0.9"

Little Glassywing   *Pompeius verna*

Dorsal male: 6/20/99, Pound Ridge Reservation, Westchester Co, NY

Dorsal female: 6/25/95, Pound Ridge Reservation, Westchester Co, NY

A familiar meadow skipper, the Little Glassywing is common in damp grasslands in many parts of our region. The only representative of its genus in the United States, it has several characteristic pattern features (e.g., the male's forewing markings). Yet the Glassywing's occurrence here is not all that exotic. Taxonomists have long regarded its genus (*Pompeius*) as being very close to *Polites*. And the Little Glassywing was itself originally classified as a *Polites*. The six currently recognized members of *Pompeius*, mostly neotropical, are distinguished by brown color and large antennal structures (Howe, 1975).

Ecologically, the Little Glassywing is interesting for its use of a single, predominant hostplant, a grass called Purpletop (*Tridens flavus*). Opinions vary as to whether this is the sole, unique host, but the Little Glassywing's range in the United States does conform very closely to that of Purpletop (*see* Pohl, 1978). Curiously, there are places along the coastal plain where Purpletop is common but Little Glassywings are not (Gochfeld & Burger, 1997); in particular, Purpletop is often found in dry, sandy woods, as well as in moister settings preferred by Little Glassywings. Perhaps the grass is more fire-tolerant than the butterfly. Such issues remain to be clarified.

Little Glassywings can be found actively nectaring throughout the day. May wander from primary habitats to feed. Males perch on low foliage to await females. Courtship is reported around noon.

Identification	An average-sized meadow skipper. The female is one of the "three witches," so-called by old-time lepidopterists because of their dark coloring, similarity of appearance, and tendency to fly together in mid-summer. Females should also be compared with Crosslines, Tawny-edges, etc. Glassberg (1999) noted a *white area on each antenna, just below the club*, which is distinctive in most populations. Fringes are buff-toned or gray, usually lighter than in other "witches."
Dorsal	Dark, often *blackish brown* on all surfaces. *Slight* tawny edging may be visible in fresh individuals. Male: lacks an orange FW field, instead has a series of separate, often yellowish glassy marks, including a long, transverse mark abutting the "keyhole" at the end of the stigma. (*Note*: Locating the stigma on such a dark wing can be challenging!) Female: prominent glassy FW marks, including a *large, central squarish hyaline spot*.
Ventral	Brown or rusty brown, with a purplish sheen when fresh. A band of thin, *inconspicuous, dashlike marks* forms a crescent on the postmedial HW; these marks usually *unevenly aligned*. Note a separate white mark on costal HW.

Habitat	Usually moist, open areas (fields, clearings, roadsides, etc.), often near shady wood edges, and streams or wetlands may be nearby. Also sometimes drier, in more sandy habitats with hostplant.
Hostplants	Mainly Purpletop (*Tridens flavus*), possibly other grasses. Caterpillars live in leaf nests.
Occurrence	Variably common. Central N. Eng. to n. FL, west to Northeast, se TX. Generally rarer on coastal plain (especially outer plain), yet common on e. LI. One brood in North (mid-Jun to early Aug), two in South, mainly Apr-Jun and Jul-early Sep. Early larva seem to overwinter (Allen, 1997).
Ecology	Generalist. Adaptable and generally widespread.

Ventral female: 6/22/96, Pound Ridge Reservation, Westchester Co, NY

av wingspan: 1.1″

Northern Broken-Dash   *Wallengrenia egeremet*

Dorsal male: 6/23/94, Cunningham Park, Queens, NY

Dorsal female: 7/4/95, Pound Ridge Reservation, Westchester Co, NY

Our knowledge of the broken-dashes (genus *Wallengrenia*) is relatively detailed, thanks to a landmark monograph published in the 1980s (Burns, 1985). Burns's focus was selective: he mainly addressed taxonomy and demographics, as opposed to behavioral style or hostplant use. Yet the resulting information base is highly comprehensive, especially in comparison with that available for most other meadow skippers.

The genus *Wallengrenia* is closely allied to *Polites* (Howe, 1975). The most obvious difference is the two-part stigma (or broken dash) on the male forewing. In this genus, the extension of the dark forewing line is part of the stigma itself, and not simply a patch of dark scaling (*compare* Long Dash).

Broken-dashes have a checkered taxonomic history. Originally described as two separate species, they were "lumped" together by Edwards in the late 19th century. At times during the decades that followed, the Northern Broken-Dash was not even considered a full subspecies. By the mid-20th century, however, clear evidence had accumulated to support the recognition of two separate entities. Burns in fact found them so well-differentiated that they were not even a good example of an emerging species-pair (Burns, 1985).

We have not noticed the reported tendency for male Northern Broken-Dashes to perch higher than Southerns (except perhaps as a function of habitat structure). Nor, to us, does their flight seem characteristically slow. Active nectarers, they usually confine their activity to rich, open habitats, where they both feed and mate.

Identification   Size varies by latitude, smallest in the North, largest in the South. But slightly smaller just south of the line where the species first becomes double-brooded (*see* Introduction, p. 20). The Northern Broken-Dash is larger than the Southern (on average) where the two species occur together (Burns, 1985). *Fringes are buffy.*

Dorsal   Similar to several related meadow skippers. Male: *stigma divided into two parts.* FW costa tawny or yellowish-orange (usually paler than Southern), with a *yellow-orange "keyhole" mark* at end of inner stigma. Female: second of the "three witches." Mostly dark brown, with a transverse series of yellowish FW marks, variably tawny on costa.

Ventral   Brown base (normally lighter than Dun or Little Glassywing), purple-violaceous sheen when fresh, also a *vague, "backward 3" pattern on HW.* Dun usually lacks this, has a more concentric-shaped crescent (if present).

Habitat   Grasslands, often near woods, usually moist, but not always, and seldom extremely wet or dry. Fields, meadows, roadsides, gardens, etc.

Hostplants   Large panic grasses, including Deertongue (*Panicum cladestinum*), Switchgrass (*Panicum virgatum*) in NJ. Also *P. dichotomum, sanguinale.* Eggs laid singly on or near hostplant.

Occurrence   Common in North and Appalachians (not high peaks); more local on coastal plain. S. ME to central FL, west to se ND and e. TX. One brood north of Philadelphia (mid-Jun to Aug), two to three in South (Apr to early Jul and late Jul to Sep). Middle instar larvae overwinter.

Ecology   Generalist. Adapted to a range of habitats and host grasses, can be very common.

Ventral: 7/11/98, Pound Ridge Reservation, Westchester Co, NY

av wingspan: 1.1″

Southern Broken-Dash *Wallengrenia otho*

Dorsal male: 8/10/97, Great Dismal Swamp NWR, Nasemond Co, VA

Dorsal female: 5/27/95, Great Dismal Swamp NWR, Nasemond Co, VA

Along with Whirlabout, the Southern Broken-Dash is the only eastern "meadow skipper" with no established population base in the Northeast (it ranges as far north as Maryland, and that only rarely). It is nevertheless among the most common and widespread of grass skippers within its core region, i.e., along the southeastern U.S. coastal plain and in Florida.

It was earlier thought that our two eastern broken-dashes (Northern and Southern) might have originated from a single, common ancestor. This was supposed to have occurred when northern, single-brooded populations gradually became isolated from southern, multibrooded ones, because of nonoverlapping flight periods. But this model does not seem to fit the facts, since Northern Broken-Dashes also have multiple broods in the Southeast, and the timing of these broods (phenology) is similar to that of Southern Broken-Dashes living in the same areas (Burns, 1985). Further, the southeastern form of Southern Broken-Dash is but one of many varied races; others occur in the West Indies and South America. Thus, a more comprehensive developmental theory is needed. The Texas form (*W. o. curassavica*), living outside our region, is distinctively yellow-toned below.

Males perch low to the ground. But individuals seen on low foliage in early morning may in some cases be basking to warm up (as with early-rising Clouded Skippers), rather than displaying. Further observation is needed.

Identification	A medium-sized meadow skipper, slightly smaller than Northerns where they occur together. Wing fringes are differently colored (gray on FW, buff or rusty on HW); in Northern, fringe color is uniform.
Dorsal	Male: *stigma divided into two parts*. FW costa tawny orange or red-orange, generally brighter and more reddish than in Northern, but some Northerns can be confusingly bright-toned, e.g., in central NJ (John Burns, pers. comm.). Several orange marks occur behind inner portion of stigma. Female: similar to several other female meadow skippers; spots generally more orange-toned than in Northern.
Ventral	Similar to Northern, with *pale, "backward 3" mark on HW*, but usually *much rustier*. S. FL males in Feb-Aug are especially reddish (Smith et al., 1994).

Habitat   Many varied open habitats. Often clearings or edges near swamps or streams, lakes, pocosins, moist prairies, etc. But also sandhills, rimrock pines, hammocks, gardens. Less common in mountains.

Hostplant   Not fully reported. St. Augustine Grass (*Stenotaphrum secundatum*) in captivity, probably also in wild. *Paspalum* grasses (*P. caespitosum* on Keys, *P. setaceum* in VA). Larva cuts circular bit of leaf for shelter, carries it from place to place (Minno & Emmel, 1993).

Occurrence   Common to abundant, coastal MD south to Keys, west to e. TX, south to ARG. Two broods in VA (May-Jun and Aug-Sep), two to three continuous broods in s. FL. Overwintering stage not reported.

Ecology   Generalist. Widespread and adaptive, uses a range of habitats and hostplants.

Ventral: 3/21/94, Corkscrew Swamp Sanctuary, Collier Co, FL

av wingspan: 1.05"

Arogos Skipper   *Atrytone arogos*

Dorsal male: 7/29/98, central NJ

Dorsal female: 7/4/01, n. NJ

This unusual skipper exists in just a handful of widely dispersed regional colonies in the East. Little studied until recently, it still needs further investigation—and dedicated protection.

The only species in its genus (*see* Burns, 1994), Arogos is one of several grass skippers with disjunct ranges in the East and Midwest. The prairie race (*A. a. iowa*) is generally smaller and yellower than the nominate eastern form.

Recent fieldwork, by Dale Schweitzer and others, has greatly improved our understanding of this elusive skipper. Some dwell in habitats with little nectar, and must travel significant distances to feed. Notoriously shy, they are prone to vanish when disturbed. Yet Arogos can at times be incredibly docile when feeding, ignoring direct touches while sipping intently at a single flower for long periods. Males display in mid- to late afternoon (5:30 P.M. in photo above/left), seldom revealing their dorsal surface at other times (except briefly when cold).

Ventral: 9/30/95, Riverside Island, Ocala NF, Marion Co, FL

Identification	Medium-sized. Males smaller, with pointier, more triangularly shaped wings. Conspicuous tuft of orange shoulder hairs contrasts with greenish thorax. Outer wing margins *pale or gray-buff, not orange* as in Delaware.
Dorsal	Male: yellow-orange with *wide, dark trailing band*, no stigma. Large orange patch on the upper HW. Female: tawny orange FW patch is comparatively small, with a *straight dark, stigma-like line* in the middle. Orange patch on upper HW smaller than male's.
Ventral	Plain orange or tan-orange below with *white scaling on HW veins*.

Habitat	Seasonally wet pine savannas, especially at regenerating burn sites. In n. NJ, grassy utility easements and transport corridors, roadside fields. Also sandhills with suitable hostplants. Not on mid-Atlantic serpentine barrens. Prairie swales in West.
Hostplants	Little Bluestem (*Schizachyrium scoparius*) in n. NJ/NY. Pine Barrens Reed Grass (*Calamovilfa brevipilis*) in s. NJ/NC, Lopsided Indian Grass (*Sorghastrum secundum*) in FL. Seems to use a single host in any locale (Dale Schweitzer, pers. comm.).
Occurrence	Isolated colonies in n.-central NJ, se NC, central SC/GA; scattered populations in n.-central FL. Formerly s. FL, NC sandhills, Staten Island, NY. Also Gr. Plains. One brood in North, mainly Jul. Two in South, Apr-Jun and Aug-Oct. Larva overwinters.
Ecology	**Specialist.** Colonizes burn sites, but may be intolerant of direct fire contact—Pyle (1981) reported pupae on grass three feet above ground.

Wet pine savanna, central NJ

av wingspan: 1.1"

Delaware Skipper *Anatrytone logan*

Dorsal male: 7/3/99, Morris Co, NJ

Dorsal female: 6/27/94, Van Cortlandt Park, Bronx, NY

Among our most colorful grass skippers—and also rather easy to identify in most situations—the Delaware is a welcome addition to early summer meadows in the northeastern states. Farther south, it is a regular feature of the skipper community throughout the warmer months.

The Delaware Skipper was recently reclassified, from the genus *Atrytone* (with Arogos Skipper) to the neotropical genus *Anatrytone* (Burns, 1994). The sole local emissary of its new genus, the Delaware still has nearby relatives, in the genus *Problema* (Byssus and Rare Skippers), which is visually as well as biologically similar. But unlike its more specialized cousins, the Delaware is widespread and successful, familiar to most local observers.

There is evidence that the Delaware Skipper may actually represent a "species complex," consisting of two or more sibling forms (Dale Schweitzer, pers. comm.). At present, however, only the western prairie race (*A. l. lagus*) is formally described, and one former subspecies (*A. l. mazai*) is now recognized as a separate entity—the Glowing Skipper (*Anatrytone mazai*)—a stray to south Texas.

Delaware males maintain a lookout for females during most of the day from low perches. Rapid fliers, they are often detected when their bright colors are glimpsed in midair. Individuals sometimes rest on the ground, and males are known to take minerals. Mating has been observed in early afternoon.

Identification	Medium-sized. Males smaller, wings more triangular than females'. Smaller on average than Rare or Byssus Skippers, notably larger than European, slightly bigger than Arogos. *Wing fringes are orange to tan-orange* (*compare* Arogos, Byssus, European, and Berry's, all with lighter fringe)—but be aware that worn individuals may have no fringe at all! Very long proboscis, 1.5 times length of FW (Opler & Krizek, 1984).
Dorsal	Bright, tawny orange with wide, dark borders. Male: FW mostly clear orange with *dark veins* near apex ("crow's-feet") and a *pointed, hook-shaped mark* at cell-end. Female: wider dark wing margins, more extensive dark veining, *two large, dark FW patches* giving some observers the impression of a snowshoe (Gochfeld & Burger, 1997). Resembles female Rare Skipper and male Byssus Skipper.
Ventral	Bright golden-orange with orange HW veins (*compare* Berry's). Orange on central HW may be paler than edges.

Habitat	Extremely varied. In our experience, fields (wet and dry), savannas, pond edges and marshes (fresh, sawgrass, and brackish), bogs and fens, dry pine bush, parks and gardens, etc. Prairies in West.
Hostplants	Varied grasses, including Big Bluestem (*Andropogon gerardi*), Switchgrass (*Panicum virgatum*), Maidencane (*P. hemitomon*), Redtop Panicum (*P. rigidulum*), beard grass (*Erianthus* sp.). Possibly *Carex* sedges (Iftner et al., 1992), based on habitat associations in IL.
Occurrence	Mostly common, se ME south to s. FL (not Keys), mostly on coastal plain in Southeast. Also west to the Rockies, n.-central TX. One brood in North, late Jun-Aug, two near-continuous broods in South, May-Jul, late Jul-Sep. Feb-Nov in s. FL. Diapause stage larva or pupa.
Ecology	Generalist. Seldom abundant, but a widespread "everyday" species, extremely adaptable.

Ventral: 6/27/94, Van Cortlandt Park, Bronx, NY

av wingspan: 1.2"

Byssus Skipper   *Problema byssus*

Dorsal male: 9/21/96, Gainesville, Alachua Co, FL

Dorsal female: 9/27/97, Bunnell, Flagler Co, FL

In our region, the Byssus Skipper inhabits thick, tall-grass wetlands, where it forms dispersed, local colonies. (Another, entirely separate population lives in remnant prairie grasslands on the Great Plains.)

The genus *Problema*, to which Byssus belongs, contains just two species. Both live in our area, and one is an East Coast endemic (Rare Skipper). *Problema* skippers are closely related to Delaware (genus *Anatrytone*). They are separated mainly by technical characters (e.g., lack of a male stigma). But both are visibly larger than Delaware.

Byssus Skippers may wander extensively during the day. Males display near colonies in early morning and late afternoon, but shift to various nectar sites at other times, often well away from display territories.

Ventral: 9/21/96, Gainesville, Alachua Co, FL

Identification	Large. Most easily confused with Delaware Skipper (which is smaller), or with Rare Skipper (which is larger, also rare, and highly local). The *antennal club is nearly all-black.* Wing fringes *white to tan.*
Dorsal	Male: *bright, yellow-orange field* with wide, dark borders on both wings. Note *long, dark FW patch with a central, orange, cell-end mark.* Resembles *female* Delaware or Rare. Female: much darker, with light color confined to a central FW mark (at cell-end) and a continuous outer FW band (two outer band elements sometimes absent).
Ventral	HW varies from yellow-orange to rust-orange, with a *faint, variable crescent-shaped HW band.* FW dark at base.

Habitat	Marsh edges (inland and coastal), grassy stream or slough edges, wet savannas, damp easements, etc., all with hostplant. Frequently at ecotones between wet woodlands and open areas. Also at nearby nectar sites during the day. Moist prairie remnants on the Gr. Plains.
Hostplants	The predominant, perhaps unique hostplant is Eastern Gamma Grass (*Tripsacum dactyloides*), a tall, broad-leaved grass that grows in dense clumps. May use other species as well, e.g., plume grass (*Erianthus* sp.), but all colonies we have seen were near Gamma Grass.
Occurrence	Rare to locally common in variably scattered colonies. NC to FL, mainly on the coastal plain, absent far se FL and Keys. Only in coastal NC, despite presence of hostplant "throughout" the state (LeGrand & Howard, 2002). Also in w. Gulf states, and separately on prairies, in MO and adjoining areas, east to ne IN. Two broods in our area, Apr-Jun, Aug-Oct. First brood less numerous in NC, possibly due to climate stress. Overwinters as a mostly grown larva. Pupa located in a dense, silk nest at the base of host plant.
Ecology	**Medium specialist.** More common and more widespread than some of our highly restricted marsh skippers. Still, Byssus is selective in its habitat, hostplant, and geography. Some midwestern populations are conservation-listed.

Eastern Gamma Grass, or Bunchgrass
(*Tripsacum dactyloides*)

av wingspan: 1.4″

Rare Skipper *Problema bulenta*

Dorsal male: 7/16/95, Cumberland Co, NJ

Dorsal female: 7/14/96, Cumberland Co, NJ

The Rare Skipper has an interesting entomological history. It was first discovered in Georgia or South Carolina in the early 1800s by the pioneering naturalist John Abbott, and was included in one of his early drawings. A formal description was published in 1834.

But *Problema bulenta* then seemingly vanished. Undetected in the field for more than 100 years, and largely given up for lost, it was rediscovered in 1925 near Wilmington, North Carolina. Based on this long absence, it was given the common name of "Rare Skipper." Eventually, more populations were found in the brackish upper reaches of four East Coast river estuaries, in Maryland, Virginia, North Carolina, and Georgia. Lastly, in the 1980s, it was reported at several sites in south coastal New Jersey (Cromartie & Schweitzer, 1993).

There is some dispute as to whether the designation of "rare" is appropriate for this butterfly (Lost Skipper might be better), since it can occur in large numbers in the marshy backwaters where it dwells, and many such sites are difficult to survey. But large tracts of likely seeming habitat do *not* host the skipper, based on diligent investigation, and it would thus be misleading to imagine them lurking just out of sight in estuaries generally. The currently known sites are geographically confined.

It is dramatic to see Rare Skippers gathered en masse on roadside nectar plants along a tidal estuary. A dozen or more skippers are sometimes seen over a short period (occasionally more than 100) when individuals are drawn in from large tracts of nectarless marsh. Partial to milkweeds, pickerelweed, morning glories, etc.

Identification	Large (especially females), with pointed wings (especially males). Note *orange antennal clubs* (below dark tip). The abdomen is rather long; *white lines encircling abdominal segments*. Fringe is light.
Dorsal	Male: bright, golden-orange above, with wide dark FW borders (narrower on HW) and a "harpoon-shaped" mark on the FW, similar to male Delaware. Male Yehl has a stigma. Female: orange field on FW proportionately smaller than in female Delaware, and black border wider. Otherwise similar, except that female Rare is *much larger*. Also larger than the surprisingly similar male Byssus.
Ventral	HW *plain brassy yellow-orange*, duller than Delaware. Dark FW mark on male visible from below (often hidden).

Habitat	Highly restricted. Tall-grass areas of upper tidal marshes, fresh to slightly brackish, in the backwaters of large, coastal rivers. Sometimes old rice paddies. Strays for nectar, e.g., to garden at Cape May Bird Observatory in Goshen, NJ.
Hostplants	Tall Cordgrass (*Spartina cynosuroides*) in NJ; Marsh Millet (*Zizaniopsis millacea*) and Wild Rice (*Zizania aquatica*) reported to the south.
Occurrence	East Coast endemic. Atlantic coast from s. NJ to GA, e.g., on Cape Fear, Chickahominy, Maurice, Santee, Savannah, and Wicomico Rivers. One to two broods, May and July-Sep. Early stages unreported.
Ecology	Specialist. Extremely restricted habitat selection makes Rare Skipper populations vulnerable. Protection is warranted at all sites.

Ventral: 7/5/98, Cumberland Co, NJ

av wingspan: 1.5″

Hobomok Skipper *Poanes hobomok*

Dorsal male: 7/16/00, Pondicherry Wildlife Reserve, Coos Co, NH

Dorsal female: 6/9/01, Pound Ridge Reservation, Westchester Co, NY

Dorsal female (*pocahontas* form): 6/12/99, Pound Ridge Reservation, Westchester Co, NY

Hobomok, a chief of the Wampanoag tribe, aided the Pilgrims after their landing at Plymouth in 1620 (Opler & Krizek, 1984). But the chief's name has no particular descriptive significance as regards the butterfly, and Pyle (1981) suggested Northern Golden Skipper instead (with Zabulon being Southern Golden). The nominate race of Hobomok occurs in our area, and our female has two highly distinct color morphs.

A denizen of northern climes, the Hobomok is one of few eastern grass skippers closely tied to mixed deciduous or transitional woods (others include the Arctic and Pepper and Salt Skippers). This hardy, prolific species emerges not long after the first spring *Hesperia*s.

Hobomok males perch on broad leaves in sunny woodland openings, aggressively pursuing intruders. They perch on woodland foliage, higher than meadow skippers, which display from grass. They often feed on bird droppings (Gochfeld & Burger, 1997), a trait common in woodland butterflies (*see* Introduction, p. 12). Both sexes stray regularly to open fields and gardens to seek nectar, but seldom far from woodlands.

Ventral female (*pocahontas* form): 6/21/97, Pound Ridge Reservation, Westchester Co, NY

Identification	Medium-sized. *Wings rounder than Zabulon* (especially males'). Can be confused with Peck's Skipper (from below) or Zabulon (especially males from above). Two isolated dots on outer FW can suggest *Hesperia*.
Dorsal	Male: bright yellow-orange field with dark, irregularly shaped border. Variable dark markings near FW apex often form an *S* shape. Female (normal): similar to male, but orange more restricted, becomes brownish near FW base. Female (dark *pocahontas*): dark with purplish overtones. Sparse FW markings include an *isolated mid-FW dot* near costa (absent in Zabulon). Dark form occurs in 25%-50% of eastern females, rare or absent farther west.
Ventral	Normal: note *continuous, blocklike yellow HW patch*. All-brown basally, with *one small, yellow spot*. Brown trailing borders *frosted purplish-white* on both wings (broader on FW in female). Male's FW apex is all brown, vs. yellow in Zabulon. Pocohontas: dark purplish-brown, with dark overscaling that obscures (but does not fully conceal) the distinctive, pale HW patch.

Habitat	As noted, likely to be seen in or near woodlands, especially in sunny clearings or along forest roads or trails. Also wanders to open fields, parks, and gardens for nectar.
Hostplants	Favors panic grasses (*Panicum*), including Small-fruited (*P. microcarpon*), Deertongue (*P. cladestinum*), etc., also bluegrasses (*Poa*). But accepts many others grasses in captive rearing situations (Allen, 1997).
Occurrence	Common in northern habitats, ME to VA, south in mountains to n. GA. Absent s. NJ pinelands, Delmarva. Also s. CAN, west to Gr. Plains, CO/NM. One long brood, late May-Jul in North, late Apr-Jun in South. Overwinters as larva.
Ecology	Generalist. Common and widespread.

Ventral male: 5/16/93, Nottingham Co Park, Chester Co, PA

av wingspan: 1.2″

Zabulon Skipper   *Poanes zabulon*

Dorsal male: 5/19/02, Clay Co, NC

Dorsal female: 9/2/96, Cape May Co, NJ

Ventral female: 9/3/95, Cape May Co, NJ

To present-day observers, armed with illustrated field guides, close-focusing binoculars, and a library of web photos, it is difficult to understand how the Hobomok and Zabulon Skippers were regularly confused by serious investigators in the late 1800s and early 1900s. Yet as late as the 1950s, Klots could not describe the Hobomok's early stages "because of the confusion of hobomok with zabulon" (Klots, 1951).

Meanwhile, any Easterner seeing a Taxiles Skipper (*P. taxiles*) in the West is sure to note its close similarity to the Zabulon. Some investigators formerly classified Taxiles as a Zabulon subspecies (e.g., Scott, 1986). The two have since been shown to be distinct (Burns, 1992).

Male Zabulons are noted for holding a single display perch for substantial periods, probably throughout their short adulthood (Opler & Krizek, 1984). Perching occurs all day. Courtship is usually in the afternoon, but occasionally as early as 8:20 A.M. Females perch lower to the ground than males, and are more likely to remain in shaded areas. As with Hobomok Skippers, Zabulon males often perch higher than typical meadow species, if only as a function of habitat structure.

Identification	Medium-sized. *Wings thin and triangular* compared with Hobomok (especially males). Female has a single dark morph, vs. two in Hobomok. A southern species, often separable from Hobomok by range or flight period.
Dorsal	Male: *very bright yellow-orange* with somewhat narrow black borders. Note variable dark patch near FW apex and often unconnected black line. No stigma. Female: all-dark with glassy, angular outer FW markings (no mark along mid-FW costa, as in *pocahontas* form of Hobomok).
Ventral	Male: *broad, ill-defined yellow patch with rust-brown dots.* HW brown at base, but with yellow "shoulder." Female: *rich, blotchy mahogany-brown* with broad, marginal frosting; distinctive *white outline on costal edge of HW.*

Habitat	Variable. Mainly brushy, second-growth habitats near woods—usually grassy, often damp. Also woodland clearings or roads, especially near streams. Adapts to suburbs, parks, gardens, etc.
Hostplants	Numerous grasses, especially lovegrass (*Eragrostis*), Purpletop (*Tridens flava*). Range of hostplants not fully described.
Occurrence	Usually common, CT and MA (local) south to n. FL. Most frequent on Piedmont in NC. Also west to central plains, e. TX, and south to Panama. Two to three broods, 1st Mar/Apr in South, late May-Jul in North; second Aug-Sep/Oct. Diapause stage not reported, probably larva.
Ecology	Generalist. Common and adaptable.

Ventral male: 6/4/94, Cape May Co, NJ

av wingspan: 1.2″

Broad-winged Skipper   *Poanes viator*

Dorsal male: 7/14/03, Cumberland Co. NJ

Dorsal female: 8/8/92, Jamaica Bay NWR, Queens Co, NY

This large, floppy skipper is becoming ever more common along the Atlantic coast, and in some freshwater marshes inland, as it capitalizes on the invasive spread of its main hostplant, Giant Reed (or Phragmites).

The Atlantic coastal race (*P. v. zizaniae*) occurs in most of our area. Its extension inland may eventually bring it into contact with the nominate Great Lakes race, which occurs sporadically in western New York and Pennsylvania.

Male Broad-winged Skippers patrol for mates, working their way deliberately through reedbeds with a slow, jerky flight, dropping visibly between wingbeats. Perches head-up on a reed stalk.

Much remains to be learned about this familiar skipper. Caterpillars overwinter, but debate continues as to where—some suggest they may hibernate underwater.

Ventral female: 7/25/98, Jamaica Bay NWR, Queens Co, NY

Identification	Among our largest grass-feeding grass skippers. Wings round and full-cut, female's somewhat more pointed.
Dorsal	Male: no stigma, resembles female of some species. FW dark with a small orange field, segmented into *discrete, orange patches*, including a central, cell-end mark and a wrist "bracelet." HW field is bright orange, with *scalloped trailing pattern*. Female: similar to male, but FW marks smaller and *cream-white*, HW marks reduced, mirror ventral pattern.
Ventral	HW variably dull to rusty tan, depending partly on wear. *Long, pale ray extends out from wing base*, crossed by a *variable spot band*, giving effect of a cross. (Spot band fainter in Gr. Lakes race, ray yellower.)

Habitat	Mainly coastal salt marshes with tall grass. Also freshwater marshes inland, including fens, sedge bogs, shrub marshes. Edges of swamps and water bodies in FL. Wanders to gardens, roadsides, etc. for nectar.
Hostplants	Giant Reed (*Phragmites communis*), Wild Rice (*Zizania aquatica*), Marsh Millet (*Zizaniopsis miliacea*); Big Cordgrass (*Spartina cynosuroides*) in NC. *Carex* sedges inland. Caterpillars reportedly rest between host leaves and stem, using little silk (mostly before molt).
Occurrence	Local but often hugely abundant. Coastal ME to n. FL, spotty inland colonies (increasing). Also Gulf Coast to e. TX, Gr. Lakes region. One long brood north of VA, July-Aug/Sep; two in South, Mar/Apr-May (or May-Jul in NC) and Aug-Oct. Larvae overwinter, details not reported.
Ecology	Medium generalist. Formerly more specialized, but has spread widely with reed host, adapting to disturbed as well as pristine settings.

Giant Reed, or Phragmites (*Phragmites communis*)

av wingspan: 1.6″

Yehl Skipper *Poanes yehl*

Dorsal male: 9/8/98, Palm Point Park, Gainesville, Alachua Co, FL.

Dorsal female: 8/10/97, Great Dismal Swamp NWR, Nasemond Co, VA

This bright-colored skipper is closely associated with southern swamps and canebrakes. Because it is wary and retiring, its life habits are not fully reported.

Yehl exhibits an unusual combination of anatomical traits. It is mostly *Poanes*-like, yet has a prominent male stigma, which once caused it to be placed in a different genus (*Paratrytone*). Such confusion is not uncommon among *Poanes*: to a nonspecialist, the entire genus seems a bit heterogeneous, with wide variations in size, lifestyle, wing pattern, habitat selection, and hostplant use.

Unlike the Broad-winged Skipper, male Yehls perch when seeking a mate, usually two to three feet off the ground on a sunny leaf surface. Loss of canebrakes and other southern wetlands has reduced local populations of this habitat-restricted butterfly.

Ventral male: 9/8/98, Palm Point Park, Gainesville, Alachua Co, FL.

Identification	Fairly large. Resembles *Euphyes* in many aspects; dorsal male could be confused with Black Dash, except that the two rarely occur together geographically. Greatest practical source of confusion is dorsal female Yehl versus female Broad-winged.
Dorsal	Male: FW *rich, reddish-orange* with wide, dark borders and a *prominent black stigma*. Female: FW brown with a variable series of spots—a transverse band, cell-end marks, a wrist "bracelet." The outer spots may be white.
Ventral	Male: tawny-orange HW with *three small, yellowish dots* forming an incomplete crescent (may be faint, especially in worn individuals). Female: darker brown HW, with more *conspicuous, white dots*. Occasionally a faint HW ray.

Habitat	Typically sunlit openings in swamps, canebrakes, wet hammocks. Likes partially shaded situations, not open marshes. Also drier sites in NC sandhills (LeGrand & Howard, 2002). May wander some distance for nectar.
Hostplants	Not definitely established throughout the range. Cane (*Arundinaria* sp.) is cited most often, but this is doubted in NC (LeGrand & Howard, 2002).
Occurrence	Local and generally uncommon, though peak flights do occur (e.g., 60 in 3 hours at Gr. Dismal Swamp, 8/10/97). Mainly on coastal plain, se VA to n. FL, also west to MO, se TX. Two broods, late May-Jun and Aug-Oct/Nov. Diapause stage not reported.
Ecology	Specialist. Breeding colonies usually found in undisturbed southern swamps or wooded wetland areas.

Ventral female: 8/10/97, Great Dismal Swamp NWR, Nasemond Co, VA

av wingspan: 1.3"

Mulberry Wing   *Poanes massasoit*

Dorsal male (unmarked): 7/20/96, Pound Ridge Reservation, Westchester Co, NY

Dorsal (marked): 7/20/96, Pound Ridge Reservation, Westchester Co, NY

Easily identified by its striking underwing pattern, this small, elfish skipper is a favorite with observers. It ranges from the north-central Atlantic Coast to the eastern Dakotas, residing in temperate wetlands over a thin, discontinuous territorial band. Although it often occurs in close association with Black Dash—a species with nearly identical habitat preferences and flight dates—the two species are actually very distinctive, each with its own characteristic physical traits and behaviors.

Mulberry Wings show significant individual pattern variation, from nearly all black above and below to quite well-marked. One regional subspecies, the dark *chermocki* race, has been described from southern Maryland (Scott, 1986).

Male Mulberry Wings patrol for mates, meandering low through thick marsh vegetation, pausing at intervals to rest. The species seldom quits its sedge marsh habitat, preferring to take nectar from wetland plants like Swamp Milkweeds. (They will take nectar at marsh edges as needed, however.)

Ventral: 7/10/93, Pound Ridge Reservation, Westchester Co, NY

Identification	Our smallest *Poanes*. Its short, rounded wings create a distinctive profile, especially when basking. Males lack stigmas.
Dorsal	Male: usually *rich blackish-brown above*, with subtle, *reddish-purple overtones* ("mulberry"), most evident in good light. May have a FW wrist "bracelet," and two small central FW marks. Fresh individuals have a bit of rufous shoulder scaling. Female: similar, but *FW spots are white, more prominent.* Upper HW variably spotted.
Ventral	Distinctive, rusty-brown base color with a *prominent, yellow cross-shaped pattern*. In rare cases, rust-colored scales overlay the HW, obscuring the cross pattern (form *suffusa*).

Habitat	Weedy, often low-growth wetlands with tussocks, such as sedge bogs, fens, marsh edges, and wet meadows (the latter usually with some standing water, not deep). As these wetlands decline in our area, so does available Mulberry Wing habitat.
Hostplants	Narrow-leaved sedges, such as Tussock Sedge (*Carex stricta*). Hostplant identification is difficult, since colonies are often found in stands of sterile sedge, lacking flower parts (Layberry et al., 1998).
Occurrence	Sometimes locally common. MA south to MD along coastal plain (rare NJ pine barrens). Also disjunct Gr. Lakes/n. plains range. One brood, late Jun–early Aug. Diapause stage not reported.
Ecology	Specialist. Seldom strays from chosen habitat.

Sedge marsh, Westchester Co, NY

av wingspan: 1.1"

Black Dash *Euphyes conspicua*

Dorsal male: 7/2/95, Amity Marsh, Orange Co, NY

Dorsal female: 7/12/96, Pound Ridge Reservation, Westchester Co, NY

Formerly known as the Pontiac Skipper, the Black Dash is the only *Euphyes* limited to the northern half of the East Coast area. (Others are either southeastern or else live in both regions.) We elected to discuss this species here—slightly out of genus order—because of its regular, close association with the Mulberry Wing. Though placed in different genera, the two species share common habitats, hostplants, flight dates, and even ranges (their distribution maps are remarkably similar). As noted below, the Black Dash is a bit more expansive in habitat selection than the Mulberry Wing, but even here the differences are secondary. The nominate race of Black Dash (*E. c. conspicua*) occurs in our region. There is also a large subspecies in Nebraska (*E. c. bucholzi*).

The genus *Euphyes* comprises a group of about two dozen skippers. It attains a high level of diversity on the East Coast, where its members embody the "marsh skipper" lifestyle. These butterflies are mainly sedge-feeders, linked to marshy lowlands with emergent, low-profile plant growth. Taxonomically, the genus is distinguished by the male's "broken" (two-part) stigma, the lack of certain middle-leg spines, and notably short antennae (less than half the length of the forewing, but with long clubs). In the field, *Euphyes* tend to be large, wary skippers, seldom common, with a classic, sexually dimorphic color patterning.

Inevitably, not all members of a genus fit the group profile exactly, and this is true of Black Dash. It is neither especially large nor difficult to approach, and it can be quite common in appropriate habitats. Only in appearance and habitat choice is its lineage highly apparent. Males perch on low vegetation when seeking mates.

Identification	A small marsh skipper. Compared in some accounts to Sachem or Long Dash, but Black Dash is unlikely to be found with these species in the field, because of differences in habitat, range, and/or flight periods. Easily confused with Dion Skipper, on the other hand, at least dorsally.
Dorsal	Note *small light marks on upper HW* in both sexes, mirroring ventral pattern. Male: bright, reddish-orange field with a bold, black stigma and wide and *very dark* borders. Female: dark with a FW wrist "bracelet" and a transverse row of elongate, creamy-orange markings.
Ventral	Rich, rusty brown with a *vague, pale spot-band* beyond the midwing, thickest at the center.

Habitat	Similar to Mulberry Wing, but more likely to include true bogs, as well as fens, sedge marshes, weedy streamsides, wet meadows. More likely to leave these habitats for nectar.
Hostplants	Narrow-leaved sedges. Tussock Sedge (*Carex stricta*) usually cited, but others are possible, and some caterpillars may refuse this species (Iftner et al., 1992).
Occurrence	Locally fairly common. Mainly central NH south to mid-Atlantic states; isolated population in se VA; scattered colonies/strays inland. Also Gr. Lakes region west to e. NE. One brood, Jun-Aug. Early instar caterpillars overwinter.
Ecology	**Medium specialist.** Only slightly less specialized than Mulberry Wing.

Ventral: 7/2/95, Amity Marsh, Orange Co, NY

av wingspan: 1.2″

Aaron's Skipper   *Poanes aaroni*

Dorsal male: 9/6/97, Cumberland Co, NJ

Dorsal female: 6/17/95, Mackay Island NWR, Currituck Co, NC

The scattershot distribution of Aaron's Skippers along the Atlantic and Gulf coasts seems to reflect a combination of habitat specialization and sheer chance. Mainly a denizen of coastal marshes, Aaron's has an unpredictable distribution. It is found in many suitable habitats along the coasts but not, inexplicably, in others. Even as new colonies are found (e.g., recently on the Gulf Coast), "real gaps" in the distribution are verified (LeGrand & Howard, 2002).

Two major subspecies of Aaron's Skipper are the northern *aaroni* race (traditionally New Jersey to Virginia) and the southern *howardi* race (coastal South Carolina to Florida peninsula). Additional populations with distinctive traits have been described, including isolated groups in Texas and inland South Carolina (Orangeburg County).

Little systematic adult life history information is recorded for this species.

Ventral male: 3/30/02, Yankeetown, Levy Co, FL

Identification	A fairly large hesperiine, but only modest-sized by marsh skippers standards (this group runs big). Wings are rather full-cut and rounded.
Dorsal	Male: note bright *orange FW field*, largely unbroken, with *ragged trailing edges*. There is a small "whisker mark" toward the FW apex and a long, thin, stigma (usually not visible in the field). Female: orange FW patch is somewhat reduced, crossed by *dark veins*. Dark patches extending from FW base and toward the apex suggest males of other species.
Ventral	FW orange near base (a dark inner patch is usually concealed by the HW); olive-brown toward margin. HW is a dull olive-tan or orange-tan with a *pale but conspicuous ray extending most of the way to the margin*. The outer part of this ray is flanked by dots in the *howardi* race, reminiscent of Broad-winged Skipper.

Habitat	Mainly coastal marshes (salt and brackish); also inland freshwater wetlands in SC and s.-central FL. Found well inland on occasion even in the North, but then seemingly in search of nectar.
Hostplants	Not clear. Smooth Cordgrass (*Spartina alterniflora*) proposed, but doubted in NC. Possibly Salt Grass (*Distichlis spicata*). Maidencane (*Panicum hemitomon*) cited in FL. Inland hosts unknown.
Occurrence	Locally fairly common, coastal mid-NJ south to e.-central NC. Separate populations near SC/GA border and in FL. Also west on Gulf Coast to e. TX. Two broods, first ranges from Feb/Mar–May in FL vs. Jun in NJ, second Aug–Oct.
Ecology	Medium specialist. Significant habitat affinity.

Salt marsh canal with Pickerelweed (*Pontederia cordata*), Currituck Co, NC

av wingspan: 1.3″

Dukes' Skipper *Euphyes dukesi*

Dorsal male: 8/20/95, Virginia Beach City, VA

Dorsal female: 6/17/95, Virginia Beach City, VA

Shadowy in habits as well as appearance, the somber-toned Dukes' Skipper is most often seen at known colony sites—which are relatively scarce, usually tucked away in soggy coastal backwaters or river swamps. The inaccessibility of these haunts may explain why the species was not described until 1923, quite late for such an obvious species. Some colonies undoubtedly still remain to be discovered, but this does not diminish Dukes' status as one of the region's notable rarities—nor lessen the need for careful protection.

Male Dukes' Skippers patrol for mates, an unusual behavior among *Euphyes*. They are relatively slow-flying and approachable—especially compared with some of their very jumpy congeners. Still, even when casually meandering they are likely to traverse some highly soggy wetlands, where they cannot easily be pursued. Dukes' are best observed at nectar plants (pickerelweed, etc.) along swampy roadsides.

Ventral: 6/17/95, Virginia Beach City, VA

Identification	A medium-large skipper. FWs are short and rounded, consistent with the species' habit of patrolling for mates (rather than perching). Said to resemble Dion, but only ventrally, and still readily distinguishable. Florida race (*calhouni*) is larger and darker than northern forms.
Dorsal	Male: rich, sooty black above, FW unmarked except for slight, rufous "shoulder fringe" and a dark stigma (which is nearly invisible against the dark wing). Upper HW shows slight tawny flush. Female: similar, but with a transverse row of *small, pale, or creamy dots* on the FW, also somewhat more orange visible on upper HW.
Ventral	*Rich tawny orange-brown* below, when fresh. Some blackish flecking with wear, and fades to a pale, brownish-beige when highly worn. HW with one to two *distinct, longitudinal yellow rays* (central one usually better defined). FW has a central dark patch (often not visible).

Habitat	Shaded backwater swamps, inland from the coast, along slow, meandering rivers. Often where swampy woods meet open marshes. Nectars in wet, roadside ditches.
Hostplants	Broad-leaved sedges. Shoreline Sedge (*Carex hyalinolepis*) usually cited.
Occurrence	Rare except in colonies (80 at one VA site, 6/17/95). Atlantic coast, se VA to n. peninsula of FL. Also Mississippi River Valley, Gr. Lakes. Two to three broods, Jun-Sep in VA, May-Oct in FL. Larvae overwinter.
Ecology	Specialist. Reclusive, colonial, and particular.

Tidal backwater habitat, Virginia Beach City, VA

av wingspan: 1.4″

Dion Skipper   *Euphyes dion*

Dorsal male: 7/2/95, Amity Marsh, Orange Co, NY

Dorsal female: 7/2/95, Amity Marsh, Orange Co, NY

This quintessential *Euphyes* embodies all the principal traits of its genus. It is geographically widespread, yet uncommon almost everywhere (indeed, it is conservation-listed in most northern states). It displays "classic" *Euphyes* coloring: the dorsal male forewing has a restricted yellow-orange patch, the female's is mostly dark with spots. Behaviorally, it is shy and reclusive—challenging to locate, harder still to approach. Dions diverge from the *Euphyes* model, in fact, only in the pale "rays" on their underwings, which are more common in the closely related genus *Poanes*.

Strong fliers (at least over short distances), male Dions perch prominently. Contrary to some reports, we have seen males apparently displaying in the morning.

Ventral: 7/2/95, Amity Marsh, Orange Co, NY

Identification	A largish grass skipper. Dark southern individuals once considered a separate subspecies (*alabamae*).
Dorsal	Highly similar above to Black Dash (both sexes), but note that upper HW pattern includes *long, rectangular stripe mirroring ray on ventral surface* (*compare* Black Dash). Male's stigma forms distinctive curved "hook" near FW apex, which reaches dark margin. *Compare* Berry's Skipper.
Ventral	HW brownish-orange to brick red, with *two longitudinal pale yellowish rays*; upper ray is shorter (does not reach margin), but is more visible. No dots on HW. *Compare* Duke's, Aaron's, and Broad-winged Skippers.

Habitat	Open freshwater wetlands, marshes, swamp edges, bogs, and fens; sometimes in brackish wetlands. Northern colonies use calcareous fens, whereas southern populations prefer bogs and other acidic wetlands (Glassberg, 1993a). Seldom found far from suitable habitat; yet seems to vanish when disturbed, may hide in adjoining habitats temporarily. Nectars on milkweeds, Button-bush, Pickerelweed, etc.
Hostplants	Various sedges (*Carex lacustris, C. stricta,* etc.) and rushes (*Scirpus cyperinus*).
Occurrence	Fairly common in local colonies, across n. FL and northward on coastal plain; rare in NC sandhills; inland across s. NY. Also Gulf Coast to e. TX, up Mississippi Valley to Gr. Lakes region. Single-brooded from central NJ north; two to three broods southward. 3rd stage caterpillars overwinter.
Ecology	Medium specialist. Difficult to classify in specialist/generalist terms. Uses widespread hostplants, yet limited to small colonies with low population density, often in niche habitats.

Calcareous fen, Orange Co, NY

av wingspan: 1.4"

Berry's Skipper *Euphyes berryi*

Dorsal male: 9/4/00, Dare Co, NC

Dorsal female: 9/28/97, Tiger Bay WMA, Volusia Co, FL

In the secluded world of marsh skippers, none is more secretive than Berry's. Klots (1951) could manage only a few sentences on this butterfly. He thought its range was limited to central Florida, where it was first studied by an investigator named Dean Berry.

Little was learned about the species' life history in the decades following its description in 1941, apart from the recording of widely scattered colonies in the Southeast. Even now, it is among the rarest and most elusive of all East Coast skippers, with few known, reliable colony sites where it can be regularly observed.

We have seen Berry's nectaring in the morning and late afternoon, usually in mixed-species groups (often with other *Euphyes*), in Florida and North Carolina. They may withdraw to inaccessible display sites in mid-afternoon.

Ventral: 9/28/97, Tiger Bay WMA, Volusia Co, FL

Identification	A medium-sized *Euphyes*, smaller than Dion or Palatka (larger than Delaware). North Carolina individuals distinctly smaller than those in Florida. *Fringes light* (orange in the Delaware). Dorsal male easily confused with Dion.
Dorsal	Male: *smallish orange FW field*, with wide, dark borders and a straight, narrow stigma. Note *gap between end of stigma and beginning of outer dark patch* (a continuous line in Dion). Dark outer patch may be smudgy (as shown) or rather thin, with a finger-like projection pointed toward the stigma. Female: similar to Dion, but *innermost spot on transverse band (i.e., the spot nearest the body) nearly touches the large spot next to it* (there is a visible gap in Dion); the band appears to have an extra spot.
Ventral	HW variably brownish-orange or coppery with *white wing veins*, no spots or rays. White on veins may fade with wear.

Habitat	Weedy edges of sedge marshes, ponds, canals, swamps, etc. Reported from sawgrass marshes in Big Cypress (Monroe Co, FL). Usually seen nectaring, probably away from breeding habitat.
Hostplants	Unreported. Almost certainly *Carex* sedges.
Occurrence	Widely distributed but very rare and local. Coastal NC (on mainland), coastal SC/GA, FL peninsula and panhandle. Not on Keys. Two broods, Mar-May and Sep-Oct in FL, not clearly determined farther north. Diapause stage not reported.
Ecology	**Specialist.** Ecology little-known, but the Berry's requirements are evidently very specific.

Weedy bank along backwater channel, Dare Co, NC

av wingspan: 1.35″

Palmetto Skipper   *Euphyes arpa*

Dorsal male: 4/15/01, Jonathan Dickenson SP, Martin Co, FL

Dorsal female: 3/24/02, Kissimmee Prairie SP, Okeechobee Co, FL

A question that comes to mind immediately with this stunning skipper is why it should be so rare and restricted, given its hyperabundant hostplant. And why should it now be in decline over much of its small range? Once locally common in south Florida, it is now greatly reduced; it has not been seen on the Keys since the early 1980s.

This is our only *Euphyes* that does not feed on sedges. Its early stages have been well-studied—perhaps because the hostplant grows in relatively accessible surroundings!

Adults are difficult to observe, given their extensive and unpredictable daily movements. Males perch in the morning and late afternoon; nectaring has been noted from late morning to mid-afternoon. Extremely skittish when not feeding. The slightest disturbance may launch it into a frenzied escape routine, never to return.

Ventral: 4/16/01, Jonathan Dickenson SP, Martin Co, FL

Identification	Among our bigger grass skippers. Wings are large and angular (sail-like). Leading tips of antennal clubs are orange. Fringes snowy white.
Dorsal	Male: head and "shoulder fringe" *bright, reddish- or golden-orange.* FW field comparatively small, with a long, thin stigma and a dark patch toward the apex (*compare* Palatka). Female: mostly dark, with a transverse band of cream-to-golden spots on FW. *Reddish head and variable costal edging*, especially when fresh.
Ventral	Both sexes bright, unmarked golden-orange or ochraceous-orange. Extremely handsome.

Habitat	Open pine savanna with palmetto understory, including sandhills, dry prairie, pine rocklands in s. FL (possibly extirpated there). Strays to nectar concentrations, such as thistle, mistflowers, etc.
Hostplants	Sole hostplant is Saw Palmetto (*Serenoa repens*), a common, low-growing member of the palm family. Caterpillars eat fresh leaves, rest in silkened rolled-leaf tubes at base of hostplant. Pupate at same site, after covering over the entrance.
Occurrence	Near-endemic in our area. Local and thinly distributed; usually rare, but may concentrate at nectar sites. N.-central FL; current southern limit unclear. Also rarely to se GA (where first described), west on Gulf Coast to AL and MS. Two to three broods, Mar/Apr-May, Jul-Oct. Recent large counts in early July at Kissimmee Prairie SP. Larvae overwinter.
Ecology	Specialist. Population-limiting factors are not clear, but very effective!

Long-leaf Pine savanna with palmetto understory, Martin Co, FL

av wingspan: 1.5″

Palatka Skipper *Euphyes pilatka*

Dorsal male: 9/12/99, Dare Co, NC

Dorsal female: 9/12/99, Dare Co, NC

Convention has it that the Palatka Skipper was named for the Florida city of Palatka—originally called Pilatka, after a Seminole village that stood on the same site (Opler & Krizek, 1984). (The Post Office changed the spelling to "Palatka" in 1875 to "avoid confusion" with another Florida town, "Picolata.") But the skipper was actually described from a specimen taken at St. Augustine—a city that was originally called "Wacca Pilatka" (or "cow crossing"). This raises an alternate possibility as to the origin of the name.

Two races of Palatka Skipper occur in our area, the common, nominate race and a specialized, highly threatened form found only in the lower Keys (*E. p. klotsi*). The *klotsi* race is darker and more rufous than the nominate. Both are strict hostplant specialists.

Male Palatka Skippers perch on or near Sawgrass, flying out to inspect intruders. They often cluster at marshside nectar sites.

Ventral: 9/4/00, Swanquarter NWR, Hyde Co, NC

Identification	Our largest grass- or sedge-eating grass skipper (*compare* Monk, Brazilian). FWs are large and triangular. Somewhat similar to a number of other marsh skippers above. Below, most like a huge Tawny-edged Skipper.
Dorsal	Male: bright orange FW field with dark borders. A *short, prominent stigmal segment* near the body, no obvious blackening near apex. Female: *extensive orange FW field* with darker veins. Note *large, oblong dark patch* toward the apex. Both sexes have much orange on upper HW.
Ventral	HW *variably dull brown below*, sometimes with orange infusion at base or near body. FW orange basally, brown-rimmed outwardly. May have some faint, pale markings past mid-FW.

Habitat	One of few butterflies to regularly inhabit open, brackish, and freshwater Sawgrass marshes, usually near coast. *Klotsi* race in pine rocklands, where Sawgrass is established in woodland clearings (generally in eroded limestone solution holes, or "pocks"). Strongly drawn to marshside nectar, such as Pickerelweed, mistflower, etc.
Hostplants	Sole host is Sawgrass (*Cladium jamaicensis*), a sharp-edged sedge that creates a near monoculture in preferred environments. Caterpillars live in silk shelters made from hostplant leaves. They pupate at same location, sealing off chamber to avoid tachinid fly parasitoids.
Occurrence	Locally fairly common in proper habitat, se VA to FL Keys (stray on upper Keys only, absent middle Keys). Also west on Gulf Coast to MS; rare stray to MD, sw LA. Two broods in North, May-Jul and Aug-Oct; two to three broods in South, Jan-Nov. Overwintering stage not reported.
Ecology	**Medium specialist.** Single habitat and host, but widespread along coast.

Sawgrass (*Cladium jamaicensis*)

av wingspan: 1.65″

Two-spotted Skipper   *Euphyes bimacula*

Peter W. Post

Dorsal male: 6/23/96, Lakehurst Bog, Ocean Co, NJ

Dorsal female: 6/24/95, Lakehurst Bog, Ocean Co, NJ

This species was reportedly described from female specimens only (Gatrelle, 1999a), but even so, in our experience the "dot count" on the dorsal forewing of most females is three, not two. The name is best supported by the two spots that are visible on the *underside* of the forewing—and by some individuals in which one of the dorsal dots is reduced. Still, this is not an outstanding example of helpful nomenclature.

Widely included on state endangered lists, this skipper is rare and sparsely distributed. It is quite possibly a relict species, more common in postglacial times, now gradually losing ground, like many other skippers with widespread but low-density distribution patterns. But such historic scenarios remain conjectural.

The nominate race of this mainly northern skipper (*E. b. bimacula*) occurs in most of our area. There is also a large prairie form (*E. b. illinois*) with grayer underwings. And it has also been suggested that the small, dark southeastern population is taxonomically differentiable. The southern form is very scarce.

We have studied this species' evasion techniques while approaching to obtain photos. Males perched in the open, at the tip of a tall strand of wetland foliage, where they could discern any form crossing the horizon line. They took flight when this happened, usually while we were still quite distant. They first hovered in place, like tiny hummingbirds, oscillating back and forth almost hypnotically. But then they burst straight upward, flying *over* our heads, settling 20 yards or so *behind* our position. They generally would not return to their original perch for a considerable period (15 to 30 minutes).

Identification	Fairly small for a marsh skipper. FW pointed, body underparts *snowy white*, also fringes when fresh.
Dorsal	Male: FW orange field is *comparatively small*, sometimes little more than an orange highlight on either side of the stigma. The stigma itself is dark and slightly crooked. Female: dark with a small wrist "bracelet" and usually *three transverse FW spots* (inner dot sometimes reduced). Leading edge of FW variably tawny, may be entirely dark. Upper HW variably tawny, often nearly unmarked (especially if worn).
Ventral	Uniform orange-brown or dull brown (especially when worn), often with *pale wing veins*. Note *strong white line on inner margin of HW* (running along the body).

Habitat	Open bogs, fens, and marshes, wet sedgy areas, often near swampy woods. May stray to adjoining habitats to find nectar. Remnant undisturbed prairie patches in the West.
Hostplants	Sedges, especially Hairy-fruited Sedge (*Carex trichocarpa*) and Tussock Sedge (*Carex stricta*).
Occurrence	Rare, low density. (Large population spike in 1968; max. eight seen by us Ocean Co, NJ, 7/7/01). Central ME south to VA, WV in mountains; rarely on se coast to GA. Also Gr. Lakes and n.-central plains, to ne CO. One brood in North, Jun-Jul; two in South, May-early Jun and late July-Aug. Mid-stage larva overwinters.
Ecology	**Medium specialist.** Although extremely local, habitat and hostplant requirements are not especially strict.

Peter W. Post

Ventral: 6/27/93, Lakehurst Bog, Ocean Co, NJ

av wingspan: 1.25"

Dun Skipper *Euphyes vestris*

Dorsal male: 7/2/95, Amity Marsh, Orange Co, NY

Dorsal female: 7/12/96, Pound Ridge Reservation, Westchester Co, NY

This smallest and most adaptable of our *Euphyes* skippers spurns the stoic lifestyle of a typical marsh skipper, opting instead for a more convivial and prolific existence—closely resembling the style of a meadow skipper. Indeed, the female is considered one of the "three witches," along with two bona fide female meadow skippers, the Little Glassywing and Northern Broken-Dash. These "three witches" regularly fly together in early summer.

Several races of Dun Skipper are recognized across the species' broad range. The dark *metacomet* form occurs in the East. A lighter form (*ruricola*) is found in the West, and the threatened *harbisoni* race in central California.

It is not clear how one member of a homogeneous biological group such as *Euphyes* skippers comes to differ so greatly from its relatives. Duns do betray their lineage to some extent—e.g., by retaining a clear affinity for wetlands over much of their range. But they have also adapted to drier habitats than their congeners, especially in the West. And they pursue a prolific, generalist life strategy, in contrast to other *Euphyes*, feeding on a wide variety of sedges (*Scirpus, Cyperus,* and *Carex*, and among the *Carex* sedges: *brevior, gracillima, heliophila, lacustris, lupulina, spissa, stricta*, and probably others). They may even stray occasionally to grasses (Layberry et al., 1998).

Male Dun Skippers station themselves on low perches, often late in the day, to await females. Fresh males regularly visit muddy ground seeking minerals.

Identification	Smaller than other *Euphyes*. FW triangular, fringes mostly brown. Note male's *yellow-orange frons* ("head") and upper thorax. It takes some practice to separate this from other, similar species with confidence.
Dorsal	Dark brown with purplish overtones when fresh. Male: *all dark*. Slight hint of a faint, orange field, but this and the dark stigma usually difficult to see. Female: similar to male, but note *very small spots* on upper FW (two central spots and usually a two-dot wrist "bracelet"). Females of similar species have more extensive white on FW; Clouded Skipper has wrist marks sweeping backward, toward apex.
Ventral	*Plain, dull brown to tan.* Female may have vague postmedian spot band. Note dark bare spot at wing attachment.

Habitat	Widely variable, but generally favors wet or damp settings. Fields and meadows, roadsides, sometimes weedy upland open spaces. Often in transitional habitats, less so in truly disturbed areas.
Hostplants	*See* above. Numerous sedges cited in the literature. Some efforts to rear caterpillars on grass have failed (Scott, 1986), but use of grass is reported in w. CAN (Layberry et al., 1998). If even a secondary host, could account for adaptability and success of the species. Ovipositing seen in late morning.
Occurrence	Widespread, usually common to abundant. Much of temperate No. Amer., all of our area except far s. FL and Keys. One extended brood in North, Jun-Aug/early Oct. Two broods to south, May-Jun and Aug-Sep; multiple flights, Mar-Oct in FL. 3rd instar larvae overwinter; pupae in spring in plugged silk tube at base of host.
Ecology	**Generalist.** No other *Euphyes* is as broadly adaptable or widespread.

Ventral male: 7/6/96, Pound Ridge Reservation, Westchester Co, NY

av wingspan: 1.1"

Dusted Skipper *Atrytonopsis hianna*

Dorsal male: 5/22/93, Pound Ridge Reservation , Westchester Co, NY

Ventral ('typical' *hianna*): 5/22/93, Pound Ridge Reservation , Westchester Co, NY

Dusted Skippers belong to a small genus, centered in Mexico and the arid Southwest. But East Coast observers should not be faulted if they regard them as honorary *Hesperia*. Seasonally targeted broomgrass specialists, they often fly at the same locations as Cobweb Skippers, emerging slightly later but with overlapping broods.

Dusted Skippers are taxonomically complex (*see* box essay). Currently, two very different-looking races are recognized, the northern *hianna* and the southeastern *loammi*. Some argue these two variable forms are sibling species. In any case, the habitats of both are widely at risk due to development pressure and unchecked succession.

Male Dusted Skippers perch on or near the ground throughout the day, in dry short-grass patches. They fly actively when disturbed and can be located by stepping carefully through their habitat.

Ventral (*loammi*): 4/15/01, Jonathan Dickenson SP, Martin Co, FL

Identification	Fairly large, especially *loammi* form; much smaller in NC mountains. Long, pointed wings, mostly brown fringed. Note *distinctive "face mask,"* framed by white palps and eyebrow. Superficially similar to cloudywings.
Dorsal	Male: dingy brown above with a small, very inconspicuous stigma and *a few small, white FW marks.* Female: more patterning above, with variable but distinct white or off-white spots.
Ventral	Rich brown with heavy marginal frosting (*hianna*) or mousy brown, some frosting (*loammi*); variable white marks. *Single basal white dot* on HW (*hianna*), or a *basal arc of three white dots* (*loammi*). *Loammi* highly patterned.

Habitat	A variety of open, usually dry habitats with host-plant. Sandy barrens and grasslands, pine-oak scrub, dry open fields, and hillsides. Favors early successional areas, e.g., recent burn sites. In the West, dry virgin prairie remnants and narrow canyons.
Hostplants	Big Bluestem (*Andropogon gerardi*), Little Bluestem (*Schizachyrium* [*Andropogon*] *scoparius*). Dune Bluestem (*S. littorale*) reported on NC coast. Larva makes silk nest at base of hostplant.
Occurrence	Uncommon to locally common in North, generally rarer in South. Se NH/MA along coastal plain to central FL (formerly s. FL); west across central and Gulf states to Gr. Lakes, w. Gr. Plains. One brood in North, May-June; two in South, Mar-Apr and Oct-Nov. Mature larva overwinters.
Ecology	Medium specialist. Colonies limited to particular sites.

Ventral (*hianna* variant): 5/31/98, Voluntown, New London Co, CT

av wingspan: 1.35"

"Seaside" Dusted-Skipper: Good Species?

It began (as such things often do) in a moment of serendipity. American Museum of Natural History entomologist Eric Quinter was poring over dune vegetation in 1983 at Ft. Macon State Park, near Moorehead City, North Carolina, looking for the larvae of borer moths (genus *Papaipema*), a group in which he specializes. The visit was unproductive as far as *Papaipemae* were concerned, but Quinter did happen to notice a large, unfamiliar hesperiine skipper ovipositing on Dune Bluestem (*Schizachyrium littorale*), and he took a small series for later study.

Little transpired for two to three years, until Smithsonian Institution skipper expert John Burns visited Quinter in New York City. Burns's attention was drawn immediately to the North Carolina specimens. He recognized them as *Atrytonopsis* skippers, similar to the *loammi* race of the common Dusted Skipper. But to Burns's experienced eye, these large, boldly marked specimens were different from any others he had seen.

For still another decade, the tale progressed at a deliberate pace, as Burns and a small cadre of local butterfly specialists (initially Bo Sullivan) built up a knowledge base concerning "the beach thing" at Ft. Macon State Park. It also occurred, they learned, at scattered public and private sites around the outlet of the Newport River, living on dunes and grassy dredge spoil flats. Still, it had a highly restricted range.

In 2000, Burns announced to a gathering of the Lepidopterists Society in Wake Forest, North Carolina, that he believed the Ft. Macon dusted skippers might be a new, isolated species, distinct from either race of the Dusted Skipper (*hianna* or *loammi*).

How could such a thing have happened? The theory most often proposed is that, at some point in the glacial past, the original Dusted Skipper became marooned in isolated territories, or refugia, by advancing ice fields. The separation must have continued long enough to induce morphological change, creating the *hianna* and *loammi* races, but not long enough to result in fully separate species. When the glaciers retreated, the two races were able to interbreed freely as their ranges became reconnected, accounting for the many intermediate forms presently seen.

Under this scenario, the "beach form" would have developed in a later glacial episode, perhaps as recently as the Pleistocene Ice Age, 10,000 years ago, when two populations of *loammi* Dusted Skippers became isolated—in the same general way as before, but this time the populations crossed the line and became separate species.

Alternatively, the "beach form" could simply be a third race of Dusted Skipper, not well differentiated enough to merit full species status. These possibilities, among others, are currently being explored with the assistance of Steve Hall (doing field surveys to further clarify the new form's range and its overlap, if any, with *loammi*) and Paul Goldstein of the Field Museum in Chicago (performing DNA analysis). In the end, a reasoned determination will be made, based on distribution evidence, host-plant use, and varied genetic and morphological comparisons.

Whatever the result, the real-world stakes are high. Developers eyeing scarce patches of coastal land are already cognizant of the implications of species designation, since the "beach form" would be an immediate candidate for federal endangered species protection once formally described (even as a full subspecies).

In the meantime, the Ft. Macon skipper flits about its tiny range, oblivious to our machinations, unconcerned that no one can say exactly what it is.

Dorsal female: 4/22/00, Cartaret Co, NC

Ventral male: 4/22/00, Cartaret Co, NC

Low-growth coastal flats (on dredge spoils), Cartaret Co, NC

Common Roadside-Skipper   *Amblyscirtes vialis*

Dorsal: 4/28/97, Flintstone, Allegheny Co, MD

Ventral: 5/22/94, Albany Pine Bush, Albany Co, NY

This inconspicuous skipper is widely distributed in the United States and Canada, but it is generally rare or uncommon, and its rapid, darting flight makes it difficult to follow once found. Although the species was described more than 140 years ago, our knowledge of its adult ecology is still limited. Why is such a durable generalist so scarce over most of its range? What conditions are critical to its success? What are its hostplant strategies?

The Common Roadside-Skipper is the most familiar *Amblyscirtes* on the East Coast—and indeed across most of the continent north of Mexico. Roadside-skippers as a group are a favorite with active observers, exerting an attraction much like that of small songbirds on birdwatchers (warblers come to mind, but *Empidonax* flycatchers are probably a better example, with their plain colors, technical identification challenges, and specialized habitats). Most roadside-skippers are identifiable in the field when fresh, especially with the aid of new, close-focusing binoculars. But once they are worn the task becomes much harder, regardless of optics, since identifying field marks then begin to fade, leaving only common base patterns that are rather similar across the genus.

Males perch all day on or very near the ground, scouting for females and taking moisture. Regular nectaring has been reported, but we have rarely seen this species at flowers ourselves. Late-instar caterpillars exude a waxy, white secretion that apparently allows them to feed in the open more safely.

Identification	Very small. Resembles a miniature Clouded Skipper, but is more likely to be confused in practice with Dusted or maybe Zabulon—or most of all with Dusky Roadside-Skipper, in the few places where the two occur together (notably in the NC sandhills). Wing shape is triangular and swept-back. (Duskys, by comparison, tend to hold the FW apex more upwardly pointed at rest.) *Fringes conspicuously checkered black-and-tan.* Abdomen circled with white rings at segment joins. Sexes similar.
Dorsal	All *dark brown to flat black*, with some light, yellowish overscaling when very fresh. Two faint, pale spots show variably on the central FW (sometimes not apparent; usually better defined in Dusky). The only conspicuous FW mark is a white wrist "bracelet."
Ventral	Base color same as dorsal. *Heavy violet-white frosting on outer FW and HW* (less confined on HW, but still more concentrated near margin; frosting is lighter, more dispersed on both wings in Dusky). Key FW mark is *prominent, tooth-like subapical dot-row*, widest near the wing margin. Note also white dash marks along the costal edge of FW, as if brushed against fresh paint.

Habitat	Variable. Colonies most reliably found on barrens (sand, shale, and serpentine); also sandhills, dry disturbed areas, canyons (in West), etc. But also occurs regularly—if less dependably—in clearings and edges of rich, mixed deciduous and transitional woodlands. Tendency to take minerals at bare soil along forest edges may be source of "roadside" reference in the common name. Little affinity, if any, for wetlands, though not averse to moderately moist surroundings. Oddly, we have often seen the species on somewhat sloped terrain.
Hostplants	Various grasses reported. Oats (*Avena*), bent grass (*Agrostis*), Kentucky Bluegrass (*Poa pratensis*), River Oats or Indian Woodoats Grass (*Chasmanthium* [*Uniola*] *latifolia*), Bermuda Grass (*Cynodon dactylon*), if available, etc. Specific patterns of hostplant use in the field require further investigation.
Occurrence	Local and uncommon (declining), but still widespread. Most of temperate US and CAN except deserts and se coastal plain. ME south to central NC, in mountains/upper Piedmont to n. GA. Isolated FL panhandle population. One brood in far North, Jun-Aug; two to three southward, Apr-May and Jun/Jul to Aug/Sep. Pupa (possibly larva) overwinters.
Ecology	Medium generalist. Tolerant in terms of climate, habitat, and hostplant, but usually few in numbers.

av wingspan: 0.9"

Pepper and Salt Skipper *Amblyscirtes hegon*

Dorsal: 6/8/96, Mt. Greylock, Berkshire Co, MA

Ventral: 6/8/96, Bekshire Co, MA

Distinctive in appearance, the Pepper and Salt Skipper is more limited in both distribution and habitat than the Common Roadside-Skipper. Yet the two are often found together in the wild, since Pepper and Salts occupy a subset of the Common's life space—specifically, damp to medium-dry northern woodlands. (Commons also frequent dry barrens and other more xeric habitats.) In the northern part of their range, Pepper and Salts are often found with Arctic Skippers. And in western South Carolina, they can occur in close proximity with as many as five other roadside-skippers (Common, Lace-winged, Carolina, Reversed, and Bell's).

As a group, the roadside-skippers (genus *Amblyscirtes*) include about two dozen species, mostly concentrated in the southwestern United States and Mexico. Extremely small, they have checkered antennal shafts and wing fringe, also white rings encircling the abdomen. The sexes are basically similar. Perched roadside-skippers tend to hold their forewings more upright than other grass skippers, especially when warm (note posture of dorsal Carolina Roadside-Skipper, p. 309). This forewing positioning is variable, however, and not always diagnostic.

Male Pepper and Salt Skippers perch on flat leaf surfaces in sunny spots to await females, usually along woodland trails or in forest openings. Frequently seen nectaring or on damp ground taking minerals. Local population declines in this species have coincided with Gypsy Moth spraying (Gochfeld & Burger, 1997).

Identification	Similar in size to Common Roadside-Skipper. Note *faint yellow-green cast* to the frons ("head") and upper thorax. Wing fringes *checkered brown-and-light* (the light color ranging from beige to white). Sexes similar.
Dorsal	Dark, reddish-brown base color with *greenish-yellow flecking toward FW base*. Outer FW is dark with a small FW wrist "bracelet" and a few white midwing marks (the latter being more defined than in Common). The male has a black stigma, but this is not easily observed in the field.
Ventral	*Numerous creamy spots* on both wings, including a *concentric dot-band* on the postmedian HW. Note the *dense, gray-green flecking on both wings* when fresh. Once this flecking wears off, however, only the dark basal wing color remains, and the Pepper and Salt may then resemble other roadside-skippers—including the Bell's. Ventral FW is dark basally, frosted toward the margin. The shape of the dark area is similar to that of the FW, as if the outer area had been sunbleached while the basal area was shielded.

Habitat	Less varied than Common Roadside-Skipper. Mainly clearings, glades, and dirt trails, all within rich hardwood forests (mixed deciduous or transitional). Often near wet meadows or low-lying boggy stream banks.
Hostplants	Kentucky Bluegrass (*Poa pratensis*), Indian Grass (*Sorghastrum nutans* and *S. secundum*), River Oats or Indian Woodoats Grass (*Chasmanthium* [*Uniola*] *latifolia*), Fowl Mannagrass (*Glyceria striata*) in WV.
Occurrence	Most common in North, scattered colonies southward. ME to MD, south in Appalachians to n. GA; separately FL panhandle; absent on most of se coastal plain. Also west to se MAN, e. TX. One brood in North, May-early Jul; two broods from VA and south, Apr-May and Jul-Aug (partial). Young larvae overwinter in leaf rolls.
Ecology	Medium generalist. Fairly adaptable and widespread.

Underside forewing: 6/8/96, Mt. Greylock, Berkshire Co, MA

av wingspan: 0.9″

Dusky Roadside-Skipper   *Amblyscirtes alternata*

Dorsal: 8/15/98, Green Swamp, Brunswick Co, NC

Ventral: 4/11/99, Sandhills Game Land, Scotland Co, NC

Of all the secretive and retiring roadside-skippers in our region, this is easily the most reclusive. It was recently described as "a skipper in search of a good common name" (LeGrand & Howard, 2003), having been variously dubbed Least Florida Skipper, Blue-dusted Roadside-Skipper, and Dusky Little Skipper. We might also recommend Ephemeral or "Swamp Fox" Roadside-Skipper—for its ability to whisk untraceably through coastal habitats of the Southeast.

Originally discovered in the Great Dismal Swamp in the 1800s, the Dusky's continued presence there is debatable (Roble et al., 2000). Its apparent absence may reflect actual extirpation, but this butterfly is difficult to locate, even in its widely scattered strongholds. We know of few exact sites where the species can be reliably found year after year. On the day that the dorsal photo (above) was taken in Green Swamp, North Carolina, Reversed Roadside-Skippers were displaying persistently on small, easily monitored territories near stands of cane. But the lone Dusky of the day skirted the perimeter of a wet savanna clearing, perching only occasionally, and always pressed ahead, bent on some compelling errand that quickly carried it into the far distance. Perhaps it was patrolling for females across a vast expanse of potential hostplant (*see* below).

Duskys fly close to the ground, resting on small, very low perches. They rarely seem to nectar.

Identification   Our smallest *Amblyscirtes* on average. Can be challenging to distinguish from Common and Bell's Roadside-Skippers (though Bell's and Dusky rarely if ever occur together in our region). *Wing fringes are checkered black-and-white,* although the black markings are often smallish, sometimes virtually absent. *Antennal clubs blunt, without a tapered apiculus* (*compare* Common and Bell's). Short FW at rest often appears upward-pointed, vs. swept-back in Common.

Dorsal   Dark, blackish-brown with *variable FW spotting* (individual shown is well-marked, others may have only a minor wrist "bracelet," or no markings at all; females seem less well-marked).

Ventral   Underside *strewn with small, bluish-gray flecks*, only somewhat concentrated toward the outer margin (vs. distinctly concentrated in the Common). Wrist "bracelet" on FW is small and delicate, not large and tapering ("toothlike"), as in Common.

Habitat   Savannas, including both dry sandhills and low-lying coastal pine savanna. In FL, also pine flatwoods and prairies.

Hostplants   Not reported. A botanic element that links many of the Dusky's varied habitats is wiregrass (*Aristida*), a dominant understory plant ubiquitous in sandhills, etc. Use of a locally abundant hostplant could account for the Dusky's thinly dispersed occurrence. By all evidence, not a cane-feeder.

Occurrence   Very local and rarely encountered. Southeast VA (historically?) and se NC, along coastal plain to n.-central FL, west to e. TX. A single, old report from Everglades seems anomalous. Two broods, Apr-Jun and Jul-Aug in North; Mar-May and Jul-Oct in FL. Overwintering stage not reported.

Ecology   Specialist. Though replete with unknowns, the Dusky's lifestyle is closely matched to specialized southern habitats.

Dry sandhill pine savanna with wiregrass (*Aristida* sp.), Riverside Island, Ocala NF, Marion Co, FL

av wingspan: 0.85"

Bell's Roadside-Skipper   *Amblyscirtes belli*

Dorsal: 5/23/99, Pickens Co, SC

Ventral: 5/8/99, Pickens Co, SC

The most locally restricted of our roadside-skippers, Bell's enters the East Coast region only at the eastern edge of its range. It is known here mainly from northern Georgia and western South Carolina. Its range may be more extensive than once thought, however, and the species might ultimately be found to have a number of additional, scattered colonies in the Carolinas, Georgia, or northern Florida, where suitable habitats exist that have not been fully explored. But it would seem a sparse inhabitant of this extended range at best.

In the taxonomically difficult world of roadside-skippers, Bell's has engendered a fair amount of discourse. It was once lumped with the visually similar Celia's Roadside-Skipper (*A. celia*), a Texas ally (*see*, e.g., Scott, 1986). The current recognition of Bell's as a "good species," based on distributional and genitalic evidence, dates back only a little more than a decade (Burns, 1990).

Males perch in open clearings or sunny corridors to await females. Interspecific competition may be one factor limiting their success in our region: in May 1999, we noted multiple instances of perching Bell's in South Carolina seeming to be displaced by Pepper and Salt Skippers. (In each case, after watching a Bell's stake claim to a strategic leaf surface, we later found a Pepper and Salt occupying the same spot, with no Bell's in sight. No other *Amblyscirtes* in the area seemed to suffer such displacements.) Active nectaring has been observed at midday.

Identification	Slightly larger than Pepper and Salt, with fairly long, rounded FWs. *Fringe prominently checkered black-and-cream.* Antennal club has a *pointed apiculus* (*compare* Dusky Roadside-Skipper). White rings visible on the abdomen at segment joints, as in other *Amblyscirtes*.
Dorsal	Dark brown base color, with a variable series of *transverse white dots* on the FW; *the innermost FW dot is shaped like a V (or chevron).* Upper HW may show vague white markings, partially reflecting the ventral pattern. The male's stigma is small and inconspicuous.
Ventral	Dark blackish, often with a purplish cast. Some *frosting toward outer wing margin, near the body* (not reaching HW apex). Also a series of poorly defined *gray-white marks* over the HW surface.
Habitat	Open areas near woodland streams with host. Also grassy wood edges with nectar, city gardens (in TX). Often seen on bare ground near creeks, taking minerals. Prefers rich, moist deciduous woodlands.
Hostplants	Sole known host in our region is River Oats or Indian Woodoats Grass (*Chasmanthium* [*Uniola*] *latifolia*). Johnson Grass (*Sorghum halepense*) reported in TX (Glassberg, 1999). Single egg laid on underside of host-plant leaf.
Occurrence	See above. W. SC and n. GA. Scattered records from n. FL; also recently found in Barnwell Co, SC (LeGrand & Howard, 2002). West across Gulf states and Ohio River Valley to se NE, e. TX. Two broods, Apr-Jun and Jul-early Sep. Overwinters as 4th instar larva. Pupates in silk-lined leaf case in soil debris.
Ecology	Specialist. Confined ecology in our region.

River Oats, or Indian Woodoats Grass (*Chasmanthium* [*Uniola*] *latifolia*)

av wingspan: 1.05"

Reversed Roadside-Skipper   *Amblyscirtes reversa*

Dorsal: 8/10/97, Great Dismal Swamp NWR, Nasemond Co, VA

Ventral: 5/23/99, Clemson Experimental Forest, Isaqueena Lake, Pickens Co, SC

For many years, the existence of the Reversed Roadside-Skipper was obscured by the presence of its close local relative, the Carolina Roadside-Skipper. Early lepidopterists considered the species we now call *Amblyscirtes reversa* to be nothing more than an "interesting color variety" of the Carolina (Klots, 1951). And even when it was finally acknowledged as a distinct species in the 1970s, *reversa* was saddled with a set of names (both common and scientific) that have cast it in the role of a biological also-ran. It is now locked in a perpetual comparison with the Carolina, which has the "original" hindwing pattern next to which the new species' markings are "reversed."

Overlooking this small—but distinct—taxonomic slight, our concern for the Reversed Roadside-Skipper should be directed toward loss of habitat within its extremely limited range, most of which falls within our region. Widespread disturbance of grassy southern wetlands (by development, spraying, unchecked succession, etc.) can eliminate key habitats, on which many as yet undocumented populations may depend.

Given its limited distribution and often uninviting habitats, relatively few observers have the opportunity to study this species closely. Local flights can be prolific at times, but this is not typical. We have seen males perch on upright grass stalks in the center of open pine savanna clearings in mid-afternoon, and on broad tree leaves (sometimes quite high) in deciduous swamplands, e.g., at Great Dismal Swamp.

Identification	A medium-sized roadside-skipper. Worn individuals can be difficult to identify. Fringes are light, and often lack the strong checkering of other *Amblyscirtes*. Head and shoulders greenish-gold when fresh (*compare* Carolina).
Dorsal	Dark brown with a series of light markings on both wings. Male's stigma, though comparatively large, is not obvious against the dark FW base color.
Ventral	Rusty brown with light veins and a series of light marks on both wings; lacks extensive yellow overscaling of Carolina, but may have a weakly defined yellow HW ray (not in the individual shown above). *Dark HW base color with light spots considered the "reverse" of color pattern on Carolina.*

Habitat	Varied settings, but always near hostplant. Favors open Longleaf or Pond Pine savannas and flatwoods on the s. NC coastal plain, where Reversed is probably most common (LeGrand & Howard, 2002), but also at the edges of southern hardwood swamps, e.g., in se VA.
Hostplants	Cane (*Arundinaria* sp.); Switch Cane (*A. tecta*) most often cited. Caterpillars live in rolled leaf nests.
Occurrence	Rare to locally uncommon, with occasional boom flights. Limited range, se VA to n. GA on the coastal plain and Piedmont (the latter sparingly). Also s. LA; rare strays in Mississippi Valley to IL and AK. Two to three broods on Atlantic coast, Apr-Jun and Jul-early Aug. Overwintering stage not reported. Pupation in tight leaf roll, almost severed from hostplant, which sways in the breeze.
Ecology	**Specialist.** Ecologically focused, little tolerance for disturbed settings.

Wet coastal plain Longleaf Pine savanna, Green Swamp, Brunswick Co, NC

av wingspan: 1.0"

Carolina Roadside-Skipper *Amblyscirtes carolina*

Dorsal: 8/19/95, Great Dismal Swamp NWR, Nasemond Co, VA

Ventral: 8/20/95, Northwest River Park, City of Chesapeake, VA

The Carolina is one of several East Coast roadside-skippers with a confined worldwide distribution, limited in this case almost entirely to the southeastern United States. Its distinctiveness as a species is diminished only slightly by its placement in the genus *Amblyscirtes*, along with several of our rarest and most exotic grass skippers. In its home territory, the Carolina is intermediate between the two other cane-feeders, Reversed and Lace-winged. The Carolina is more common than the Reversed over most of its range, and also a bit more bright and colorful. But it is considerably less widespread—and to most viewers less impressive-looking—than the Lace-winged.

It is difficult to understand why the Carolina Roadside-Skipper, which is absent from southern Georgia and all of Florida, does not range farther south than it does. Adequate supplies of wet bottomland with ample stands of cane certainly exist in areas it does not colonize. Lace-winged Roadside-Skippers do inhabit this southern tier, and even the Reversed reaches the Gulf Coast (albeit west of our area). Some unknown combination of historic chance and specialized needs has limited the Carolina to a narrow, east-west distributional band.

Male Carolina Roadside-Skippers perch in sunny areas to await females, often resting directly on hostplant leaves. They are avid nectarers (we have seen up to seven individuals on a single milkweed flowerhead in southeastern Virginia), but as with many other *Amblyscirtes*, nectaring seems to occur in focused episodes.

Identification	Size similar to Reversed. Fringes usually more checkered (still less so than in some other *Amblyscirtes*). Note *dark spots along abdomen*. Below, sometimes mistaken for a Fiery Skipper, or even a Whirlabout, although Carolina is smaller and the similarity is not all that close.
Dorsal	Dark brown with a *yellowish head and "shoulders"* and a number of *yellow dots* on both wing surfaces.
Ventral	HW *dull, mottled yellow with the appearance of dark spotting*. To see this as the "reverse" of the HW pattern on the Reversed Roadside-Skipper can take a bit of squinting, but the Carolina's overall base color is yellower.

Habitat
Somewhat varied, but typically sunny clearings or edges of hardwood swamps, bottomlands, cane-brakes, or pocosins, always with hostplant. Less likely away from shady habitats than Reversed, generally not in open pine woods. May seek nectar along ditches or edges of wood road.

Hostplants
Cane (*Arundinaria* sp.). Once again Switch Cane (*A. tecta*) cited most often. Caterpillar invariably found on or near the host.

Occurrence
Can be locally fairly common. Upwards of 60 seen in a day at Croatan NF during peak flights (LeGrand & Howard, 2002). Se VA to SC and n. GA. Also west in n. Gulf Coast states to AK, s. IL. Rarely to DE. Three broods, mainly Apr, Jun-early Jul, late Jul-e Sep; often flies slightly later than Reversed when sympatric. Diapause stage unreported.

Ecology
Specialist. If anything, a bit more restricted in habitat selection than *reversa*, although more common overall.

Southern hardwood swamp, edge habitat, Great Dismal Swamp NWR, Nasemond Co, VA

av wingspan: 1.0"

Lace-winged Roadside-Skipper   *Amblyscirtes aesculapius*

Dorsal: 5/25/96, Great Dismal Swamp NWR, Nasemond Co, VA

Ventral: 8/22/92, Great Dismal Swamp NWR, Nasemond Co, VA

No August trip to Great Dismal Swamp would be complete without sighting a Lace-winged Roadside-Skipper. This alluring southern specialty has the unusual power to draw enthusiasts into such steamy climes in mid-summer.

The Lace-winged Roadside-Skipper is not the most common *Amblyscirtes* in our area, nor the most widespread, yet it was described more than 200 years ago (in 1793), almost 70 years before any other species in its genus. This early recognition is likely a tribute to the Lace-winged's dramatic ventral hindwing pattern, which draws quick attention for a butterfly of its small size. Still, systematic study of the Lace-winged's life history did not proceed with equal dispatch: to our knowledge, the early stages were not reported until quite recently (Tveten & Tveten, 1996), and a full discussion of adult ecology is yet to follow.

In conducting fieldwork on Lace-wingeds in Texas, the Tvetens made several interesting discoveries. One involved the value of the caterpillar's defensive strategy during early stages. Lace-winged caterpillars, like those of many other skippers, live in tight leaf shelters. They pupate in dangling leaf bundles, nearly detached from the main plant stem. Of 12 caterpillars the Tvetens located, only one had been parasitized by a tachinid fly. Yet of 47 caterpillars and pupae of Southern Pearly Eye they found in the same vicinity (all living in the open), only three were *not* parasitized. Such astute field observations can generate themes for follow-up research.

Identification   Slightly larger than our other roadside-skippers, and by far the easiest to identify, at least from the ventral surface. *Wing fringes are prominently checkered black-and-white*. As in other *Amblyscirtes*, sexes are mostly similar.

Dorsal   Dark brown above with a *full row of curved, white FW spots*, plus a cell-end white mark (in females). Male has a long, narrow, two-part stigma, inconspicuous in the field.

Ventral   Rich, uniform dark brown with a highly distinctive "cobweb" pattern of HW marks with light wing veins. *Compare* Cobweb Skipper.

Habitat   A variety of southern swamp and wetland habitats. Clearings and openings in bottomlands, canebrakes, forest paths, ditches, etc. Seldom more than 50 yards from hostplants or suitable, moist woodland habitats.

Hostplants   Cane (*Arundinaria*). Giant Cane (*A. gigantea*) has been documented in TX, Switch Cane (*A. tecta*) also cited. Isolated young caterpillars rest in leaf folds near the tip of a hostplant leaf. Mature caterpillars form rolled leaf tubes (Tveten & Tveten, 1996). After consuming their protective sheath, they quickly crawl to another leaf and form a new shelter.

Occurrence   Uncommon to locally fairly common across most of the Southeast. Se VA to n. FL, mostly along the coastal plain and Piedmont, also broadly west through Gulf states and lower Ohio River Valley to AK and e. TX. Rare in Appalachians. Brood sequence confusing. Two to three broods depending on location, Feb-Jun (South) or June (North) and Jul-Sep. Diapause stage not reported.

Ecology   **Medium specialist.** Restricted to southern "canelands," but these are fairly varied and diverse in this region, and the Lace-winged has adapted to many, both geographically and ecologically.

Cane (*Arundinaria* sp.)

av wingspan: 1.15″

Twin-spotted Skipper   *Oligoria maculata*

Dorsal: 5/5/96, Ocala SF, Marion Co, FL

Ventral: 5/5/96, Ocala SF, Marion Co, FL

Placed in its own, monotypic genus, the Twin-spotted Skipper is an unassuming singleton among East Coast grass skippers. Klots (1951) suggested that its nearest local relatives are longwings (*Panoquina*), dusted skippers (*Atrytonopsis*), and Clouded Skipper (*Lerema*). Yet none of these resemble it very closely, in either general appearance or lifestyle. Indeed, its distribution and field behavior superficially suggest *Poanes* or *Euphyes*. It is frankly difficult to understand exactly where this anomalous skipper fits in the taxonomic order.

The Twin-spotted's genus name, taken from the Greek *oligaria* (meaning "light esteem" or "contempt") (Opler & Krizek, 1984), is ironically apt, for despite being common in much of the Southeast, the "light esteem" in which it is held is readily apparent from the lack of attention Twin-spotteds have received in the literature since first described in 1865 (the last year of the Civil War!). We know exceptionally little even today about their hostplant use, adult behaviors, or ecology. A fondness for thistle nectar is one of the species' few well-documented traits, but even here we know that mistflower, pickerelweed, bidens, etc., are all readily used as well.

The Twin-spotted's common name has been a point of frequent comment, since three spots are plainly visible on the hindwing. It is just that two of them are located close together, as "twins." Scott (1986) once proposed calling this species "Three-spotted Skipper," while renaming the current Three-spotted Skipper (*Cymaenes tripunctus*) as "Dingy Dotted Skipper," but this suggestion has not been adopted.

We have seen Twin-spotted Skippers nectaring at virtually all times of day, a trait often associated with patrolling species. On one occasion, courtship was observed between 9:30 and 10:00 in the morning.

Identification	Fairly large, with ample, rounded wings (also consistent with patrolling, although the Twin-spotted flies rapidly, some distance off the ground; it appears very dark in flight). Wing fringes are plain brown. Sexes similar. In spite of its generally dull coloration, the Twin-spotted is actually rather easy to identify because of its diagnostic ventral pattern. *Compare* with Dun Skipper and Little Glassywing (both are smaller, lack distinct ventral HW marks); Clouded and female Zabulon Skippers (both have marginal frosting, no spots on underside); Dukes' Skipper and Monk (both also dark above, Monk much larger, Dukes' slightly larger). Tip of antennal club (apiculus) is long and pointed, a trait that helps define the genus, although it is not a useful mark in the field.
Dorsal	Plain, uniform blackish brown, with little or no shading. *Three to four small, white FW spots* are visible on most individuals, but these can be much reduced, sometimes absent. Male lacks a stigma.
Ventral	A *warm, reddish (or chestnut) brown*, develops slightly brassy tones near the margins with wear. *Three conspicuous spots on the HW, with the lower two paired (the "twins")*. Spotting can be less conspicuous in some populations, e.g., on the Florida Keys (Minno & Emmel, 1993).

Habitat	Quite variable. Edges of freshwater and brackish or tidewater marshes, sawgrass marshes, canal edges, and dikes. But also drier habitats, such as sandhills, pine rocklands, Florida scrub, and dry prairies. Mainly along the coastal plain except in FL, where it flies inland, across the peninsula. Moves actively in search of nectar.
Hostplants	Largely unreported, but bluestem grasses recently mentioned (Glassberg et al., 2000).
Occurrence	Generally common, sometimes very common; does not appear colonial. E. NC, south along coast to peninsular FL; in pine rocklands on lower Keys. Also along Gulf Coast to e. TX. Two broods from mid-FL north, May-Jun (NC) or Mar-Apr (n. FL) and Aug-Oct; three-plus nearly continuous broods in s. FL. Diapause stage not reported.
Ecology	**Medium specialist.** Little as we know about this species, it has adapted to a limited range of habitats and remains successful throughout its southeastern range.

av wingspan: 1.3″

Monk   *Asbolis capucinus*

Dorsal male: 11/25/00, Big Pine Key, Monroe Co, FL

Ventral: 2/18/95, Everglades NP, Miami-Dade Co, FL

Here is a species that seems quite at home in a monotypical genus of its own. Beyond its distinctive size and bold appearance, it has made an important adaptive leap, to a dedicated diet of palm leaves. The only other grass skipper to make this specific jump is the Palmetto Skipper, but its menu is limited to one plant, the Saw Palmetto (*Serenoa repens*), whereas Monks feed on a wide range of both native and introduced hosts, spanning the Palm Family (Arecaceae). Brazilian Skippers have also shifted to nongrass monocotyledons, but to cannas, not palms.

A native of Cuba and the Isle of Pines (where it is very common), the Monk appeared in Miami in the late 1940s. It is not an aggressive colonist elsewhere, so the assumption has been that Monks probably arrived in the United States with human assistance (e.g., attached to imported botanical goods), although it is impossible to know for sure. Whatever the port of entry may have been, Monks are now well-established in central and southern Florida, making a home for themselves even in some of "the most unpromising areas" of central Miami (Smith et al., 1994).

Males perch to await females, darting out to accost intruders. Fast-flying (commensurate with their large size), Monks remain active into the late afternoon, even at times after sunset. Generally avid nectarers, they use their exceptionally long proboscises to reach nectar in large or tubular flowers, such as hibiscus.

Identification	Our second-largest grass skipper, robust and thick-bodied, with long, pointed wings. Males are somewhat smaller than females. Monks are most easily mistaken (in undisturbed marshland habitats) for Palatka Skippers, but are larger, have *light wing fringes (white to tawny)*, especially on the HW near the body, and have uniform-colored wings below (not two-toned). After actively scavenging for pollen, their heads may be dusted any color!
Dorsal	Male: dark, velvety black-brown with dull, fulvous tones toward the FW base, especially when fresh. The divided FW stigma is visible, but difficult to pick out. Female: similar, but somewhat less blackish, with *two pale, elliptically shaped patches* on the FW.
Ventral	Unmarked, but *flecked with chestnut- or mahogany-brown and white scales*. Given their muscular flight style, Monks' wings quickly abrade, losing their colorful overscaling to become plain brown.

Habitat	Widely variable, based on availability of suitable palms. (Yet, since palms are more ubiquitous in FL than Monks, additional constraints must be at work.) Regular in open parks and suburbs, rimrock pinelands, coastal hammocks, sometimes savannas.
Hostplants	A wide variety of palm species, Coconut Palm (*Cocos nucifera*), Paurotis Palm (*Acoelorrhaphe wrightii*), Silver Palm (*Coccothrinax argentata*), many others. Caterpillars feed on mature leaf tips, pulling frond edges around to form nests.
Occurrence	South and central FL, range seems to be gradually expanding. Three to four broods, Mar-Dec in North, all year in successive broods, s. FL and Keys.
Ecology	**Medium generalist.** This is still a transitional species in our area. Yet while "newly arrived" in ecological terms, it seems positioned for continued expansion.

Silver Palm (*Coccothrinax argentata*)

av wingspan: 1.75"

Brazilian Skipper   *Calpodes ethlius*

Dorsal: 9/9/01, Hendry Co, FL   **Ventral**: 9/24/95, Miami-Dade Co, FL

Unlike many skippers, whose life histories are scarcely explored, the Brazilian is exceptionally well-studied, thanks mainly to an acquired (but persistent) taste for ornamental cannas. So entwined is its life story with this popular genus of garden flowers that we often cannot tell whether its range extensions are the result of migratory movements or recent botanical shipments.

Pest control authorities refer to this species as "Larger Canna Leafroller," comparing it to the Lesser Canna Leafroller (*Geshna cannalis*), which is actually a moth. In the West Indies, Brazilian Skipper larvae can defoliate a local food plant (*Maranta arundinacea*). In our region, the aesthetic impact of young skippers on canna leaves has caused the Miami-Dade Parks Authority to stop using the flowers in ornamental plantings (Smith et al., 1994). This may be our only skipper that is easier to locate as a larva than as an adult.

Brazilian Skippers are known to "trap-line" during the day, traveling between established nectar sites. They are reportedly crepuscular, even at times nocturnal, but we have also seen them behaving actively during the day, especially in natural habitats. With large proboscises, Brazilian Skippers tend to visit large flowers. They nectar actively. Studies have claimed that ovipositing females prefer cannas with different leaf and flower colors, but much of the evidence seems contradictory and whatever rules apply in this respect may be very subtle.

Identification	Our largest grass skipper. (Note that three of the four largest, including also Palmetto Skipper and Monk, use non-grass hostplants.) With a *wide head and long, pointed FWs*, Brazilians can resemble a giant horsefly. The HWs are much shorter and more rounded. Sexes similar.
Dorsal	Wings blackish-brown with *three to five conspicuous translucent markings* ("hyaline spots"). Hammock Skipper lacks hyaline on the HW (upper or lower). Scaling on the head gives it a grayish appearance.
Ventral	Uniform, reddish-brown when fresh (fades later to pale brown), with *three to four conspicuous, but variable, light spots on the HW.*

Habitat	Extremely dichotomous. Best known from parks, gardens, and residential areas, wherever canna is planted. But also swamp edges, wet sloughs, flooded fields, and ditches, where the native host can be abundant.
Hostplants	Many species and varieties of garden canna (*Canna indica, flaccida, edulis, generalis*). Also Arrowroot, also called Alligator Flag (*Thalia geniculata*). Caterpillars dwell in folded-over leaf tents from which they emerge at night to feed. With many larvae on a single plant, their chewing can be audible. Feeding waste (frass) is ejected forcibly from the nest.
Occurrence	Variably common. Permanent range FL west to CA, south to ARG. Also W.I. Emigrates northward along coast in the fall, and south to Keys. Durable northern colonies currently at least to se VA. Earlier northern populations probably linked to canna trade. Flies all year in far South, late summer in far North. Warm periods elsewhere. Overwintering stage not reported.
Ecology	**Medium specialist.** Ecologically choosy despite taste for garden canna.

Arrowroot or Alligator Flag (*Thalia geniculata*)

av wingspan: 1.9"

Ocola Skipper   *Panoquina ocola*

Dorsal: 9/28/95, Ocala NF, Marion Co, FL

Ventral: 9/28/95, Ocala NF, Marion Co, FL

Ocola Skippers engage in long-distance, mass emigrations late each summer, making them one of the very few grass skippers fully dedicated to this lifestyle. The Ocola's seasonal movements are seldom obvious, evidenced only by a few stray individuals nectaring out of range, in some northern park or garden. Yet on rare occasions, bona fide mass movements have been witnessed. In October 1950, most notably, many thousands passed through southern Louisiana in a giant swarm, clouding the windshields of cars crossing Lake Pontchartrain.

The origin of the name "Ocola" is somewhat mysterious. If the butterfly's namesake is the city in central Florida, it should be spelled "Ocala." There was also a early Timucuan village and district in north-central Florida called "Ocali"—still a misspelling, if this was the inspiration. "Ocola" itself is a relatively uncommon suname, also an acronym for Overseas Cost of Living Adjustment, but neither has any discernible link to the butterfly.

There is strong evidence that Ocola Skippers migrate both north and south in the fall, emanating from strongholds in the Deep South. Incursions reach as far north as Massachusetts (even rarely to Pt. Pelee in Ontario), also southward on the Florida peninsula. Ocolas fly rapidly, and are very active nectarers, visiting a wide range of flowering plants.

Identification	Like other long-winged skippers (genus *Panoquina*), Ocolas have *extended FWs, almost exaggerated in appearance*. The apparent length of the FWs is accentuated at rest, when they are nested between the HWs, protruding directly behind them. As regular migrants, with frequent gene mixing, Ocolas show little regional pattern variation. *Dark lines running along abdomen* are characteristic of the genus.
Dorsal	Dark, yellowish-brown when fresh, fading to a flat, dark brown (usually darker overall than the other two *Panoquinas*). Note a variable series of *off-white to yellowish FW spots*, the largest shaped like an arrowhead. Spots generally larger on female. Center of upper HW lighter.
Ventral	Brown with somewhat lighter veins. Usually *unmarked*, except for a *faint, postmedian band of tiny spots* in some individuals. Female has a *bluish-purple iridescent sheen* when fresh.

Habitat   Nearly any flower bed during migration, especially disturbed coastal sites with nectar (parks, gardens, etc.). Residents found in varied open settings, old fields, roadsides, wood edges. Many writers suggest that moist habitats are preferred, but we have not noticed this to any appreciable extent.

Hostplants   Not well-reported. Sugarcane (*Saccharum officinarum*) and rice (*Oryza sativa*) in agricultural areas. Rice Cutgrass (*Leersia oryzoides*) in WV, Torpedo Grass (*Panicum repens*) in FL; undoubtedly others.

Occurrence   Numbers fluctuate seasonally. Common to abundant in GA/SC breeding zone, regularly emigrates (*see* above); but accidental on Keys. Also W.I., west to AZ and south to ARG. Flies all year in s. FL. Emigrants reach northern range limits in Sep-Oct.

Ecology   Generalist. A consummate generalist, constantly dispersing, adapting to new habitats and circumstances. Readily accepts disturbed habitats.

Sugar Cane (*Saccharum officinarum*)

av wingspan: 1.3″

Obscure Skipper *Panoquina panoquinoides*

Dorsal: 10/28/99, San Padre Island, Cameron Co, TX

Ventral: 3/22/94, Valhalla, Monroe Co, FL

The longwinged skippers (genus *Panoquina*) are a small group, with just three species in our area, yet they encompass a diverse spectrum of lifestyles, almost as broad as that of the entire grass skipper subfamily (*Hesperiinae*) combined. Their ranks include both a highly sedentary niche specialist and a consummate, wide-ranging emigratory generalist. The Obscure Skipper—mainly a tropical species, reaching its range limit in the southern United States—is intermediate between these two extremes. It is closer in most respects to the Salt Marsh Skipper than to the Ocola, but has some affinities with both.

Two races of Obscure Skipper occur in the United States, the eastern nominate race (*Panoquina p. panoquinoides*), which reaches as far south as Colombia, and the threatened Wandering Skipper (*P. p. errans*) of southern California. Other races occur in the West Indies and South America. Some of these may ultimately prove to be full, sibling species.

It is sometimes suggested that the Obscure Skipper "replaces" the Salt Marsh Skipper in the Florida Keys and along the Gulf Coast, but in fact the ranges of the two species overlap extensively. Further study is needed to determine the specific factors that differentiate their distributions locally (they are seldom seen flying together).

Obscure Skippers dart quickly through beachside foliage. Males perch on grass stalks or other vantages to await females. They are jumpy and usually difficult to approach.

Identification	The smallest East Coast Panoquin. The FW is less narrow and angular than in other genus members. Wing fringes range from buff to grayish. *Dark abdominal stripes* are typical of the genus. This skipper can at times be challenging to identify. Its ventral pattern is similar to the Twin-spotted's, although that species is much larger and darker, with rounded wings. And when its ventral pattern is not seen (or is not well-expressed), confusion is possible with Eufala, Neamathla, or Three-spotted—to say nothing of other Panoquins.
Dorsal	*Dull brown or yellowish-brown* with some fulvous scaling on the upper body and FW costa. *Pale spots* variable, often faint or absent in males, better-expressed in females. Upper HW unpatterned.
Ventral	*Plain brown with somewhat lightened veins.* When present, *three spots on postmedian HW, two close together*, are diagnostic. But these may be lost with wear, or simply not be present. Individuals lacking clear ventral HW markings are of course the most difficult to identify confidently.

Habitat	Somewhat variable locally. In s. FL and the Keys, mostly beachside scrub, grassy uplands along mangrove edges, coastal prairies, dunes, salt marshes. Salt marshes and dunes seem preferred in TX, but perhaps merely because of their greater prevalence. Also groves, open fields and second-growth scrub away from the immediate coast, usually in search of nectar, but seems somewhat less habitat-restricted than salt marsh.
Hostplants	In our area, Seashore Saltgrass (*Distichlis spicata*), a tall grass of marshes (salt or brackish) and beaches, also Seaside Dropseed (*Sporobolus virginicus*), found on drier ground near salt marshes. Sugarcane (*Saccharum officinarum*) and Bermuda Grass (*Cynodon dactylon*) also reported. Caterpillars live in rolled leaf nests, feeding at night.
Occurrence	Populations are small and usually very local, often in ephemeral colonies. Spraying is likely responsible, however, for some of the vagaries of the Obscure Skipper's unpredictable occurrence. Coastal FL and the Keys, from mid-Atlantic coast south. Also W.I. and along Gulf Coast (w. of FL panhandle) into So. Amer. Multiple broods, all year in s. FL; inactive in mid-winter farther north. Caterpillars overwinter in CA.
Ecology	**Medium specialist.** Mainly restricted to coastal, subtropical habitats. Local and retiring, not prone to stray widely.

av wingspan: 1.0″

Salt Marsh Skipper   *Panoquina panoquin*

Dorsal: 4/22/00, Ft. Macon, Cartaret Co, NC

Ventral: 3/21/94, Collier-Seminole SP, Co, FL

Until the 1980s, this sprightly denizen of coastal marshes was unrecorded west of Louisiana, making it seem a near-endemic in our area (Tveten & Tveten, 1996). But today we know that its range extends much farther west along the Gulf Coast, reaching as far as Texas, with a substantial colony in Rockport and individual records from as far down the coast as South Padre Island. (Unlike our other two Panoquins, however, the Salt Marsh Skipper does not occur in the West Indies.)

The Salt Marsh Skipper's ecological style is distinctive. It has long been described as "more strikingly limited to salt and brackish marshes than any other butterfly in our area" (Klots, 1951, p. 269). The biological adaptations necessary to succeed in an open salt marsh habitat are substantial (yet we have barely any idea of the specific adaptations that allow Salt Marsh Skippers to live here).

Identification of this species is not especially challenging, and is made even easier by the fact that it is often the only butterfly of any kind to be seen in its core habitat. Many sightings occur along dikes and marsh edges, where colony members congregate to find nectar (along with other species). But males can be seen perched alone on strands of marsh grass away from such feeding sites. Obtaining photographs of perched individuals is challenging, since they are quite wary and the grass on which they perch is often riotously windblown.

Identification	Intermediate in size between the other two Panoquins. The FW is long and pointed. The body has *dark stripes*, indicative of the genus. Similar above to Obscure Skipper (usually darker), but Obscure is usually more difficult to distinguish from Ocola than from Salt Marsh. Fringe color is a varying combination of buff, tan, and gray.
Dorsal	Brown, variably dark or yellowish. FW has *pale, off-white or yellowish markings*, including a small wrist "bracelet," a transverse spot band, and a cell-end mark at midwing. These marks are generally better defined in females. *Upper HW shows faint reflection of white, ventral dash mark.*
Ventral	Light brown with yellowish wing veins and a diagnostic *white, rectangular bar or dash mark* on the outer HW.

Habitat	As already noted, mostly confined to salt or brackish marshes near the immediate coast. But individuals stray for nectar to marsh edges, and sometimes well inland. Strays seen far from the shore may represent dispersing emigrants.

Hostplants	Uncertain. Colonies are often in salt marshes with Saltmarsh Cordgrass (*Spartina alterniflora*), or brackish marshes with Black Needlerush (*Juncus roemerianus*). But the true host may be Seashore Saltgrass (*Distichlis spicata*), growing with them.
Occurrence	Locally common in colonies, often wanders singly to nectar. Saltmarsh spraying and drainage impacts populations. NY to e. TX along coast; rare in s. FL and Keys. Two broods in North, Jun-Jul and Aug-Sep, three or more southward, nearly all year in s. FL.
Ecology	Specialist. A paradigm of the specialist lifestyle.

Don Riepe/American Littoral Society

Maritime grasses in salt marsh

av wingspan: 1.15"

Giant-Skippers: Subfamily Megathyminae

Giant-Skippers are among the most specialized—and intriguing—of East Coast butterflies. Their mystique among observers is enhanced by their rarity, and the fact that many aspects of their life histories remain poorly known. The Megathyminae is a small subfamily, consisting of four genera with about 20 species (13 found in the southern United States, the rest limited to Mexico and Central America). The two regular genera occurring in the United States are *Agathymus* (agave-feeders, tribe Aegialini) and *Megathymus* (yucca-feeders, tribe Megathymini). Both East Coast giant-skippers are Megathymini.

Giant-skippers are heavy and muscular, capable of rapid flight over short distances (up to 60 mph by one report). Swirling aerial dogfights among territorial competitors are truly breathtaking. Displaying males can make a clicking noise with their wings.

Giant-skipper bodies contain high levels of stored fat. This is an important attribute, since Megathyminae do not feed as adults. Male *Agathymus* skippers sometimes do puddle, but *Megathymus* males reportedly do not (Bailowitz & Brock, 1991).

Giant-skippers probably evolved from the closely related grass skippers (Hesperiinae). Both groups share key taxonomic traits (antennal club with bent tips, distinctive wing venation, the familiar "jet fighter" basking posture, etc.). But giant-skippers differ in their large size, thick, cigar-like shape (head narrower than thorax), and—significantly—in their highly specialized lifestyle.

Giant-Skipper caterpillars burrow into the roots of agaves and yuccas to feed. Eggs are usually laid singly on the undersides of leaves and hatch in about 10 days. Young caterpillars initially spin silken webs and eat leaves (just as their grass skipper relatives do), but older ones burrow through the plant's growth crown, down as far as 1/2 meter into the thick rootstock. They create a narrow tunnel as they chew their way along.

Megathymus caterpillars return to the surface periodically to deposit frass and other waste. A silk-lined tent or "chimney" is gradually erected on these trips to the surface, leaving a telltale clue that larvae are present, or recently have been. Caterpillars overwinter partially grown, attaining full development in the following spring or summer.

Megathymus pupae are as specialized as their larvae. Their abdominal segments are jointed more flexibly than in typical butterflies, allowing them to wiggle up and down in their silk-lined passageways. In addition, the giant-skippers' cremaster (an appendage that usually attaches a pupa securely to its chosen substrate) has broad bristles, providing traction for movement. Relocation within the tunnel may be useful for thermoregulation, defense, positioning for a final exit, etc.

At the end of the pupal stage, the emerging adult climbs from the top of its tent, clinging to the outer surface for several hours while its wings harden. This process reportedly begins in the morning, with flight achieved by afternoon.

There is some inconsistency in the literature concerning adult behavior. As noted earlier, adult *Megathymus* skippers are believed not to feed at all, or even to engage in puddling. Yet many accounts routinely suggest that giant-skipper males in our area sometimes visit mud. It is not clear if these statements are based on general literature references (which may refer to *Agathymus* giant-skippers), or if they result from specific, local observations.

Descriptions of daily activity cycles also vary. We have seen Yucca Giant-Skippers (often described as crepuscular) behave much as ordinary skippers in both Florida and North Carolina, establishing territorial perches in the morning, then becoming less active during mid-afternoon. But the pattern is quite different, as others have noted, with Cofaqui Giant-Skippers. Cofaquis are active almost exclusively at sunset, when they engage in highly structured, ritual arena displays (*see* species account). These displays need much further investigation, and the possibility of nocturnal activity should not be discounted. We have not observed dawn (or morning) Cofaqui display flights, despite careful searching at active colonies. Ovipositing and emergence are possible early in the day, however.

Northern Cofaqui display arena (hostplant in nearby woods)

Larval "tent" of Yucca Giant-Skipper, 7/31/99, De Kalb County, GA

Ventral Cofaqui Giant-Skipper (northern race) during display, 8/7/99, deKalb Co, GA

Yucca Giant-Skipper *Megathymus yuccae*

Dorsal male: 4/10/99, Pender Co, NC

Ventral: 4/10/99, Pender Co, NC

In early spring, the tranquility of southeastern forests is disrupted—in certain select locations, at least—by the barging, erratic flight of male Yucca Giant-Skippers. These large, unlikely looking butterflies hurtle headlong through open woodlands at high speed, usually returning afterward to their original perch. They are the most widespread of the Megathyminae (giant-skippers) in the United States. Yet despite their wide geographical distribution and diverse habitat tolerance, they are quite rarely seen, living in isolated colonies with only a single, early flight.

Male Yucca Giant-Skippers perch territorially near yucca hostplants, often basking in sunny glades. We have seen them actively perched in the morning, withdrawing in the afternoon. Since they do not feed as adults, giant-skippers can remain immobile when not engaged in courtship, mating, or egg-laying. We have not witnessed crepuscular behavior for this species, though it may occur (*compare* next species account). Nor can we verify the common view that male Yuccas take moisture from mud. This possibility should be corroborated by current field reports.

Like other *Megathymus* skippers, older Yucca caterpillars (3rd instar and following) burrow into hostplant roots, forming a conspicuous tent at the outlet of the burrow. The Yucca's tent is comparatively large, usually emerging from the central crown of the hostplant.

Identification	Large-bodied and thick-set, can resemble a large moth on casual view. FWs long and pointed. Upper thorax hoary gray.
Dorsal	Dark brown with a *variable, yellow band across the outer FW*; also two white to yellowish cell-end FW spots, *not connected to main yellow band*. Outer HW rimmed in bright yellow (inner edge of band usually not scalloped). Female has a row of small yellow dots across the outer HW.
Ventral	Dark with variable white frosting toward the margin; note c*onspicuous white "wedge" on leading edge of HW.*

Habitat	Pine flatwoods and sandhills, open bottomland woods, old fields, rock outcrops, coastal dunes, and maritime forest—all where yucca occurs.
Hostplants	Various yucca species, including Spanish Bayonets (*Yucca filamentosa* and *aloifolia*), sometimes Small's Yucca (*Y. smalliana*) or Spanish Dagger (*Y. gloriosa*). Said to favor small yuccas as hosts (Butler & Covell, 1957).
Occurrence	Extreme se VA, across the se coastal plain to the FL peninsula (absent s. FL). Also westward along the US southern rim, to Colorado Plateau, lower Gr. Basin, s. CA. Western populations were once regarded as separate species, vs. one regionally diverse species complex. One brood, Mar-May most places; mid-Feb to early May in FL. Normally does not fly for more than a month at any one site.
Ecology	**Medium specialist.** Though Yucca Giant-Skippers feed on common species of yucca, which occur in several widespread habitats, the precise conditions required for successful colonization are evidently quite specific. The species is absent in a great many suitable-seeming locations. Almost never found far from an established colony.

Open riparian woods with yucca, Pender Co, NC

av wingspan: 2.5″

Cofaqui Giant-Skipper *Megathymus cofaqui*

Dorsal male (northern race) in dusk display: 8/7/99, DeKalb Co, GA

Ventral male (peninsular race): 3/29/02, Pasco Co, FL

A southeastern endemic, the Cofaqui has two distinct subspecies: a single-brooded, northern race and a paler, double-brooded Florida peninsular form. Rare and elusive, Cofaquis may be seen less often as free-flying adults than any butterfly in our region (particularly the northern race).

With luck, Cofaquis can be "kicked up" from low perches or shaded trunks by day, but in our experience both races are active mainly at dusk, in 15- to 30-minute display flights. Displays are staged in flat, open "arenas" with late sun, away from hostplants. Participants fly in low circles of decreasing diameter; if "crossed" they initiate high-speed aerial dogfights. These arena displays begin and end with almost clocklike regularity each evening.

Ventral female (peninsular race): 3/30/02, Pasco Co, FL

Identification	Smaller on average than the Yucca Giant-Skipper, with shorter wings. But these subtle differences are of limited diagnostic value during rapid display flights in low light. Male's FW is more pointed than female's.
Dorsal	Dark brown above (northern race darker, peninsular form paler). *Yellow FW band broadened into a wide patch,* visible in fluttery display flight, *which touches yellowish cell-end dots.* The inner edge of yellow HW border is *more scalloped* than in Yucca Giant-Skipper. Females have additional yellow dot-row on the upper HW.
Ventral	Dark brown below with substantial white frosting. *Two white dots on leading HW edge* (not a bold triangular wedge); these dots are most prominent in peninsular females; reduced or absent in northern race.

Habitat	Similar to Yucca Giant-Skipper, but peninsular race is most common in Gulf coastal habitats. Northern race is best known from granite outcrops along the southern Piedmont (though it is not confined to such locales).
Hostplants	Yuccas, including some or all species cited for Yucca Giant-Skipper. (*Y. filamentosa* seen at colonies we have visited.) Larval tents relatively inconspicuous. Older larvae tunnel away from hostplant (in n. race at least), making a small, well-camouflaged tent on the surface nearby.
Occurrence	Range poorly documented. Southern: n.-central FL, especially along coast. Northern (*harrisi*): w. NC; n. and e. GA; central SC; FL panhandle; w. TN. One brood in North, Jul to mid-Aug; two in South, Mar-Apr, Aug-Nov (peak early Oct).
Ecology	**Specialist.** Complex habitat requirements and lifestyle, despite use of common hostplants. Should be protected wherever it occurs.

Sandy live oak savanna, southern race breeding habitat, Pasco Co, FL

av wingspan: 2.0"

Sources List

Abrahamson, W.G. & W.G. Hartnett. 1990. Pine flatwoods and dry prairies, in Myers, R.L. & J.J. Ewel, Ecosystems of Florida, University of Central Florida Press, Orlando

Ackery, P.R. 1985. Systematic and faunistic studies on butterflies, in Vane-Wright, R.I. & P.R. Ackery, The Biology of Butterflies: Symposium of the Royal Entomological Society of London Number 11, Academic Press, Harcourt Brace Jovanovich, London, 2nd ed.

Ackery, P.R. & R.I. Vane-Wright. 1984. Milkweed Butterflies: Their Cladistics and Biology, Being an Account of the Natural History of the Danainae, a Subfamily of Lepidoptera, Nynphalidae, Comstock Publishing Associates, Cornell University Press, Ithaca

Adler, L.S. 2000. Alkaloid uptake increases fitness in a hemiparasitic plant via reduced herbivory and increased pollination, The American Naturalist, 156(1):92-99

Agrawal, A.A., J.K. Conner, M.T.J. Johnson & R. Wallsgrove. 2002. Ecological genetics of an induced plant defense against herbivores: Additive genetic variance and costs of phenotypic plasticity, Evolution, 56(11):2206-2213

Alden, P., R. Cech & G. Nelson. 1998. National Audubon Society Field Guide to Florida, Knopf, Chanticleer Press, Inc., New York

Allen, T.J. 1997. The Butterflies of West Virginia and Their Caterpillars, University of Pittsburgh Press, Pittsburgh, PA

Altizer, S.M., K.S. Oberhauser & L.P. Brower. 2000. Associations between host migration and the prevalence of a protozoan parasite in natural populations of adult monarch butterflies, Ecological Entomology, 25:125-139

Anderson, R.C. et al. 1999. Savannas, Barrens, and Rock Outcrop Plant Communities of North America, Cambridge University Press, Cambridge

Anderson, S.L. 2001. Butterfly flight patterns: How flight and color attribute palatability, http://www.colostate.edu/Depts/Entomology/courses/en507/papers_2001/anderson.htm

Andersson, J., A-K. Borg-Karlson & C. Wiklund. 2000. Sexual cooperation and conflict in butterflies: A male-transferred antiaphrodisiac reduces harassment of recently mated females, Proceedings of the Royal Society of London, 267:1271-1275

Arms, K., P. Feeny & R.C. Lederhouse. 1974. Sodium: Stimulus for puddling behavior by tiger swallowtail butterflies, Papilio glaucus, Science, 185:372-374

Ayres, M.P. & J.M. Scriber. 1994. Local adaptation of regional climates in Papilio canadensis (Lepidoptera: Papilionidae), Ecological Monographs, 64:465-482

Bailowitz, R.A. & J.F. Brock. 1991. Butterflies of Southeastern Arizona, Sonoran Arthropod Studies, Inc., Tucson, AZ

Barros-Bellanda, H.C.H. & F.S. Zucoloto. 2001. Influence of chorion ingestion on the performance of Ascia Monuste and its association with cannibalism, Ecological Entomology, 26:557-561

Barth, F.G. 1991. Insects and Flowers: The Biology of a Partnership, Princeton University Press, Princeton, NJ

Barton, A.M. 1986. Spatial variation in the effect of ants on an extrafloral nectary plant, Ecology, 67(2):495-504

Bates, M. 1960. The Forest and the Sea: A Look at the Economy of Nature and the Ecology of Man, Vintage Books, Random / Knopf, New York

Baum, N.M. & K. Baum. 1999. An experimental test of corridor effects on butterfly densities, Ecological Applications, 9:623-633

Beldade, P. & P.M. Brakefield. 2002. The genetics and evo-devo of butterfly wing patterns, Nature Reviews, Nature, 3:442-452

Bell, C.R. & B.J. Taylor. 1982. Florida Wild Flowers and Roadside Plants, Laurel Hill Press, Chapel Hill, NC

Bernath, R.F. 1982. Pupal polymorphism in the cabbage white butterfly, Pieris rapae L., in Stock, M.W. & A.C. Bartlett, Proceedings Symposium at National Meeting, Entomological Society of America, 57-63

Bernzweig, A. 1999. Satyr-Eyes, American Butterflies, 7(4):37

Betts, C.R. & R.J. Wootton. 1988. Wing shape and flight behaviour in butterflies (Lepidoptera: Papilionoidea and Hesperioidea): A preliminary analysis, Journal of Experimental Biology, 138:271-288

Beutenmuller, W. 1902. The butterflies of the vicinity of New York City, in Supplement to American Museum Journal, American Museum of Natural History, Vol. II, No. 5, Guide Leaflet No. 7

Bitzer, R.J. & K.C. Shaw. 1979. Territorial behavior of the red admiral, Vanessa atalanta (L.) (Lepidoptera: Nymphalidae), Journal of Research on the Lepidoptera, 18:36-49

Bitzer, R.J. & K.C. Shaw. 1983. Territorial behavior of Nymphalis antiopa and Polygonia comma (Nymphalidae), Journal of the Lepidopterists' Society, 37:1-13

Blest, A.D. 1957. The function of eyespot patterns in the Lepidoptera, Behavior, 11:209-255

Boggs, C.L. 1998. The Y files: Salt, sex, and the single butterfly, American Butterflies, 6(3):4

Boggs, C.L., W.B. Watt & P.R. Ehrlich. 2003. Butterflies: Ecology and Evolution Taking Flight, University of Chicago Press, Chicago

Bosworth, S.B. 2003. Cerulean Warbler Relative Abundance and Frequency of Occurrence Relative to Large-scale Edge, master's thesis, West Virginia University, http://etd.wvu.edu/templates/showETD.cfm?recnum=3116

Bowden, S.R. 1971. American white butterflies (Pieridae) and English food-plants, Journal of the Lepidopterists' Society, 25:6-12

Bowers, M.D. 1980. Unpalatability as a defense strategy of Euphydryas phaeton (Lepidoptera: Nymphalidae), Evolution, 34:586-600

Bowers, M.D. 1983. Mimicry in North American checkerspot butterflies Euphydryas phaeton and Chlosyne harrisii (Nymphalidae), Ecological Entomology, 8:1-8

Boyd, H.P. 1991. A Field Guide to the Pine Barrens of New Jersey: Its Flora, Fauna, Ecology and Historic Sites, Plexus Publishing, Medford, NJ

Brakefield, P.M., J. Gates, D. Keys, F. Kesbeke, P. J. Wijngaarden, A. Monteiro, V. French & S.B. Carroll. 1996. Development, plasticity and evolution of butterfly eyespot patterns, Nature, 384:236-242

Brakefield, P.M. & T.B. Larsen. 1984. The evolutionary significance of dry and wet season forms in some tropical butterflies, Biological Journal of the Linnaean Society, 22:1-12

Braunschweig, S.H., E.T. Nilsen & T.F. Wieboldt. 1999. The Mid-Appalachian shale barrens, in Anderson, R.C., J.S. Fralish & J.M. Baskin, Savannas, Barrens, and Rock Outcrop Plant Communities of North America, Cambridge University Press, Cambridge

Britten, H.B. & P.F. Brussard. 1993. Isozyme data and the taxonomy of checkerspot butterflies (Euphadryas), Journal of Research on the Lepidoptera, 32:124-134

Brockman, C.F. 1979. Trees of North America: A Field Guide to the Major Native and Introduced Species North of Mexico, Golden Press, Western Publishing Company, Racine, WI

Brodsky, A.K. 1994. The Evolution of Insect Flight, Oxford University Press, Oxford

Brooks, R.R. 1987. Serpentine and Its Vegetation: A Multidisciplinary Approach, Dioscorides Press, Portland, OR

Brower, A.V.Z. & K.R. Sime. 1998. A reconsideration of mimicry and aposematism in caterpillars of the Papilio machaon group, Journal of the Lepidopterists' Society, 52(2):206-212

Brower, L.P. 1985. Chemical defenses in butterflies, in Vane-Wright, R.I. & P.R. Ackery, The Biology of Butterflies: Symposium of the Royal Entomological Society of London Number 11, Academic Press, Harcourt Brace Jovanovich, London, 2nd ed.

Brower, L.P., B.S. Albert & S.C. Glazier. 1970. Observational learning in the feeding behavior of blue jays, American Zoology, 10:475-476

Brower, L.P. & J.V.Z. Brower. 1962. The relative abundance of model and mimic butterflies in natural populations of the Battus philenor mimicry complex, Ecology, 43:154-158

Brower, L.P., W.H. Calvert, L.E. Hendrick & J. Christian. 1977. Biological observations of an overwintering colony of monarch butterflies (Danaus plexippus, Danaidae) in Mexico, Journal of the Lepidopterists' Society, 31:232-242

Brown, C. & E. Woods. 1999. Nature in Cities, Sustainable Cities Program, Open Space Network, New York

Brown, J.W. & P.A. Opler. 1990. Patterns of butterfly species density in peninsular Florida, Journal of Biogeography, 17(6):615-622

Brown, K.S., Jr. 1981. The biology of Heliconius and related genera, Annual Review of Entomology, 26:427-456

Brown, K.S., Jr. 1984. Adult-obtained pyrrolizidine alkaloids defend ithomiine butterflies against a spider predator, Nature, 309:707-709

Brown, K.S., Jr. & R.B. Francini. 1990. Evolutionary strategies of chemical defense in aposematic butterflies: Cyanogenesis in Asteraceae-feeding American Acraeinae, Chemoecology, 1:52-56

Brown, L. 1985. The Audubon Society Nature Guides: Grasslands, Knopf, New York

Brown, W.D. & J. Alcock. 1990. Hilltopping by the red admiral butterfly: Mate searching alongside congeners, Journal of Research on the Lepidoptera, 29(1-2):1-10

Bryson, C.T. & D.L. Sudbrink. 1999. Investigations for the biological control of congongrass, Proceedings of the X International Symposium on Biological Control of Weeds, Montana State University

Burger, J. & M. Gochfeld. 2001. Smooth-billed Ani (Crotophaga ani) predation on butterflies in Mato Grosso, Brazil: Risk decreases with increased group size, Behavioral Ecology and Sociobiology, 49:482-492

Burns, J.M. 1968. Mating frequency in natural populations of skippers and butterflies as determined by spermatophore counts, Journal of the Lepidopterists' Society, 6(3):852-859

Burns, J.M. 1975. Isozymes in Evolutionary Systematics, Academic Press, San Francisco

Burns, J.M. 1983. Queen of the Carolinas (Lepidoptera: Nymphalidae: Danainae: Danaus Gilippus), Journal of the Lepidopterists' Society, 85(2):388-396

Burns, J.M. 1984. Evolutionary Differentiation: Differentiating Gold-banded Skippers—Autochton cellus and More (Lepidoptera: Hesperiidae: Pyrginae), Smithsonian Institution Press, Smithsonian Contributions to Zoology, Washington, DC, Number 405

Burns, J.M. 1985. Wallengrenia otho and W. egeremet in Eastern North America (Lepidoptera: Hesperiidae: Hesperiinae), Smithsonian Institution Press, Smithsonian Contributions to Zoology, Washington, DC, Number 423

Burns, J.M. 1990. Amblyscirtes: Problems with species, species groups, the limits of genus, and genus groups beyond—A look at what is wrong with the skipper classification of Evans (Hesperiidae), Journal of the Lepidopterists' Society, 44(1):11-27

Burns, J.M. 1992. Genitalic recasting of poanes and paratrytone (Hesperiidae), Journal of the Lepidopterists' Society, 46(1):1-23

Burns, J.M. 1994. Genitalia at the generic level: Atrytone restricted, anatrytone resurrected, new genus Quasimellana—And yes! We have no mellanas (Hesperiidae), Journal of the Lepidopterists' Society, 48(4):273-337

Burns, J.M. 2000. Pyrgus communis and Pyrgus albescens (Hesperiidae: Pyrginae) are separate transcontinental species with variable but diagnostic valves, Journal of the Lepidopterists' Society, 54:52-71

Burns, R.M. & B.H. Honkala. 1990. Silvics of North America: 1. Conifers; 2. Hardwoods. Agriculture Handbook 654, U.S. Department of Agriculture, Forest Service, Washington, DC

Butler, R.B. & C.V. Covell. 1957. Megathymus yuccae in North Carolina, Lepidoptera News, 11:137-141

Calhoun, J.V. 1993. A 'Heathii' aberration of Mitoura grynea sweadneri (Lycaenidae theclinae), Journal of the Lepidopterists' Society, 47(3):247-248

Calhoun, J.V. 1996. Conquering soldiers: The successful invasion of Florida by Danaus eresimus (Lepidoptera: Nymphalidae), Holarctic Lepidoptera, 3(1):7-18

Calhoun, J.V. 2002. Sibling rivalry in Florida: The displacement of Pyrgus communis by Pyrgus albescens (Hesperiidae), Journal of the Lepidopterists' Society, 56(2):98-103

Calhoun, J.V. 2003. The history and true identity of Melitaea ismeria (Nymphalidae): A remarkable tale of duplication, misinterpretation, and presumption, Journal of the Lepidopterists' Society, 57(3):204-219

Calhoun, J.V., J.R. Slotten & M.H. Salvato. 2000. The rise and fall of tropical blues in Florida: Cyclargus ammon and Cyclargus thomasi bethunebakeri (Lepidoptera: Lycenadae), Holarctic Lepidoptera, 7(1):13-20

Camara, M.D. 1997. A recent host range expansion in Junonia coenis Hubner (Nymphalidae): Oviposition preference, survival, growth and chemical defense, Evolution, 51:873-884

Cappuccino, N. & P. Kareiva. 1985. Coping with a capricious environment: A population study of a rare pierid butterfly, Ecology, 66(1):152-161

Carde, R.T., A.M. Shapiro & H.K. Clench. 1970. Sibling species in the eurydice group of Lethe (Lepidoptera: Satyridae), Psyche, 77:70-103

Carroll, S. 1997. Genetics on the wing: Or how the butterfly got its spots, Findings, Natural History, 28-32

Casagrande, R.A. & J.E. Dacey. 2001. Monarch butterfly (Danaus plexippus) oviposition on black swallowwort (Vincetoxicum nigrum), The Newsletter of the Rhode Island Natural History Survey, 8(1):2-3

Cassie, B. 1994. Initial foray: Bogging down in northern New England, American Butterflies, 2(2):10

Caterino, M.S., S. Cho & F.A.H. Sperling. 2000. The current state of insect molecular systematics: A thriving Tower of Babel, Annual Review of Entomology, 45:1-54

Caterino, M.S., R.D. Reed, M.M. Kuo & F.A.H. Sperling. 2001. A partitioned likelihood analysis of swallowtail butterfly phylogeny (Lepidoptera: Papilionidae), Systematic Biology, 50(1): 106-127

Caterino, M.S. & F.A.H. Sperling. 1999. Papilio phylogeny based on mitochondrial cytochrome oxidase I and II genes, Molecular Physics and Evolution, 11(1):122-137

Catling, P.M. 1997. Notes on the ecology of Phyciodes batesii batesii (Lepidoptera: Nymphalidae), Holarctic Lepidoptera, 4(1):35-36

Catling, P.M. & V.R. Brownell. 1999. Alvars of the Great Lakes basin, in Anderson, R.C., J.S. Fralish & J.M. Baskin, Savannas, Barrens, and Rock Outcrop Plant Communities of North America, Cambridge University Press, Cambridge

Catling, P.M. & J.V. Calhoun. 1997. Genus Megisto in Florida and the taxonomic status of Megisto viola (Lepidoptera: Nymphalidae: Satyrinae), Holarctic Lepidoptera, 4(1):27-33

Catling, P.M. & R.A. Layberry. 1998. Distribution and biology of Chlosyne gorgone carlota (Nymphalidae) at its northeastern limit, Journal of the Lepidopterists' Society, 52(1):98-104

Cech, R.B. 1993. A Distributional Checklist of the Butterflies and Skippers of the New York City Area (50-mile Radius), New York City Butterfly Club

Cech, R.B. 1995. Wheeling through West Virginia: Definitely not in Brooklyn anymore, American Butterflies, 3(2):12

Cech, R.B. 2002. The care and feeding of Dina yellows, American Butterflies, 10(1):24-26

Chamblin, H.D. 2002. Small Mammal Communities on a Reclaimed Mountaintop Mine/Valley Fill Landscape in Southern West Virginia, master's thesis, West Virginia University, http://etd.wvu.edu/templates/showETD.cfm?recnum=2480

Chapman, W.K., V.A. Chapman, A.E. Bessette, A. Rainis Bessette & D.R. Pens. 1998. Wildflowers of New York in Color, Syracuse University Press, Syracuse, NY

Chermock, F.H. 1947. Appalachian brown, Entomological News, 58:33

Chermock, R.L. & O.D. Chermock. 1947. Notes on the life histories of three Floridian butterflies, Canadian Entomologist, 79: 142-144

Chew, F.S. 1975. Coevolution of pierid butterflies and their cruciferous foodplants I. The relative quality of available resources, Oecologia, 20:117-128

Chew, F.S. & R.K. Robbins. 1985. Egg-laying in butterflies, in Vane-Wright, R.I. & P.R. Ackery, The Biology of Butterflies: Symposium of the Royal Entomological Society of London

Number 11, Academic Press, Harcourt Brace Jovanovich, London, 2nd ed.

Clarke, C.A. & P.M. Sheppard. 1962. The genetics of the mimetic butterfly Papilio glaucus, Ecology, 43:159-161

Clench, H.K. 1972. Celastrina ebenina, a new species of Lycaenidae from the eastern United States, Carnegie Museum Annuals, 44:33-34

Clewis, S.B. 1999. Cogongrass (Imperata cylindrica), in WSSNC, Weed Science in a Changing Environment: Proceedings of the WSSNC Seventeenth Annual Meeting, Weed Science Society of North Carolina, http://www.wssnc.ncsu.edu/pro1999.htm

Codella, S.G., Jr. 1986. Intersexual Comparison of Mimetic Protection of Black Swallowtail Butterfly (Papilio polyxenes asterius Stoll), master's thesis, Rutgers University, Newark, NJ

Colby, D.M. 2002. Effects of Fire Frequency and the Red Imported Fire Ant on Insects in a Louisiana Longleaf Pine Savanna, doctoral dissertation, Louisiana State University

Connecticut Butterfly Association. Not dated. Connecticut Butterfly Atlas Project, http://george.peabody.yale.edu/cbap/

Connors, R. 2002. Kinky Diana, American Butterflies, 8(3):45

Cordero, C. 2000. Is spermatophore number a good measure of mating frequency in female Callophrys xami (Lycaenidae)?, Journal of the Lepidopterists' Society, 53(4):169-170

Courant, A.V. & F.S. Chew. 1995. Latent polyphenism and direct development in Pieris virginiensis (Pieridae), Journal of the Lepidopterists' Society, 49(1):84-87

Courant, A.V., A.E. Holbrook, E.D. Van der Reijden & F.S. Chew. 1994. Native pierine butterfly (Pieridae) adapting to naturalized crucifer?, Journal of the Lepidopterists' Society, 48(2):168-169

Courtney, S.P. 1984. The evolution of egg clustering by butterflies and other insects, The American Naturalist, 123(2):276-281

Courtney, S.P. 1985. Habitat versus foodplant selection, in Vane-Wright, R.I. & P.R. Ackery, The Biology of Butterflies: Symposium of the Royal Entomological Society of London Number 11, Academic Press, Harcourt Brace Jovanovich, London, 2nd ed.

Courtney, S.P. 1988. Oviposition on peripheral hosts by dispersing Pieris napi (L.) (Pieridae), Journal of Research on the Lepidoptera, 26(1-4):58-63

Cromartie, W.J., Jr. 1975. Influence of habitat on colonization of collard plants by Pieris rapae, http://www.eap.mcgill.ca/CPCM_4.htm

Cromartie, W.J. & D.F. Schweitzer. 1993. Biology of the rare skipper, Problema bulenta (Hesperiidae) in southern New Jersey, Journal of the Lepidopterists' Society, 47(2):125-133

Crozier, L. 2002. Climate change and its effects on species range boundaries: A case study of the sachem skipper butterfly, Atalopedes campestris, in Schneider, S.H. & T.L. Root, Wildlife Responses to Climate Change: North American Case Studies, Island Press, Washington, DC, pp. 57-91

Culbert, D.F. 1995. Florida Coonties and Atala Butterflies, Florida Cooperative Extension Service, University of Florida / Institute of Food and Agricultural Sciences, Gainesville

Cunningham, J.P. & S.A. West. 2001. Host selection in phytophagous insects: A new explanation for learning in adults, Oikos, 95(3):537-543

Damman, H. 1986. The osmeterial glands of the swallowtail butterfly Eurytides marcellus as a defense against natural enemies, Ecological Entomology, 11:261-265

Damman, H. & P. Feeny. 1988. Mechanisms and consequences of selective oviposition by the zebra swallowtail butterfly, Animal Behavior, 36(2):563-574

Daniels, J.C. 2000. Your Florida Guide to Butterfly Gardening: A Guide for the Deep South, University of Florida Press, Gainesville

Davies, N. & E. Bermingham. 2002. The historical biogeography of two Caribbean butterflies (Lepidoptera: Heliconiidae) as inferred from genetic variation at multiple loci, Evolution, 56(3):573-589

Davies, N.B. 1978. Territorial defence in the speckled wood butterfly (Pararge aegeria): The resident always wins, Animal Behavior, 26:138-147

Davis, R.H. & A. Nahrstedt. 1987. Biosynthesis of cyanogenic glycosides in butterflies and moths, Insect Biochemistry, 17:689-693

Davis, W.T. 1909. List of the Macrolepidoptera of Staten Island, New York, Proceedings of the Staten Island Association of Arts and Sciences, 3:1-30

Dempster, J.P. 1985. The natural enemies of butterflies, in Vane-Wright, R.I. & P.R. Ackery, The Biology of Butterflies: Symposium of the Royal Entomological Society of London Number 11, Academic Press, Harcourt Brace Jovanovich, London, 2nd ed.

Dethier, V.G. & R.H. MacArthur. 1964. A field's capacity to support a butterfly population, Nature, 201:728-729

DeVries, P.J. 1987. The Butterflies of Costa Rica and Their Natural History: Papilionidae, Pieridae, Nymphalidae, Princeton University Press, Princeton, NJ

DeVries, P.J. 1997. The Butterflies of Costa Rica and Their Natural History: Volume II, Riodinidae, Princeton University Press, Princeton, NJ

Deyrup, M.A. 1997. Learning the language of insects and flowers, Wings, Xerces Society, 20:12-15

Dietrich, S. 2001. Cycads—Palmferns—and their Insects (Herbivores and Pollinators), European Symposium for Insect Taste and Olfaction, http://www.esito-symp.org/talks.html

Dirig, R. 1997. Karner blue sing your purple song, American Butterflies, 5(1):14

Dirig, R. & J.F. Cryan. 1991. The status of silvery blue subspecies (Glaucopsyche lygdamus lygdamus and G.l. couperi: Lycaenidae) in New York, Journal of the Lepidopterists' Society, 45: 272-290

Dobb, E. 1998. Reality Check: The debate behind the lens, Audubon, National Audubon Society, New York, 45

Dombrowski, D.S. & R.R. Mills. 1996. Fennel (Foeniculum vulgare), primary host plant for the eastern black swallowtail (Papilio polyxenes asterius) (Papilionidae) on the eastern shore of Virginia, Journal of the Lepidopterists' Society, 50(1):87-89

Dos Passos, C.F. & A.B. Klots. 1969. The systematics of Anthocharis mida Hubner (Lepidoptera: Pieridae), Entomologica America, 45:34

Douglas, M.M. 1989. The Lives of Butterflies, University of Michigan, Ann Arbor, MI

Douglas, M.M. & J.W. Grula. 1978. Thermoregulatory adaptations allowing ecological range expansion by the pierid butterfly, Nathalis iole Boisduval, Evolution, 12(4):776-783

Douglas, M.S. 1997. The Everglades: River of Grass, Pineapple Press, Inc., Sarasota, FL, 50th anniv. ed.

Duncan, W.H. & M.B. Duncan. 1988. Trees of the Southeastern United States, University of Georgia Press, Athens

Duncan, W.H. & M.B. Duncan. 1999. Wildflowers of the Eastern United States, University of Georgia Press, Athens

Dyck, H. Van. 2003. Mate location: A matter of design? Adaptive morphological variations in the speckled wood butterfly, in

Boggs, C.L., W.B. Watt & P.R. Ehrlich, Butterflies: Ecology and Evolution Taking Flight, University of Chicago Press, Chicago, pp. 353-366

Dyck, H. Van & E. Matthysen. 1998. Thermoregulatory differences between phenotypes in the speckled wood butterfly: Hot perchers and cold patrollers?, Oecologia, 114:326-334

Dyer, L.A. 1997. Effectiveness of caterpillar defenses against three species of invertebrate predators, Journal of Research on the Lepidoptera, 34:46-48

Dyer, L.A. & M.D. Bowers. 1996. The importance of sequestered iridoid glycosides as a defense against an ant predator, Journal of Chemical Ecology, 22:1527-1539

Edgar, J.A. 1985. Parsonsieae: Ancestral larval foodplants of the Danainae and Ithomiinae, in Vane-Wright, R.I. & P.R. Ackery, The Biology of Butterflies: Symposium of the Royal Entomological Society of London Number 11, Academic Press, Harcourt Brace Jovanovich, London, 2nd ed.

Edwards, W.H. 1868. The Butterflies of North America, vol. 1, American Entomological Society, Philadelphia.

Edwards, W.H. 1884. The Butterflies of North America, vol. 2, Houghton Mifflin, Boston.

Ehrlich, P.R. 1985. The structure and dynamics of butterfly populations, in Vane-Wright, R.I. & P.R. Ackery, The Biology of Butterflies: Symposium of the Royal Entomological Society of London Number 11, Academic Press, Harcourt Brace Jovanovich, London, 2nd ed.

Ehrlich, P.R. & P.H. Raven. 1964. Butterflies and plants: A study in coevolution, Evolution, 18:586-608

Eliot, J.N. & A.K. Kawazoe. 1983. Blue butterflies of the Lycaenopsis group, British Museum (Natural History), London

Ellers, J. & C.L. Boggs. 2002. The evolution of wing color in Colias butterflies: Heritability, sex linkage, and population divergence, Evolution, 56:836-840

Ellison, A.M. & E.J. Farnsworth. 1996. Spatial and temporal variability in growth of Rhizophora mangle saplings on coral cays: Links with variation in insolation, herbivory, and local sedimentation rate, Journal of Ecology, 84:717-731

Emmel, T.C. 1973. Dispersal in a cosmopolitan butterfly species (Pieris rapae) having an open population structure, Journal of Research on the Lepidoptera, 11(2):95-98

Emmel, T.C. & P.J. Eliazar. 2002. Pesticide problems in Paradise, American Butterflies, 11(1):28-35

Emmel, T.C., M.C. Minno & B.A. Drummond. 1992. Florissant Butterflies: A Guide to the Fossil and Present-Day Species of Central Colorado, Stanford University Press, Stanford, CA

Emmel, T.C. & J.C. Tucker. 1991. Mosquito Control Pesticides: Ecological Impacts and Management Alternatives, Scientific Publishers, Inc., Mariposa Press, Gainesville

Encyclopaedia Britannica. 2002. Ultimate Reference Suite DVD, Encyclopaedia Britannica, Inc.

Engler, H.S., K.C. Spencer & L.E. Gilbert. 2000. Insect metabolism: Preventing cyanide release from leaves, Nature, 406:144-145

Etheredge, J.A., S.M. Perez , O.R. Taylor & R. Jander. 1999. Monarch butterflies (Danaus plexippus L.) use a magnetic compass for navigation, Proceedings of the National Academy of Sciences, 96:13845-13846

Ewel, K.C. & H.T. Odum (eds.). 1984. Cypress Swamps, University Presses of Florida, University of Florida Press, Gainesville

Fasulo, T.R. (coord.). 1996. Featured Creatures, University of Florida, Department of Entomology and Nematology, http://creatures.ifas.ufl.edu/

Fernandez, D.M. 2001. New oviposition and larval hostplant

records for twenty-three Cuban butterflies, with observations on the biology and distribution of some species, Caribbean Journal of Science, 37(1-2):122-125

Ferris, C.D. (ed.). 1989. Supplement to: A catalogue/checklist of the butterflies of America north of Mexico, The Lepidopterists' Society, Memoir No. 3

Ferriter, Amy (ed.). 1997. Brazilian Pepper Management Plan for Florida, Florida Exotic Pest Plant Council

Florida Fish and Wildlife Conservation Commission. 1999. Land Cover of Florida, http://www.floridaconservation.org/viewing/landcovers/index.htm

Folsom, W. 2000. Art and Science of Butterfly Photography, Amherst Media, Inc., Buffalo, NY

Food and Agriculture Organization (FAO) of the United Nations. 2002. Proceedings: Second Expert Meeting on Harmonizing Forest-related Definitions for Use by Various Stakeholders, http://www.fao.org/DOCREP/005/Y4171E/y4171e00.htm#TopOfPage

Foote, L.E. & S.B. Jones, Jr. 1989. Native Shrubs and Woody Vines of the Southeast: Landscaping Uses and Identification, Timber Press, Portland, OR

Ford, E.B. 1977. The New Naturalist Butterflies, Bloombury Books, William Collins Son's & Co. Ltd, London, 4th ed.

Fordyce, J.A. 2003. Aggregative feeding of pipevine swallowtail larvae enhances hostplant suitability, Oecologia, 135(2):250-257

Fordyce, J.A., C.C. Nice, M.L. Forister & A.M. Shapiro. 2002. The significance of wing pattern diversity in the Lycaenidae: Mate discrimination by two recently diverged species, Journal of Evolutionary Biology, 15:871-879

Forys, E.A., A. Quisrorff & C.R. Allen. 2001. Potential fire ant (Hymenoptera: Formicidae) impact on the endangered Schaus' swallowtail (Lepidoptera: Papilionidae), Florida Entomologist, 84(2):254-258

Fraser, A.M., A.H. Axén, & N.E. Pierce. 2001. Assessing the quality of different ant species as partners of a myrmecophilous butterfly, Oecologia, 129:452-460

Freitas, A.V.L. & P.S. Oliveira. 1992. Biology and behavior of the neotropical butterfly Eunica bechina (Nymphalidae) with special reference to larval defence against ant predation, Journal of Research on the Lepidoptera, 31(1-2):1-11

Friedlander, T.P. 1987. Taxonomy, phylogeny and biogeography of Asterocampa Rober (Lepidoptera, Nymphalidae, Apaturinae), Journal of Research on the Lepidoptera, 25(4):215-338

Fullard, J.H. & N. Napoleone. 2001. Diel flight periodicity and the evolution of auditory defences in the Macrolepidoptera, Animal Behavior, 62:349-368

Futuyma, D.J. & G. Moreno. 1988. The evolution of ecological specialization, Annual Review of Ecology and Systematics, 19:207-233

Gannon, A.T. 1986. Oviposition site selection by the gulf fritillary Agraulis vanillae nigrior (Michener), master of science thesis, University of Florida, Gainesville

Gatrelle, R.R. 1971. Notes on the confusion between Lethe creola and Lethe portlandia (Satyridae), Journal of the Lepidopterists' Society, 25:145-146

Gatrelle, R.R. 1998a. An addendum to Anthocharis midea Dos Passos and Klots 1969. (Description of a New Subspecies from Texas), The Taxonomic Report, 1(1):1-5

Gatrelle, R.R. 1998b. Subspecific status of southeastern U.S. Megathymus Cofaqui and M. yuccae: Renaming of the Florida subspecies of M. Cofaqui, The Taxonomic Report, The International Lepidoptera Survey, Goose Creek, SC, 1(2)

Gatrelle, R.R. 1998c. The rediscovery, taxonomy, and biology of Chlosyne gorgone gorgone and Chlosyne ismeria (Nymphalidae) in Burke County, Georgia, The Taxonomic Report, The International Lepidoptera Survey, Goose Creek, SC, 1(2)

Gatrelle, R.R. 1998d. Two new Nymphalidae from western North Carolina: New subspecies of Speyeria aphrodite and Phyciodes batesii, The Taxonomic Report, The International Lepidoptera Survey, Goose Creek, SC, 1(2)

Gatrelle, R.R. 1999a. A comment on Friedlander's Asterocampa (Nymphalidae, Apaturinae): Designation of neotypes for A. celtis and A. clyton, The Taxonomic Report, The International Lepidoptera Survey, Goose Creek, SC, 1(5)

Gatrelle, R.R. 1999b. An evolutionary subspecific assessment of Deciduphagus henrici (Lycaenidae) based on its utilization of Ilex and non-Ilex hosts, The Taxonomic Report, The International Lepidoptera Survey, Goose Creek, SC, 1(6)

Gatrelle, R.R. 1999c. Hubner's Helicta: The forgotten Neonympha, The Taxonomic Report, The International Lepidoptera Survey, Goose Creek, SC, 1:(8)

Gatrelle, R.R. 1999d. Three new Hesperioidea (Hesperiinae) from South Carolina: New subspecies of Euphyes bimacula, poanes, aaroni, and Hesperia attalus, The Taxonomic Report, The International Lepidoptera Survey, Goose Creek, SC, 1(10)

Gatrelle, R.R. 2000. A new North American swallowtail butterfly: Description of a relict subspecies of Pterourus troilus (Papilionidae) from the southern tip of Florida, The Taxonomic Report, The International Lepidoptera Survey, Goose Creek, SC, 2(4)

Gatrelle, R.R. 2001a. An examination of southeastern U.S. Satyrium (Lycaenidae: Theclinae). Part one: an obscure new subspecies of Satyrium edwardsii, The Taxonomic Report, The International Lepidoptera Survey, Goose Creek, SC, 3(2):1-9.

Gatrelle, R.R. 2001b. An examination of southeastern U.S. Satyrium (Lycaenidae: Theclinae). Part two: The identification and delimitation of nominate Satyrium liparops and the description of a new subspecies from west central peninsular Florida, The Taxonomic Report, 3(3):1-8

Gatrelle, R.R. 2001c. An examination of the Mitoura (Lycaenidae) in the southeastern United States: With the description of a new subspecies of Mitoura hesseli, The Taxonomic Report, 3(4):1-10

Gatrelle, R.R. 2001d. Colias ID'ing, unpublished, http://www.carolinabutterflysociety.org/Articles/colias01.htm

Gatrelle, R.R. 2001e. Initial survey of the butterflies and skippers in the vicinity of the Buck Creek Serpentine Barrens, Clay County, North Carolina, The Taxonomic Report, The International Lepidoptera Survey, Goose Creek, SC, 3(5):1-6

Gatrelle, R.R. 2001f. Thorybes clarification, Newsletter, The International Lepidoptera Survey, Goose Creek, SC, 3(1):1-2

Gatrelle, R.R. & R.T. Arbogast. 1974. A new subspecies of Lethe appalachia (Satyridae), Journal of the Lepidopterists' Society, 28(4):359-363

Geiger, H.J. 1988. Enzyme electrophoreses and interspecific hybridization in Pieridae (Lepidoptera)—The case for enzyme electrophoresis, Journal of Research on the Lepidoptera, 26(1-4):64-72

Geiger, H.J. & A.M. Shapiro. 1992. Genetics, systematics and evolution of holarctic Pieris napi species-group populations (Lepidoptera: Pieridae), Zeitschrift fuer Zoologische Systematik und Evolutionsforschung, 30:100-122

Gerberg, E.J. & R.H. Arnett, Jr. 1989. Florida Butterflies, The Sandhill Crane Press, Inc., Gainesville

Gibo, D.L. & J.M. McCurdy. 1993. Lipid accumulation by migrating monarch butterflies (Danaus plexippus L.), Canadian Journal of Zoology, 71:76-82

Gibson, D.J., R.A. Zampella & A.G. Windisch. 1999. New Jersey pine plains: The "true barrens" of the New Jersey pine barrens, in Anderson, R.C., J.S. Fralish & J.M. Baskin, Savannas, Barrens, and Rock Outcrop Plant Communities of North America, Cambridge University Press, Cambridge

Gilbert, L.E. 1972. Pollen feeding and reproductive biology of Heliconius butterflies, Proceedings of the National Academy of Sciences, New York, 69(6):1403-1407

Gilbert, L.E. 1982. The coevolution of a butterfly and a vine, Scientific American, 247:110-121

Gilbert, L.E. 1985. The biology of butterfly communities, in Vane-Wright, R.I. & P.R. Ackery, The Biology of Butterflies: Symposium of the Royal Entomological Society of London Number 11, Academic Press, Harcourt Brace Jovanovich, London, 2nd ed.

Gilbert, L.E. 2003. Adaptive novelty through introgression in Heliconius wing patterns: Evidence for shared genetic "toolbox" from synthetic hybrid zones and a theory of diversification, in Boggs, C.L., W.B. Watt & P.R. Ehrlich, Carol L., Butterflies: Ecology and Evolution Taking Flight, University of Chicago, Chicago

Gilchrist, G.W. 1990. The consequences of sexual dimorphism in body size for butterfly flight and thermoregulation, Functional Ecology, 4:475-487

Gimbutas, M. 1989. The Language of the Goddess, Harper & Row, New York

Glassberg, J. 1993a. Butterflies Through Binoculars: A Field Guide to Butterflies in the Boston New York Washington Region, Oxford University Press, Oxford

Glassberg, J. 1993b. First foray: The Great Dismal Swamp of Virginia: How many great purple hairstreaks can dance on the head of a Hercules Club?, American Butterflies, 1(1):10

Glassberg, J. 1999. Butterflies Through Binoculars: The East, Oxford University Press, Oxford

Glassberg, J. 2000. Eyeing the greater fritillaries, American Butterflies, 8(3):14-17

Glassberg, J. 2003. Miami blue story, American Butterflies, 11(1):16-20

Glassberg, J., M.C. Minno & J.V. Calhoun. 2000. Butterflies Through Binoculars: Florida, Oxford University Press, Oxford

Gleason, H.A. & A. Cronquist. 1991. Manual of Vascular Plants of Northeastern United States and Canada, New York Botanical Garden, New York

Gochfeld, M. & J. Burger. 1997. Butterflies of New Jersey: A Guide to Their Status, Distribution, Conservation and Appreciation, Rutgers University Press, New Brunswick, NJ

Gonzalez, A., C. Rossini, M. Eisner & T. Eisner. 1999. Sexually transmitted chemical defense in a moth (Utetheisa ornatrix), Proceedings of the National Academy of Sciences, 96:5570-5574

Goulson, D., J. Ollerton & C. Sluman. 1997. Foraging strategies in the small skipper butterfly, Thymelicus flavus: When to switch?, Animal Behavior, 53:1009-1016

Grossmueller, D.W. & R.C. Lederhouse. 1987. The role of nectar source distribution in habitat use and oviposition by the tiger swallowtail butterfly, Journal of the Lepidopterists' Society, 41:159-165

Gupton, O.W. & F.C. Swope. 1979. Wildflowers of the Shenandoah Valley and Blue Ridge Mountains, University Press of Virginia, Charlottesville

Gupton, O.W. & F.C. Swope. 1981. Trees and Shrubs of Virginia, University of Virginia Press, Charlottesville

Haddad, N.M. 2000. Corridor length and patch colonization by a butterfly: Junonia coenia, Conservation Biology, 14(3):738-745

Haddad, N.M. & K.A. Baum. 1999. An experimental test of corridor effects on butterfly densities, Ecological Applications, 9(2):623-633

Haddad, N.M. & W.M. Hicks. 2000. Host pubescence and the behavior and performance of the butterfly Papilio troilus (Lepidoptera: Papilionidae), Environmental Entomology, 29:299-303

Hagen, R.H., R.C. Lederhouse, J.L. Bossart & J.M. Scriber. 1991. Papilio canadensis and P. glaucus (Papilionidae) are distinct species, Journal of the Lepidopterists' Society, 45:245-258

Hall, D. & J. Butler. 2000. Atala, Featured Creatures, University of Florida, Gainesville

Hall, S.P. 1996. Butterflies of Xanadu, American Butterflies, 4(2):4

Halpern, S. 2001. Four Wings and a Prayer: Caught in the Mystery of the Monarch Butterfly, Knopf, New York

Hammer, R.L. 2002. Everglades Wildflowers: A Field Guide to Wildflowers of the Historic Everglades, Including Big Cypress, Corkscrew, and Fakahatchee Swamps, Globe Pequot Press, Guilford, CT

Hammer, R.L. 2004. Florida Keys Wildflowers: A Field Guide to Wildflowers, Trees, Shrubs, and Woody Vines of the Florida Keys, Globe Pequot Press, Guilford, CT

Hammond, P.C. 1990. Patterns of geographic variation and evolution in polytypic butterflies, Journal of Research on the Lepidoptera, 29(1-2):54-76

Hancock, D.L. 1983. Classification of the Papilionidae (Lepidoptera): A phylogenetic approach, Smithersia, 2:1-48

Hanski, I. 1999. Metapopulation Ecology, Oxford University Press, Oxford

Hanson, B., G.F. Garifullina, S.D. Lindblom, A. Wangeline, A. Ackley, K. Kramer, A.P. Norton, C.B. Lawrence & E.A.H. Pilon-Smits. 2003. Selenium accumulation protects Brassica juncea from invertebrate herbivory and fungal infection, New Phytologist, 159:461-469

Haribal, M., P. Feeny & C.C. Lester. 1998. A caffeoylcyclohexane-1-carboxylic acid derivative from Asimina triloba, Phytochemistry, 49(1):103-108

Haribal, M. & J.A.A. Renwick. 1998. Isovitexin 6"-O-b-D-glucopyranoside: A feeding deterrent to Pieris napi oleracea from Alliaria petiolata, Phytochemistry, 47:1237-1240

Harrar, E.S. & J.G. Harrar. 1962. Guide to Southern Trees, Dover Publications, Inc., New York, 2nd ed.

Harris, L., Jr. 1972. Butterflies of Georgia, University of Oklahoma, Norman

Harris, S.K., J.H. Langenheim, F.L. Steele & M. Underhill. 1964. AMC Guide to Mountain Flowers of New England, Appalachian Mountain Club, Boston

Hauptfleisch, J.O. & R.B. Kinnel. 2001. Enantiomeric Composition of Germacrene D in Asters, unpublished, http://www.chem.hamilton.edu/research/research_kinnel2001.html

Hazel, W.N. & D.A. West. 1996. Pupation site preference and environmentally cued pupal colour dimorphism in the swallowtail butterfly Papilio polyxenes Fabr. (Lepidoptera: Papilionidae), Biological Journal of the Linnaean Society, 57:81-87

Hebert, P.D.N., E. Penton, J.M. Burns, D.H. Janzen & W. Hallwachs. 2004. Ten species in one: DNA barcoding reveals cryptic species in the neotropical skipper butterfly Astraptes fulgerator, Proceedings of the National Academy of Sciences USA, 10.1073/pnas.0406166101

Heinrich, B. 1986. Thermoregulation and flight activity of a satyrine, Cenonympha inornata (Lepidoptera: Satyridae), Ecology, 67(3):593-597

Heitzman, J.R. & C.F. dos Passos. 1974. Lethe portlandia (Fabricus) and L. anthedon (Clark), sibling species, with descriptions of new subspecies of the former (Lepidoptera: Satyridae), Transactions of the American Entomological Society, 100:52-99

Hereau, H. & M. Scriber. 2003. Interspecific copulation of a dark morph Papilio glaucus female and a male P. Polyxenes (Papilionidae): Observations and significance, Journal of the Lepidopterists' Society, 37(1):71-74

Hicks, P. 1997. Photographing Butterflies and Other Insects, Fountain Press, Kingston-on-Thames

Higgins, L.G. 1981. A revision of Phyciodes hübner and related genera, with a review of the classification of the Melitaeinae (Lepidoptera: Nymphalidae), Bulletin of the British Museum of Natural History, 43:77-243

Hill, J.K., C.D. Thomas & O.T. Lewis. 1999. Flight morphology in fragmented populations of a rare British butterfly, Hesperia comma, Biological Conservation, 87:277-283

Hirota, T. & Y. Obara. 2000. The allocation of the reproductive and feeding behaviors in the male cabbage white butterfly, Zoological Science, 17:323-327

Hoffman, R.J.H. 1973. Environmental control of seasonal variation in the butterfly Colias eurytheme. I. Adaptive aspects of a photoperiodic response, Evolution, 27(3):387-397

Holland, W.J. 1899. The Butterfly Book: A Popular Guide to a Knowledge of Butterflies of North America, Doubleday & McClure Co., New York

Holland, W.J. 1931. The Butterfly Book, Doubleday & Company, Garden City, NY, 1931 ed.

Holmgren, N.H. 1998. Illustrated Companion to Gleason and Cronquist's Manual, New York Botanical Garden, New York

Hovanitz, W. 1963. The relation of Pieris virgiensis Edw. to Pieris napi L.: Species formation in Pieris?, Journal of Research on the Lepidoptera, 1(2):124-134

Hovanitz, W. & V.C.S. Chang. 1963. Change of food plant preference by larvae of Pieris rapae controlled by strain selection, and the inheritance of this trait, Journal of Research on the Lepidoptera, 1(2):163-168

Howe, W.H. 1975. The Butterflies of North America, Doubleday & Company, Inc., Garden City, NY

Huxley, J.S. 1942. Evolution: The Modern Synthesis, Hafner Press, New York, 3rd ed.

Iftner, D.C., J.A. Shuey & J.V. Calhoun. 1992. Butterflies and Skippers of Ohio, College of Biological Sciences, Ohio State University, Columbus

Janz, N. 1999. Ecology and Evolution of Butterfly Host Plant Range, doctoral dissertation, http://www.zoologi.su.se/research/ihp/abstracts/JanzThesis_sum.html

Janz, N. 2003. Sex linkage of host plant use in butterflies, in Boggs, C.L., W.B. Watt & P.R. Ehrlich, Butterflies: Ecology and Evolution Taking Flight, University of Chicago Press, Chicago, pp. 229-240

Janz, N., K. Nyblom & S. Nylin. 2001. Evolutionary dynamics of host-plant specialization: A case study of the tribe Nymphalini, Evolution, The Society for the Study of Evolution, 55(4):783-796

Jewell, S.D. 1997. Exploring Wild South Florida: A Guide to Finding the Natural Areas and Wildlife of the Southern Peninsula and the Florida Keys, Pineapple Press, Inc., Sarasota, FL, 2nd ed.

Jhee, E.M., K.L. Dandridge, A.M. Christy, Jr. & A.J. Pollard. 1999. Selective herbivory on low-zinc phenotypes of the hyperaccumulator Thlaspi caerulescens (Brassiceae), Chemoecology, 9:93-95

Johnson, A.F. & M.G. Barbour. 1999. Dunes and maritime forests, in Anderson, R.C., J.S. Fralish & J.M. Baskin, Savannas, Barrens, and Rock Outcrop Plant Communities of North America, Cambridge University Press, Cambridge

Johnson, C.W. 1985. Bogs of the Northeast, University Press of New England, Hanover

Johnson, K. & S. Coates. 1999. Nabakov's Blues: The Scientific Odyssey of a Literary Genius, Zoland, Cambridge, MA

Jones, M.T., I. Castellanos & M.R. Weiss. 2002. Do leaf shelters always protect caterpillars from invertebrate predators?, Ecological Entomology, 27(6):753

Joron, M. 2003. Mimicry, in Carde, R.T. & V.H. Resh, Encyclopedia of Insects, Academic Press, New York

Kapan, D.D. 2001. Three-butterfly system provides a field test of mullerian mimicry, Nature, Letters to Nature, 409:338-340

Kawahara, A.Y. 2003. Behavioral observations of Libytheana carinenta Cramer, News of The Lepidopterists' Society, 45(4):107

Kelly, L. & D.M. Debinski. 1998. Relationship of hostplant density to size and abundance of the regal fritillary Speyeria idalia drury (Nymphalidae), Journal of the Lepidopterists' Society, 52(3):262-276

Kiel, W.J. 2003. The Butterflies of the White Mountains of New Hampshire, Globe Pequot Press, Guilford, CT

Kilduff, T.S. 1972. A population study of Euptychia hermes in northern Florida, Journal of Research on the Lepidoptera, 11:219-228

Kimball, C.P. 1965. Lepidoptera of Florida, State of Florida Department of Agriculture, Division of Plant Industry, Gainesville

Kingsolver, J.G. 1983. Thermoregulation and flight in Colias butterflies: Elevational patterns and mechanistic limitations, Ecology, 64(3):534-545

Kingsolver, J.G. 1989. Weather and the population dynamics of insects: Integrating physiological and population ecology, Physiological Zoology, 62:314-334

Kingsolver, J.G. 1995. Fitness consequences of seasonal polyphenism in western white butterflies, Evolution, 49(5):942-954

Kingsolver, J.G. & M.A.R. Koehl. 1994. Selective factors in the evolution of insect wings, Annual Review of Entomology, 39:425-451

Kingsolver, J.G. & W.B. Watt. 1983. Thermoregulatory strategies in Colias butterflies: Thermal stress and the limits to adaptation in temporally varying environments, The American Naturalist, 121(1):32-55

Kingsolver, J.G. & D.C. Wiernasz. 1991. Seasonal polyphenism in wing-melanin pattern and thermoregulatory adaptation in Pieris butterflies, American Naturalist, 137:816-830

Kirby, J.T. 1995. Poquosin: A Study of Rural Landscape and Society, University of North Carolina Press, Chapel Hill

Klass, C. & R. Dirig. 1992. Learning About Butterflies, Cornell Cooperative Extension

Klots, A.B. 1951. A Field Guide to the Butterflies of Eastern North America, Houghton Mifflin, Boston

Klots, A.B. & H.K. Clench. 1952. A new species of Strymon Huebner from Georgia (Lepidoptera: Lycaenidae), American Museum Novitiates, 1600:1-19

Koch, P.B., D.N. Keys, T. Rocheleau, K. Aronstein, M. Blackburn, S.B. Carroll & R.H. ffrench-Constant. 1998. Regulation of dopa decarboxylase expression during colour pattern formation in wild-type and melanic tiger swallowtail butterflies, Development, 125:2303-2313

Konvicka, M., O. Nedved & Z. Fric. 2002. Early-spring floods decrease the survival of hibernating larvae of wetland-inhabiting population of Neptis rivularis (Lepidoptera: Nymphalidae), Acta Zoologica Academiae Scientiarum Hungaricae, 48(2):79-88

Kopper, B.J., D.C. Margolis & R.E. Charleton. 2001b. Notes on the behavior of Speyeria idalia (Drury) (Nymphalidae) larvae with implications that they are diurnal foragers, Journal of the Lepidopterists' Society, 54(3):96-97

Kopper, B.J., S. Shu, R.E. Charlton & S.B. Ramaswamy. 2001a. Evidence for reproductive diapause in the fritillary Speyeria idalia (Lepidoptera: Nymphalidae), Annals of the Entomological Society of America, 94(3):427-432

Krebs, R.A. 1988. The mating behavior of Papilio glaucus (Papilionidae), Journal of Research on the Lepidoptera, 26(1-4):27-31

Kricher, J.C. 1988. A Field Guide to Ecology of Eastern Forests, North America, The Peterson Field Guide Series, Houghton Mifflin, Boston

Krizek, G.O. & P.A. Opler. 1986. Observations on Problema bulenta, Journal of Research on the Lepidoptera, 25(2):146-148

Kronforst, M.R. & T.H. Fleming. 2001. Lack of genetic differentiation among widely spaced subpopulations of a butterfly with home range behavior, Heredity, 86:243-250

Kroutov, V., R.L. Reep & T. Fukuda. 2002. Experience-related changes in the brain of Agraulis vanillae (L.) (Nymphalidae), Journal of the Lepidopterists' Society, 56(4):193-198

Kukal, O., M.P. Ayres & J.M. Scriber. 1991. Cold tolerance of pupae in relation to the distribution of swallowtail butterflies, Canadian Journal of Zoology, 69:3028-3037

Kyhl, J.F. & J. Hahn. 2002. Yard and garden brief: Woolly aphids on trees and shrubs, University of Minnesota Extension Service, http://www.extension.umn.edu/projects/yardandgarden/ygbriefs/e453woollyaphid.html

Langan, A.M., C.P. Wheater & P.J. Dunleavy. 2001. Does the small white butterfly (Pieris rapae L.) aggregate eggs on plants with greater gas exchange activity?, Journal of Insect Behavior, 14(4):459-468

Lawrence, D.A. & J.C. Downey. 1966. Morphology of the immature stages of Eeres comyntas Godart (Lycaenidae), Journal of Research on the Lepidoptera, 5(2):61-96

Layberry, R.A., P.F. Hall & J.D. Lafontaine. 1998. The Butterflies of Canada, University of Toronto Press, Toronto

Leather, S.R., K.F.A. Walters & J.S. Bale. 1993. The Ecology of Insect Overwintering, Cambridge University Press, Cambridge

Lederhouse, R.C. 1982. Territorial defense and lek behavior of the black swallowtail butterfly, Papilio polyxenes, Behavioral Ecology and Sociobiology, 10:109-118

Lederhouse, R.C. 1993. Territoriality along flyways as mate locating behavior in male Limenitis arthemis (Nymphalidae), Journal of the Lepidopterists' Society, 47:22-31

Lederhouse, R.C., M.P. Ayres & J.M. Scriber. 1990. Adult diet affects male virility in Papilio glaucus, Functional Ecology, 4:743-751

LeGrand, H.E., Jr. 1995. Letter to Editor, in Cech, R. ed., The Anglewing, North American Butterfly Association, 2(2):3

LeGrand, H.E., Jr. 2000. Definitive destination: The North Carolina sandhills, American Butterflies, 8(2):4

LeGrand, H.E. Jr. & T.E. Howard. 2002. Notes on the Butterflies of North Carolina, http://www.ncsparks.net/butterfly/nbnc.html

Levins, R. 1968. Evolution in Changing Environments: Some Theoretical Explorations, Princeton University Press, Princeton, NJ

Li, W., M.A. Schuler & M.B. Berenbaum. 2003. Diversification of furanocoumarin-metabolizing cytochrome P450 monooxygenases in two papilionids: Specificity and substrate encounter rate, in National Academy of Sciences, Chemical Communication in a Post-Genomic World, http://www.nap.edu/openbook/030909089X/html/81.html#pagetop

Lind, E.M., M.T. Jones, J.D. Long & M.R. Weiss. 2001. Ontogenetic changes in leaf shelter construction by larvae of the silver-spotted skipper, Epargyreus clarus, Journal of the Lepidopterists' Society, 54(3):77-82

Lindroth, R.L., J.M. Scriber & M.T. Stephen Hsia. 1988. Chemical ecology of the tiger swallowtail: Mediation of host use by phenolic glycosides, Ecology, 69:814-822

Line, L. 2001. Swaths of flowering dogwood fall prey to fungus blight, New York Times, p. C5

Lipke, C. & R.B. Kinnel. 2001. Investigation of the Chemical Ecology of Asters, unpublished, http://www.chem.hamilton.edu/research/research_kinnel2001.html

Little, E.A. 1980. National Audubon Society Field Guide to North American Trees: Eastern Region, Knopf, Chanticleer Press, Inc., New York

Loader, C. & H. Damman. 1991. Nitrogen content of food plants and vulnerability of Pieris rapae to natural enemies, Ecology, 72(5):1586-1590

MacNeill, C.D. 1964. The skippers of the genus Hesperia in Western North America with special reference to California (Lepidoptera: Hesperiidae), University of California Publications in Entomology, 35:1-230

Manos-Jones, M. 2000. The Spirit of Butterflies, Harry N. Abrams, Inc., New York

Marsh, N., M. Rothschild & F. Evans. 1985. A new look at Lepidoptera toxins, in Vane-Wright, R.I. & P.R. Ackery, The Biology of Butterflies: Symposium of the Royal Entomological Society of London Number 11, Academic Press, Harcourt Brace Jovanovich, London, 2nd ed.

Martin, J. M., S. R. Madigosky, Z. Gu, J. Wu, D. Zhou & J. L. McLaughlin. 1999. Chemical defense in the zebra swallowtail butterfly, Eurytides marcellus, involving annonaceous acetogenins, Journal of Natural Products, 62 (1):2-4

Masters, J.H. 1972. Habitat: Oeneis jutta ascerta Masters & Sorensen, Journal of Research on the Lepidoptera, 11(2):94

Mattiacci, L., M. Dicke & M.A. Posthumus. 1995. B-Glucosidase: An elicitor of herbivore-induced plant odor that attracts host-searching parasitic wasps, Proceedings of the National Academy of Sciences, 92:2036-2040

McDonald, A.K. & H.F. Nijhout. 2000. The effect of environmental conditions on mating activity of the buckeye butterfly, Precis coenia, Journal of Research on the Lepidoptera, 35:22-28

McFarland, K.P. 2002. Conservation Assessment for White Mountain Butterfly (Oeneis Melissa semidea) and White Mountain Fritillary (Boloria montinus montinus), USDA Forest Service, Eastern Region

McNeeley, C. & M.C. Singer. 2001. Contrasting the roles of learning in butterflies foraging for nectar and oviposition sites, Animal Behavior, 61:1-6

McPhee, J. 1967. The Pine Barrens, Ballantine Books, New York

Mendoza, M. & R.L. Rutowski. Date unknown. Evolution of Ultraviolet Reflectance Patterns in Sulphur Butterflies, http://lsweb.la.asu.edu/rrutowski/poster.pdf

Menges, E.S. 1999. Ecology and conservation of Florida scrub, in Anderson, R.C., J.S. Fralish & J.M. Baskin, Savannas, Barrens, and Rock Outcrop Plant Communities of North America, Cambridge University Press, Cambridge

Mennechez, G., N. Schtickzelle & M. Baguette. 2003. Metapopulation dynamics of the bog fritillary butterfly: Comparison of demographic parameters and dispersal between a continuous and a highly fragmented landscape, Landscape Ecology, 18: 279-291

Milius, S. 1998. How bright is a butterfly?, Science News, 153:233-235

Miller, D.G. 1995. An isolated population of bog elfin, Incisalia lanoraieensis Sheppard (Lepidoptera: Lycaenidae), in Central New York, Journal of the New York Entomological Society, 103(1):122-123

Miller, J.S. 1987. Phylogenetic studies in the Papilioninae (Lepidoptera: Papilionidae), Bulletin of the American Museum of Natural History, 186:365-512

Miller, J.Y. 1992. The Common Names of North American Butterflies, Smithsonian Institution Press, Washington, DC

Minno, M.C. & M. Minno. 1999. Florida Butterfly Gardening: A Complete Guide to Attracting, Identifying, and Enjoying Butterflies of the Lower South, University of Florida Press, Gainesville

Minno, M.C. and T.C. Emmel. 1993. Butterflies of the Florida Keys, Mariposa Press, Scientific Publishers, Gainesville

Moerman, D.E. 1998. Native American Ethnobotany, Timber Press, Portland, OR

Morrison, H.K. 1873. Notes on North American Lepidoptera, Bulletin of the Buffalo Society of Natural Science, 1:186-187

Myers, R.L. 1990. Scrub and high pine, in Myers, R.L. & J.J. Ewel, Ecosystems of Florida, University of Central Florida Press, Orlando

Myers, R.L. & H. Beles. 1998. Fire Ecology: Dynamics & Restoration of Old Growth High Pines, Archbold Biological Station, http://www.archbold-station.org/abs/research/fireecology.htm

Myers, R.L. & J.J. Ewel. 1990. Ecosystems of Florida, University of Central Florida Press, Orlando

Nachtigall, W. 1968. Insects in Flight, McGraw-Hill Book Company, New York

Nagano, C. et al. 1991. A Bibliography of the Monarch Butterfly, Bioquip Products, Rancho Dominguey, CA

Neck, R.W. 1983. Significance of visits of hackberry butterflies (Nymphalidae: Asterocampa) to flowers, Journal of the Lepidopterists' Society, 37(4):269-274

Nekola, J.C. & C.E. Kraft. 2002. Spatial constraint of peatland butterfly occurrences within a heterogeneous landscape, Oecologia, 130:53-61

Nelson, G. 1994. The Trees of Florida: A Reference and Field Guide, Pineapple Press, Inc., Sarasota, FL

Nelson, G. 1995. Exploring Wild Northwest Florida: A Guide to Finding the Natural Areas and Wildlife of the Panhandle, Pineapple Press, Inc., Sarasota, FL

Nelson, G. 1996. The Shrubs and Woody Vines of Florida: A Reference and Field Guide, Pineapple Press, Inc., Sarasota, FL

Neumann, E. 1955. The Great Mother: An Analysis of the Archetype, Pantheon Books, New York, Bollingen Series XLVII

Newcomb, L. 1977. Newcomb's Wildflower Guide: An Ingenious New Key System for Quick, Positive Field Identification of the Wildflowers, Flowering Shrubs and Vines of Northeastern and North-central North America, Little, Brown and Company, Boston

Nice, C.C. & A.M. Shapiro. 1999. Molecular and morphological divergence in the butterfly genus Lycaeides (Lepidoptera: Lycaenidae) in North America: Evidence of recent speciation, Journal of Evolutionary Biology, 12:936-951

Nice, C.C. & A.M. Shapiro. 2001. Population genetic evidence of restricted gene flow between host races in the butterfly genus Mitoura (Lepidoptera: Lycasnidae), Annals of the Entomological Society of America, 94(2):257-267

Nielsen, E.T. 1961. On the habits of the migratory butterfly Ascia montuse, Biologiske Meddellser, 23:1-81

Niering, W.A. 1979. National Audubon Society Field Guide to North American Wildflowers, Knopf, Chanticleer Press, Inc., New York

Nijhout, H.F. 1991. The Development and Evolution of Butterfly Wing Patterns, Smithsonian Institution Press, Washington, DC

Nitao, J.K., M.P. Ayres, R.C. Lederhouse & J.M. Scriber. 1991. Larval adaptation to lauraceous hosts: Geographic divergence in the spicebush swallowtail butterfly, Ecology, 72:1428-1435

North American Butterfly Association. 1995. Checklist & English Names of North American Butterflies, NABA Names Committee, 1st ed.

North American Butterfly Association. 2001. Checklist & English Names of North American Butterflies, NABA Names Committee, 2nd ed.

Nylin, S., K. Nyblom, F. Bonquist, N. Janz, J. Belicek & M. Kallersjo. 2001. Phylogeny of Polygonia, Nymphalis and related butterflies (Lepidoptera: Nymphalidae): A total evidence analysis, Zoological Journal of the Linnaean Society, 132:441-468

Odendaal, F.J., M.D. Rausher, et al. 1987. Predation by Anolis lizards on Battus philenor raises questions about butterfly mimicry systems, Journal of the Lepidopterists' Society, 41:141-144

Ohsaki, N. & Y. Sato. 1994. Food plant choice of Pieris butterflies as a trade-off between parasitoid avoidance and quality of plants, Ecology, 75(1):59-68

Ohio Valley Environmental Coalition (OHVEC). 2003. Mountaintop Removal Website, http://www.ohvec.org/issues/mountain top_removal

Olafsdottir, E.S., J.W. Jaroszewski & M. Mercedes Arbo. 1990. Cyanohydrin glucosides of Turneraceae, Biochemical Systematics and Ecology, 18(6):435-438

Oliveira, E.G., R.B. Srygley & R. Dudley. 1998. Do neotropical migrant butterflies navigate using a solar compass?, Journal of Experimental Biology, 201:3317-3331

Oliver, C.G. 1982. Distinctiveness of Megisto c. cymela and M. c. viola (Satyridae), Journal of the Lepidopterists' Society, 36:153

Opler, P.A. 1983. County Atlas of Eastern United States Butterflies (1840-1982), U.S. Fish and Wildlife Service, Washington, DC

Opler, P.A. 1992. A Field Guide to Eastern Butterflies, Houghton Mifflin, Boston

Opler, P.A. & G.O. Krizek. 1984. Butterflies East of the Great Plains, Johns Hopkins University Press, Baltimore

Opler, P.A. & A.D. Warren. 2004. Scientific Names for Butterfly Species of North America, North of Mexico, Gilette Publications, Ft. Collins, CO

Oudemans, J.T. 1903. Etude sur la position de repos chez les Lepidoptera, Verhandelingen der Koninklijke Akademie van Wetenschappen, Amsterdam, (2)10

Panzer, R. 2002. Compatibility of prescribed burning with the conservation of insects in small, isolated prairie reserves, Journal of the Society for Conservation Biology, 16(5):1296-1307

Parmesan, C., N. Ryrholm, C. Stefanescus, J.K. Hill, C.D. Thomas, H. Descimons, B. Huntley, L. Kaila, J. Kullberg, T. Tammaru, W.J. Tennent, J.A. Thomas, & M. Warren. 1999. Poleward shifts in geographical ranges of butterfly species associated with regional warming, Nature, Nature Publishing Group, New York, London, 399:579-583

Parshall, D.K. 2002. Conservation Assessment for Olympia Marble Butterfly (Euchloe olympia), USDA Forest Service, Eastern Region

Pavulaan, H. 1998. A new subspecies of Incisalia henrici Grote & Robinson (Lynaenidae) from the Outer Banks of North Carolina, Maryland Entomologist, 4(2):1-16

Pavulaan, H. 2000. A revised look at red-banded hairstreak caterpillar hostplants, The Virginia Butterfly Bulletin, p. 3

Pavulaan, H. & R.R. Gatrelle. 1999. A new subspecies of Brephidium isophthalma (Lycaenidae: Polyommatinae) from coastal South Carolina, The Taxonomic Report, The International Lepidoptera Survey, Goose Creek, SC, 1(7)

Pavulaan, H. & D.M. Wright. 2000. The biology, life history, and taxonomy of Celastrina neglectamajor (Lycaenidae: Polyommatinae), The Taxonomic Report, The International Lepidoptera Survey, Goose Creek, SC, 2(5)

Pavulaan, H. & D.M. Wright. 2002. Pterourus appalachiensis (Papilionidae: Papilioninae), a new swallowtail butterfly from the Appalachian region of the United States, The Taxonomic Report, 3(7)

Penz, C.M. & D. Peggie. 2003. Phylogenetic relationships among Heliconiinae genera based on morphology (Lepidoptera: Nymphalidae), Systematic Ecology, 28:451-479

Peterson, R.T. & M. McKenny. 1968. A Field Guide to Wildflowers of Northeastern and North-central North America, Houghton Mifflin, Boston

Petr, D. 1998. Hawaiian White Monarch, Danaus plexippus form nivosus, http://biology.swau.edu/faculty/petr/ftphotos/hawaii/monarch/

Petrides, G.A. 1972. A Field Guide to Trees and Shrubs: Field Marks of All Trees, Shrubs, and Woody Vines that Grow Wild in the Northeastern and North-central United States and in Southeastern and South-central Canada, Houghton Mifflin, Boston, 2nd ed.

Phillips, W.C. 1940. Butterflies of New Jersey: A list of the Lepidoptera suborder rhopalocera occurring in the State of New Jersey; giving time of flight, food plants, records of capture with locality and date, Journal of the New York Entomological Society, 48:47-84

Pierce, N.E. 1987. The evolution and biogeography of associations between Lycaenid butterflies and ants, in Harvey, P.H. & L. Partridge, Oxford Surveys in Evolutionary Biology, Oxford University Press, Oxford, 4:89-116

Pivnick, K.A. & J.N. McNeil. 1985. Effects of nectar concentration on butterfly feeding: Measured feeding rates for Thymelicus lineola (Lepidoptera: Hesperiidae) and a general feeding model for adult lepidoptera, Oecologia, 66:226-237

Pivnick, K.A. & J.N. McNeil. 1986. Sexual differences in the thermoregulation of Thymelicus lineola adults (Lepidoptera: Hesperiidae), Ecology, 67(4): 1024-1033

Platt, A.P. & J.F. Allen. 2001. Sperm precedence and competition in doubly-mated Limenitis arthemis-astynax butterflies (Rhopalocera: Nymphalidae), Ecology and Population Biology, 94(5):654-663

Platt, S., C. Brantley & T. Rainwater. 2002. Canebrakes: Bamboo forests of the Southeast, Wild Earth, 39-45

Platt, W.J. 1999. Southeastern pine savannas, in Anderson, R.C., J.S. Fralish & J.M. Baskin, Savannas, Barrens, and Rock Outcrop Plant Communities of North America, Cambridge University Press, Cambridge

Pohl, R.W. 1978. How to know the grasses, WCB McGraw-Hill, Boston, 3rd ed.

Polcyn, D.M. & M.A. Chappell. 1986. Analysis of heat transfer in Vanessa butterflies: Effects of wing position and orientation to wind and light, Physiological Zoology, 59:706-716

Pollard, E. 1988. Temperature, rainfall and butterfly numbers, Journal of Applied Ecology, 25:819-828

Pollock, D.D., W.B. Watt, V.K. Rashbrook & E.V. Iyengar. 1997. Molecular phylogeny for Colias butterflies and their relatives (Lepidoptera: Pieridae)., Heredity, 88(5):335-342

Porter, A.H. 1994. Implications of introduced garlic mustard (Allaria petiolata) in the habitat of Pieris virginiensis, Journal of the Lepidopterists' Society, 48(2):171-172

Porter, A.H. & J.C. Mueller. 1998. Partial genetic isolation between Phyciodes tharos and P. cocyta (Nymphalidae), Journal of the Lepidopterists' Society, 52(2):182-205

Pratt, G.F., D.M. Wright & H. Pavulaan. 1994. The various taxa and hosts of the North American Celastrina (Lepidoptera: Lycaenidae), Proceedings of the Entomological Society of Washington, 96(3):566-578

Profant, D. 1989. The Lepidoptera of a central Florida sand pine scrub community, Journal of Research on the Lepidoptera, 28(1-2):37-74

Prudic, K.L., A.M. Shapiro & N.S. Clayton. 2002. Evaluating a putative mimetic relationship between two butterflies, Adelpha bredowii and Limenitis lorquini, Ecological Entomology, 27(1):68

Pyle, R.M. 1981. The Audubon Society Field Guide to North American Butterflies, Chanticleer Press, Random / Knopf, New York

Pyle, R.M. 1984. Handbook for Butterfly Watchers, Houghton Mifflin, Boston

Pyle, R.M. 1999. Chasing Monarchs: Migrating with the Butterflies of Passage, Houghton Mifflin, Boston

Ragusso, R.A. & M.A. Willis. 2003. Hawkmoth pollination in Arizona's Sonoran Desert: Behavioral response to floral traits, in Boggs, C.L., W.B. Watt & P.R. Ehrlich, Butterflies: Ecology and Evolution Taking Flight, University of Chicago Press, Chicago, pp. 43-66

Rapoport, E.H. 1982. Areography: Geographical Strategies of Species, Pergamon Press Ltd., Oxford

Rausher, M.D. 1979. Egg recognition: Its advantage to a butterfly, Animal Behavior, 27:1034-1040

Rawlins, J.E. 1980. Thermoregulation by the black swallowtail butterfly, Papilio polyxenes (Lepidoptera: Papilionidae), Ecology, 61:345-357

Rickett, H.R. 1973. Wild Flowers of the United States: The Northeastern States, New York Botanical Garden, McGraw-Hill Book Company, New York

Riley, N.D. 1975. A Field Guide to the Butterflies of the West Indies, Dempter Press, Quadrangle/The New York Times Book Co., Boston

Riley, T.J. 1988. Effect of larval photoperiod on incidence of adult seasonal forms in Anaea andria (Lepidoptera: Nymphalidae), Journal of the Kansas Entomological Society, 61:224-227

Ritland, D.B. 1994. Variations in palatability of queen butterflies (Danaus gilippus) and implications regarding mimicry, Ecology, 73(3):732-746

Ritland, D.B. & L.P. Brower. 1991. The viceroy is not a Batesian mimic, Nature, 350:497-498

Ritland, D.B. & L.P. Brower. 2002. Mimicry-related variation in wing color of viceroy butterflies (Limenitis archippus): A test of the model-switching hypothesis (Lepidoptera: Nymphalidaw), Holarctic Lepidoptera, 7(1):5-11

Robbins, R.K. 1981. The "false head" hypothesis: Predation and wing pattern variation of Lycaenid butterflies, The American Naturalist, 118:770-775

Robbins, R.K. 1993. False heads: The real tale, American Butterflies, 1(4):19-22

Robbins, R.K. 1994. Naming hairstreaks, American Butterflies, 2(3):28-32

Robbins, R.K. 1997. These legs were made for walking—and so much more!, American Butterflies, 5(1):28-32

Robbins, R.K. & S.S. Nicolay. 1998. Taxonomy and nomenclature of Strymon istapa and S. columella (Lycaenidae: Theclinae: Eumaeini), Journal of the Lepidopterists' Society, 52(3):318-327

Roble, S.M., W.D. Hartgroves & P.A. Opler. 2000. The butterflies and skippers (Lepidoptera) of the Great Dismal Swamp and vicinity, in Rose, R.K. The Natural History of the Great Dismal Swamp, Omni Press, Madison, WI

Roff, D.A. 1983. Phenological adaptation in a seasonal environment: A theoretical perspective, in Brown, V. K. & I. Hodek, eds., Diapause and Life Cycle Strategies in Insects, Junk, The Hague, pp. 253-270

Ross, G.N. 1997. Goddesses fly again: Butterfly images, American Butterflies, 5(4):14

Ross, G.N. 1999. Sleepers, American Butterflies, 7(2):26

Ross, G.N. 2001. New hostplant for the silvery checkerspot, Chlosyne nycteis, News of The Lepidopterists' Society, 43(4):101

Ruffin, J. & J. Glassberg. 2000. Miami blues still fly, American Butterflies, 8(1):28-29

Russwurm, A.D.A. 1978. Aberrations of British Butterflies, E.W. Classey Ltd, Oxon, UK

Rutkowski, F. 1971. Observations on Papilio aristodemus ponceanus (Papilionidae), Journal of the Lepidopterists' Society, 25(2):126-136

Rutowski, R.L. 1980. Courtship solicitation by females of the checkered white butterfly, Pieris protodice, Behavioral Ecology and Sociobiology, 7:113-117

Rutowski, R.L. 1981. Courtship behavior of the dainty sulphur butterfly, Nathalis iole with a description of a new, facultative male display (Pieridae), Journal of Research on the Lepidoptera, 20:161-169

Rutowski, R.L. 1983. Courtship leading to copulation in the cloudless sulphur, Phoebis sennae (Pieridae), Journal of Research on the Lepidoptera, 22(4):249-253

Rutowski, R.L. 1991. Temporal and spatial overlap in the mate-locating behavior of two species of Junonia (Nymplahidae), Journal of Research on the Lepidoptera, 30(3-4):267-271

Rutowski, R.L. 1998. Mating strategies in butterflies, Scientific American, 64-69

Rutowski, R.L. 2003. Visual ecology of adult butterflies, in Boggs, C.L., W.B. Watt & P.R. Ehrlich, Butterflies: Ecology and Evolution Taking Flight, University of Chicago Press, Chicago, pp. 9-26

Rutowski, R.L. & M.B. Kimball. 2000. Seeing the world through butterfly eyes, American Butterflies, 8(4):18-25

Ruxton, G.D. 1998. Mimicry: Sheep in wolves' clothing, Nature, News and Views, 394:833-834

Salvato, M.H. 1998. The Florida Keys, a paradise endangered, American Butterflies, 6:26-35

Salvato, M.H. 2001. Influence of mosquito control chemicals on butterflies (Nymphalidae, Lycaenidae, Hesperiidae) on the Lower Florida Keys, Journal of the Lepidopterists' Society, 55: 8-14

Schappert, P. 2000. A World for Butterflies, Firefly Books, Buffalo, NY

Schultz, J.K. & P.V. Switzer. 2001. Pursuit of heterospecific targets by territorial amberwing dragonflies (Perithemis tenera Say): A case of mistaken identity, Journal of Insect Behavior, 14(5): 607-620

Schwanwitsch, B.N. 1924. On the ground-plan of wing-pattern in Nymphalids and certain other families of Rhopalocerous Lepidoptera, Proceedings of the Zoological Society of London, 34:509-528

Schwartz, A. 1987. The butterflies of the Lower Florida Keys, in Milwaukee Public Museum, Contributions in Biology and Geology, 73:1-35

Scoble, M.J. 1995. The Lepidoptera: Form, Function and Diversity, The Natural History Museum, Oxford University Press, Oxford, 2nd ed.

Scott, J.A. 1974. Mate-locating behavior of butterflies, American Midland Naturalist, 91:103-117

Scott, J.A. 1975. Mate-locating behavior in western North American butterflies, Journal of Research on the Lepidoptera, 14:1-40

Scott, J.A. 1978. The identity of Rocky Mountain lycaena dorcas-helloides complex (Lycaenidae), Journal of Research on the Lepidoptera, 17(1):40-50

Scott, J.A. 1986. The Butterflies of North America: A Natural History and Field Guide, Stanford University Press, Stanford

Scott, J.A. 1992. Direction of spring migration of Vanessa cardui (Nymphalidae) In Colorado, Journal of Research on the Lepidoptera, 31(1-2):16-23

Scott, J.A. & R.E. Stanford. 1981. Geographic variation and ecology of Hesperia leonardus (Hesperiidae), Journal of Research on the Lepidoptera, 20(1):18-35

Scott, J.A. & D.M. Wright. 1991. Celastrina nigra and its synonym C. ebenina (Lepidoptera: Lycaenidae), Journal of Research on the Lepidoptera, 30(3-4):257-260

Scriber, J., J.M. Keefover & S. Nelson. 2002. Hot summer temperatures may stop movement of Papilio canadensis butterflies and genetic introgression south of the hybrid zone in the North American Great Lakes region, Ecography, 25:184-192

Scriber, J.M. & M.H. Evans. 1988. Bilateral gynandromorphs, sexual and/or color mosaics in the tiger swallowtail butterfly, Papilio glaucus (Lepidoptera: Papilionidae), Journal of Research on the Lepidoptera, 26(1-4):39-57

Scriber, J.M., N. Margraf & T. Wells. 2000. Suitability of four families of Florida "bay" species for Papilio Palamedes and P. Glaucus (Papilionidae), Journal of the Lepidopterists' Society, 54(4):131-136

Scriber, M., Y. Tsubaki & R.C. Leder. 1995. Swallowtail Butterflies: Their Ecology & Evolutionary Biology, Scientific Publishers, Inc., Gainesville

Scudder, S.H. 1889. The Butterflies of the Eastern United States and Canada with Special Reference to New England, self-published, Cambridge, 3 volumes

Shapiro, A.M. 1974. Butterflies and Skippers of New York State, Search, Cornell University Agricultural Experiment Station, Ithaca, 4(3)

Shapiro, A.M. 1979. Erynnis baptisiae (Hesperiidae) on Crown Vetch (Leguminosae), Journal of the Lepidopterists' Society, 33:258

Shapiro, A.M. 1991. Extirpation and recolonization of the buckeye, Junonia coenia (Nymphalidae) following the northern California freeze of December 1990, Journal of Research on the Lepidoptera, 30(3-4):209-220

Shapiro, A.M. 2002. The California urban butterfly fauna is dependent on alien plants, Diversity and Distributions, 8:31-40

Shapiro, A.M. & A.R. Shapiro. 1973. Butterflies of Staten Island, Journal of Research on the Lepidoptera, 12(2):65-128

Shapiro, A.M., R. von Buskirk, G. Kareofelas, & W.D. Patterson. 2003. Phenofaunistics: Seasonality as a property of butterfly faunas, in Boggs, C.L., W.B. Watt & P.R. Ehrlich, Butterflies: Ecology and Evolution Taking Flight, University of Chicago Press, Chicago, pp. 111-148

Shepard, J.H. 1998. The correct name for the Boloria chariclea/titania complex in North America (Lepidoptera: Nymphalidae), in Emmel, T.C. (ed.), Systematics of Western North American Butterflies, Mariposa Press, Gainesville, pp. 727-730

Shuey, J.A. 1997. Conservation status and natural history of Mitchell's satyr, Neonympha mitchellii French (Insecta: Lepidoptera: Nymphalidae), Papers in Conservation Issues, Natural Areas Journal, 17(2):153-163

Shull, E.M. 1987. The Butterflies of Indiana, Indiana Academy of Science, Indiana University Press, Bloomington

Shure, D.J. 1999. Granite outcrops of the SE United States, in Anderson, R.C., J.S. Fralish & J.M. Baskin, Savannas, Barrens, and Rock Outcrop Plant Communities of North America, Cambridge University Press, Cambridge

Silberglied, R.E. & O.R. Taylor, Jr. 1978. Ultraviolet reflection and its behavioral role in the courtship of the sulphur butterflies Colias eurytheme and C. philodice (Lepidoptera, Pieridae), Behavioral Ecology and Sociobiology, Springer-Verlag, 3:203-243

Sime, K.R. 2002. Chemical defence of Battus philenor against attack by the parasitoid Trogus pennator, Ecological Entomology, 27(3):337

Sime, K.R., P.P. Feeny & M.M. Haribal. 2002. Sequestration of aristolochic acids by the pipevine swallowtail, Battus philenor (L.): Evidence and ecological implications, Chemoecology, 10(4):169-178

Simpson, B. 1990. The Great Dismal Swamp: A Swamp Memoir, Henry Holt and Company, New York

Singer, M.C. 1972. Complex components of habitat suitability within a butterfly colony, Science, 176:75-77

Smedley, S.R., F.C. Schroeder, D.B. Weibel, J. Meinwald, K.A. Lafleur, J.A. Renwick, R. Rutowski & T. Eisner. 2002. Mayolenes: Labile defensive lipids from the glandular hairs of a caterpillar (Pieris rapae), Proceedings of the National Academy of Sciences, 99(10):6822-6827

Smith, D.S., L.D. Miller & J.Y. Miller. 1994. The Butterflies of the West Indies and South Florida, Oxford University Press, Oxford

Smith, K.C. 1991. The effects of temperature and daylength on the rosa polyphenism in the buckeye butterfly, Precis coenia (Lepidoptera: Nymphalidae), Journal of Research on the Lepidoptera, 30(3-4):225-236

Snyder, J.R., A. Herndon & W.B. Robertson, Jr. 1990. South Florida Rockland, in Myers, R.L. & J.J Ewel, Ecosystems of Florida, University of Central Florida Press, Orlando

Sourakov, A. 1998. Elliott Key's endangered swallowtail, Letters from the Field, California Wild, http://www.calacademy.org/calwild/sum98/field.htm

Spencer, K.C. 1988. Chemical mediation of coevolution in the

Passiflora-Heliconius interaction, in Spencer, K.C, Chemical Mediation of Coevolution, Academic Press, Harcourt Brace Jovanovich, San Diego, pp. 167-240

Sperling, F. 2003. Butterfly molecular systematics: From species definition to higher-level phylogenies, in Boggs, C.L., W.B. Watt & P.R. Ehrlich, Butterflies: Ecology and Evolution Taking Flight, University of Chicago Press, Chicago, pp. 431-458

Spomer, S.M., L.G. Higley, T.T. Orwing, G.L. Selby & L.J. Young. 1993. Clinal variation in Hesperia leonardus (Hesperiidae) in the Loess Hills of the Missouri River Valley, Journal of the Lepidopterists' Society, 47:291-302

Spomer, S.M. & W.W. Hoback. 1998. New ant associations for Glaucopsyche lygdamus Doubleday (Lycaenidae), Journal of the Lepidopterists' Society, 52(2):216-217

Sprandel, G.L. 2001. Fall dragonfly (Odonata) and butterfly (Lepidoptera) migration at St. Joseph peninsula, Gulf County, Florida, Florida Entomologist, 84(2):234-238

Srygley, R.B. 2001. Sexual differences in tailwind drift compensation in Phoebis senna butterflies (Lepidoptera: Pieridae) migrating over seas, Behavioral Ecology, International Society for Behavioral Ecology, 12(5):607-611

Srygley, R.B. & R. Dudley. 1993. Correlations of the position of center of body mass with butterfly escape tactics, Journal of Experimental Biology, 174:155-166

Srygley, R.B. & E.G. Oliveira. 2001. Sun compass and wind drift compensation in migrating butterflies, Journal of Navigation, 54:405-417

Srygley, R.B. & A.L.R. Thomas. 2002. Unconventional lift-generating mechanisms in free-flying butterflies, Nature, 420:660-664

Stanton, E.J. 2001. Status of Pieris virginiensis (Pieridae) in New York State, The Lepidopterists' Society, 55(3):122-123

Stanton, M.L. 1984. Short-term learning and the searching accuracy of egg-laying butterflies, Animal Behavior, 32:33-40

Steele, F.L. 1982. At Timberline: A Nature Guide to the Mountains of the Northeast, Appalachian Mountain Club, Boston

Stevenson, G.B. 1969. Trees of the Everglades National Park and the Florida Keys, Banyan Books, Inc., Miami, FL

Stiling, P. 1999. Butterflies of the Caribbean and Florida, Macmillan Education Ltd., London

Strahler, A.N. 1966. A Geologist's View of Cape Cod, The Natural History Press, American Museum of Natural History, Garden City, NY

Struttmann, J.M. 2001. Butterflies of North America, Northern Prairie Wildlife Research Center, U.S. Geological Survey, http://www.npwrc.usgs.gov/resource/distr/LEPID/bflyusa/bflyusa.htm

Suffert, F. 1927. Zur vergleichenden Analyse der Schmetterlingszeichnung, Biologisches Zentralblatt, 47:385-413

Sutton, A. & M. Sutton. 1985. Eastern Forests, Chanitcleer Press / Borzoi Books, Knopf, New York

Swanson, H.F. & J. Monge-Najera. 2000. The effects of methodological limitations in the study of butterfly behavior and demography: A daily study of Vanessa atalanta (Lepidoptera: Nymphalidae) for 22 years, Review of Tropical Biology, 48(2/3):605-614

Sweeney, A., C. Jiggins & S. Johnsen. 2003. Insect communication: Polarized light as a butterfly mating signal, Nature, 423:31-32

Swengel, A.B. 1996. Effects of fire and hay management on the abundance of prairie butterflies, Biological Conservation, 76:73-85 Swengel, A.B. 1997. Habitat associations of sympatric violet-feeding fritillaries (Euptoieta, Speyeria, Boloria) (Lepidoptera: Nymphalidae) in tallgrass prairie, The Great Lakes Entomologist, 30(1&2):1-18

Swengel, A.B. & S.R. Swengel. 2001. A ten-year study of the status and trend of the Regal Fritillary (Speyeria idalia) (Lepidoptera: Nymphalidae) in Wisconsin, U.S.A., The Great Lakes Entomologist, 34(1):111-128

Swengel, S.R. & A.B. Swengel. 1999. Correlations in abundance of grassland songbirds and prairie butterflies, Journal of the Lepidopterists' Society, 90:10-11

Taylor, W.K. 1992. The Guide to Florida Wildflowers, Taylor Publishing Company, Dallas, TX

Taylor, W.K. 1998. Florida Wildflowers in Their Natural Communities, University Press of Florida, Gainesville

Tekulsky, M. 1985. The Butterfly Garden, The Harvard Common Press, Harvard, MA

Thomas, J. & R. Lewington. 1991. The Butterflies of Britain & Ireland, The National Trust, Dorling Kindersley, London

Thompson, J.N. 1988. Evolutionary genetics of oviposition preference in swallowtail butterflies, Evolution, 42(6):1223-1234

Tietz, H.M. 1936. The Lepidoptera of Pennsylvania: A Manual, School of Agriculture, The Pennsylvania State College

Timmermann, S. 1999. Uric acid deposition in larval integument of black swallowtails and speculation on its possible function, Journal of the Lepidopterists' Society, 53(3):104-107

Tolman, T. 1997. Collins Field Guide: Butterflies of Britain & Europe, HarperCollins, London

Tranefors, T. 1999. Variation among androconial substances from three Hesperiidae species, 16th annual meeting, International Society of Chemical Ecology, http://www.chemecol.org/meetings/99/p-133.html

Turner, T.W. & J.R. Parnell. 1985. The identification of two species of Junonia hubner (Lepidoptera: Nymphalidae); J. evarete and J, genoveva in Jamaica, Journal of Research on the Lepidoptera, 24(2):142-153

Tveten, J. & G. 1996. Butterflies of Houston & Southeast Texas, University of Texas Press, Austin

Tyndall, R.W. & J.C. Hull. 1999. Vegetation, flora, and plant physiological ecology of serpentine barrens of eastern North America, in Anderson, R.C., J.S. Fralish & J.M. Baskin, Savannas, Barrens, and Rock Outcrop Plant Communities of North America, Cambridge University Press, Cambridge

Urquart, F.A. & N.R. Urquart. 1976. The overwintering site of the eastern population of monarch butterfly (Danaus p. plexippus) in southern Mexico, Journal of the Lepidopterists' Society, 30:153-158

U.S. Environmental Protection Agency. 2003. Mid-Atlantic Mountaintop Mining: Draft Environmental Impact Statement

U.S. Fish and Wildlife Service. 1999. Schaus swallowtail butterfly, Heraclides aristodemus ponceanus, Department of the Interior, southeast.fws.gov/vbpdfs/species/inverts/ssbu.pdf

U.S. Fish and Wildlife Service. 2003. Schaus swallowtail butterfly, Department of the Interior, https://ecos.fws.gov/species_profile/SpeciesProfile?&spcode=I016

Vawter, A.T. & J. Wright. 1986. Genetic differentiation between subspecies of Euphadryas phaeton (Nymphalidae: Nymphalinae), Journal of Research on the Lepidoptera, 25(1):25-29

Vlieger, L., P.M. Brakefield & C. Muller. 2004. Effectiveness of the defence mechanism of the turnip sawfly, Athalia rosae (Hymenoptera: Tenthredinidae) against predation by lizards, Bulletin of Entomological Research, 94:283-289

Vukusic, P., J.R. Shambles & C.R. Lawrence. 2004. Structurally assisted blackness in butterfly scales, Proceedings of the Royal Society of London B, published online, doi:10.1098/rsbl.2003.0150

Wade, N. 2000. In death-defying act, butterfly thrives on poison vine, New York Times, p. F6

Wagner, D.L. 1995. Rearing regals for reintroduction: Playing the odds but still losing ground, American Butterflies, 3(2):19-23

Wagner, D.L., V. Giles & R.C. Reardo. 1999. Caterpillars of Eastern Forests, U.S. Forest Service, U.S. Department of Agriculture

Wagner, D.L., M.W. Nelson & D.F. Schweitzer. 2003. Shrubland Lepidoptera of southern New England and southeastern New York: Ecology, conservation, and management, Forest Ecology and Management, 185:95-112

Wagner, H., Jr. & T. Mellichamp. 1978. Foodplant, habitat and range of Celastrina ebenina (Lycaenidae), Journal of the Lepidopterists' Society, 32(1):20-36

Wahlberg, N. 2000. The Ecology and Evolution of Melitaeine Butterflies, http://ethesis.helsinki.fi/julkaisut/mat/ekolo/vk/wahlberg/

Wahlberg, N. 2001. The phylogenetics and biochemistry of host plant specialization in melitaeine butterflies (Lepidoptera: Nymphalidae), Evolution, 55:522-537

Waldbauer, G. 1998. The Birder's Bug Book, Harvard University Press, Cambridge, MA

Walker, T.J. 1978. Migration and re-migration of butterflies through north peninsular Florida: Quantification with Malaise traps, Journal of the Lepidopterists' Society, 32:178-190

Walker, T.J. 1991. Butterfly migration from and to peninsular Florida, Ecological Entomology, 16:241-252

Walker, T.J. 2001. Butterfly migrations in Florida: Seasonal patterns and long-term changes, Environmental Entomology, Entomological Society of America, 30(6):1052-1060

Watanabe, M. & K. Sato. 1993. A spermatophore structured in the bursa copulatrix of the small white Pieris rapae (Lepidoptera, Pieridae) during copulation, and its sugar content, Journal of Research on the Lepidoptera, 32:26-36

Watson, F.E. & F.E. Lutz. 1926. Our Common Butterflies, in American Museum of Natural History, Guide Leaflet Series, No. 38, 5th rev. ed.

Watt, W.B. 1969. Adaptive significance of pigment polymorphism in Colias butterflies. I. Variation of melanin pigment in relation to thermoregulation, Evolution, 22:437-458

Weber, L. 2002. Butterflies of New England, Kollath-Stensaas Publishing, Duluth, MN

Webster, R.P. & M.C. Nielsen. 1984. Myrmecophily in the Edwards' hairstreak butterfly Satyrium edwardsii (Lycaenidae), Journal of the Lepidopterists' Society, 38:124-133

Weed, C.M. 1917. Butterflies Worth Knowing, Doubleday, Page & Company, Garden City, NJ

Weiss, M.R. 1991. Floral colour changes as cues for pollinators, Nature, 354:227-229

Weiss, M.R. 1997a. Brains on the wing: Learning ability of butterflies, American Butterflies, 5(2):28-32

Weiss, M.R. 1997b. Innate colour preferences and flexible colour learning in the pipevine swallowtail, Animal Behavior, 53:1043-1052

West, D.A. 1988. Suppression of the black phenotype in females of the P. glaucus group (Papilionidae), Journal of Research on the Lepidoptera, 26(1-4):187-200

West, D.A., W.M. Snellings & T.A. Herber. 1972. Pupal color dimorphism and its environmental control in Papilio polyxenes aasterius (Lepidoptera: Papilionidae), Journal of the New York Entomological Society, 80:205-211

Whittaker, P.L. 1984. Population biology of the great purple hairstreak, Atlides halesus, in Texas (Lycaenidae), Journal of the Lepidopterists' Society, 38(3):179-185

Wiklund, C. 1972. Pupal coloration in Papilio machaon in response to wavelength of light, Die Naturwissenschaften, 1-2

Wiklund, C. 1975. Pupal colour polymorphism in Papilio machaon L. and the survival in the field of cryptic vs. non-cryptic pupae, Transactions of the Royal Entomological Society of London, 127:73-84

Wiklund, C. 2003. Sexual selection and the evolution of butterfly mating systems, in Boggs, C.L., W.B. Watt & P.R. Ehrlich, Butterflies: Ecology and Evolution Taking Flight, University of Chicago Press, Chicago, pp. 67-90

Williams, E.H. 2002. Harris' checkerspot: A very particular butterfly, American Butterflies, 10(2):18-25

Wilsman, L.A. & D.F. Schweitzer. 1991. A rangewide status survey of Mitchell's satyr Neonympha mitchellii mitchellii (Lepidoptera: Nymphaliae), unpublished report for U.S. Forest and Wildlife Service, Twin Cities Regional Office, NJ

Wiser, S.K. & P.S. White. 1999. High-elevation outcrops and barrens of the southern Appalachian Mountains, in Anderson, R.C., J.S. Fralish & J.M. Baskin, Savannas, Barrens, and Rock Outcrop Plant Communities of North America, Cambridge University Press, Cambridge

Woodbury, E.N. 1994. Butterflies of Delmarva, Tidewater Publishers, Centreville, MD

Wright, A.B. 1993. Peterson First Guides: Caterpillars, Houghton Mifflin, Boston

Wright, D.M. 1983. Life history and morphology of the immature stages of the bog copper butterfly Lycaena epixanthe (Bsd. & Le C.) (Lepidoptera: Lycaenidae), Journal of Research on the Lepidoptera, 22(1):47-100

Wright, D.M. 1995. The American azures: Our blue heaven, American Butterflies, 3(1):20-28, 30

Wright, D.M. & H. Pavulaan. 1999. Celastrina idella (Lycaenidae: Polyommatinae): A new butterfly species from the Atlantic coastal plain, The Taxonomic Report, International Lepidoptera Survey, Goose Creek, SC, 1(9)

Yack, J.E., M.L. Smith, & P.J. Weatherhead. 2001. Caterpillar talk: Acoustically mediated territoriality in larval Lepidoptera, Proceedings of the National Academy of Sciences, 98(20):11371-11375

Yack, J.E. & J.H. Fullard. 2000. Ultrasonic hearing in nocturnal butterflies, Nature, 403:265-266

Index